HETEROSIS BREEDING IN VEGETABLE CROPS

NIPA® GENX ELECTRONIC RESOURCES & SOLUTIONS P. LTD.
New Delhi-110 034

HETEROSIS BREEDING IN VEGETABLE CROPS

Dr. Nagendra Rai
Senior Scientist (Horticulture)
Indian Institute of Vegetable Research
Shahanshahpur, PB 01, P.O. - Jakhani
Varanasi - 221 005 Uttar Pradesh

Dr. Mathura Rai
Director
Indian Institute of Vegetable Research
Shahanshahpur, PB 01, P.O. - Jakhani
Varanasi - 221 005 Uttar Pradesh

nipa®
NIPA® GENX ELECTRONIC RESOURCES & SOLUTIONS P. LTD.
New Delhi-110 034

NIPA® GENX ELECTRONIC
RESOURCES & SOLUTIONS P. LTD.

101,103, Vikas Surya Plaza, CU Block
L.S.C. Market, Pitam Pura, New Delhi-110 034
Ph : +91 11 27341616, 27341717, 27341718
E-mail: newindiapublishingagency@gmail.com
web: www.nipabooks.com

For customer assistance, please contact
Phone: + 91-11-27 34 17 17
Fax: + 91-11- 27 34 16 16
E-Mail: feedbacks@nipabooks.com

ISBN: 978-81-19215-25-6

This is the Paperback edition of Harback edition

Composed and Designed by NIPA®.

Dr. G. KALLOO
Deputy Director General
(Horticulture & Crop Science)

भाकृ अनुप
ICAR

INDIAN COUNCIL OF AGRICULTURAL RESEARCH
KRISHI ANUSANDHAN BHAVAN-II,
PUSA, NEW DELHI 110 012
Phone : (Off.) 2584 2068 (Resi.) 2584 0191
Fax : 91-11-2584 1976 E-mail: kalloog@icar1.nic.in

FOREWORD

Production of hybrid vegetable is gaining momentum due to increased quality, productivity, uniformity, better desirable traits and managment of biotic stresses. The next decade will be for hybrid vegetable crops for sustainable development of horticulture as well as for higher economic returns to farmers and vegetable industry. Substantial increase in area under hybrid in vegetable crops has been witnessed during last decade indicating its popularity and adoption by vegetable growers. The area under hybrids in tomato, brinjal, chillies, cabbage, cauliflower and cucurbits is in increasing trend due to realization of high yield, uniformity and better return to the farmers. There is an ample scope for the development of hybrids in several vegetables, especially okra, capsicum, cabbage, cauliflower and chilli to improve the production and productivity to growers. The seeds of hybrid varieties of carrot, cabbage and chilli are in high demand and requirement is met through import of seeds. Dissemination of knowledge on hybrid technology for seed production is key in achieving the objective of increasing area under vegetable hybrids. To achieve this, the efforts made by Dr. N. Rai, Senior Scientist (Horticulture) and Dr. Mathura Rai, Director, Indian Institute of Vegetable Research, Varanasi in bringing out the valuable publication *'Heterosis Breeding in Vegetable Crops'* is praiseworthy. This publication contains both theoretical and practical aspects for vegetables hybrid seed production along with their package of practices. I am sure this book will be of great use to vegetable breeders, students, seed industries and farmers of the country. I wish to compliment the authors of this book for their endeavour in compilation of the information in the form of book.

G. KALLOO
Deputy Director General
(Horticulture and Crop Sciences)
Indian Council of Agricultural Research
New Delhi

Preface

In India, vegetables are grown from dry temperate to humid tropics between the altitudes from sea level to snow line. The vegetables are rich source of vitamins, minerals, antioxidants, phytochemicals and provide protection against health hazards. A diet rich in vegetables ensures nutritional security and productive life. Most of vegetables are short duration, cultivated as sole, mix, relay, catch crops provides ample scope in crop diversification, provides opportunity for higher employment generation consequently higher economic return. It also provides raw material for establishment of processing industries and value addition, the the recent past tremendous progress has been made in the vegetable production and India has emerged second largest producer of vegetable in the world after China. It shares about 14% of world vegetable production. The total production of vegetable has increased from 23.4 mt in the year 1961-62 to the tune of 90 mt in the year 2003-04 with annual increase of about 3.5%. The export of vegetable is in increasing trends. The export of vegetable in the year 2000-01 was Rs.1480 crores which has increased to Rs.1950 crores in the year 2004-05 registering a growth rate of 17.9%. However, considering the demand to meet the recommended requirement of 300g vegetables per caput per day for ever growing population with annual growth rate of 1.93% and for export market India needs to produce about 150 mt of vegetables by 2011. The focus of research for increasing the national productivity of vegetable crops by the year 2011 encompasses many research areas including popularisation of hybrid varieties, integrated nutrient management, disease/pest management, diversification, value addition and post harvest management. In this direction the sincere efforts made by the authors will be of immense use in popularising the hybrid technology. The book, *Heterosis Breeding in Vegetable Crops,* has two sections, the first section deals with basic principles of heterosis and the second section includes techniques for development of hybrids in 16 vegetables along with improved production technology. This would be useful to teachers, scientists, students, extension workers, seed industries and farmers.

Authors express their gratitude to Dr. Mangala Rai, Secretary, DARE, Govt. of India and Director General, Indian Council of Agricultural Research, New Delhi, India for encouragement and suggestions; Dr G. Kalloo (DDG, Horticulture and Crop Sciences) for technical guidance and for writing the foreword. The authors are grateful to Dr. K.M. Bujarbaruah, Director, ICAR Research Complex for NEH Region, Shillong for permission to write and publish this book. Thanks to Dr. Omvir Singh, Principal Scientist, NBPGR Regional Station, Srinagar and Dr. N. Ahmed, Professor and Head (Vegetable Crops), Sher-E-Kashmir University of Agriculture and Technology, Srinagar for providing important literature. The authors expresses sincere thanks to Late Dr. C.B.S. Rajput, Ex.-Director, Institute of Agricultural Sciences, Banaras Hindu University, Varanasi for encouragement and motivation. The help extended by Mr. K.K. Patel, Mr. P. Sharma, Mr. B. Dey, Mr. B.S. Asati and Mr. R.M. Yadav is highly appreciated.

Thanks are also due to Dr. D. S. Yadav, Dr. N. P. Singh, Dr. Y. P Sharma, Dr. V. K. Mishra, Dr. A. N. Sylesha, Dr. B. R. Some, Dr. Rajesh Kumar, Er. R. K. P. Singh and Er. K. N. Agrawal of ICAR Research Complex for NEH Region for their valuable comments and suggestions for improvement of the manuscript.

Though, every effort has been made to ensure the authenticity in the content of the book. The views and facts presented are solely those of the authors of various books and journals consulted. The authors do not take any responsibility of the correctness, authenticity or views of previous authors. The book has been edited for language and content improvement as well as length management by a professional editor.

We would appreciate the receipt of suggestions/views from readers for improvement.

December, 2005

Nagendra Rai
Mathura Rai
Indian Institute of Vegetable Research
Varanasi - 221 005

Contents

Foreword *v*
Preface *vii*

SECTION I
PRINCIPLES OF HETEROSIS

Chapter 1: Introduction **1-13**
History, *2*
Concept of Heterosis, *4*
General Features, *5*
Types of Heterosis, *5*
Inbreeding Depression, *8*
Vegetable Hybrids in India, *9*

Chapter 2: Basis of Heterosis **15-29**
Genetic Basis of Heterosis, *15*
Physiological Basis of Heterosis, *19*
Biochemical Basis of Heterosis, *20*
Biometrical Basis of Heterosis, *26*

Chapter 3: Manifestation of Heterosis **31-64**
Basic Requirements, *31*
Types of Hybrids Used for Commercial Cultivation, *31*
Types of Crossing, *32*
Procedures of Hybrid Seed Production, *32*
Manifestation of Hybrid Vigour, *34*

Chapter 4: Reproduction, Pollen Biology and Pollen Biotechnology **65-120**
Reproduction, *65*
Pollen and Pollen Biology, *68*
Pollen Biotechnology, *70*
Mechanism of CMS, *82*
Cytoplasmic Genetic Male Sterility, *83*
Self- incompatability, *85*
Comparasion Between Gametophytic and Sporophytic Self- incompatability, *94*
Diallelic Sporophytic SI and Heteromorphy, *94*
Problems in Exploiting Self-incompatibility, *96*
Barriers to Hybridization, *98*
Methods for Overcoming Interspecific Crossing Barriers, *101*
Techniques for Overcoming Post-fertilization Barriers, *103*
Techniques for Overcoming F_1 Sterility, *104*
Storage of Pollen, *105*

Causes in Losses of Pollen Viability, *106*
Methods of Storage of Pollen, *107*
Estimation of Pollen Viability After Storage, *109*
Pollination Constraints and Management of Pollination Insects for Crop Production, *110*
Pollen Efficiency of Insects, *116*

Chapter 5: Population Improvement of Parents/ Hybrid Varieties **121-124**
Concept, *121*
Methods Used for Population Improvement, *121*

Chapter 6: Hybrid Varieties Testing and Release Procedures in India **125-141**
Evaluation, *125*
Hybrid Testing Centers in India, *130*
Identification of Hybrids for Release, *132*
Release of Hybrids by the Central Sub-committee on Crop Standards and Release of Varieties/ Hybrids, *138*
Release of a Hybrid by a State Variety Release Committee, *141*

Chapter 7: Genetic Purity Testing of Hybrid Seeds **143-156**
Meaning and Importance, *143*
Advantages of Genetic Purity, *143*
Methods of Testing, *144*
Application of Molecular Markers- Proteins/ Isozymes/ DNA based Markers, *146*
Advantages, *150*
Applications, *151*
Comparative Advantages of Common Molecular Techniques, *155*
Sample Size for Genetic Purity Testing, *155*

Chapter 8: Biotechnology and Vegetable Improvement **157-223**
Origin, History and Definition, *157*
Applications, *162*
Approaches, *163*
Solanaceous Vegetables, *164*
Cucurbitaceous Vegetables, *167*
Somaclonal Variation, *170*
Distant Hybridization, *172*
Limitations of Distant Hybridization, *183*
Somatic Hybridization, *184*
Anther Culture, *190*
Achievements of Tissue Culture, *192*
Genetic Engineering, *194*

Engineering for Disease Resistance, *201*
Engineering for Insect-pest Resistance, *203*
Engineering for Parthenocarpic Plants, *205*
Engineering for Post-Harvest Traits, *205*
Engineering for Male Sterility and Fertility Restoration, *205*
Engineering for Herbicide Tolerance, *207*
Engineering for Salt Tolerance, *208*
Engineering for Abiotic Stress, 208
Use of Molecular Markers, *212*
Synthetic Seed Technology, *218*
Use Polymerase Chain Reaction (PCR), *222*

Chapter 9: Intellectual Property Rights-Related Issues in Hybrid Technology **225-246**
Definition and History, *225*
Benefits From IPR, *226*
Problems From IPR, *226*
Management of IPR, *227*
Patent, *227*
Plant Breeders' Rights (PBRs), *228*
Related Issues in Hybrid Technology, *238*
Institutional Committees, *241*
Terminator and Verminator Technology, *243*
Verminator and Triator, *246*

Chapter 10: Economics, Import and Export **247-258**
Introduction, *247*
Factors Affecting Production and Popularization of Hybrid Seeds, *250*
Cost of Hybrid Seed Production, *250*
Method of Hybrid Seed Production and their Production Cost Analysis, *252*
Seed Import and Export, *258*

SECTION II
TECHNIQUES FOR HYBRID SEED PRODUCTION

Chapter 11: Solanaceous Crops **259-293**
Tomato, *259*
Brinjal, *272*
Chilli, *281*
Bell Pepper, *289*

Chapter 12: Malvaceous Crop **295-301**
Okra, *295*

Chapter13: Cole Crops **303-316**
Cabbage, *303*
Cauliflower, *313*

Chapter 14: Cucurbitaceous Crops **317-356**
Bitter Gourd, *317*
Bottle Gourd, *322*
Cucumber, *327*
Muskmelon, *335*
Watermelon, *340*
Pumpkin and Squash, *347*
Ridge Gourd and Smooth (Sponge) Gourd, *353*

Chapter 15: Root Crop **357-362**
Carrot, *357*

Chapter 16: Bulb Crop **363-372**
Onion, *363*

Chapter 17: Hints for Cultivation **373-410**
Tomato, *373*
Brinjal, *378*
Chilli, *381*
Bell Pepper, *383*
Okra, *383*
Cabbage, *386*
Cauliflower, *389*
Bitter Gourd, *391*
Bottle Gourd, *395*
Cucumber, *396*
Muskmelon, *398*
Watermelon, *400*
Pumpkin, *401*
Ridged Gourd and Smooth Gourd, *402*
Carrot, *403*
Onion, *406*

Glossary *411*
Appendices *425*
Colour Plates (1-15) *Between Pages 436-437*
References *437*
Index *513*

SECTION I

PRINCIPLES OF HETEROSIS

(PAGES 1-258)

1

INTRODUCTION

Vegetables are rich source of nutrients, vitamins and minerals. Phytochemical contents in vegetables and their strong antioxidant potential in scavergering free radicals have generated tremendous attention among the consumers. Consumption of vegetables has significant health-promoting effect and can reduce the incidence of cardiovascular diseases, cancer, Acquired Immuno Deficiency Syndrome (AIDS) and various other degenerative diseases. Human health, in recent years, has assumed an unprecedented status. The rising population levels, increase in mental stress and changes in dietary habits have resulted in various forms of degenerative diseases including cardiovascular, cancer and ageing. As a result, there has been an alarming explosion in consumers' interest in health. In recent past, tremendous progress has been made for increasing vegetable production in the country. India has emerged as the second largest producer of vegetables next to China. During 2000-2001, India produced 93.92 million tonnes of vegetables from 6.24 million ha of land, accounting a productivity of 15.0 tonnes/ha. During 1991-1992, total vegetable production was 58.54 million tonnes from 5.59 million ha of land, with productivity of 10.5 tonnes/ha. Thus, in last one decade country has multiplied its vegetable production one-and-a-half times. Presently, India's share is more than 13.60%. Our country will need to produce 215 million tonnes vegetables by 2015 to meet the demand of domestic and export markets. The increase in vegetable production by increasing area under vegetable cultivation is limited due to continued decrease of land holdings. Hence, it is essential to increase vegetable productivity to fulfill country's requirements for food and nutritional security and poverty alleviation.

Hybrid vegetable technology is one of the better options particularly due to that full potential of hybrids in vegetable crops, has not been completely exploited as compared to several cereals, despite growing vegetable crops is four times more remunerative than cereals. In world agriculture, hybrid varieties are available in about 50 field and horticultural crops. Hybrid breeding has helped in overcoming yield barriers through accelerating the increase in productivity initially and subsequently. Presently, F_1 hybrid breeding method is commonly utilized to exploit heterosis in several economically important vegetable crops like tomato, brinjal, peppers (both hot and sweet), cabbage, cauliflower, radish, carrot, melons, onion *etc.* Vegetable breeders prefer to select hybrid breeding because it is comparatively easy to incorporate resistant genes for biotic and abiotic stresses in F_1 hybrid and right of the breed variety is protected in terms of parental lines. Moreover, despite high cost of hybrid seeds, there has been an increasing concern among farmers on cultivation of hybrids, because under optimum crop production and protection management, crops raised from seeds of F_1 hybrids have several advantages. These are:

- High-yielding due to large size or weight of fruits, bulbs or heads as in cabbage, onion, melons and increased number of fruits in okra and tomato.

- Early-maturing (cabbage, tomato and brinjal), hence, fetch premium price in market.
- Can fit in multiple cropping sequences, hence, increase the cropping intensity.
- By virtue of buffering capacity, they are more uniform than varieties, thereby, reducing cost in grading, harvesting and other allied operations.
- Perform better under stress environments.
- In general, they have wider adaptability.
- They are stable in performance.
- The chance of cross-pollination and admixtures are practically nil, as fresh seed is to be used every year.
- Ensure extended availability of vegetables.
- Better transportability for reaching to markets.
- Better resistance to insects-pest and diseases (onion, spinach, tomato, cucumber *etc.*).
- Better fruit quality (tomato, watermelon *etc.*).
- Better monetary returns to the growers per unit area.
- They promote seed industry.

History

J.G. Koelreuter was the first to describe plant hybrids during eighteenth century. Although many of the early attempts at hybridization lacked scientific depth, well-known classical by Koelreuter (1766) put plant hybridization upon a more scientific basis. His investigations of hybrid vigour in interspecific crosses in *Nicotiana, Dianthus, Verbascum, Mirabilis, Datura,* and other genera are intensive, systematic, and scientific. Charles Darwin discussed hybrid vigour in plants in his book, *The Effects of Cross and Self-fertilization in the Vegetable Kingdom,* which appeared in 1876. Darwin demonstrated that cross-fertilization frequently resulted in increased size, vigour, and productiveness compared with self-fertilization or other close-breeding within the same species. He did not, however, recognize this increased vigour as *hybrid vigour*, nor did he attribute it to the differences between uniting gametes. Beal (1880) described how he had planted in alternate rows, two stocks of the same type of corn (that had been grown 100 miles apart for a number of years). All of the plants of one stock of corn were detasselled, and the hybrid seeds produced on these plants were then planted. Yield on these hybrid plants exceeded the parental stocks by more than 50%. Beal not only demonstrated the increased vigour from hybridizing two different stocks, but also was probably one of the first to utilize the detasselling method that is regularly used by corn breeders and seed companies for production of commercial hybrid seed. It is interesting to note that maize researchers working in late nineteenth century and early twentieth century recognized the injurious effects of inbreeding, but did not emphasize the positive gains resulting from cross-breeding. For example, Shamel (1905) reported that prime objective in a breeding programme is the prevention of injurious effects of cross-fertilization between closely-related plants, or inbreeding. Investigations by Shull (1908 and 1909) provided a change in emphasis from negative effects of self-fertilization and inbreeding to the positive effects from hybridization. Therefore, foundations for a more comprehensive understanding of heterosis were formed.

In his article, "*The Composition of a Field of Maize*", he concluded that a variety was a complex mixture of hybrids, with each plant of a different genotype. Another conclusion was that (within a variety), the variability among strains subjected to inbreeding, as well as the reduction in vigour, was a consequence of segregation with resulting homozygosity of both desirable and deleterious alleles. He also reported that yield of F_1s between certain lines exceeded that of the parental variety from which the lines were derived. His 1909 paper provided many of the considerations that later became standard operating procedures in maize breeding programmes as they capitalized the concept of heterosis. Investigations by East (1908 and 1909) on the effects of inbreeding and cross-breeding of maize also contributed to the development of the concept of heterosis. These studies by Shull and East demonstrated that inbreeding and hybridization appeared to offer an alternative route for improving yields over mass selection, which had been deemed to be ineffective. However, marked loss in vigour due to inbreeding of parents tend to preclude the commercial use of hybrid vigour. Jones (1918) suggested the use of double cross hybrids for commercial planting. In double cross production, commercial seed is produced on high yielding single cross parents, so that limitations imposed by low-yielding inbred parental lines are removed.

In vegetable crops, first suggestion to exploit hybrid vigour was made by Hayes and Jones (1916a) in cucumber. F_1 hybrid brinjal was commercially used in Japan before 1925, (Kakizaki, 1931). Demonstration of the value of hybrids in corn in early 1920's prompted a few exploratory studies in other crops, but methods of its application to field scale operations were first proposed in 1932, using self-incompatibility system in cabbage (Pearson, 1931), and using cytoplasmic male sterility mechanism in onion (Jones and Clarke, 1943). Programmes to exploit the sex systems to produce hybrids were developed in spinach (monoecious forms), cucumber (genetic controlled), squash (response to chemical), and carrot (cytoplasmic male sterility). Identification and measurement of expression of heterosis — the per cent increase over mid-parent value—of components of economic yield have been attempted. Educated guesses as to the value of parental combinations have been the base used in choice of parents. Different studies of the same crop often differ when analyzed for heterosis effects because the number, identity and effectiveness of multiple gene systems involved vary with in varieties used as parents. Diallel studies can be used to locate sources of desirable additive factors and guide the development of more efficient hybrids. In India, first report of hybrid vigour appeared in 1933 in chilli at Indian Agricultural Research Institute (IARI), New Delhi. However, first hybrid 'Pusa Meghdoot' of bottle gourd was developed at IARI, Regional Station, Katrain, and released in 1971. Immediately after two years, F_1 hybrids of summer squash 'Pusa Alankar' and cucumber 'Pusa Sanyog' were also developed and released for cultivation. From the beginning of vegetable hybrid research, private sector is actively associated with the development of hybrids. Indo-American Hybrid Seed Company, Bangalore, took the lead in release of first tomato hybrid 'Karnataka' and first capsicum hybrid 'Bharat' in 1973 for commercial cultivation (Kalloo and Singh, 2001). From 1975, intensive research work has been started in public sector organizations for development of F_1 hybrids, releasing a number of hybrids. According to Chadha (1997), growing of hybrids became popular with the introduction of new seed policy declared in 1988, which liberalized seed imports under Open General License (OGL) with the objective to provide the farmers best quality planting material. During 1994, first products of biotechnological application in vegetables was introduced in pilot test markets. Vine ripe tomatoes with extended shelf life, processing tomatoes with superior quality and deep red colour, squash with novel virus resistance and potatoes genetically modified to produce an insect killing protein are examples of traits introduced into commercial vegetable varieties with the tools of biotechnology. These first products such as *Flavr Savr* tomato received both public visibility and regulatory scrutinies. For promotion of hybrid technology in India, Indian Council of Agricultural Research (ICAR), New Delhi, initiated research work on 'Promotion of Hybrid Research in Vegetable Crops' in 1995-96. Initially, research work was initiated in nine vegetable crops, *viz*. tomato, brinjal, chilli, capsicum, okra, onion, cabbage, cucumber and bitter gourd at 15 Centers including four national institutes and 10 State Agricultural Universities (SAUs) and National Horticultural Research and Development Foundation, Nasik. In 1999, a World Bank funded mission mode project under National Agriculture Technology Project (NATP) was formulated. The project aims for development of specific quality hybrids in tomato, brinjal, chilli and onion at 12 Centers in different parts of the country.

Concept of Heterosis

Hybrid vigour is used as a synonymous of heterosis. The term heterosis was first used by Shull in 1910 who defined this phenomenon as the stimulus of heterozygous and in his word it has been the 'interpretation of increased vigour, size, fruitfulness, speed of development, resistance to diseases and insect pest or to the climatic regions of any kind manifested by outbreeding organism as compared with corresponding inbreds as a specific result of the unlikeliness in the constitution of the uniting parental gametes'. This definition is often interpreted as not implying a genetic basis for heterosis, because the definition basically describes the phenotypes that results from crossing two different inbred lines. Generally, heterosis is manifested as an increase in vigour, size, growth rate, yield or some other characteristics. But in some cases, the hybrid may be inferior to the weaker parent. This is also represented as heterosis (average heterosis) and if heterosis is estimated over superior parent, such an estimate is sometimes referred to as heterobeltiosis. Hence, authors feel 'heterosis refers a biological phenomenon in which F_1 population obtained by crossing of two genetically dissimilar gametes or individual shows increased or decreased vigour over better parent or over mid-parental value'. Power (1944) also suggested that the term heterosis should be used only when the hybrid is either superior or inferior to both the parents. Other situations should be regarded as partial or complete dominance. It can be easily explained with the help of Table 1.

Table 1: Heterosis and dominance in relation to parental values

Position and value of parents	Value of F_1 hybrids	Phenomenon
Parent A (10)	>10	Heterosis
	10	Complete dominance
	<10 but >8	Partial dominance
Mid-parent	8	No dominance
	<8 but >6	Partial dominance
Parent B (6)	6	Complete dominance
	<6	Heterosis

Source: Singh (2002).

General Features

The general features of phenomenon of heterosis are as follows:

- Heterosis is a widely occurring biological phenomenon in both the plant and animal species. In plants, it has been reported to occur more frequently in a number of naturally cross-pollinated crop species as compared with self-pollinated ones. The specific ones heterotic crosses usually show increase in size, vigour, seed producing capability, usually better resistance to insect pests or diseases, increased metabolic activity and better stability and thus ultimately result in better performance of hybrids than parents. Usually, these hybrids show better fitness and breeding value as compared to parents from which they are made.
- It is a genetically governed phenomenon.
- Heterosis is confined only to specific F_1 or first hybrid generation and considerably diminishes in F_2 and the later segregating generations. This reduction in performance in later generations is because of the breakdown of specific gene combination building up the expression of heterosis.
- It has been observed that the expression of heterosis is usually more in hybrids obtained from genetically related lines. The expression of heterosis in a crop species shows that some degree of genetic diversity, heterozygosity and or dominance exists in parents with respect to character or characters in which this phenomenon has been expressed.
- The expression of heterosis is highly associated with specific combining ability of a cross.
- Heterosis is usually unfixable from generation to generation as heterozygote advantage is confined to F_1 population only. But some natural contrivances, which favour the existence of heterozygotes, or those, which preserve the heterozygote condition from generation to generation, could fix it. Some examples for the existence of such mechanism in nature are homozygous lethal factors, balanced heterozygotes, apomixis or the use of vegetative reproduction.
- Heterosis has been adequately reported in vegetable crops like brinjal, brassica, onion, tomato, radish, carrot, cucumber, squashes, legumes, *etc.*

Types of Heterosis

Heterosis can be classified into two classes: (i) euheterosis (true heterosis), and (ii) pseudoheterosis (luxuriance), depending upon the nature of origin, adaptability and reproducibility or non-reproducibility. Euheterosis is further divided into: (a) mutational euheterosis, and (b) balanced euheterosis; depending upon the nature and origin of genes, causing heterosis in a biological system (Dobzhansky, 1952). A brief descriptions of various types of heterosis is given :

Mutational Euheterosis

It is the result of natural consequences of the phenomenon of dominance. This is the simplest type of euheterosis derived from the sheltering or overshadowing the deleterious, unfavourable often lethal, recessive mutants by their adaptively superior dominant alleles in population of sexually reproducing cross- fertilizing organisms (Dobzhansky, 1952). Most of the mutations, which occur in natural conditions, are genetically recessive and adaptively unfavourable. The phenomenon of mutational euheterosis shelters the deleterious recessive mutants and for them works as protective device against deleterious mutation pressure. For newly-evolved mutants, this would perhaps be necessary. Because a large number of such mutations which occur under natural conditions are deleterious to organic evolution and are easily liable to be wiped out by natural selection. Mutational heterosis shelters protects them from their quick elimination.

Balanced Euheterosis

It is based on the theory of over-dominance and a type of true heterosis which arises out of a balanced, specifically-adapted gene combination or from the occurrence of rather special type of adapted mutations which confer on the heterozygote, a better adaptive value or a higher agricultural usefulness, is usually found in corresponding homozygotes. There are number of recessive genes in nature which are comparatively less favourable for their natural adaptation than their dominant counterparts and are consequently liable to be lost. This phenomenon of specific in gene recombination functions as a built-in evolutionary mechanism, which maintains the multiplicity of the adapted recessive genes in a population, providing additional vigour to the heterozygote and saving it from being lost. This makes the cross-pollinated species virtually and potentially self-improving ones under natural conditions.

Luxuriance

Luxuriance is basically by the crossing of two parental forms bringing in an accidental, excessive, usually unadaptable and an unbalanced expression of an attribute. Some such examples are seen in excessive vegetative growth of foliage, stems, flower buds, *etc*; and are for more encountered in case of domesticated plants than wild species. This is found in two parental species crossed. Though, this phenomenon resembles like heterosis and has been noticed in many domesticated plants species, but it has not been found to be agronomically and economically feasible to utilize. This is primarily because of the reasons of adaptability under natural conditions of crop production. Usually, there is no indication whatever or any conclusive proof, the luxuriant hybrids would also simultaneously prove adaptively superior in comparison or in competition with parental forms in natural habitat of the latter.

Positive and Negative Heterosis

The superiority of F_1 over its superior parent is referred as positive heterosis, whereas, its inferiority over inferior parents is negative heterosis.

Adaptive Heterosis

This heterosis refers to increased fitness of hybrid that is similar to euheterosis.

Selective Heterosis

This heterosis refers to competitive ability. Its usage is restricted to studies of population. It takes into consideration to phenotypic reaction of individual and not the expression of any single character.

Liable Heterosis

In heterosis under heterozygous, genetic situation breakes down in subsequent generations and thus is not fixable, such heterosis is termed as liable heterosis.

Fixed Heterosis

The heterosis that can be fixed by different mechanisms such as vegetative propagation, apomixes or chromosomal changes is called fixed heterosis.

Useful Heterosis

The term useful heterosis was used by Meredith and Bridge (1972). It refers to superiority of F_1 over the standard commercial check variety. It is also called economic heterosis. This type of heterosis is of direct use in plant breeding.

Standard Heterosis

Sometimes heterosis is worked out over standard commercial hybrid. It is estimated in those crops where hybrids are already available for comparison. This type of heterosis is termed as standard heterosis. This is also of direct practical importance in plant breeding.

In a particular cross, heterosis is usually measured in terms of two parameters, namely (i) heterosis over mid-parental (MP) value and (ii) heterosis over better parent (BP). In plant breeding programmes, however, it is also estimated in terms of heterosis over the check variety. For commercial exploitation, this value is, perhaps, most important one. The procedure for estimation of these values is as follows:

Heterosis over mid parental value (relative heterosis):

$$= \frac{F_1\text{-MP}}{\text{MP}} \times 100$$

where F_1 and MP are average performances of F_1 and mid-parental values, respectively.

Heterosis over better parent (heterobletiosis):

$$= \frac{F_1\text{-BP}}{\text{BP}} \times 100$$

where BP is the average performance of the better parent.

Heterosis over the check parent (standard heterosis):

$$= \frac{F_1 - \text{Check variety}}{\text{Check variety}} \times 100$$

where check variety denotes its average performance.

Based on the estimates of gene effects obtained from six components model of the analysis of quantitive gene action, *i.e.* additive (a), dominance (d) and its various interaction components like (a x a), (a x d) and (d x d). Jinks and Jones (1958) have suggested the following formula for calculation of heterobeltiosis at genotypic level.

Heterobeltiosis = (dominance-additive x additive)-(additive + 1/2 additive x dominance)

Symbolically, it could be expressed as:

$$F_1 - BP = (d - a \times a) - (a + 1/2\ a \times d)$$

Inbreeding Depression

The inbreeding depression refers to decrease in fitness and vigour due to inbreeding. The degree of inbreeding is measured by inbreeding coefficient. The main differences between inbreeding and heterosis are as follows:

- The inbreeding results from mating between related individuals, whereas heterosis results from crossing unrelated strains.
- Inbreeding depression is the decline in fitness and vigour with decreased heterozygosity, whereas heterosis is increase in fitness and vigour with increased heterozygosity.
- The inbreeding depression results due to fixation of unfavourable recessive genes in F_2, while in case of heterosis, unfavourable recessive genes of one line (parent) are covered by favourable dominant genes of other parent. The fixation of all favourable dominant genes in one homozygous line is impossible due to linkage between some unfavourable recessive and favourable dominant genes.

Estimation of Inbreeding Depression

The inbreeding depression is estimated when both F_1 and F_2 populations of the same cross are available. It is measured as follows.

Inbreeding depression = $(F_1 - F_2 / F_1) \times 100$

where, F_1 and F_2 are the mean values of F_1 and F_2 progeny, respectively of the same cross for a given character. The inbreeding depression may be high, medium, low and nil depending on the crop species. The following inferences can be drawn from the estimates of heterosis and inbreeding depression.

- If high heterosis is followed by inbreeding depression, it indicates the presence of non-additive gene action (dominance and epistasis).
- If performance is the same in F_1 and F_2, it reveals presence of additive genes.
- If heterosis is negative in F_1 and there is increase in F_2, it again indicates presence of additive genes.
- The heterosis will be greatest when some alleles are fixed in one parent and other alleles in other parent.
- The genes with lack of dominance will not exhibit heterosis in F_1 but may show increase in performance in F_2 due to fixation of genes, *i.e.* additive action. If some genes have dominance in one direction and some in other direction there will be no heterosis due to mutual cancellation effects of such genes.

Test of Significance

The significance of heterosis and inbreeding depression is tested with the help of critical difference. The general formula for the estimation of critical difference is given below:

Critical difference (CD) = SE difference x t at 5% or 1%.

Mean error variance from combined analysis of variance of parents, F_1s and F_2s are used for calculating the SE of difference. The mean values over replications are used for comparison. For heterosis, difference between F_1 and the parent used for estimation of heterosis (mid-parent or better parent or standard commercial cultivar) is taken into account cross-wise. For inbreeding depression, mean values of F_1s and F_2s are compared. When the differences are greater than CD, it is considered significant and *vice-versa*.

Vegetable Hybrids in India

Status

Vegetable production of our country is still dominated by the locally available genotypes or open-pollinated varieties. This is mainly due to farmers ignorance and poor extension activities. Hybrid varieties in India are of recent origin. Hybrid technology has been most successful in tomato, resulting in a large area coverage in Karnataka, Maharashtra, Gujarat and some parts of Andhra Pradesh. Hybrid cabbage is next in area coverage to tomato. The entire cabbage production in southern parts of Maharashtra and West Bengal are under F_1 hybrids. In capsicum, brinjal, cucumber and watermelon, area under F_1 hybrids are small (Table 2). With awareness of advantages for cultivation of F_1 hybrids, the area is bound to extend.

Some Success Stories

In spite of several constraints, there are some success stories of hybrid vegetable technology in certain specific regions of this country. These achievements were made possible by utilization of genetic male sterile lines, producing more economic F_1 seeds and adoption

Table 2: Estimated share, open-pollinated and local or unidentified varieties in total cropped area in India (1997-98)

Crop	Total area (ha)	Area under F_1 hybrids (ha)	Share of hybrid (%)	Share of open-pollinated (%)	Share of unidentified varieties (%)
Brinjal	4,79,095	85,300	17.80	50.00	32.20
Cabbage	2,42,140	76,000	31.39	68.61	-
Cauliflower	3,04,156	10,000	3.29	50.00	46.71
Chilli	5,73,529	14,000	2.44	37.56	60.00
Gourds	4,09,270	10,000	2.44	20.00	77.56
Melons	1,69,350	68,000	4.02	25.98	70.00
Okra	3,71,665	20,000	5.38	80.00	14.62
Tomato	4,85,520	153000	31.51	50.00	18.49

Source: Singh (1998).

of improved agro-techniques for hybrid seed production of vegetable crops. The Punjab Agricultural University (PAU), Ludhiana, has taken lead in developing muskmelon hybrid 'Punjab Hybrid' from Cross MS-1 x Hara Madhu. In Punjab, many farmers are trained for utilization of male-sterility for F_1 hybrid seed production. The PAU, Ludhiana and National Seed Corporation (NSC), New Delhi supply seeds of both male and female lines to farmers and F_1 seeds are then produced. This programme is highly successful in muskmelon and gaining momentum of chilli hybrid 'CH1' developed from cross MS-12 x ELS. MS-12 is a genic male sterile line in the background of 'Punjab Lal', developed by incorporating male sterile gene introduced from France. 'CH1' hybrid exhibited 160% heterosis for yield besides having tolerance to some diseases. Utilizing self-incompatibility and genetic cytoplasmic male sterility system, two hybrids, *viz*. 'BRH5' and 'H64', respectively have been developed in cabbage at Katrain (Himachal Pradesh) and are being evaluated under monitoring of Indian Institute of Vegetable Research (IIVR), Varanasi, at several centres. In cucumber, 'Pusa Sanyog' (Japanese Gynoecious line x Green Long) has been developed at IARI, New Delhi, and recommended for cultivation through All India Coordinated Vegetable Improvement Project. Genetic cytoplasmic male sterility system has been used to develop 'Arka Pitambar' and 'Arka Kirtiman' hybrids in onion at Indian Institute of Horticultural Research (IIHR), Bangalore. In carrot cytoplasmic male sterility and in muskmelon genic male sterility are being used to develop low cost hybrid seeds at many places.

Major Constraints

High cost of F_1 hybrid seed is one of the major handicaps of successful hybrid vegetable technology. This is because under-utilization of available genetic tools for economic F_1 seed production. In the entire country, mostly hand-emasculation and hand-pollination techniques is being followed (with very few exception) to develop F_1 hybrid seeds in most of the vegetable crops. Besides, there are several other constraints prevailing, which act as barriers in the rapid spreading of hybrid vegetable technology in our country. Some of these constraints are:

- Lack of awareness among farmers about hybrid crop production techniques.
- Unorganized marketing system for fresh vegetables.
- Lack of post-harvest management techniques.
- Non-availability of other inputs at proper time.
- Non-availability of biotic stress resistant hybrids.

Future Strategies

In coming years, aim should be to increase vegetable production not only to meet requirement of our population but also to increase per capita income of marginal farmers. In the context of decreasing land holding of Indian farmers and liberalized economic policy of Indian Government, hybrid vegetables seem to be most remunerative. By growing F_1 hybrids, production of quality vegetables can be increased manifold, which will ensure surplus production for fresh export and post-harvest industry. Therefore, followings the areas where, there is urgent need to take attention for vegetable hybrids are :

General Areas

- There is a need to develop crop-and location-specific effective production technique for hybrid variety and while developing these, poor resources of small and marginal farmers should be taken into account.
- In order to meet F_1 seed of popular hybrids developed through private organization, its parental line should be available from private organization to public sector. Already, public organization, like IARI, New Delhi, PAU, Ludhiana (Punjab) and IIHR, Banglore have taken lead in providing the parental lines of popular hybrids to the seed producers, *e.g.* in brinjal (Pusa Hybrid-5 and Pusa Hybrid-6), tomato (Pusa Hybrid-1 and Pusa Hybrid-2) and muskmelon (Punjab Hybrid).
- In order to streamline the development, infrastructure like greenhouse, nethouse and polyhouse with drip irrigation system are required. In several vegetable crops, especially in cucurbits, hybrid varieties can be grown successfully in off-season with extraction of full potential of hybrid variety. Therefore, development of such infrastructure may boost hybrid vegetable technology.
- Development of sound post- harvest management techniques and established network of local chains and cold rooms will be a boon for vegetable growers.
- Promotion of cooperative society will ensure adequate supply of quality F_1 seeds as well as other inputs at proper time. Such varieties are hot spot for the adoption of hybrid vegetable technology. A successful example of cooperative sector involvement is the launching of a modern fruits and vegetables shops in Chennai during 1994. It was set up with the aim to establish stronger linkage between growers and consumers. In Raipur (Chattisgarh), progressive farmers are also growing vegetables under the umbrella of Cooperative Society.

- Training programme on hybrid seed and crop production for state government officials and farmers need to be strengthened. On the pattern of Punjab farmers, hybrid seed production at farmers fields is feasible in many regions of the country by imparting training to vegetable growers.Techniques for emasculation, pollination, proper bagging and tagging, identification of proper stage of hybridization, male sterile plants, harvesting and extraction of F_1 seeds, and testing of genetic tools are important aspects for training programme.

Crop-specific Research Priorities

Vegetable crops are highly prone to biotic stress, namely diseases, insect-pests and nematodes. Judicious use of pesticides is hazardous to the health and environment. Considering growing concern about residue-free vegetables and export of fresh canned vegetables, it is high time when major aim should be to develop multiple resistant hybrid varieties against biotic stresses in vegetables (Table 3). This will serve the purpose of producing residue-free quality vegetables, meeting standards of competitive international market. In general, productivity is major criteria to get maximum return but in vegetable crops, its quality and availability during lean periods are also equally important in order to get better return in the vegetable market.

Table 3: Research priorities for developing multiple resistant hybrids against biotic stresses (Kalloo and Singh, 2001)

Crop	Target biotic stresses
Tomato	Tomato leaf curl virus + early blight + bacterial wilt + roots knot nematode.
Brinjal	Phomopsis + bacterial wilt + fruit and shoot borer.
Chilli	Leaf curl + thrips + mites + anthracnose.
Capsicum	Phytophthora + thrips + mites.
Okra	Yellow vein mosaic virus + fruit borer.
Onion	Stemphyllium blight +purple blotch + thrips.
Cucumber	Downey mildew + mosaic.
Muskmelon	Powdery mildew + downey mildew + anthracnose + fusarium.
Watermelon	Powdery mildew + downey mildew + anthracnose.
Cabbage	Black rot + diamond back moth.
Cauliflower	Black rot + diamond back moth.

Therefore, off-season vegetable hybrid varieties (having better adaptability) may be developed in tomato, brinjal, capsicum, cauliflower, cabbage, onion and other vegetables. Further, in order to decrease hybrid seed production cost, it would be appropriate that we must take initiative to utilize available genetic mechanisms more effectively for hybrid seed production (Table 4).

Table 4: Research priorities for hybrid seed production system and specific traits (Kalloo and Singh, 2001)

Crop	F_1 seed production system*	Heterosis breeding aim
Tomato	Use of genetic cytoplasmic male sterile (CMS).	Tolerant to high temperature (off-season), high TSS and high lycopene.
Chilli	Use of cytoplasmic male sterile or genetic male lines.	High oleoresin.
Capsicum	Standardization of techniques for plains (use of protected infrastructure).	Adapted to plains of north India.
Cauliflower	Use of strong self-incompatible of CMS lines.	Heat tolerance (off-season).
Cabbage	Use of strong self- incompatible of CMS lines.	Tolerant to high temperature (tropical cabbage).
Onion	Development and use of short day CMS lines.	Photoperiod insensitive (off-season).
Carrot	Use of cytoplasmic male sterile lines.	High carotene per cent.
Watermelon	Use of male sterile lines.	High TSS and seedlessness.
Muskmelon	Use of male sterile lines.	High TSS.
Cucumber	Use of gynoecious line.	Pickling type (for export).

*Initiative may be taken to identify, characterize and utilize conditional mutants of male sterility in all vegetables.

2

BASIS OF HETEROSIS

Genetic Basis of Heterosis

There are three theories to explain the genetic basis of heterosis: (1) dominance hypothesis, (2) over-dominance hypothesis and (3) epistasis.

Dominance Hypothesis

Davenport (1908) first proposed this hypothesis and he considered that heterosis is a consequence of action and interaction of favourable dominant genes brought together in a hybrid. Bruce, Keeble and Pellew explained it in 1910. This hypothesis suggests that at each locus the dominant allele have favourable effect, while recessive allele has unfavourable effect. In heterozygous state, deleterious effects of recessive alleles are masked by their dominant alleles. Thus, heterosis results from masking of harmful effects of recessive alleles by their dominant alleles. The harmful effects of recessive alleles, which becomes homozygous due to inbreeding. Therefore, according to dominance hypothesis, heterosis is not the result of heterozygosity; it is the result of prevention of expression of harmful recessive allele by their dominant alleles (Fig.1). In this hypothesis, there are two objections (i) the first objection relates to the failure in isolation of homozygous lines for all the dominant genes, and (ii) the second objection is directed at the symmetrical distribution in F_2. Both the objections are discussed below:

Failure in Isolation of Inbreds as Vigorous as Hybrids

According to dominance hypothesis, it should be possible to isolate inbreds with all dominant genes. Such inbreds would be as vigorous as F_1 hybrids. However, such inbreds have not been isolated. It can clearly understand by Fig1:

Cross I	**Cross II**
Inbred A x Inbred B	Inbred A x Inbred D
↓	↓
AabbCCdd *AabbCCdd*	*AabbCCdd* *aaBBccDd*
F_1 hybrid	F_1 hybrid
AabbCCdd	*AaBbCcDd*
NO HETEROSIS	HETEROSIS
(The same number of genes has dominant alleles as in two parents. Harmful effects of b and d genes are not masked)	(Twice as many genes have dominant alleles leading to heterosis because harmful effects of a, b, c and d are hidden by A, B, C and D)

Figure 1: Heterosis as a function of genotypes of two parents (dominance hypothesis)

Source: Singh, 1983, p.187.

Symmetrical Distribution in F_2

In F_2, dominant and recessive characters segregate in the ratio of 3:1. Quantitative characters, according to dominance hypothesis, there should not exists symmetrical distribution in F_2. This is because dominant and recessive phenotypes would segregate in proportion $(^3/_4+^1/_4)^n$, where n is the number of genes segregating. However, F_2s nearly always show a symmetrical distribution.

These objections were over ruled by Jones (1917) and Collins (1921). Jones (1917) suggested that since many genes govern quantitative characters, they are likely to show linkage. It may be expected that dominant and recessive genes governing a character would be linked together. In such case, inbreds containing all the dominant genes cannot be isolated because this would require several precisely placed crossovers. It would also explain the symmetrical curves in F_2. This is often known as the *Dominance of Linked Genes Hypothesis*. However, Collins (1921) showed that if the number of genes governing a quantitative character is large, symmetrical distribution would be obtained even without linkage. Further, it is unlikely that a plant containing all the dominant genes would be recovered if the number of genes is large even if they are not linked. The distribution curve would further become symmetrical due to the effects of environment, that is, due to less than 100% heritability. Research work on *Nicotiana* species, the conclusion drawn by Spargue (1983) after reviewing several experiments in maize, and more recent work on rice using molecular markers, have strongly supported the dominance hypothesis. Barabas (1991), however, opined that dominance certainly plays an important role in heterosis, but is hardly acceptable as a fundamental cause *per se.*

Dominant traits are not always linked to productivity. Wild species possessing more dominant traits than cultivated crops plants are generally low yielder, indicating that accumulation of dominant genes alone can not explain the heterosis phenomenon. Barabas (1991) further argued that if only the accumulation of dominant genes dispersed in parents, required for heterosis, plant breeder would have been successful in assembling such dominant blocks in a single variety long ago. Although, the objection raised by Barabas may have some relevance, it is still important to note that plant breeders have been successfully in assembling dominant building blocks in a single variety / inbred, leading to some highly vigorous and productive inbreds, that are quite superior to even hybrids of yesteryears. Also, in the recent years, researches have been remarkably successful in transferring favourable dominant quantitative trait loci (QTLs) for yield and other agronomically important characters from wild species to tomato and rice cultivated crop plants, leading to significant improvement of the target traits.

Over-dominance Hypothesis

East and Shull independently proposed this hypothesis in 1908. This is known by several names, *viz. pseudo over-dominance* (tightly linked complementary genes) *single gene heterosis, super dominance, cumulative action of divergent alleles* and *stimulation of divergent alleles*. This hypothesis explains that heterozygotes at least some of the loci are superior to both the relevant homozygotes. Thus, heterozygote *Aa* would be superior to both the homozygotes *AA* and *aa*. Consequently, heterozygosity is essential for and is the cause of heterosis, while homozygosity resulting from inbreeding produces inbreeding depression. It would, therefore, be impossible to isolate inbreds as vigorous as F_1 hybrids. Milborrow (1998) proposed an alternative hypothesis for over-dominance hypothesis. He suggested that genes that regulate certain metabolic pathways to lower levels than maximum possible might limit the growth of a plant. Heterozygotes may partially escape this regulation since they have two slightly different alleles for these genes, allowing greater flow on these pathways. This is not over-dominance; but, like over-dominance hypothesis, it predicts that heterozygotes have an inherent advantage in vigour that cannot be duplicated by any amount of selection in open- pollinated homozygous lines. The most interesting part of Milborrow's theory is the implication as to why there are so many *weak allele*s in a population to start with. These are not sub-lethal mutants that have accumulated under the protection of closely linked genes that are strongly selected, but necessary parts of the genetic adaptation of the population.

Over-dominance has been experimentally observed from some monogenic characters (Berger, 1974), but could not be substantiated for polygenic traits (Jink, 1983). Gallais (1988), however, opines that contribution of over-dominance is low relative to that of dominance, but that it is difficult to exclude a role of marginal over-dominance in both autogamous and allogamous plants.

Comparison Between Dominance and Over-dominance Hypotheses

Both hypotheses lead to similar expectations. But they differ from each other with respect to one expectation. The similarities and differences between them are given below:

Similarities

The two hypotheses have following similarities.

- Inbreeding would produce inbreeding depression.
- Outcrossing would restore vigour and fertility.

The degree of heterosis would depend upon the genotypes of two parents. In general, greater the genetic diversity between the parents, the higher magnitude of heterosis.

Differences

The chief differences between both hypotheses are as follows :

- According to dominance hypothesis, inbreeding depression is due to homozygosity of harmful recessive alleles, while, as per over-dominance hypothesis, it is due to homozygosity itself.
- According to dominance hypothesis, heterosis is the consequence of homozygosity *per se*. But as per over-dominance hypothesis, it is the result of dominant alleles masking the deleterious effects of their recessive alleles, and heterozygosity itself is not the cause of heterosis.
- Heterozygotes are superior to the two homozygotes according to over-dominance hypothesis, while according to dominance hypothesis, they are as good as the

dominant homozygote. Therefore, inbreds as vigorous as F_1 hybrid can be isolated according to dominance hypothesis, but it will be impossible according to over-dominance hypothesis.

In spite of the large experimental evidence accumulated, it is not possible to conclusively accept or reject one of the two hypotheses. There are definitely some genes that show heterozygote superiority. But the number of such genes appears to be rather small, and even these cases could be due to linkage in repulsion phase or epistasis or both. It is generally accepted that heterosis, to a large extent, is due to dominance gene action, but epistasis and over-dominance are also involved. The relative importance of these phenomena is, however, not clearly understood. Recent evidence accumulated with maize seems to suggest that over-dominance may not be the primary cause of heterosis. Over-dominance is easily imitated by epistasis and linkage, that most reported cases of over-dominance may not represent true over-dominance (Singh, 2002).

Epistasis Hypothesis

This hypothesis takes into consideration the influence of one locus on the expression of another for heterozygote adaptive superiority (Gowen, 1952). Evidence for positive interaction of non-alleles as contributors to heterosis is available from many single-cross hybrids in maize (Otsuka *et al.*, 1972; Stuber *et al.*, 1973). Jinks and Jones (1958) observed a significant association between epistasis and heterosis in 80% of the crosses in *Nicotiana rustica.* However, they pointed out that not all heterotic crosses showed non-allelic interactions, and conversely, not all crosses showing non-allelic interactions, were heterotic. Wright (1968) visualized a *'net-like'* structure of population genotypes such that the variations of most characters are affected by many loci and that each gene replacement may have effects on many characters. Based on this perspective, epistasis should be one of the most important genetic components in the inheritance of quantitative characters. Stuber (1994) reported that epistasis effect could be dramatic when such an allele is masked by its corresponding dominant allele in a hybrid.

Cheverud and Routman (1995) identified two concepts of epistasis: physiological epistasis and statistical epistasis. Physiological epistasis occurs when genotypic values at one locus vary depending on alleles present at other loci. Statistical epistasis occurs when there are quantitative genetic variance components that can be attributed to interactions among genes. Physiological epistasis is a property of genotype that is unaffected by population gene frequencies, while statistical epistasis is a property of populations that does changes with gene frequency (Goodnight, 1999). Strinfield (1950) found that in many cases backcrossed showed in consistently higher yields than F_2. This suggests interaction, as if the gene combinations selected during inbreeding are broken up in F_2, but partially retained in backcrosses. Similarly, double cross hybrids in maize usually yield somewhat lower than the three-way crosses (Troyer, 1991). Schnell and Cockerham (1992) reported multiplicative interaction as a cause of heterosis and epistatic effects are non-linear function of one- locus effect and the mean.

Experimental evidences also suggest that intralocus genetic interactions or epistasis should also be considered (Pray and Goodnight, 1995). Yu *et al.* (1997) analyzed the genetic components underlying yield and its component traits by using an F_{2-3} populations derived from a highly heterotic rice cross, and detected a large number of digenic interactions involving loci distributed throughout the entire rice genome. They suggested that epistasis plays an important role as the genetic basis of heterosis. Li *et al.* (2001) and Luo *et al.* (2001) also suggested that epistasis and over-dominance are primary genetic basis of inbreeding depression and heterosis.

Physiological Basis of Heterosis

Physiological basis of heterosis is in general, concerned with metabolic superiority of hybrid over inbred parents and also the resultant effect on growth size and other developmental characteristics. Earlier, studies on physiological basis of heterosis related to embryo and seed sizes, growth rates in various stages of development, rates of reproduction and of various assimilation activities. It has been suggested that hybrid vigour resulted from large embryo and endosperm size of hybrid seeds as compared to those of inbreds. As a result, the rate of growth in seedling stage may be expected to be greater in hybrids than in inbreds. But these relationships are clearly demonstrated in some cases, while in other cases they are not detectable. There is evidence that increased size of hybrids is a result of an increase in both in size and number of cells. This and other observations indicate a basic difference in metabolic activities of hybrids and inbreds. Earlier, Whaley (1952) concluded that primary heterotic effect concerns growth regulators and enzymes. He suggested that hybrid embryo would be able to mobilize stored food materials earlier than those of inbreds due to a more efficient enzyme system. This would lead to superiority of hybrids in early seedling stages. Most of these studies concerned with the gross expressions of heterosis, *e.g.* growth rate, cell number and size *etc.* and their conclusions about the physiological causes of these expressions are highly speculative. There is no evidence to point out clearly the possible reasons for heterozygote advantage. Various studies demonstrate that heterozygote superiority may be produced in several different ways, which may differ considerably from one case to another. These mechanisms are briefly described below:

Reduced Amount of a Single Gene Product

In some cases, heterozygote produces an intermediate amount of a gene product, which may lead to the increased vigour and growth rate. Such a situation would be encountered when homozygote for dominant allele produces more and homozygote for the recessive allele produces less than the optimum level of a product. An example of such a situation is found in the bread mould (*Neurospora crassa)*. Gene *pab* is responsible for the production of p-aminobenzoic acid. The heterozygote *pab+* pab produces intermediate amounts of p-aminobenzoic acid and is faster in growth than two homozygotes, that is, *pab+* pab+ and *pab pab*.

Separate Gene Products

The two alleles may produce proteins with slightly different functions and properties. The heterozygote would possess the proteins produced by both the alleles, while homozygotes would have only one of the two proteins. Thus, heterozygotes may be expected to be more adapted to changes in the environment. Consequently, they would be superior to the two homozygotes. An example of such a situation is the sickle cell anemia in human beings (*Homo sapiens*). The homozygote *ss* does not survive because the defective haemoglobin β^s is unable to perform the functions of normal haemoglobin β^s. The heterozygote *Ss* is more resistant to malaria than normal homozygote *SS*. This is because the defective haemoglobin β^s produced by recessive alleles, is not suitable for the multiplication of malarial parasite, which multiplies in red blood cells. Thus, heterozygotes survive due to normal haemoglobin β^A produced by normal allele *S*, and are resistant to malarial parasite due to defective haemoglobin β^s produced by recessive allele.

Combined Gene Product

This is also known as hybrid product. Many enzymes are dimers or polymers, that is, they are made of two or more identical polypeptides. In such cases, products of two alleles in heterozygotes combine to produce an enzyme molecule with somewhat different properties as compared to enzymes found in two homozygotes. This new molecule is often detectable by electrophoresis in heterozygote, and is often termed as *hybrid substance.* Such enzyme molecules from heterozygotes may often differ from the enzyme found in homozygotes in their physical properties, *e.g.* heat sensitivity. The two alleles governing the enzyme *octanyl dehydogenase-I* produce heat sensitive enzymes in homozygous state. The enzyme from heterozygote, however, is less heat sensitive than enzymes produced by two homozygotes.

Effect in two Different Tissues

In some cases, heterozygote superiority may arise in a somewhat strange manner. The two alleles in homozygous state may produce high levels of an enzyme in two different tissues. The heterozygote may show intermediate levels of enzyme in both the tissues. In maize, allele Adh^f produces more activity of enzyme *alcohol dehydogenase* in the scutellar tissue of seeds, while allele Adh^s produces more activity in pollen grains. The hybrid exhibits intermediate activities in both the tissues.

Biochemical Basis of Heterosis

Strictly speaking, there is no biochemical model or explanation of heterosis, which is independent of one or the other of both genetic theories of heterosis. Nevertheless, biochemical approach to heterosis is mainly based on the basic concept of heterosis expression in traits like growth and grain yield which themselves are the end results of a series of biochemical reactions, each of which is controlled or catalyzed by one or more specific enzymes. Thus, the role of major metabolic enzymes and enzyme system in manifestation of heterosis effects has been recognized as the prime cause of heterosis. However, ultimate determination of biochemical manifestation of heterosis has been pivoted upon the genetic control. In most of the studies metabolic superiority of hybrid has been related to gene replication, gene-enzyme- information transfer, protein synthesis and physiological attributes like energy transfer for growth and photosynthetic activity. Superiority of hybrid in terms of biochemical gene products has been studied both in plants as well as lower organisms. There are some models, which explain biochemical basis of heterosis. They are :

Model of Single Gene Heterosis

Models of single gene heterosis with respect to biochemical characters have further furnished insight into the understanding of heterosis. These evidences primarily throw light on production of heterotic effect as a consequence of complementation or by providing the production of a balanced metabolic interaction, which provides the hybrid advantage. Earlier, Robbin (1941 and 1952) has reported the effect of internal biochemical factors on heterotic expression in root growth in tomato. He excised root tips from the varieties, Johannesfeur and Red Currant as well as their F_1 hybrids, and grew them in the solution supplemented with growth-promoting substances like pyridoxin and nicotinamide. He observed that Johannesfeuer showed greater response to pyridoxin while Red Currant to nicotinamide, suggesting thereby that these two varieties are not able to synthesize these growth factors on their own and their root growth is limited because of this bottleneck factor. However, hybrid is able to grow even in the absence of these two growth factors and is not deficient, *i.e.* it is able to synthesize them on its own. This observation suggested that the expression of heterotic advantages is an expression of the activity of favourable combination of biochemical growth factors produced as the consequence of complementary gene action in hybrids. Another example of interagenic complementation and production of a different effect (which in essence is analogous to over-dominance) have been given by Srb and Weyman (1966) with respect to peak gene in *Neurospora*. A number of recessive peak mutants occurs in *Neurospora* and usually give non-linear asci arrangement in ascus. But when different peak mutants are crossed, most of the combinations produce linear asci (Fig.2).

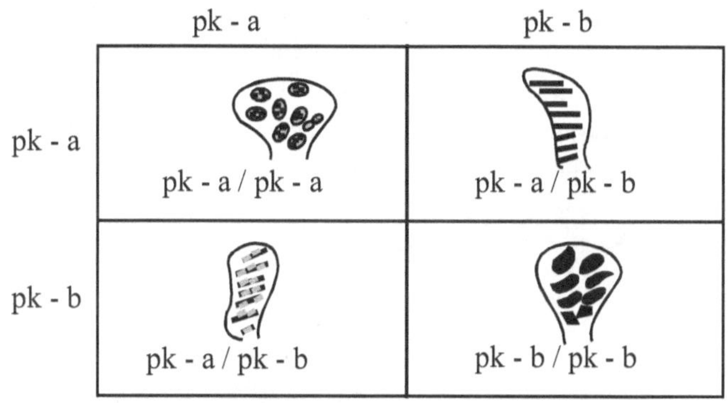

Figure 2: Intragenic complementation at the peak locus in *Neurospora* (after Srb and Weyman, 1966).

Allelic Diversity and Production of Balanced Metabolic Interaction

Heterosis due to heterozygosity at one locus as a result of balanced metabolic interaction has been reported by Emerson (1948) with respect to suppression of sulfonmide requiring character in *Neurospora*. He obtained a mutant, which required additional supply of p-aminobenzoic acid (pab) to grow on the medium. He obtained another mutant, which required the additional supply of sulfonamide (sfo) in growth medium to grow. It is known that sulfonamide usually forms a complex substance which counteracts and regulates the excessive production of p-aminobenzoic acid which, if unchecked or uncontrolled in biochemical path-way is poisonous to *Neurosora*. He crossed these two and obtained a double mutant carrying both p-aminobenzoic acid (pab), which blocks the synthesis of pab and sfo, which needs extra supply of sulfonamide. When he made heterokaryon of this double mutant pab/ sfo with a mutant which carries sfo and original type of pab which produces and does not require pab, *i.e.* (pab sfo pab+ sfo) this heterokaryon grows on minimal medium having neither sulfamide nor para-aminobenzoic acid. Emerson interpreted this instance of one gene heterosis as a consequence of 'growth resulting from a balance obtained by the production of para-aminobenzoic acid by one type of nucleus and lack of production or counteraction by other to give a tolerable optimum amount of pab thereby, giving it most favourable dose of para-aminobenzoic acid and by providing a favourable metabolic balance sufficient for growth.' In other words, the heterozygote (pab+ /pab) performed better than either of the homozygotes.

Production of Hybrid Substance

Irwine and Cole (1936) and Irwine (1952) observed the specificity in gene effects in heterozygotes and have reported the production of hybrid substances in pigeon-dove hybrids. A cross made between males of an Asiatic species, the pearlneck (*Streptopelia chinensis*), and was domesticated Ringdove female (*Strepto risoria*). The corpuseles of hybrid had almost all the substances common to each of the parental species along with a complex of antigens substances known as hybrid substance (Fig. 3). The production of hybrid substance is specific to this hybrid and is not found in other hybrids in crosses of pigeon and doves. It is suggested that production of extra hybrid substances in a hybrid is the consequence of heterozygosis and the interaction between genes. This, though, had not ultimately reflected in terms of heterosis in quantitative characters. However, it is suggested that such comparable specificity of gene effect might well have some relevance to the phenomenon of heterosis.

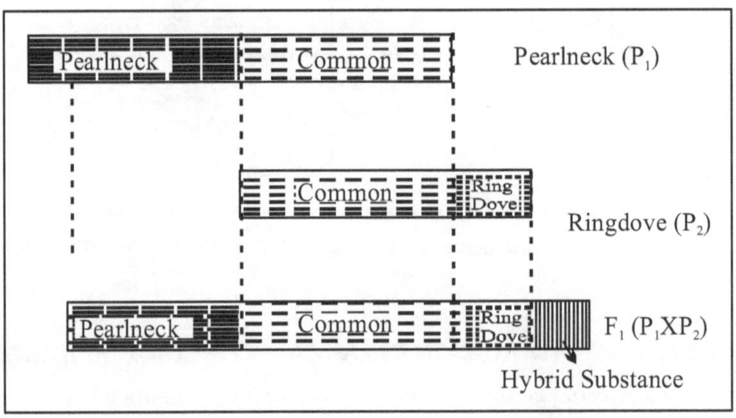

Figure 3: Production of hybrid substance in interspecific cross between Pearlneack x Ringdove (After Irwin, 1952).

Nitrate Reductase and Expression of Heterosis

Since plant breeders are basically interested in yield of crop plants which in turn often constitutes of the bulk of grain protein, carbohydrate *etc*. It is rather interesting to study the synthetic process of these biochemical products in hybrids and inbreds. The enzyme nitrate reductase (NR) is predominantly needed in nitrogen metabolism and plays a major role in the conversion of nitrates into ammonia. This is utilized in the production of various amino acids, which build up the total protein. Therefore, nitrate must be effectively reduced prior to its elaboration into amino acids. This enzyme initiates the reduction sequences. It is also a major control point for the systems that supply reduced nitrogen to the plant.

Theoretical Insight Into Complexities of Heterosis

A theoretical insight to illustrate the need for a balanced system and an integrated control of these systems to bring high production efficiency (with respect to production of a complicated automobile car, and its relevance to integration of enzymatic systems that might be operation with respect to phenomenon of heterosis) has been given by Hageman *et al.* (1967). It is clear that higher production of cars depends not only on the more raw materials or more production by one assembly line but by the properly coordinated rate of operation in all the assembly lines (Fig. 4). In a similar way, the heterotic effect in a single enzyme system in root zone, meristem, seedling or elsewhere alone could not be the basic cause of total heterotic vigour. The heterotic effect, as a matter of fact, lies in efficient, integration and timely operation of essential individual enzymes and optimum balance of enzymatically catalyzed metabolic systems.

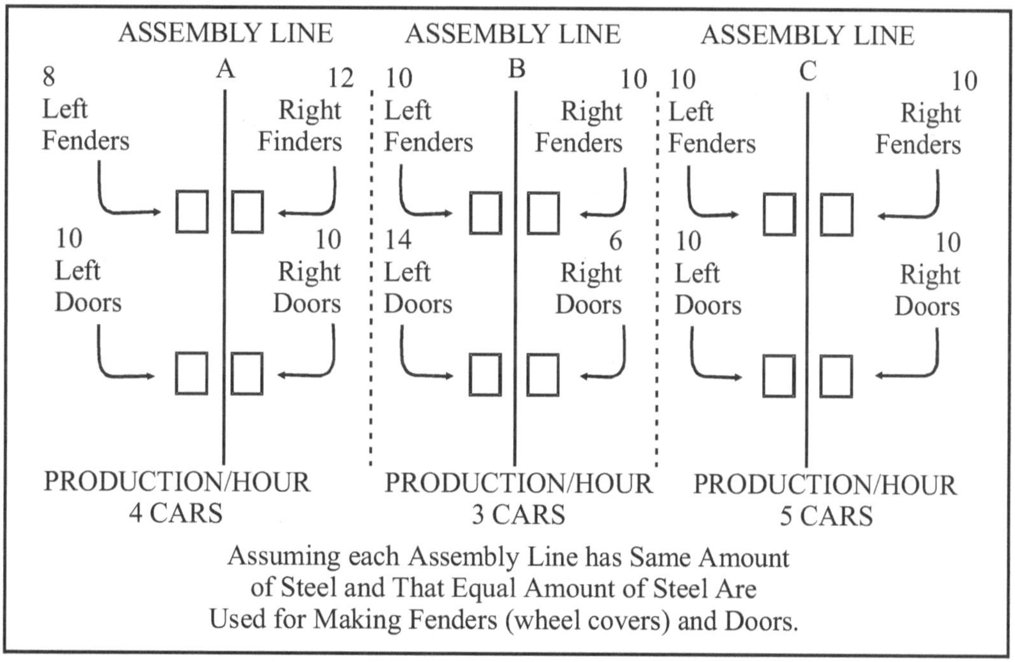

Figure 4: Illustrating the need for balanced and integrated control system of car production to explain the phenomenon of heterosis (After Hageman, Leng and Dudley, 1967).

The F_1 hybrids usually possess a more favourable genetic constitution and thus produce a better enzymatic balance for overall enzymatic efficiency than does either of the parents. Heterozygote has been observed to be usually intermediate in enzymatic activity, which as a matter of fact, is adequate for most of the major metabolic processes and may actually be more effective for overall metabolism. At enzymatic level, allele 'A' may specify one enzyme and allele 'a' a different enzyme. Thus, hybrid could possess both enzymes and enzyme forms which are available in one and separate inbreds. The inbreds on other hand rather has a fixed type of enzymatic balance due to inbreeding either at very high or at very low or at an otherwise ineffective level of activity. When these are there in hybrids, F_1 usually has better metabolic efficiency and whenever this balanced situation exists, this might be resulting in the ultimate expression of heterosis.

Cytoplasmic Basis of Heterosis

Based on the postulate that genetic polymorphism in any form is related to increased adaptability, Sarkissian (1972) and Srivastava (1975) considered organelle polymorphism or heterogeneity as a cause of hybrid vigour. Organelle polymorphism can result from recombination between organelle genomes derived from parental and maternal sources due to biparental (non- Mendelian) transmission. However, by the available evidence, biparental transmission of cytoplasmic orgenelles may be an exception rather than a rule. Cytoplasm is a limpid fluid, which is usually transmitted through the female line to each of the daughter cells taking part in the development of the organism. It is usually rich in RNA, and contains cytoplasmic bodies like mitochondria, centrioles *etc.* So far, we had been considering the role of genes exclusively for the expression of heterosis. However, a number of evidences are now available which indicate that heterosis may result from the interaction of genes and cytoplasm. Ruenbenbauer (1962) proposed that the yield of a hybrid is a function of both genome and cytoplasm. Srivastava (1981) also emphasized on chloroplast complementation causes heterosis in crop plants.

The term chloroplast complementation is generally used to indicate that greater activity is observed from a 1:1 mixture of isolated parental chloroplasts compared with mid- parental value. In parasexual hybrids, however, chloroplast from both parents may stay together and be distributed to daughter cells, with one type of chloroplast dominating the other (Melchers *et al.,* 1978 and Pelletier *et al.*, 1995). Mitochondria isolated from seedlings of hybrids and their parents showed different efficiencies of oxidative phosphorylation during ATP synthesis (Mc Daniel and Sarkissan, 1966). The mitochondria from non-heterotic hybrids had the same phosphorylative efficiency as that of parents was considered as an expression of heterosis (Srivastiva, 1983). Heterosis for mitochondrial activity in terms of respiration has also been reported in hybrids at late seedling stage in maize (Trivic and Leskovac, 1987).

Molecular Basis of Heterosis

The availability of molecular basis has made it possible to develop linkage maps with a high degree of resolution in a number of crop plants. Such linkage maps are useful for locating and manipulating quantitative traits loci (QTLs). Stuber *et al.* (1992) and Stuber

(1999) discussed several cases where heterosis has been analyzed with respect to gene and enzyme activities. Demonstrations of multimeric hybrid enzymes or isozyme spectrum due to non-allelic or intergenomic complementation provide evidence for a buffing mechanism at the subcellular level in hybrids. Through an analysis of relationship between protein amount polymorphisms and single-cross hybrid performance in maize, Leonardi *et al*. (1991) suggested that genes controlling protein amounts, particularly those with multiple effects, directly affected the expression of hybrid vigour. An important question in relation to molecular basis of heterosis is, whether the simple complementation of different slightly deleterious alleles generates a growth response that can explain heterosis.

Birchler (2003) provided a strong argument against simple complementation to explain heterosis on the basis of three observations: (i) although inbred lines have been improved greatly over the decades, the magnitude of heterosis has not diminished but has increased slightly. If heterosis is caused by the complementation of deleterious alleles and inbred lines, have been purged of the most of heterosis might be expected to decline somewhat. The slight increase in hybrid vigour over the years might have occurred by selecting alleles at the right set of loci that produce the best combinations in hybrids to bring about heterosis; (ii) progressive heterosis in tetraploids, where vigour improves with the greater number of distinct genomes present, and (iii) inbreeding depression in tetraploids of many species proceeds faster than expected based on the homozygosis of alleles. One possible molecular model of heterosis, outlined recently by Birchler (2003), was that the combination of different alleles produces an interaction that causes gene expression in hybrid to deviate relative to mid-parent predictions (*e.g.* by an up regulation of many housekeeping genes). This scenario might best be considered the result of regulatory gene allelic interactions. Song and Messing (2003) provided evidence for the altered regulatory effects in hybrids in comparison with parental lines.

Regulatory genes in multicellular organisms often function as part of complexes, and exhibit some measures of dosage dependence, where as housekeeping genes or those that encode metabolic functions are less likely to show dosage effect. Quantitative traits are expected to be controlled in large part by multiple dosage-dependent regulatory loci. This logic has led to the proposal by Birchler (2003) that heterosis is the result of different alleles being present at loci that contribute to regulatory hierarchies that control quantitative traits. Studies at protein level clearly indicate that the expression of many genes does not exhibit the expected mid-parent value (*e.g.* Leonardi *et al*., 1991 and Song and Messing, 2003). The reasons for such a deviation are still not clear. Detailed analysis of gene expression patterns using microarrays (available in some model plants, such as rice), could potentially offer interesting insights about the patterns of changes in expression of a spectrum of genes in hybrids and in which direction the changes occur. The degree of DNA methylation as a possible molecular regulator of heterosis was proposed by some researchers (Stuber, 1999).

This hypothesis suggests a gradual accumulation of methylation during inbreeding or selfing, which is then released and/or re-patterned when these lines are crossed to generate hybrids. Tsaftaris *et al*. (1997) strongly supported this hypothesis, and reported that hybrids are less methylated than inbreds. In addition, their data also showed that inbreds varied in their methylation status and improved lines are less methylated than older low-yielding lines. Under stressful environments (such as high-density planting), the F_1 hybrids are found to be more resistant to such a density induced increase in methylation than their parental inbreds. Epigenetic modification such as DNA methylation could be heritable and influence diverse developmental processes in plants. There is now an increasing appreciation of influences of such epigenetic changes on manifestation of agronomically important traits.

Biometrical Basis of Heterosis

The additive dominance model, based on the biometrical procedures developed by Mather and Jinks (1971,1982), is the easiest method for understanding heterosis. Breeding material for this procedure includes $P_{1,}$ P_2, $F_{1,}$ $F_{2.,}$ $B_1(F_1 \times P_1)$, and B_2 $(F_1 \times P_2)$, and B_2 $(F_1 \times P_2)$ of a cross. Analysis of mean values provides a test of adequacy of additive dominance model. In the event of heterobeltiosis, $F_1 - P_1$ = [h]-[d], [h] must be positive and greater than [d] for few or most of the loci for heterosis to occur. There must be over-dominance (h/d > 1) at some or all loci, and degree of association or dispersion of alleles (r_d) must be less than 1 or approaching 0. The genetic expectations of heterosis are:

$$\sum_{i=1}^{k} hi > \sum_{i=1}^{k} di,$$

where, k is number of loci for which the parents differ, di the additive effects of the ith locus, and hi the dominance effects of ith locus. Heterosis is maximum when dispersed genes show unidirectional dominance and there is no internal cancellation of h's of different signs. If over-dominance is the cause, Σhil >Σdil, the hybrid will always be superior to pure breeding line. On contrary, if heterosis results from gene dispersion (rd<1) and dominance, it is theoretically possible to extract pure lines, which equals or exceeds the hybrid. Thus, such an extreme line will deviate from mean m by as much as F_1 if Σhi >Σdi, or more than F_1 if Σhi >/Σdi. The choice between line *vs.* hybrid is a genetic enigma; the best way to solve it is by estimation of dominance ratio, √H/D which requires relative estimates of additive genetic [D=(Σdi^2)] and dominance [H=(Σhi^2)] components of variation. This ratio is a true estimate of the weighted degree of dominance overall loci. Over-dominance is indicated if the ratio is more that 1; a value equal to 1 suggests dominance, and less than 1 means incomplete dominance.

Linkage and Linkage Disequilibrium

For additive genetic component (D), the bias in genetic interpretation results from the excess of coupling or of repulsion linkages. The coupling phase increasing and repulsion decreasing it. In case of dominance (H), it means an excess of either reinforcing or opposing combinations of dominance deviations at linked loci (Mather and Jinks, 1971). The reinforcing type of linkage inflates H and opposing type deflates it. The linkage bias on D and H decreases with the number of cycles of recombination, as indicated by rank increases:

Excess of coupling linkages	:	$D_1 > D_2 > D_3 > D$
Excess of repulsion linkages	:	$D_1 < D_2 < D_3 < D$
Directional dominance	:	$H_1 > H_2 > H_3 > H$
Directional dominance	:	$H_1 < H_2 < H_3 < H$

If heterosis is due to directionally dominant genes with complete or incomplete dominance, linkages will be mainly in the repulsion phase and will deflate the estimate of D and inflate the estimate of H. This will cause an overestimate of the degree of dominance. Biometrical methods to test linkage, to estimate its predominant phase (Mather, 1949), and to detect linkage disequilibrium (Perkins and Jinks, 1970) are available. It must, however, be realized that linkage disequilibrium biases the specification of heterosis or components of heterosis only in the presence of epistasis.

Nonallelic Interactions

Involvement of nonallelic interactions in reducing or enhancing the magnitude of heterosis is well-documented (Jinks 1955; Jinks and Jones, 1958). It is possible to detect and estimate epistases by fitting models, which include their contribution towards means, variances, *etc.* through the analysis of populations derived from crosses (Mather and Jones 1971, 1982). Three types of parameters are needed for a pair of genes to infer interaction of homozygotes at both loci : (i) of homozygotes at one and heterozygote at another (j) and heterozygote at both the loci (l). All interactions can be generalized for any number of loci (k)

$$[i] = r_1 \overset{k(k-1)}{\Sigma i}$$

$$[j] = r_1 \overset{k(k-1)}{\Sigma i}$$

$$[l] = r_1 \overset{k(k-1)}{\Sigma i}$$

The coefficients r_1 and r_j accommodate the effects of distribution of increasing and decreasing alleles among pairs of loci in parental lines and are analogous to r_d (Jinks, 1983). When all the pairs are associated, r_i and r_j are equal to 1, whereas with complete dispersion of interacting pairs $r_i = 1$ and $r_j = 0$. These parameters can be estimated from the basic set of generations. After making allowance for gene interactions, heterosis is:

$$F_1 - P_1 = ([h] + [l]) - ([d] + [i]).$$

For positive heterosis to occur ([h] + [l]) must be positive and greater in magnitude than ([d] + [i]). Heterosis is higher when complementary epistasis ([h] and [l] of the same sign combine with dispersion ($r_d = 0$ and $r_i = -1$). It is minimized in the presence of duplicate epistases (different signs of [h] and [l]) and complete association ($r_d = r_i = 1$) occurring together. The relative contribution of each of these four components lies in the values of r_d and r_i and the sign and magnitude of the individual h's, l's and I's. The component of variances can be estimated as $D = \Sigma d^2$, $H = \Sigma h^2$, $I = \Sigma i^2$ and $L = \Sigma l^2$ for additive dominance, additive x additive, and dominance x dominance variances, respectively (Verma and Chahal, 1993). It is possible to differenciate trait between over-dominance and dispersion of completely or incompletely dominant genes from the ratio $\sqrt{(H/D)}$. Similarly, the ratio $\sqrt{(L/I)}$ allows differentiation between the dispersion of interacting genes and over interaction. No method is available to estimate D, H, I and L, when all are present together. Epistasis causes bias in the estimation of the degree of dominance if it is estimated from the analysis of six basic families (P_1, P_2, F_1, F_2, B_1, B_2). Dispersed complementary genes inflate the ratio, leading traits misinterpretation as over- dominance.

On the other hand, over-dominance may be misinterpreted as complete or even partial dominance if the degree of dominance is under estimated due to associated complementary or dispersed duplicate genes. In triple test cross analysis, if dispersed or associated genes show either complementary or duplicate action, the estimate of H is inflated. Similarly, the estimate of D is also inflated except for dispersed duplicates and inflation may be, more or less the same as for H component. Thus, chances of misinterpreting the degree of dominance in presence of epistasis are less. The presence of complementary epistasis (similar sign of (h) and (I) may not cause serious bias, but in the event of duplicate epistasis (opposite sign of (h) and (I), D is likely to be deflated more as compared to the inflatory effect on H, leading to spurious over-domiance. Trigenic and higher order interactions can also bias such estimates and experiments must take this into account.

Linkage Between Interacting Genes

Linkage has no effect on the mean of non-segregating generations if epistasis is absent, and thus does not affect either the specifications or the magnitude of heterosis. However, since linkage does not affect the mean of segregating generations (B_1, B_2 and F_2), it corrupts the estimates and the relative contributions of different kinds of gene action and interaction. Linkage between interacting genes results in an overestimation of dominance and an underestimation of epistasis interactions as contributory factors for heterosis. Linkage of interacting genes can be detected in triple-test cross analysis (Jinks, 1978).

Maternal Effects

Maternal effects are present if the mean performances of reciprocal crosses differ. These differences result from cytoplasmic influence on the expression of nuclear genes. Mather and Jinks (1971) symbolized the maternal effects as $/d_m/$ for P_1 maternal parent and - $/d_m/$ for P_2 mother. Therefore:

$$P_1 = m + [d] + [dm]$$
$$P_2 = m - [d] - [dm]$$
$$F_1 = m + [h] + [dm] \ (P_1 \text{ mother})$$
$$R_F = m + [h] + [dm] \ (P_2 \text{ mother})$$

A positive or negative sign of the maternal component $/d_m/$ will depend on whether the P_1 mother contributes more or less than the P_2 mother. If F_1 shows heterosis over either P_1 or P_2 as maternal parent. F_2 heterosis may be possible. In F_2/RF_2, the expected F_2 heterosis ($F_2 - P_1$ is expressed as:

$$\tfrac{1}{2}[h] + [h_m] - [d] - [d_m].$$

Therefore, F_2 and RF_2 will show greater heterosis than F_1 or RF_1 if $[h_m]$, the maternal contribution of F_1 mother is more than ½ [h]. Unbiased estimates of D and H can be obtained from family variances of basic generations and from an F_2 triple-test cross testers (P_1, P_2 and F_1) are used as maternal perents (Mather and Jinks, 1971).

Genotype X Environmental Interactions

The differential response of homozygous genotypes to environmental conditions influences the magnitude of heterosis, but no such effect is expected if both homozygous and heterozygous have similar response. Relative sensitivity is symbolized as g_d and g_h for homozygous and heterozygous genotypes, respectively, and these measures of sensitivity (g_d, g_h) are linear functions of additive environmental effect (e_j) (Bucio-Alania *et al.,* 1969), due to which:

$$g_{dj} = b_d e_j$$

$$\text{and, } g_{hj} = b_h e_j$$

where b_d and b_h are linear regression coefficients. These coefficients can be detected by raising parents P_1, P_2 along with F_1, F_2, B_1, and B_2 over several environments. The actual magnitude of heterosis in any environment (e_j) will then be:

$$F_{1j} - P_{1j} = [h] - [d] + [(b_h - b_d)\ ej]$$

Thus, if $b_h = b_d$, heterosis will remain constant across environments. Contrarily, if $b_h < b_d$, heterosis will be less, and if $b_h > b_d$, it will be more in above-average environments. In presence of interacrions, although the performance mean will be specified by m, [d] and [h], the magnitude will vary with difference in environmental component e_j, specified by b_d and b_h. It is, therefore important to regard environment sensitivity as a character because it may be under the control of the same genetic system that determines the mean performance.

3

MANIFESTATION OF HETEROSIS

Basic Requirements

Hybrid seeds could be produced at cheap rates only if the crop species have certain built-in morphological or physiological mechanisms, suitable for the control of parentage for successful hybrid seed production and its floral morphology, is amenable to it. Some of the basic requirements and techniques employed in large-scale commercial hybrid seed production of crop plants are as follows:

- Availability of a proven heterotic hybrid combinations which could distinctly and profitably surpass the yield levels of commercial variety being grown.
- Availability of an easy, economic and effective means of eliminating or rendering functionless of male part of bisexual seed parent, mechanically, genetically or even biochemically.
- Availability of a strong fertility restoration system (in case of the use of cytoplasmically governed male sterility system) or availability of a tightly linked marker gene system in genetically governed male sterility along with full and detectable expression of self-incompatibility.
- Absence of modifier genes or gene systems in case of the use of self-incompatibility for hybrid seed production.
- Complete synchronization of flowering in both seed and pollen parents.
- Free, unrestricted and natural transfer of pollen from pollen to seed parents.
- Good seed setting on seed parent.
- A skilled organized effort for large-scale seed production, certification, processing and well-knitted distribution of hybrid seeds.

Types of Hybrids Used for Commercial Cultivation

Various types of hybrids utilized in hybrid seed programme of vegetable crops are single-crosses, double-crosses, double-top-crosses, triple-crosses *etc*. An account of these hybrids are given in Table 1.

Table 1: Commercial hybrids used in vegetable crops

Hybrid	Symbolic Representation	Crops
Single-crosses	(A x B)	Brinjal, carrot,chilli cucurbits, onion, tomato and watermelon
Double-crosses	(A x B) x (C x D)	Sugarbeet
Triple-crosses	[(A x B) x (C x D) x (E x F)]	Cabbage and kale
All types of hybrids	-	Potato

A, B, C, D, E and F are constituent inbred lines of hybrids.

Types of Crossing

Five different types of crossing used in the production of seed of F_1 hybrids are as follows:

- ***Both emasculation and pollination have done by hand***

  Tomato, brinjal, capsicum and muskmelon (and romonoecious varieties).

- ***Artificial pollination alone required***

  Cucurbits: Cucumber, watermelon, muskmelon, pumpkin and squash.

  Tomato: Using male sterile lines.

  Cucumber: Using pistillate lines.

- ***Emasculation and use of wind pollination***

  Sweet corn by detasselling.

  Spinach: Emasculation by pulling out male plants.

- ***Emasculation or removal of male flowers and use of insect-pest pollination***

  Cucumber, watermelon, pumpkin, squash and bottle gourd.

- ***Free insect-pollination***

  Using self-incompatibility: Cabbage, Chinese cabbage, Radish, Turnip, brussels'sprouts, broccoli and Cauliflower.

  Using male sterility: Onion, carrot and radish.

Procedures of Hybrid Seed Production

Heterosis breeding involves three important stages, namely production of inbred lines, testing of combining ability of inbred lines, and production of seed in bulk of promising and desirable F_1 hybrids.

Inbreeding and Production of Inbred Lines

Inbreeding is the mating of closely related individual, which results in a large number of lethal weak types and a general decline in their vigour, fertility and fecundity. Recessive, lethal, and deleterious genes are usually concealed in a heterozygous system, which are increased in homozygous balance after inbreeding. Inbreeding increases homozygosity and reduces the proportion of heterozygosity in population, thus reducing the vigour of plants. For inbreeding, usually selfing and sibmating are practised, for selfing in cole crops, bud pollination is practised. For quick and efficient selfing, electrically-aided pollination and thermally-aided pollination are followed. In self-pollinated crops, *i.e.* tomatoes, brinjal, chilli and beans, pure lines maintained in collection are used as inbred lines. In cross-pollinated crops like onion, cabbage, cauliflower, brussels'sprouts, radish, turnip, carrot and beet, it is necessary to produce inbred lines. However, in cucurbits like cucumber, pumpkin, squash, bottle gourd, watermelon and muskmelon in which varieties can be maintained in pure form by inbreeding without loss of vogour. It is not necessary to produce inbred lines.

In cucurbitaceous crops, homozygous varieties can be used directly in place of inbred lines as done in self-pollinated crops like tomato, brinjal and capsicum. When inbreds are to be produced, selection is done within and between the inbred lines to obtain promising inbreds. These inbred lines are usually used in single crosses in onion, cucurbits, cabbage, cauliflower, broccoli, Brussels'sprouts, carrot and spinach, in double crosses in cabbage and sweet corn or in three way crosses in sweet corn and onion. For achieving uniformity in hybrids, it is essential to inbreed the parents. The degree of inbreeding required in parents will be determined by the extent of uniformity desired. The number of generations of artificial inbreeding or selfing depends upon the degree of inbreeding depression in a crop. A high degree of inbreeding has been observed in Brussels' sprouts, marrow-stem, kale, cabbage and broccoli and a moderate to low degree in radish, carrot, onion and beet.

In cole crops, there is a high degree of inbreeding depression, therefore it requires five generations of inbreeding, but carrot and onion require only 3-4 generations of inbreeding, as they show moderate degree of inbreeding depression. In cucurbits, inbreeding can be practised to obtained homozygous true-breeding varieties without any loss of vigour. Uniformity may not be desired so much in hybrids in which disease resistance is an important consideration as heterogeneity may offer greater insurance against diseases, particularly to a new pathogen or physiological race. Therefore, testing of uniformity of parents is essential as the uniformity of hybrids depends entirely upon the homozygosity of parents. When there is more homozygosity, there is a problem of maintenance of inbred lines. Proper maintenance of parental lines is an integral part of hybrid seed production.

Testing Combining Ability of Inbred Line

The successful exploitation of hybrid vigour depends upon combining abilities (gca and sca) of parental lines and techniques used for economical hybrid seed production. It has been observed that some varieties have better ability to transmit high yield and quality to their progeny. Sometimes there are also differences in reciprocal hybrids. Generally, sca value of the cross gives better predictive information than gca of parents. Besides, sca is also more reliable for the selection of parents for hybrid production. However, parents can also be selected on the basis of genetic diversity or crossing widely different varieties in a number of vegetable crops, which produces potential hybrids. The combining ability of inbred lines can be tested in topcross, single cross or paired cross, diallele cross and poly cross. The top-cross and poly cross tests are helpful in determining gca ability, while by single or pair crosses, sca can be determined. The diallele cross, which include a complete set of all possible combinations of crosses, determine both the general and specific combining abilities. It is convenient to follow top-cross in sweet corn and beet, poly cross in cabbage, Brussels' sprouts, and cauliflower in which inbred lines can be easily maintained by vegetative propagation, and single and diallel crosses in onion, radish, turnip, cabbage, cauliflower, Brussels'sprouts, broccoli, cucumber, watermelon, muskmelon, pumpkin, squash, spinach, tomato and brinjal. Nieuwhof (1969) had also suggested the use of paired cross, diallel cross and poly cross in cole crops.

Production of Hybrid Seed

Shull (1909) was the first to produce a single-cross hybrid in maize involving two inbred lines. The practical utilization of a single-cross hybrid suffered from the limitation of its higher cost. It is produced from inbreds, which are poor yielder, thus less quantity of hybrid seed is produced. This is the reason that single-cross hybrids are uneconomical and did not attain commercial importance. However, in tomato, pepper, cucumber, muskmelon, and in some cole crops, single-cross hybrids are being commercially used for hybrid seed production. The utilization of hybrid in maize came into practice only when Jones (1918) produced a double cross hybrid by crossing two single hybrids, and this was economical because of a heterotic hybrid was taken as the seed parent. In maize such hybrids are very popular in maize-growing countries. Production of a double-cross hybrids have also been suggested in cabbage (Odland and Noll, 1950) and marrow stem–kale (Thompson, 1959). In cole crops, particularly in kale, a triple–cross hybrid has been suggested by Thompson (1964) and in carrot, it has been reported by Erickson *et al* (1982). Single-cross and triple-cross hybrids have been reported by Thompson (1955) in spinach. Although, double-cross and triple-cross hybrids are economical but they lack uniformity. In pointed-head cabbage, single-cross hybrid has been found superior to double-cross (Nieuwhof and Garretsen , 1975). Therefore in vegetable crops, single-cross hybrids are generally preferred than double and triple-hybrids. Thus, single-cross is best alternative for horticultural quality and value.

It is comparatively easy to produce the seeds of F_1 hybrid in dioecious crops like spinach and asparagus, monoecious vegetables like cucurbits and sweet corn, and self-incompatible crops like cabbage, brussels'sprouts, broccoli, cauliflower, radish and turnip than in self-fertile species having hermaphrodite flowers, such as tomato, brinjal and capsicum. Three main mechanisms utilizing natural pollination have facilitated the procedure of crossing. These include sex expression, dioecism in spinach, asparagus, and monoecious in cucurbits and sweet corn; self-incompatibility in cabbage, Chinese cabbage, brusseles' sprouts, broccoli, cauliflower, radish and turnip; and sterility-both genetically male sterility as in tomato, brinjal, capsicum, onion, pumpkin and squash, and triploidy in seedless watermelon. In self-fertile species possessing hermaphrodite flowers, emasculation is usually done either by hand or genetically with the help of male-sterility genes.

Manifestation of Hybrid Vigour

Various workers have reported manifestation of hybrid vigour in a number of vegetables. These differ from the common agricultural crops in that almost all parts of the plant are used, depending on individual kind. The parts may be swollen root (carrot), modified stem (knol-khol), tuber (potato), corm (colocasia and amorphophallus), hypocotyl (radish, turnip and beet), leaf (spinach, beet, cabbage and lettuce), flower (cauliflower), fruit (legumes like pea, French bean, lablab bean, cowpea and cluster bean; tomato, brinjal, okra, cucurbits and rat-tail radish) and seed (green pea, lima bean and lallab bean). Therefore, investigation on hybrid vigour, the response of various parts to vigour is an important feature. Heterosis in vegetable crops is manifested in different ways like: (i) high yield due to larger size or weight of fruits, bulbs or heads (cabbage, watermelon, onion, pumpkin, squash and okra), (ii) uniformity in size (onion and cabbage), (iii) uniformity in maturity (onion and sweet corn) (iv) early maturity (cabbage, watermelon, muskmelon, onion, tomato and brinjal), (v) larger number of marketable heads or bulbs per unit area (cabbage and onion), (vi) better resistance to insect-pest (onion), (vii) better resistance to diseases (spinach, cucumber, tomato and watermelon), (viii) better fruit quality (tomato, watermelon, better flavour and more sugar) and (ix) better and wider adaptability to environmental conditions (sweet corn). Various studies on extent of heterosis in vegetable crops are summarized below:

Solanaceous Crops

Tomato

The phenomenon of hybrid vigour in tomato was first observed as early as in the beginning of 20th century (Hedrick and Booth, 1908). Wellington (1912) and Stuckey (1916) reported that F_1 hybrids considerably excelled their parents in earliness and total yields. Wellington (1922) proposed the large-scale use of F_1 hybrids of tomato for the first time. In tomato, reported range of heterosis for various yield attributes is 40-60%. Earlier in Indian conditions, the manifestation of heterosis has been reported by Pal and Singh (1943a) for fruit size; Choudhury and Khanna (1972) for earliness; Bhutani *et al.* (1973) for total yield; Kalloo *et al.* (1974) for number of locules and Reddy and Reddy (1994) for fruit quality. A work of Rai (1992) revealed that positive heterosis for plant height ranged from 5.36% to 56.29% and from 1.48 to 133.22% over superior and mid-parent, respectively, while for number of primary branches, Kumar *et al.* (1995) observed 83.37% heterosis over superior parent and Rai *et al.* (1996) also noted 98 and 133.22% heterosis for this trait over superior and mid-parent, respectively. Early in flowering expressed as negative heterosis, is desirable for early maturity. The work of Rai and Joshi (1994) revealed that negative heterosis for this attribute over mid and superior parent in the range of –2.66 to –21.05% and from –2.97 to –26.67%, respectively. High and positive heterosis is a desirable feature in tomato for number of fruits/plant. Dundi (1991) recorded the extent of heterosis from 48.38 to 93.49% over better parent. However Dev *et al.* (1994) and Kumar *et al.* (1995) found 115.7% and 193.65% heterosis for number of fruits/plant over better parent and superior parent, respectively. High and positive heterosis in tomato for number of fruits/plant has also been reported up to 79.66 and 143.68% over superior and mid-parent, respectively by Rai *et al.* (1998).

Dev *et al.* (1994) found that a cross of EC-156 x Marglobe showed high heterosis for number of fruits/plants which gave 83.18% more yield than better parent. Similarly, Kumar *et al.* (1995) reported 87.06 % heterosis over superior for yield. Singh *et al.* (1995) calculated 79.72% and 57.86% significant and positive heterosis over better parent in cross combinations NDT-120 x Kalyanpur Kuber and NDT-5 x NDT-21, respectively. Rai *et al.* (1998) observed the highest heterosis in cross HS-101 x Punjab Chhuhara for yield/plant to an extent of 133.30% and 176.40% over superior and mid-parent, respectively. Firmness of fruit is an important character for long distance transportation, which is associated with pericarp thickness and less number of locules. Anbu *et al.* (1981) have recorded a heterosis of 20.90% over the best parent. Positive heterosis for number of locules has also been reported by Kalloo *et al.* (1974). Rai *et al.* (1996) recorded 22.81% and 38.49% heterosis over better parent and mid-parent, respectively for number of locules/fruit in cross HS-101 x Pusa Ruby. However, Prabhushanker (1990) and Dundi (1991) have reported negative heterosis for same trait.

Positive and high degree of heterosis for pericarp thickness has also been recorded by Yadav *et al.* (1991). High degree of heterosis over better parent for pericarp thickness has been observed in cross S-14 x Pusa Ruby (Patil and Patil, 1988), Early Dwarf x S-12 (Dod *et al.,* 1992) and HS-101 x Flora-Dade (Rai *et al.,* 1996). The importance of fruit dry-matter content is well recognized for preparation of various products made from tomato. The extent of positive heterosis for fruit dry-matter content ranged from 36.94 to 71.80% over better parent and mid-parent, respectively (Rai *et al.,* 1996). High TSS in fruits is desirable for quality point of view and high TSS has been observed by Sonone *et al.* (1981); Daundi (1991) and Rai (1992). Like fruit dry-matter content, TSS and acidity are also major components for processed products. Rai *et al.* (1996) observed both increase and decrease in heterosis over better parent and mid-parent for fruit acidity. Some of the crosses showing high heterosis in tomato are given in Table 2.

Table 2: List of hybrids showing high heterosis

Hybrid	Attributes	References
Best of All x Pusa Early Dwarf and Pusa Ruby x Rutgers	Yield	Choudhury *et al.*, 1965
MS x 6L, MS x RD, MS x RD and MS x ZR	Yield	Singh and Nandpuri, 1970
Pusa Red Plum x Horkheff and	Yield	Khanna *et al.*, 1972
Marglobe x CR	Yield	Chaudhury and Khanna, 1972
Punjab Tropic x S-120 and Nematex x Angurlata	Yield	Bhutani *et al.*, 1973
Atkinson x HM-82782, Ostenkinskiz x HM-82782 and Atkinson x Nova	Yield	Nandpuri *et al.,* 1974 a, b
Kalyanpur Kuber x HBS	Total & early yield	Mittal *et al.*, 1974 a, b
Sel-3–7 x HM-8278	Number of fruits and fruit size	Nandpuri *et al.,* 1975
Sel-313 x Sel-1–2, Pusa Ruby x HS-101, Pusa Ruby x Sioux		Gill *et al.,* 1976 and
HS-101 x Pusa Early Dwarf and HS-101 x Pusa Early Dwarf	Yield and quality	Singh *et al.*, 1966
EC-55055 x Punjab Tropic	Pericarp thickness	Nandpuri and Tyagi, 1976
Pusa Ruby x HS-101, Pusa Ruby x Sioux, Pusa Early Dwarf x Sioux and HS-101x Pusa Early Dwarf	Yield	Singh *et al.*, 1978
Peru-2 x Roma and Mosasi x B-2747	Early yield	Swamy and Mathai, 1978
HS-101 x Sioux, Pusa Early Dwarf x HS-101 and Keckruth Ageti x Pusa Early Dwarf	Ascorbic acid	Singh *et al.*, 1979
Pusa Red Plum x Punjab Chhuhara and Best of All x HS-101	Ascorbic acid	Bhutani, 1980
Rossol x Punjab Chhuhara	Fruit weight	Virdelwala, 1981
Gamed x Chiku Grande, Roma x Punjab Chhuhara and Punjab Chhuhara x Kalyanpur Angurlata	Yield	Sidhu *et al.*, 1981

Contd...

Table 2. ***Contd...***

Hybrid	Attributes	References
Sel-152 x Roma, EC-6050 x Lalmani, EC-12156 x Lalmani and EC-121193 x Gola	Yield	Jaiswal *et al.*, 1984
Punjab Chhuhara x Sweet-72 and S-14 x Pusa Ruby	Total soluble solids, tritable acidity and pericarp thickness	Patil and Patil, 1988
Walient x French Variety and LA-402 x EC-118271	Fruit dry-matter content	Yadav *et al.*, 1988
LE-3x LE-1052	Yield	Kanthaswamy, 1989
L-15 x DWD-2 and DWD-2 x UC-2048	High yield	Mandal *et al.,* 1989
CO-3 x LE-812, CO-3 x LE 376 & PKM-1x LE-376	Processing attributes	Shoba and Aruangam, 1991
UC-204B x 79B-1390	High shelf-life	Dundi and Madalageri, 1991
HS-101 x S-12	Earliness, number of fruits/plant and plant height	Dod *et al.*, 1992
BT-1 x BT-10, BT-10 x LE-79 and BT-10 x KT-10	High yield	Bora *et al.,* 1993
Ohio-7663 x Rossol	High yield	Sidhu *et al.,* 1993
LE-75 x LE-76 and LE-76 x LE-22	Moisture stress condition	Natgragan, 1993
Punjab Chhuhara x 84-8	Yield	Singh and Singh, 1993
NT-3 x Flora-Dade	Earliness	Rai and Joshi, 1994
NT-3 x Pusa Ruby	Resistant to blotchy ripening	Rai and Joshi, 1994
Punjab Chhuhara x Roma	Yield	Pujari and Kale, 1994
EC-156 x Marglobe	Number of fruits per plant	Dev *et al.,* 1994
NDT-120 x Kalyanpur Kuber, NDT-5 x NDT-21, NDT-90 x NDT-21, NDT-5 x NDT-21 and NDT-120 x NDT-5	Number of fruits	Singh *et al.,* 1995
Hisar Arun x Sel-30, Sel-30 x Flora-Dade, Hisar Arun x Ace and Hisar Arun x Flora - Dade	Yield	Kumar *et al.,* 1995
HS-101 x S-12, Pusa Early Dwarf x S-12 and Pusa Ruby x S-12	Yield	Dod *et al.,* 1995
NT-3 x Punjab Chhuhara	Radial fruit cracking	Rai *et al.*, 1996
CO-3 x LE-113, LE-812RS x LE-1139, LE-812 RS x FFS- 120 and LE-812 RS x IIHR-709	Yield	Aruna and Veeragavatham, 1996
L-4139 x CL-246-0-3-B-9-oo	High temperature stress	Rajadhav *et al.*, 1996
20/6 Alcobasa x L-58, 20/5 Alocobasa x N-229 8MF 6 and 20/4 Alcobasa x L-58	Least infestation of TLCV and white fly	Dharmati *et al.*, 1996

Contd...

Table 2. ***Contd...***

Hybrid	Attributes	References
Flora- Dade x Punjab Chhuhara and NT-3 x Punjab Chhuhara	Less number of locules	Rai *et al.,* 1996
Flora – Dade x Punjab Chhuhara	Days to first flow.ering Number of fruits/plant	
Pant Bahar x Flora -Dade	Fruit weight	
BT-1 x Arka Vikas		Ghosh *et al.,* 1997
Arka Vikas x Flora-Dade	Equatorial fruit diameter	
HT-12 x NT-3	Pericarp thickness	
Pant Bahar x Pusa Ruby	TSS	
NT-3 x Flora-Dade	High pericarp thickness	Rai *et al.,* 1997
L-211 x S-281, C-121 x N-232	Processing purpose	Singh *et al.,* 1998
Pusa Ruby x Flora- Dade, NT-3 x HS-101, NT-3 x Pusa Ruby, Flora- Dade x Hisar Arun, and Sel-18 (Hisar Lalima) x Flora-Dade	Yield and its traits	Rai *et al.,* 1998
BTN-46 x Roma, BTL x AC-402 and BTR-49 x Roma	Yield	Sharma *et al.,* 1999
Arka Saurabh x Arka Ahuti and IIHR-709 x LE-812	Fruit weight	Mangeshwari and Natrajan, 1999
LE-812 x Arka Saurabh	TSS	
IIHR-709 x LE-812	Titratable acidity	
Arka Saurabh x Arka Ahuti	Ascorbic acid	
Edkawy x *L. pimpinellifolium*	Salinity tolerance	Hassan *et al.,* 1999
EC-4881 x EC-130204, EC-429 x Pant T-1, EC-41025 x Pant T-1, & EC-32557x Pant T-1	Yield per plant	Baishya *et al.,* 2001
Arka Saurabh x H-711492 and H-711492 x AC-402	Yield	Fageria *et al.,* 2001
BTL-33 x Roma, BTN x Roma, BTR-17 x AC-402, BTL-33 x AC-402 and BTR-49 x AC-402	Yield	Sharma *et al.,* 2001
Arka Saurabh x DT-10, Sel-7 x Flora-Dade-7421, H-24 x Flora-Dade-7421, H-24 x Agata, DVRT-1 x Flora-Dade-7421, Araka Saurabh x Flora Dade-7421, JT-99 x DT-10 and H-24 x ACC.	Yield and yield traits	Rajput *et al.,* 2003
H-711492 x 101 and 260 x V-16	Fruit firmness and pericarpthickness	Joshi *et.al.,* 2005

Brinjal

The extent of heterosis ranges from 4 to 104% for yield and its contributing characters. Nagi and Kida (1926) were probably the first to observe hybrid vigour in brinjal in crosses among some Japanese varieties. Kakizaki (1931) noted that highest-yielding F_1 hybrids gave 14.8% increased yield. Earlier, exploitation of heterosis in brinjal has been reported by Pal and Singh 1949; Samandan, 1962; Raman, 1964; Pal and Rajaki, 1966; Komochi, 1966; Thakur *et al.,* 1968; Singh *et al.*, 1971, 1974 and 1978; Papova *et al.*, 1976 a, b; Peter and Singh 1973; Vishwanthan, 1973; Singh *et al.,* 1978; Salehuzzaman and Alam, 1983; Balmohan *et al.,* 1983; Patil and Shinde 1984; Sanguineti *et al.*, 1985; Nulsari *et al.*, 1986; Dixit and Gautam,1987 and Randhawa, 1987. Saha *et al.* (1991) observed maximum heterobeltiosis for plant height and number of branches/plant in the range of 26.4 (Islampuri x 11-1-324) and 48.45% (Uttara x Shingnath Long), respectively. Plant height has been estimated to be controlled by two pairs of genes or group of genes but only one pair of gene involved in number of branches/plant. Singh and Rai (1990) recorded that cross of Pusa Purple Long x Pusa Purple Cluster and Erangre x Pusa Purple Cluster exhibited high heterosis for yield and other fruit characters, while F_2 showed inbreeding depression for all the characters. Chadha *et al.,* (1988) recorded increased yield over better parent (162.5 %) in cross Pusa Purple Cluster x Sel-5. They also reported that the crosses H-4 x Sel-5, Annamalai x Sel-5 and Sel-5 x ARU-1 exhibited high heterosis for yield. A work of Ingale and Patil (1997) revealed that magnitude of heterosis ranged between -77.9 and 82.7 % over-mid, better and top-parent values for the characters, *viz.* earliness, fruit yield, number of fruits, plant height, plant spread, primary branches and stem girth. The frequency and magnitude of heterotic hybrids have been observed more towards desired direction for days to flowering, fruit yield, plant height, plant spread and fruits/plant, indicating the presence of over-dominance. In order of merit, hybrids Gokak Local x Pusa Bhairav, Surya x Gokak Local and Gane Local x Pusa Kranti may be useful for future breeding programme. Babu and Thirumurugan (2000) noted that hybrid EP-390 x Pusa Kranti showed highest heterosis for plant height (14.2%), number of branches/plant (25.5%), fruit length (12.4%), number of fruits (16.3%), fruit weight (17.7%) and fruit yield (35.6%). Heterotic effects on yield and yield-contributing characters revealed that the crosses PR x MDU-1 for number of branches/plant, 190 x CO-2 for percentage of long-styled flowers, H-9 x MDU-1 for fruit girth, 202 x CO-2 for fruit length, 180 x AVI for fruit weight and 190 x AVI for fruits/plant possessed maximum heterosis over better parent revealing the presence of over-dominance (Prasath, 2000). Some of the crosses showing high heterosis in brinjal are given in Table 3.

Table 3: Crosses of brinjal showing high hetrosis

Cross	Attributes	References
Black Long x Pusa Purple Long and White Bunches x Pusa Purple Long	Yield and fruit weight	Mittal *et al.,* 1976
Pusa Kranti x BR-112	Earliness	Bhutani *et al.,* 1980
S-148 x 6703	Fruit length	Ram *et al.,* 1981
DMP x IMK, Pusa Purple Cluster x PK and DMP x S-2	Yield	Chadha and Sidhu, 1982
Pusa Kranti x S-16	Fruit weight	Dixit *et al.,* 1982
Pusa Purple Long x Punjab Bahar, Pusa Purple Long x S-373 and Annamalai x S-96	Yield	Kandaswamy, 1983

Contd...

Table 3. *Contd...*

Cross	Attributes	References
Ludhiana Local Long x Pusa Purple Long	Yield	Dahiya *et al.*, 1984
5623 x S-4	Fruit number	
Pusa Purple Cluster x Extra Long, Pusa Purple Long x R-34, Pusa Kranti x Pusa Purple Long x H-34, Pusa Purple Long x Aushey, BR-112 x R-34, PH-4 x Pusa Kranti, PH-4 x S-16, PH-4 x Aushey, BL-3 x Punjab Bahar, R-34 x P-8 and Punjab Bahar x Pusa Purple Long	Yield	Peter and Singh, 1976; Dixit *et al.*, 1982; Bhutani *et al.*, 1980, Dixit *et al.*, 1982, Randhawa, 1987 and Verma, 1986
BL-3 x S-16	Early yield	Randhawa, 1987
BL-3, BL-4	Fruit length	Randhawa, 1987
BL-3 x BL-4	Leaf length	Randhawa, 1987
BL-4 x Nurki	Leaf breadth	Randhawa, 1987
H-4 x Sel-5, Annamalai x Sel-5 and Sel-5 x ARU-1	Yield	Singh and Kumar, 1988
Pusa Krnati x Punjab Chamkila and Pusa Kranti x Pusa Purple Long	Chlorophyll a &b	Chadha *et al.*, 1988
Pusa Purple Cluster x Bhubaneshwar-4, Pusa Purple Cluster x Keonjhar-1, Dhumchai x Keonjuar-1, Deogarh x Bhubaneshwar-4 and Deogarh x Keonjhar-1	Yield plant	Mishra and Mishra, 1990
Pant Rituraj x Pusa Purple Long, Pant Rituraj x 7-3 and Pant Rituraj x Pusa Purple Cluster	Yield	Chadha *et al.*, 1990
Rayadonga x Pusa Kranti, Milk White x Pusa Purple Long, Arka Shirish x Rayadonga & Rayadonga x Pusa Purple Long	Plant height and plant spread	Shankaraiah and Rao, 1990
Pusa Purple Long x Pusa Purple Cluster and Erangar x Pusa Purple Cluster	Fruit yield	Singh and Rai, 1990
Manajari Gota x Local Brinjal and Arka Kusumakar x Local Brinjal	High yield and resistant to bacterial wilt	Sawant *et al.*, 1992
Pusa Purple Cluster x 178	Yield	Mandal and Dana, 1993
CH-4 x CH-82, CH-4 x CH- 95 and CH-13 x CH-82	Yield plant	Singh and Prasad, 1995
Swarnashree x CH-190 CH-190 x CH-792 CH-190 x BL-5	Yield	Singh *et al.,* 2004
CHES-309 x Hisar Shyamal Arka Keshav x Pusa Purple Cluster Punjab Barsati x Hisar Shyamal Pusa Anupam x Pusa Purple Cluster	Marketable Yield/Plant	Kumar and Pathiania, 2004
DBL-11 x PB-33, Pant Samrat x KS-331 PB-33 x PB-30 and PB-30 x KS-331	Yield	Pratibha *et al.,* 2004
RCMB-10 x RCMB-4, RCMB-10 x RCMB-1, RCMB-3 x BB-64, x RCMB-1x BB-64 and BB 40 x BB 64	Yield and its traits	Rai *et al.*, 2005

Chilli

Very little work has been done on hybrid vigour in chilli. The magnitude of heterosis for yield, maturity, plant height, fruit thickness and number of fruits are considered important characters and reported to be 20-30% average range of heterosis. In abroad, heterosis for yield has been reported up to 46% in cross California 1230 x Romanian and 38% in California 1230 x Maikop 470 over male parent in red pepper (Studentsova, 1973). Lippert (1972) reported 27.4% heterosis for dry weight and 7.3% for fruit length, whereas, Park *et al.*, (1975) found 61% heterosis for fruit yield in red pepper.

In India, first report on heterosis in chilli came from Deshpande in 1933. He observed it for earliness, plant height, fruit girth, fruits/plant and yield/plant in crosses between KAU Cluster and bell pepper varieties (Hungarian Wax, Early Calwonder and Bell Boy). Pal (1945) also observed that a hybrid between two Pusa Types gave increased yields, matured earlier and had fruit thicker than better parent. The highest heterosis of 19.15% for yield, 29.8% for fruit number, 44.7% for fruit length, 0.5% for fruit thickness and 18.5% for plant height over better found in *C. annum* (Singh *et al.,* 1973). Pious (1985) observed heterosis for earliness, plant height, fruit length, fruit diameter, average fruit weight and green fruit yield. Heterosis in chilli has also been reported by various workers for different traits (Pal and Sikka, 1956; Mishra *et al.,* 1976; Singh and Singh, 1978; Gangadhar Rao *et al.*, 1981; Pandey *et al.*, 1981; Gopalkrishnan *et al.*, 1987; Singh and Singh, 1992; Joshi and Thakur, 1995; Jagadish, 1995; Hundal 1995; Patel *et al.,* 1997; Lohithaswa, 1997; Shukla *et al.*, 1999 and Joshi and Shukla, 2000). Since hand-emasculation and hand pollination are uneconomical methods of seed production in chilli crop. Use of male sterility is the only solution to produce F_1 hybrid seed. By using genic male sterile line, Lakshmi *et al.*, (1988) were able to produce 43.80% heterosis over mid-parent. However, Hundal and Khurana (1988) recorded heterosis over 'Punjab Lal' standard variety up to 235.71% for green fruit yield and 138.01% for red ripe fruit yield. A hybrid of MS_{12} x LLS, has been identified for commercial production at PAU, Ludhiana, and gave 249.42 q/ha red ripe fruit yield which has been found 81.99% more yield than 'Punjab Lal'. In India, some of the hybrids were developed in different institutes and studied for number of attributes. Crosses performing high heterosis are given in Table 4.

Table 4: Chilli crosses performing high heterosis.

Cross	Attributes	References
Pusa Jwala x Pant C-1 and Pusa Jwala x CA-23	Early and high yield	Gopalakrishnan *et al.,* 1987
Bell Boy x Kau Cluster and 672 Hungarian Wax x Kau Cluster	Yield	Pious and Peter, 1988
J-218 x CA-586, Pusa Jwala x Sindur and BR Red x G-4	Fruit yield/plant	Mishra *et al.,* 1988
NP- 46 A x Kalyanpur Yellow	Fruits/plant	Ram and Lal, 1989
Pusa Jawala x Sindur	Fruit weight and yield	Mishra *et al.,* 1989
LCA-206 x LCA-960, LCA-206 x LCA-1079 and LCA-960 x LCA-206	Yield	Bhagyalakshmi *et al.,* 1991

Contd...

Table 4. ***Contd...***

Cross	Attributes	References
Tiwan x Jawahar-218	Yield traits	Singh *et al.,* 1992
NC-1x K -33, J-3x LAC-11, NCI x J-3 and J-3 x K-33	Yield	Rahman *et al.*, 1996
CA-133 x CA-84	Number of fruit/plant and fruit weight	Subashri and Natrajan, 1999
Jwala x Jagudan	Yield	Shukla *et al.,* 1999
H- 44 x IC-119769 and Jwala x IC-119769	Dry fruit yield	Gandhi *et al.,* 2000
S-49 x DPS- 120	Green fruit yield	Patel *et al.*, 2001
MS-12 x S-2530, LT-2 x RHCHUP, Punjab Guchnedar x S-2545, Punjab Guchnedar x LLS	Oleoresin yield Oleoresin (%) Capsaicin Colouring matter	Singh and Hundal, 2001
Bell Pepper x Round Early	Early, higher yield, uniform fruits	-
D-103 x Sivriya-600	Early ripening and higher yield	-
Jubilee x Krymskwhite 29-293	Higher ascorbic acid content	-
Early Canning 79-68, Maiko x Pazardshik Pepper	Higher ascorbic acid content	-
Arka Gaurav x VR-1	Earliness	Prasad *et al.*, 2003
VR-2 x VR-47	Fruit length	
Arka Gaurav x VR-2	Fruit width	
VR-47 x VR-55	Number of fruits/ plants	
VR x VR-55	Dry fruit yield/plants	

Bell Pepper

In bell pepper, the average range of heterosis for yield and its contributing characters has been reported as 28-47%. Earlier, heterosis for different attributes in bell pepper have been reported by Attavar and Bhatt (1973), Popova and Mikhailov (1973), Ignatov and Popova (1977) and Shifriss and Rylski (1973) and Joshi and Singh (1980), Thomas and Peter (1988), Sharma (1992), Joshi *et al*., (1995) and Thakur *et al*., (1998). Joshi (1986) observed that Bullnose x HC-201, HC-201 x Ruby King and Yolo Wonder x Bullnose showed respectively 46.79, 22.9 and 23.2% superiority in total yield over best parent HC-201. Poor x Poor yielder showed high heterosis relative to better parents but have been found low in yield compared with high x high and high x poor cross combinations. Kaul and Sharma (1988) found that Sweet Banana x California Wonder, Osh Region x California Wonder and HC-201 x California Wonder expressed 34, 33 and 25% heterosis, respectively over better parent for fruit yield. HC-201, Sweet Banana, Sel-4 and Osh Region have been found good combiners for fruit length, whereas, for fruit diameter only Early Prolific was good combiner. Significant gca has been noted for number of fruits/plant and fruit yield/

plant in 'California Wonder'. Joshi and Singh (1987) highlighted that for bell pepper improvement, parent HC-201, Golden Queen, Vinedale, Ruby King and Yolo Wonder can be used in hybridization programme. Thakur (1987) reported that predominance of non-additive gene effect and absence of non-interacting crosses for yield indicated utilization of heterosis to obtain higher production in bell peppers. Rahman *et al.*, (1996) also observed that high level of heterosis for yield and its component in the crosses NC-1 x K-33, J-3 x LAC-11, NC-1 x J-3 and J-3 x K-33. Ahmed *et al.*, (1997) recommened crosses of California Wonder x KSPA-2, California Wonder x KSPS-1 and KSPA-2 x KSPS-1 for commercial cultivation. Verma *et al.*, (1999) observed maximum heterosis (55.3%) for yield/plot over standard check Pusa Deepti (F_1). Ahmed *et al.*, (2000) reported that hybrids KSPS-461 x Oskash, KSPS-461 x KSPS-2, KSPS-461 x California Wonder, KSPS-13 x California Wonder and HC-201 x KSPS-2 can successfully be exploited under temperate-growing condition in India. Sanjeet *et al.*, (2000) showed hybrid of parentage sweet pepper (California Wonder) x hot pepper (LCA-235) exhibited high heterosis for yield (111 % and 100.2 %), days to fruit setting (-55.3 % and -40.5 %) over MP and BP, respectively. Pandey *et al.* (2002) reported that cross combination Yolo Wonder x CW-51 may be utilized for development of capsicum hybrid for hilly region as showed high heterosis with high specific combining ability. A study of Mulge *et al.* (2001) revealed that hybrid PMR-28 x IIHR-1023 expressed high heterosis over the best parent for fruit dry-matter content (26%). Linganagouda *et al.* (2003) observed, 36.94% and 35.08% for stem girth; 28.63% for plant height; -12.05 and -27.72 % for height at first branching; 42.63% and 36.47% for plant spread; 14.73% and 7.25% number of primary branches and 17.05% and 12.12% for number of secondary branches over mid and better parents, respectively in crosses of bell pepper x chilli.

Cole Crops

Cabbage

Heterosis in cabbage has been documented for yield, earliness, size of head, weight of head, marketable head and uniformity. Nieuwhof (1963), Swarup *et al.*, (1963) and Averchenkova (1975) have studied heterosis for yield and earliness. According to Nieuwhof (1963) heterosis for yield attributed to head weight and earliness, may be due to rapid growth of plants. Swarup *et al.* (1963) found highest heterosis up to 120.97% for yield over better parent and manifested in term of net heads, are the main components of yield. It has also been observed that hybrid from 'Branunschoeig' x 'Amager' exceeded the yield average of two parents by 30%, whereas, reciprocal exceeded the yield average of parents by 100% (Khimic, 1935). Myres (1942) crossed the strains of 'Pennstate Ballhead' cabbage inbred by bud pollination for 2-4 generations and found that average difference in yield between cross and high-yielding parent. It has been observed that greatest advantages in hybrid cabbages are uniformity in head. The parentage of plants forming good heads is markedly high compared with ordinary cabbage crop in which 'barren' plants are not uncommon.

Norman (1973) and Leal and Arume (1974) reported high productivity in hybrid cabbage. In Europe, especially in the Netherlands, pointed head cabbage is very important.

In this cabbage, earliness and uniformity are of major consequences. Paired crossing from the same strain without previously selfing for required generation, may lead to good results for the production of heterosis for uniformity in F_1 hybrid. In cabbage, uniformity and synchronization in maturity are associated with once-over harvesting or mechanical harvesting, recorded from hybrids TSKHA-1789, TSKHA-1781, TSKHA-1784 and TSKHA-1764 (93-100%) (Bondarenko, 1983). The uniformity of hybrids depends upon the homozygosity of inbred lines used in hybrid combination, therefore inbred lines used as parent should have a high percentage of homozygosity and uniformity. Such inbred lines are generally maintained vegetatively. Disease-resistant hybrids have been produced for commercial growing in cabbage by utilizing self-incompatibility. Sixteen self-incompatible lines have been obtained by 6-9 years of inbreeding from two varieties and three hybrids. Three lines have been found resistant to *Peronospora parasitica*, which can be used in crosses for producing *P. parasitica* resistant hybrid (Alipieva and Dimitrov, 1984).

Multiple disease-resistant F_1 hybrids Hybelle and Sanibel have been developed in cabbage, which are the classical examples of multiple disease-resistance hybrids. These hybrids are resistant to yellow, tip burn, powdery mildew, rhizoctonia and mosaic. Hybrid Hybelle is a cross between Badger Inbred-12 and Globelle and Sanibel is a cross between Badger Inbred-13 and Globelle (William *et al.*, 1968). Isozyme genes have been analysed in a large number of samples of F_1 hybrids of cole crops. Electrophones is used as suitable technique for F_1 hybrid purity (Arus *et al.*, 1985). Verma *et al.*, (1990) developed two hybrids in cabbage (Pusa Mukta x Pride of Asia and Pride of India x Pride of Asia), based on combining ability of hybrids. The F_1 hybrids are early, high- yielding and have higher head size index. F_1 hybrids have also been made by using of cytoplasmic male sterile line (EC-173419). Further, lines A (male sterile) and B (maintainer) were imported from Cornell (USA) and used to develop F_1 hybrids. On evaluation at Katrain (Himachal Pradesh) and New Delhi, hybrid 'H-11' ('Cornell 83-23' x 'Golden Acre') and 'H-46' ('Cornell 83-23' x 'Golden Acre') have been found promising. Self-incompatibility has also been utilized to develop F_1 hybrids. More *et al.*, (1991a) identified 'H-44' ('S_2S_2' x Cornell 83-6) and H-43 ('S_2S_2' x 'Pusa Mukta') to be heterotic for yield. The parents/lines, showing high heterosis for different traits in cabbage, are given in Table 5.

Table 5: Parents/lines showing high heterosis in cabbage.

Parent/cross combinations/hybrids	Attributes
Summer Monarch, Autumn Monarch, Winter Monarch, Hurt and Ideal	Large heads, long storability and high yielding
Sri Ganesh Gol, Hari Rani, Meenakshi, Green Boy and Green Express	Larger heads and high-yielding.
Nomes Purvyj x Jure 17' (Russian)	Early and high-yielding
Glory of Enkhusizen x Pudjan	Early, high-yielding and resistant to black rot
BRH 5 (Pusa Mukta x AC-204)	High-yielding
Sel-1 x Sel-36	Net head weight and ascorbic acid

Source : Peter (1998).

Cauliflower

Pal *et al*., (1952) reported hybrid vigour for plant growth, fruiting habits and early-flowering in interspecific hybrids. In cauliflower, there is the problem of uniformity, which can be increased by producing hybrid varieties, are more uniform than pure line varieties. Heterosis for yield has been reported (Singh, 1984; Singh *et al*., 1976 and Sandhu *et al*., 1977). Heterosis for yield may be mainly due to increase in curd size (Watts 1965 and Swarup and Chatterjee, 1974) and curd weight (Swarup and Pal, 1966 and Swarup and Chatterjee, 1974). Earliness is an important attribute and heterosis for earliness and its usefulness are well-documented (Jones *et al*., 1932; Weiring, 1961; Swarup and Pal, 1966 and Sandhu *et al*., 1977). Manifestation of heterosis for earliness over better parent ranged from 5 to 7 days and increase in net weight and size of curd ranged from 24.55 to 28.91% and 22.54 to 34.85%, respectively (Swarup and Pal, 1966). Swarup and Chhaterjee (1974) observed heterosis of 41.23% for yield, 16% for curd size and 23.4% for curd weight in an early mid group of cauliflower, whereas in mid-late group, it has recorded as 22.22% for curd size and 27% for curd weight. There are sufficient heterosis for yield up to 37 and 27% over better and standard parents, respectively, which can be utilized on a commercial scale (Sandhu *et al*., 1977). Kale *et al*. (1979) recorded maximum heterosis in curd weight (93.3%), followed by plant height (91.1%), curd size index (55.5%) curd to plant ratio (44.4%), early maturity (20.2%) and less number of leaves (2.2%) over better parent.

Sharma *et al*. (1988) studied heterosis and combining ability in a diallel cross self-incompatible lines from winter cauliflower to exploit heterosis in snowball group. Jamwal *et al*. (1991) recorded that cross combination K-1 x MGS 2-3 showed high heterosis for yield/plant and curd size index, with increase of 46.6 and 37. 5% over better parent, respectively. Gangopadhyay *et al*. (1997) selected five promising hybrids namely aa (395) x 820-1-3, CC-13 x Pusa Deepali, CC-14 x 820-1-3 and CC-14 x Pusa Deepali in early group of cauliflower which exhibited significant heterosis over better parent for first maturity and days to 50% maturity. Of which, hybrid aa (395) x 820-1-3 has shown significant heterosis for curd diameter and marketable weight and CC-14 x 820-1-3 hybrid has been found to show significant heterosis for marketable weight only. Appreciable heterosis for curd weight has also been observed in hybrid PG-26 x C-52 (Thakur *et al.,* 1995) and Poosi x Pant Subhra and 12-C x Improved Japanese (Singh *et al*., 1996). Thakur *et al*., (2001) observed high heterosis for days to curd maturity, gross weight, marketable yield, curd size and leaf size index. The top heterotic crosses for all round performance have been recorded in Snowball-16 x PSB-1, Janavon x PSB-1, SN-445 x PSB-1, Pusa Synthetic x PSB-1 and Kathmandu Local x PSB-1. Mahajan *et al*. (1999) found the crosses 267-6-7 x Kathmandu Local, HR-5-4 x 3-5-1-1 and Kathmandu Local x 12 C exhibited high heterosis for net curd weight. The cross CC x 452-10 yielded white good quality curd, possessing resistance to powdery mildew. Chaubey *et al*. (2002) recorded best heterobeltiotic crosses for different characters : NDC-1-I_6 x 235-S-I_6 for days to maturity, 75-IC-I_7 x 235-5-I_6 for curd diameter and NDC-1-I_6 x Early Kuwari I_7 for curd depth, 235-S-I_7 x 235-S-I_6 for curd depth and 234 -S -I_7 x 235-S-I_6 for curd weight. However,best standard heterotic crosses for different characters are 75-2C-I_7 x Early Kuwari-I_7 for days to maturity, NDC-I-I_6 x Early Kuwari I_7

for curd diameter, 75-2C-I_7 x Pusa Early Synthetic-I_6 for curd depth and 75-1C-I_7 x Pusa Early Synthetic-I_6 for curd weight. Dixit *et al.,* (2004) observed the crosses *viz.* Pusa synthetic x Pusa Sharad, Pusa Synthetic x RCJOB-1, RCJOB x Pant Shubhra, Pant Shubhra x Pusa Synthetic and Pant Shubhra x Pusa Sharad showed high heterosis for yield and yield components.

Malvaceous Crops

Okra

Vijayaraghvan and Warries (1946) observed distinct evidence of hybrid vigour with respect to fruits/plant as well as size and weight of fruits. Swarup (1948) made an attempt for interspecific hybridization among *Hibiscus esculentus* f. *fuculans* L., *H. tetraphyllus* Roxb. and *H. pungens* spp. He observed that F_1 hybrids of all the crosses are vigorous and have numerous branches. Ustomova (1949) reported that superior F_1 hybrids for stem diameter over both parents in interspecific hybridization between *Hibiscus esculentus* L. and *Hibiscus manihot.* Pal *et al.* (1952) reported hybrid vigour in okra for plant growth, fruiting habit and early-flowering in interspecific hybrids. Venkataramani (1952) also recorded an increase in yield of 50.4-14.5% over better parent in okra hybrids. The plant height of hybrids is intermediate or slightly more than that of parent. The increase in yield attributed to the production of more fruits/plant. Manifestation of heterosis for early and late flowering, plant height, number of branches, weight and size of fruits, number of ridges, marketable maturity, marketable and total yield and also fewer nodes and fewer days to first virus appearance have also been reported (El-Maksound *et al.*, 1986; Joshi *et al.*, 1958; Raman and Ramu, 1963; Sharma, 1965; Singh, 1979 and Singh, 1983).

At IARI, hybrid Pusa Makhmali x Long White Darbhanga showed vigorous growth and abundant fruiting, giving 60% increase in yield over Pusa Makhmali, the standard variety. In addition, hybrid vigour has also been found manifested in interspecific crosses for quick seed germination, plant height, number of branches, fruiting habit, early flowering and marketable maturity (Dhillon *et al.,* 1982 and Pal *et al.,* 1952). Significant heterosis has also been observed by Sharma (1965) for plant height, late flowering, fruit weight, number of fruits and fruit yield. The increase in yield over respective parent in heterotic crosses ranged from 11.8 to 101.7%. According to Sharma and Mahajan (1978), manifestation of heterosis over mid-parental value ranged from 0.03 to 68.03% for days to first flowering, marketable maturity, number of fruits/plant, fruit length, weight, diameter, number of ridges, plant height and fruit yield plant. However, over-dominance has been observed only for days to first flowering, fruit length and weight, plant height and yield/plant. Singh (1979) reported heterosis over mid-parent for fruit number, fruit weight and fruit length, higher marketable yield and total yield, plant height, less number of nodes and days to first flower appearance. Manifestation of heterosis over better parent has also been evidenced in interspecific crosses for number of branches, fast seed germination, early-flowering and marketable maturity (Dhillon and Sharma, 1982). The crosses showing high heterosis are given in Table 6.

Table 6: Crosses showing high heterosis in okra

Cross	Attributes	References
Pusa Sawani x Pusa Makhmali	Early flowering, early maturity, higher fruit weight and yield	Joshi *et al.*, 1958 and Raman and Ramu, 1963
Local 5 (Malaysian) x Emerald (American), Local 7 (Malaysian) x Gold Coast (American), Local 5 (Malaysian) x Gold Coast (American) and Local 7 (Malaysian) x Emerald	Higher seed germination, early flowering, more plant height and higher yield	Jalani and Graham, 1973
Vaishali Vadhu x Local Akola, Local Akola x Selection 2-2, IC-12934 x IC-13997, Vaishali Vadhu x MDU1, Vaishali Vadhu x IC-18553, IC-18960 x Selection-2-2, IC-12934 x IC-13997 & IC-18960 x IC-3769	Yield and its contributing characters	Kale and Dod, 1975
Pusa Sawani x Smooth Long Green	More fruits/plant and high yield	Sharma and Mahajan, 1978
American Seven Dhari x Pusa Sawani	More plant height and high yield	Sharma and Mahajan, 1978
New Selection x Red Wondered, Rashmi Lucknow x Pusa Sawani	Number of branches	Elangovan *et al.,* 1981
Sel-6-2 x Pusa Sawani and Sel-6-2 x Vaishali Vadhu	Earliness	Mishra *et al.,* 1982
IC-6663 x IC-6316	Number of fruits/plant	Pratap *et al.,* 1982
New Selection x AE-91	Yield	Poshiya and Shukla, 1986
Balady x Gold Coast	Plant height, fruits/plant and fruit weight	El-Maksoud *et al.,* 1986
Vaishali Vadhu x EC-168475, Pusa Selection-6-2 x EC-68475 and Pusa Sawani x EC-68475	Yield/plant	Korla and Sharma, 1987
IC-12205 x Parbhani Kranti and KS-310 x Pusa Sawani	Yield	Shukla *et al.,* 1989
IC-5292 x Pusa Sawani and KS-306 x Pusa Sawani	Fruit diameter	Shukla and Gautam, 1990
Pusa Sawani x Ludhiana Selection-1	Less number of seeds, more number of fruits, more fruit weight and marketable yield	Singh and Sharma, 1990
Padra 18-16 x KS-312 and Punjab Selection x KS-312	Yield	Kumbhani *et al.,* 1993
Sel-10 x Punjab Padmini, Sel-4 x Punjab Padmini	Plant height and fruits /plant	Mandal and Dana, 1993

Contd...

Table 6. ***Contd...***

Cross	Attributes	References
Sel-7 x KS-312, Parbhani Kranti x KS-312, Sel-4 x KS-312	Total yield	Singh and Mandal, 1993
PB No. 57 x Punjab Padmini	Plant height	Dayasagar, 1994
PB No. 57 x Parbhani Kranti	Stem girth	Dayasagar, 1994
Pusa Sawani x Parbhani Kranti	Number of branches/ plant, days to flowering, number of fruits/plant and total yield	Dayasagar, 1994
Pusa Sawani x Janardhan	Fruit weight	Dayasagar, 1994
Seven Ridges x Pusa Sawani	Length of fruit	Dayasagar, 1994
EC-169519 x Pusa Sawani	Number of fruits, number of branches and yield/plant	Panda and Singh, 1995
Japan x Parbhani Tillu, Japan x 168 and Parbhani Tillu x 168	Green pod yield/plant	Poshiya and Vashi, 1995
Rajasthan-12 x Parbhani Kranti	Fruit number, branches/ plant and yield	Dhankar *et al.,* 1996 a
Pusa Makhmali x Parbhani Kranti, Pusa Sawani x Parbhani Kranti, Pusa Makhmali x Pusa Sawani, Parbhani Kranti x Punjab-7	Yield	Singh *et al.,* 1996
MR-92 x MR-12-1	Yield	Dhankar *et al.*, 1998
SB-3 x Punjab -7, SB-5 x Pusa Sawani and Perkins Long Green x Parbhani Kranti	Yield	Ahmed *et al.*, 1999
P-7 x Arka Abhay, P-7 x Arka Anamika, P-7 x Parbhani Kranti and Pabhani Krant x Arka Abhay	Fruit yield	Singh and Sood, 1999
Hybrid 410 x 407, Hybrid 409 x 421 and Hybrid 409 x 408	Number of fruits/plant and yield/plant	Rattan and Bindal, 2000
HRB-55 x Arka Anamika Prabhani Kranti x HRB-9-2 and BO-1 x P-7	Resistant to yellow vein mosaic virus	Chandra Deo *et al.,* 2000
P-7 x Arka Abhay and P-7 x Arka Anamika	Fruit yield	Sood and Kalia, 2001
MR-10 x Varsha Uphar and MR-12 x Raj-12	Yield	Dhankhar & Dhankar, 2001
Arka Anamika x Arka Abhay, Shagun x Varsha Uphar, EMS8 x P-7, DC-97 x Varsha Uphar, PSB-1 x Varsha Uphar, K-21 x P-7 and Punjab Padmini x Varsha Uphar	Yield High	Chauhan and Singh, 2002

There are some interspecific hybrids, which may be used to transfer disease resistant gene (s) from related species to *Abelmoschus esculentus*. A large number of interspecific crosses have been executed in genes *Abelmoschus* (Table 7).

Table 7: Interspecific hybrids

Hybrid	References
A. esculentus x *A. manihot* var. *medicus*	Arumugham *et al.,* 1975
A. esculentus x *A. manihot* var. *adotachtylus*	Arumugham *et al.,* 1975
A. esculentus x *A. tetraphyllus*	Ugale *et al.,* 1976
A. esculentus x *A. manihot* var. *manihot*	Thakur, 1976
A. esculentus x *A. tetraphyllus*	Mamidwar *et al.,* 1979
A. esculentus x *A. manihot*	Jambhale and Nerker, 1983
A. esculentus x *A. manihot* ssp. *manihot*	Jambhale and Nerker, 1983
A. esculentus x *A. manihot* ssp. *tetraphyllus*	Dutta, 1984
A. esculentus x *A. manihot* ssp. *tetraphyllus*	Suresh Babu, 1987

Source: Peter (1998).

The wide range of heterosis observed by various workers may not be of much practical value unless it is manifested as increase over standard commercial varieties. In some cases over-dominance has been observed for days to first flowering, fruit length and weight, plant height and fruit yield. Thus, it suggests that hybrid vigour can be exploited in okra for increasing pod yield through early flowering, increased plant height, fruit length and weight. The potential of this approach is evident from the reports where more than 100% increase in yield over better parent has been recorded and also presence of dominant gene that affect for yield and yield-contributing traits.

Cucurbitaceous Crops

Bitter gourd

Bitter gourd is a cross-pollinated crop, there is ample scope for exploitation of heterosis (Munshi and Sirohi, 1993). More pronounced hybrid vigour could be observed with inclusion of diverse parents (Pal *et al*., 1983). Heterobeltiosis has been observed for many economic characters including total yield/plant (Pal and Singh, 1946 and Srivastava, 1970). The hybrid between Panipat Local x Ambala Local gave consistently higher yield. The F_1 of B-15 x J-21 showed a heterobeltiosis of only 2.41% for fruit yield (Khole, 1972). While, Green Local x White Local gave 139 % increase over better parent (Lal *et al*., 1976), F_1 hybrid of Pusa Do Mausami x S-144, Pusa Do Mausami x S-63 and Coimbatore Long x S-63 appeared promising for total yield/plant and its component characters and they showed 84.10%, 72% and 45.46% higher yield, respectively than best parent, Pusa Do Mausami. The F_1 hybrid Pusa Do Mausami x S-144 which gave 84% higher yield than better parent and recommended for commercial cultivation (Sirohi and Choudhary, 1983). Choudhary (1987) showed hybrid MD-41 x VK-1 'Priya' displayed maximum standard heterosis for yield over top parent Khandesh Mali. Vahab (1989) recorded hybrids MC-78 x MC-48, MC-49 x MC-34, MC-49 x MC-69 and Arka Harit x MC-79 exhibited heterobeltiotic for yield/plant. The F_1 CO-1 Green x Hisar Selection produced highest average yield of 86.09 kg/plant. (Lawande, 1989). The F_1 hybrids C-96 x Green Bitter Gourd, Khandesh Mali x Green Bitter Gourd, BG-114 x Coimbatore Long and Washin Local x BG-110 have been found good specific combiners for yield (Choudhury and Kale, 1991). The extent of heterosis studied by various workers is summarized in Table 8.

Table 8: Crosses showing high heterosis in bitter gourd

Hybrid	Attributes	References
Panipat Local x Ambala Local	High yield	Pal and Singh, 1946
B-15 x J-21	High yield	Kohle, 1972
Green Local x White Local	High yield	Lal *et al.,* 1976
Pusa Do Mausami x S-144, Pusa Do Mausami x S-63 and Coimbatore Long x S-63	High yield	Sirohi and Choudhury, 1977
MDU1 x VK1 Priya	High yield	Gopalkrishnan, 1986.
C-96 x Green Bitter Gourd and MC-49 x Arka Harit	High percentage of female flowers and yield	Vahab, 1989.
Pusa Vishesh x Arka Harit	Yield	Munshi and Sirohi, 1993
Pusa Do Mausami x Priya	Anthesis of first female flower, days to first harvesting and fruit length	Khartra *et al.,* 1994.
BG-14 x ACC-32	Fruit number & yield	
Pusa Vishesh x Arka Harit, Pusa Do Mausami x Kalyanpur Sona and MC-48 x Pusa Do Mausami	Yield per plant	Celine and Sirohi, 1996
Pusa Vishesh x Pusa Do Mausami	High yield	Tendulkar, 1997
ACC-23 x ARU-41	Yield	Singh *et al.*,1997
IC-50516 x VRBT-77	Number of fruits/plant	
Narendra x VRBT- 46 & IC-50516 x VRBT-77	Yield	Ram *et al.,* 1997
Pusa Hybrid 1	Yield	Sirohi, 2000
BG-11 x BG-25 and BG-5 x BG-25	Number of fruits, earliness, lowest number of node bearing first female flower	Singh *et al.,* 2000
BL-2351 x Punjab-14 and AR-4-41 x PDM	Early flowering	
HK-12 x PDM	Fruit length	Singh *et al.,* 2001
ACC-23-1 x PDM and Palwal Selection x PDM	Fruit weight	
BG4-2-1 x PDM and Palwal Selection x PDM	Number of fruits/vine	
Punjab-14 and Pusa Do Mausami	Yield and its traits	Kattra *et al.,* 2000
PBIG-1 x PBIG-2, PBIG-1 x PBIG-4, PBIG-2 x Kalyanpur Sona, PBIG-4 x Kalyanpur Sona, Kalyanpur Sona x Kalyanpur Baramasi and PBIG-2 x Priya White	Yield	Tiwari *et al.,* 1999 and 2001

Bottle gourd

Early maturity, higher yield, better quality and fruit uniformity are major advantages of bottle gourd hybrids. Rajendran (1961) probably was the first to report 266.5% heterosis for yield over better parent in bottle gourd (Sirohi, 2004). There is considerable scope to exploit heterosis in bottle gourd. Earlier, heterosis to the extent of 100.6% for yield has been reported by Anon (1962), Choudhury and Roy (1964), Tyagi (1973), Pal *et al.* (1984), Sharma *et al.* (1993) and Absaleha and Dutta (1995) Sirohi *et al.* (1985) recorded 115% higher yield/plant in best F_1 hybrid over commercial cultivar Pusa Summer Prolific Long, high standard heterosis over top parent for standard check (Pusa Summer Prolific Long) as well as parental line, respectively. According to Verma (1997) in bottle gourd, two lines, Sel-2 in long type and Sel-11 in round type, have been found to combine well the IARI recommended varieties. Two hybrids have released as Pusa Meghdoot (Pusa Summer Prolofic Long x Sel-2) and Pusa Manjari (Pusa Summer Prolific Round x Sel-11). Pusa Meghdoot recorded 75% increase in total yield, while Pusa Manjari increased 106%. Janakiram and Sirohi (1989) reported 80-85% heterosis over better parent for yield trait. Sit (1999) found 113.47% heterosis in the best hybrid combination over commercial check Pusa Naveen. The extent of heterosis studied by various workers is summarized in Table 9.

Table 9: Crosses showing high degree of heterosis in bottle gourd

Hybrids	Attributes	References
S-46 x S-47	High yield	Sirohi *et al.*, 1985
S-46 x S-54	High yield	Jankiram and Sirohi, 1989.
Pusa Naveen x S-10	High yield	Pichaimuthi, 1991
S 36-1 x NC-59812-1	High yield	Jankiram and Sirohi, 1992
Green Sel-2 x Faizabadi Local	Number of fruits	Sharma *et al.,* 1993.
Hisar Local Sel-2 x PSPL	High fruit length	
BTG-1 x Pusa Summer Prolific Long	Yield	Singh *et al.*, 1996
Pusa Summer Prolific Long x NDBG-1	Yield/plant	Kumar *et al.,* 1998
ARBG-7 x LC2-1	Fruit weight and days to first fruit harvest	Singh *et al.,* 1998.
L-13 x T1	Yield	Kumar *et al.,* 1999.
Pusa Hybrid -3	Yield	Sirohi, 2000
Pusa Naveen x Pusa Summer Prolific Long, KLG x Pusa Naveen, 9503 x Pusa Naveen and KLG x Pusa Summer Prolific Long	Yield	Pandey *et al.*, 2004

Cucumber

Hayes and Jones (1916a,) were the first to observe heterosis in cucumber (*Cucumis sativus* L.). Hybrid vigour expressed itself in total yield and the increased yield is due to larger number of fruits/plant. The highest yielding-hybrids, which are early-fruiting, gave higher yield and showed an increase in fruit size compared with corresponding parents. Hutchins (1939), crossed pickling variety 'Mincu' with 9 slicing varieties and showed that F_1 generation exhibited vigour for increased yield, resulting from increased number of fruits/plant. Shifriss (1945b) developed Burpee hybrid cucumber that was particularly suitable as an early, all-purpose variety besides being a heavy yielder, it showed considerable resistance to downy mildew and mosaic. Cizov (1945) also observed hybrid vigour for earliness, yield and higher fruit weight. Alexandrove (1952), Carlsson (1952), Axelsan (1956) and Yurnia (1977) reported hybrid vigour for earliness, increased yield and disease resistance. In India, Gill *et al.* (1973) developed hybrid Pusa Sanyog by crossing a Japanese variety Kaga Aomoga Fushinari with Green Long Naples. This hybrid gave 23.05-128.75% more yield as out yielded variety Japanese Long Green. Development of gynoecious lines suitable for the tropical conditions hold great promise in giving a fill up to the evolution of F_1 hybrids.

Musmade and Kale (1986) reported that hybrid Poona Khira x Japanese Long Green, White Long Cucmber (Pandhari Lambi Kakari) x Poinsett and Kalyanpur Ageti x Panvel Local have been found promising, since these hybrids showed highest sca effects and recorded 135.47, 56.52 and 54.72% higher yield/vine over better parent in that order. Hormuzadi and More (1989) recommended hybrids WL-2757 x RKS-295 and Poinsett x RKS-300 for rainy season and SR-551Fx Balam Khira, SR-551F x Japanese Long Green and 551F x Poona Khira for summer season. At IARI, New Delhi, three gynoecious lines, 304-C, 319-12 and 338, developed by More (1993) are also being utilized in hybrid breeding programme. Vijayakumari *et al.* (1993) recorded hybrids Poinsett x Japanese Long Green, Line-304 x RKS-296, Line 304 x EC-129110, Gynoecious Japanese Long Green x Japanese Long Green, Gynoecious Japanese Long Green x RKS-296 and SR-551 F x Japanese Long Green; SR-551 F x EC-129110 and SR 551F x RKS-295 showed high heterosis for earliness, yield and its components. Dogra *et al.* (1997) observed heterosis to the extent of 51.34 and 51.15% in cross K-75 x Gyn and K-90 x Gyn-1, respectively.

Singh *et al.* (1999) observed that heterotic response in maternal seeds on out- crossing in cucumber F_1 seed weight influenced more than the seed size with a maximum heterosis of 25.78 and 23.90% over mid-and better-parents in cross GY 338 x C-6. While, maximum heterosis (10.30% over mid-parent) for seed size exhibited in cross JLG x Poinsette. Bairagi *et al.* (2002) observed that crosses C-31 x PCUC-28 (121.96%), PCUC-98-25 x C-31 (111.80%) and PCUC-98-25 x DC-1 (97.92%) showed high heterosis over their mid-parents in term of number of fruits/plant. However, for yield /plant best heterobeltiotic hybrids have been found in PCUC-98-25 x DC-1 (123.82%), followed by PCUC-98-25 x PCUC-9-5 (108.69%) and C-31 x PCUC-28 (19.42%). Sarkar (2004) suggested DC-2 x PCUC-28, PCUC-28 x VRC-11 and DC-2 x VRC-11 are most promising crosses as they exhibited 50.43, 43.61 and 35.32% heterosis, respectively for yield over top-parent DC-2 (Pusa Uday), a newly-released variety of cucumber.

Heterosis for yield and quality has also been reported by Delaney and Lower (1987), Vijay Kumari *et al.* (1993) and Dogra *et al.* (1997). Cucumber has been the subject for production of multiple disease resistance. An attempt has been made to utilize multiple disease resistant gynoecious lines and pollen parents for developing hybrids resistant to downy mildew, anthracnose, angular leaf spot, cucumber mosaic and scab (Barnes, 1961). For that, Hybrid Rodrichok has been developed which is resistant to powdery mildew, *Pseudomonas ascochyta*, *Colletotrichum lagenarium* and *Cladosporium cucumerinum* (Strelnikova *et al.*, 1981). There are a number of crosses which showed high heterosis. They are given in Table 9.

Table 9: Promising crosses of cucumber showing high heterosis

Cross	Attributes
VIR-501 x VIR-505	Intense photosynthetic activity and high-yielding.
Conqueror x Vijazhiki-37	Early and high-yielding.
Don-175 x Boston	High germination, growth rate and better pollen viability.
MSU-713-5 x Spartan-27	Early and high-yielding.
Koga Aomoga Fishinavi (Japanese gynoecious Line) x Green long of Naples), Pusa Sanyog	Early and high-yielding.
Udipi x Kasargod	High-yielding.
Nileswar x Udipi	High-yielding.
CH-2-3-1-2 x CH-1-3-5-2	Number of fruits.

Source: Prasad and Singh (1992).

Muskmelon

Heterosis in muskmelon has been expressed for total yield, earliness and vegetative characters such as number and length of stem, leaf size, and number of male and female flowers, quality (generally TSS) and resistance to diseases. Earliness is important to fetch better market price. There are several first-generation hybrids, which outyielded parents (Nandpuri *et al.*, 1974; Foster, 1967; Dixit and Kalloo, 1983). Heterosis for yield has been found 84% more than commercial parent. In certain cases, there is improvement in fruit quality in addition to an increase in the yield and also hybrid combined crown blight resistance along with good shipping quality of fruits (Foster, 1967). Munger (1942) observed increased vigour, more fruits, higher percentage of flesh and much more sugar in hybrids of muskmelon (*Cucumis melo* L.) than adapted standard varieties. Some of the hybrids also produced early fruits. Bohn and Davis (1957) reported that hybrid exhibited responses ranging from no dominance to apparent heterosis for earliness in total growth period and requirement to attain sexual maturity. All the hybrids showed early flowering and fruit ripening than the average of their parents. Dixit and Kalloo (1983) observed highest heterosis over better parent for number of fruits/plant (54.3%) in cross Punjab Sunehari x Sel-1, while for fruit yield, heterosis to an extent of 46.70% and 38.10% has been recorded in crosses, Pusa Sharbati x Sarola Melon and Pusa Sharbati x Punjab Sunehari, respectively. Heterotic hybrids *viz.* Arka Jeet x Durgapura Madhu and Arka Jeet x Sarola Melon for fruit quality (TSS content) have also been identified.

Nandpuri *et al.* (1974) found that Arka Rajhans and Hara Madhu among males and MS-1 and MS-2 among female have been noted good combiners for maximum number of economic characteristics. They also reported that among superior combinations at least one parent involved with high general combining ability and segregants with better fruit weight and yield can be obtained from the crosses MS_1 x Arka Rajhans and Edisto x Arka Rajhans. In 1983, Nandpuri and coworkers have also identified inbreds as well as superior cross combinations like FMS_{1-2} and Arka Rajhans for fruit weight, MS-1 and Hara Madhu for TSS and Pusa Sharbati, Home Garden and Sarda for earliness, and FMS_{1-2} x Sarda, MS-1 x Hara Madhu and MS-1 x Pusa Sharbati for exploiting heterosis. Favourable heterosis for yield and its component characters has also been reported by Pandey and Kalloo (1976) in monoecious hybrids. Mishra and Seshadri (1985) observed early yield expressed as percentage of total yield. F_1 hybrids also exhibited heterosis for total yield and total soluble solids. For TSS of early fruits and proportion of flesh, hybrids are generally intermediate between the parents. Kesavan and More (1991) recorded that for days to first harvest, hybrids FM-5 x Lucknow Safeda (11.2%), FM-5 x Pusa Madhuras (11.0) and M-4 x Lucknow Safeda (8.8%) showed high heterotic responses. FM-5 x Gulf Stream, FM-1 x Pusa Mudhuras, FM-5 x Harela, FM-5 x Lucknow Safeda and M-4 x Pusa Sarati have emerged as hetrotic for yield and its components especially average fruit weight.

Heterosis did not show much promise for quality components such as TSS and F:C ratio. Munshi and Verma (1997) also observed high heterosis in cross Pusa Madhuras x Ravi which showed 28.15% higher yield over top-parent as well as best commercial cheek Pusa Madhuras, may be used as one of the parent for development of muskmelon hybrid. Moon (2002) reported Pusa Madhuras x Hara Madhu, a most promising F_1 hybrid, which showed significantly higher yield over top-parent and commercial check Hara Madhu and Punjab Hybrid, respectively.

Male sterility can also be used for hybrid seed production. Bohn and Whitaker (1949), Bohn and Principe (1964), McCreight and Elmstorm (1984) and Le-Couvior (1990) have reported male sterility in muskmelon. Male sterile mutants in muskmelon are monogenic recessive. Male sterility has also been utilized in accertaining the magnitude of heterosis and combining ability (Nandpuri *et al.*, 1974, a, b; Nandpuri *et al.*, 1983; Mishra and Sheshadri, 1980 and 1988). Male sterile line MS-1, identified by Bohn and Whitaker (1949), has proved boon to hybrid seed industry. The line showed good nicking ability when tested in a series of hybrid combinations (Nandpuri *et al.*, 1974, a, b), resulting Punjab Hybrid F_1 MS-1 x Hara Madhu has been released (Nandpuri *et al.*, 1982) in Punjab and subsequently at national level in 1985. This hybrid still occupies a large acreage of muskmelon. Male sterile line MS-3, discovered by McCreight and Elmstorm (1984) has been found superior to MS-1 and MS-2, due to that male sterile plants are very easy to identify and possess superior horticultural traits. In another male sterile line MS-5, male flower buds abort prematurely, hence male sterile plants can easily be identified. Use of marker genes to simplify the procedure of identification, thus improving the quality of hybrid seed, have been attempted by Foster (1968) and Mishra (1981). But 100% pure hybrid seed could not be produced on commercial scale even by using marker genes. This indicated that linkage of genetic markers with *ms* gene is poor.

Muskmelon in general is an andromonoecious but some genotypes show monoecious sex expression. Since, use of monoecy excludes emasculation, it can reduce time required for a given number of pollination by 5% and enhance fruit setting by 40-70 per cent as compared to 5-10 per cent in andromonoecious parent (More *et al.*, 1982). Monoecious lines M_1, M_2, M_3 and M_4 have been developed by More *et al.* (1980) for use in heterosis breeding. More and Sheshadri (1980) examined 50 hybrids and found that M_1 x Mix Top Mark, M_1 x Golden Perfection, M_2 x Yanco Treat, M_2 x PMR-6 and M_2 x Top Mark have been found superior than better parents for earlier yield and quality attributes. A F_1 M_3 x Durgapura Madhu (Pusa Rasraj) has been found to be promising (Kesavan *et al.*, 1987) and released for commercial cultivation in 1990. In muskmelon, it has been found that a single pair of recessive gene, controls monoecious character. Korzeniewska *et al.* (1995) reported that use of ethephon (300-550 ppm) for F_1 hybrid seed production with monoecious lines. Foliar applications of ethephon induce temporary gynoecious stage (averaging 7-19 days) during which development of male bud inhibited and bees can be used to pollinate the flowers. The problem in using monoecy in heterosis breeding is undesirable, linkage between gene controlling monoecy sex expression and fruit shape (Bains and Kang, 1963; Risser, 1984 and Kubicki, 1979), consequently, F_1 combinations with round fruits can not be easily obtained. This problem is with hybrid Pusa Rasraj, which is not acceptable commercially, because of its undesirable fruit shape and poor external appearance.

Foster (1968) and Rowe (1969) argued that it is a case of linkage rather than pleiotropy. Because they were able to identify monoecious lines with round fruit shape. Even P1-124111, a source of resistance to powdery and downy mildew, is monoecious, yet has oblate fruits. So the linkage is not absolute and hybrid combinations can be obtained with desirable fruit shape. But commercial hybrid seed production using monoecy under open field conditions is handicapped on account of large mixture of selfed or sib seed in hybrid progeny (More *et al.*, 1991b).

The muskmelon hybrid seed is 12-30 times costly than that of open-pollinated cultivar (McCreight and Elmstorm, 1984). The gynoecious lines having natural bias of producing only female flowers ensure 100% hybrid seed production, thus having potential to lower cost of hybrid seed considerably. By using gynoecy in heterosis breeding, tedious job of identifying and roguing of male fertile from the mixed population is avoided. Probably, for this reason, Frankel and Galum (1977) and Loy *et al.* (1979) advocated the use of gynoecious lines in hybrid seed production. Kubicki (1966) described the method producing gynoecious lines from a cross of hermaphrodite (AAGG) with monoecious (AAGG) and extracting gynoecious (AAGG) and hermaphrodite lines (AAGG). Wisconsin 998 is the first gynoecious line in muskmelon developed by Peterson *et al.* (1983). Cohen *et al.* (1993) selected Gylon, is resistant to pathotype 3 of *Pseudoperonospora cubensis* and races 1 and 2 of *Sphaerotheca fuliginea*. Good hybrid fruit with andromonoecious varieties Vedranta and Tam-Uvalde can be produced. Wisconsin 998, when in hybrid production, has been found to show good combining ability for yield and earliness (Dhaliwal and Lal, 1996). A cross between Wisconsin 1998 x Punjab Sunehari (MH-10) has been found promising and released for commercial cultivation in Punjab during 1995. It is not only early and high-yielding but also has good shipping quality (Lal and Dhaliwal, 1996). The gynoecious lines are maintained by stimulation of male flowers. Use of $AgNO_3$ @ 100-200 ppm (Owen *et al.*, 1980) has been found to induce production of 12 perfect flowers on first 20 nodes, which can be used to self or sib for maintenance of gynoecious lines. Use of Ethrel and Alar (Rudich *et al.*, 1970) and silver thiosulphate (More and Sheshadri, 1987) have also been suggested for inducing maleness. Sandha and Lal (1999) reported that future research on heterosis breeding needs attention to develop high TSS and disease resistant gynoecious lines.

Watermelon

A considerable degree of heterosis has been observed in watermelon for yield, number of fruits, size of fruit, weight of fruit, female flowers, earliness, TSS and number of seeds. Yanagisawa and Hosono (1951) reported first time an outstanding hybrid of watermelon evolved by crossing 'Asahi Yamato' and 'Miyako', characterized by earliness, high yield, disease resistance and transport ability. Heterosis for fruits yield and number of fruits has also been extensively studied by Daskalov and Popova, (1958), Ivanoff and Albritton (1962), Nath and Dutta, (1970) and Sidhu and Brar (1977). Eighty hybrids have been studied and out of these, three hybrids namely IHR-6 x Charleston Gray, IHR-20 x Crimson Sweet and Sugar Baby x Crimson Sweet, produced more than 50% heterosis for yield and also exhibiting superiority in fruit quality. Hybrid IHR-6 x Charleston Gray produced attractive red-fleshed fruit with granular flesh texture (Nath and Dutta, 1970). Sachan and Nath (1974) reported high heterosis for yield (87%), number of fruits/plant (41.8%), fruit weight (16%) and TSS (21%). Heterosis for yield has been found to increase in number, length and breadth of fruits. Sidhu and Brar (1977) recorded appreciable heterosis for fruit yield, fruit weight, fruit number, flesh weight, 100-seed weight and number of female flowers. Among these, highest has been found for yield/ plant. They also observed that crosses, Sugar Baby x Shipper, Durgapura Meetha x Black Diamond, Sugar Baby x Black Diamond and Shipper x 10-6-5, exhibited high heterosis for yield, fruit weight, fruit number and female flowers, respectively. Fruit number as best yield component for heterosis has been well documented (Nath and Dutta, 1970; Sidhu and Brar, 1977; Brar and Sukhija, 1978).

Sidhu and Brar (1977) observed negative heterosis for number of seeds/fruit, which is important to increase the quality of fruits. They also noted that cross combinations of low x high and medium x high gca are more effective for production of superior hybrids and for expression of heterosis in F_1 at least one of the parents should be high-yielding and varieties should be of diverse geographical origin. According to Brar and Sukhija (1978) American x Indian and American x Japanese varieties produced better hybrids than American x American and Indian x Indian varieties. They also suggested that one of the parents involved in the production of hybrids should have a marker characteristic to identify the purity of hybrids and non-dissected leaf characteristic has been found suitable recessive marker. This trait has been introduced by backcrossing in Sugar Baby (Feher, 1985). Seshadri *et al.* (1972) produced Pusa Bedana a seedless watermelon by crossing of tetraploid line 'Tetra-2' from USA and variety 'Pusa Russel' from India. In a study of 20 cultivars and selections and 32 hybrids derived from them for yield and fruit characteristics were evaluated. Sixteen hybrids outyielded the mean yield of their parents, highest yield and number of fruits being obtained from Shipper x Striped Klondik, Sugar Baby x Verona have been found the earliest to bear female flowers with the highest total soluble solids (Nandpuri *et al.*, 1974).

The heterosis for sugar content in various hybrid combinations has also been reported (Brar and Sidhu, 1977). Kale and Seshadri (1981) also observed crosses, Pusa Rasal x Asahi Yamato, Sugar Baby x Pusa Rasal, Russian Sel. x Shikkarchini, exhibited high heterosis for yield-contributing characters. Sidhu and Brar (1985) showed non-consistency between parental divergence and magnitude of heterosis in F_1 hybrids. Gill and Kumar (1988) identified superior hybrid combinations like Sugar Baby x Durgapura Meetha for yield and number of fruits/plant and also good performer for earliness. Reddy *et al.* (1987) observed that crosses namely Sugar Baby x Tirupati and Durgapura Mitha x Arka Manik showed high heterosis for yield and quality characters. There are some crosses, which exhibited high heterosis for yield and quality traits, are given in Table10.

Table10: Crosses showing high heterosis.

Cross	Attributes
Special 1 (Round x Charleston Gray (long)	Round-fruited, high-yielding and high TSS.
Shipper (round) x Harleston Gray' (long)	Round-fruited, high-yielding and high TSS.
Sweet Princess (long) x Special 1 (round)	Round-fruited and high-yielding.
Crimson Sweet x New Hampshire	High-yielding and high TSS.
Midget IHR 6 x Charleston Gray	High-yielding and high TSS.

Source: Peter (1998).

Pumpkin and Squashes

In pumpkin, heterosis has been reported for earliness, fruits/plant, fruit weight and weight of seeds/plant. Hutchins and Croston (1941) found in their study of 10 F_1 hybrid lines of pumpkin (*Cucurbita maxima* Duch.) that increased yield over that of higher-yielding parent in seven cases. In such high-yielding hybrids, the yield ranged from 88.7 to 532.7 q/ha. The weight per fruit was more in hybrids, which also possessed larger number of fruits compared with the parents. Diojode *et al.* (1982 and 1984) studied heterosis and combining ability for yield related traits in 21 F_1 derived from a 7 x 7 half diallel, CM-12 x Arka Chandan and CM-12 x IHR-8 gave significant positive heterobeltiosis for fruit weight. Sirohi *et al.* (1986) reported that the cross S- 93 x CM-12 showed high heterosis for yield / plant, fruits/plant and fruit weight. Sirohi (1994) also recorded appreciable heterosis in cross Pusa Vishwas x S-122, Pusa Vishwas x S-93 and S-93 x S-122 at 36.9, 25.8 and 24.6% heterosis, respectively for yield over Pusa Vishwas, a best parental line. Mohanty and Mishra (1998 and 1999) recorded heterosis to the extent of 37.7, 28.6, 76.9, 142.9, 96.3, 48.9 and 188.7% over better parent and 52.1, 36.9, 89.6, 161.5, 109.9, 68.9 and 197.1% over mid-parent for vine length, number of primary branches, female flowers, fruits/ plants, average fruit weight, flesh thickness and yield / plant, respectively. The cross Ambile x BBS-10 has been found the best performing hybrid which showed 134.5 and 165.6% heterosis over better parent and check parent, respectively. Sirohi (2000) reported 30% higher yield of newly released Pusa Hybrid-1 as compared to commercial check Pusa Viswas. Sirohi *et al.*

(2002) observed that cross Pusa Vishwas x S-124-10 showed significantly higher heterosis for number of fruits and yield/plant over commercial check, Pusa Vishwas, whereas Pusa Vishwas x S-12 and S-107 B x S-17 showed heterosis for short vine and early maturity, respectively. For exploitation of hybrid vigour for yield and yield components in pumpkin have also been reported by Diojode (1981), Suneal Kumar (1984) and Sirohi *et al.* (1985).

Most of the improvement work in summer squash has been reported from USA and other European countries. In India, very little improvement work in this crop has been done. The high early yield and attractive green fruit with desirable fruit shape and resistance to powdery mildew, virus and red pumpkin beetle are the objectives of hybrid breeding programme. Pusa Alankar is the only F_1 hybrid available so far, developed from a cross of EC-27-50 x Sel-1 PI-81 and released by the IARI Regional Station, Katrain (Gill *et al*; 1972).

Ridge and Smooth Gourds

In spite of wide genetic variability (Irulappan, 1992), studies on extent of heterosis in this crop are limited. Singh and Pal (1949) observed that different interspecific crosses made between ridge gourd and sponge gourd produced F_1 hybrids which have been found both normal and bitter or abnormal and non-bitter type, depending on varieties of two species entering the cross. The abnormal type hybrid showed a decrease of 28.8% in yield as compared to ridge gourd. The normal type hybrid showed an increase of 57.5% in yield. Abusualeha and Dutta (1995) found the cross combination IIHR-9 x IIHR-23 exhibited 90.78% over better parent and 72.78% over top-parent. Thus, IIHR-76 can be utilized for exploitation of hybrid vigour. Chen *et al.* (1996) highlighted a ridge gourd hybrid Feng King which showed 33.2% higher yield than commercial cultivar Dongguan of Shuang Qing. Rao (2000) showed appreciable heterosis over better parental lines consistently over the seasons for yield and its contributing characters. The F_1 hybrids, LA-46 x LA-99 and LA-12 x LA-76, showed 51.8 and 125.9, 81.8 and 121.1% higher yield over the better parent in kharif and summer seasons, respectively. The best F_1 hybrid LA-46 x LA-99, which gave 51.8 and 125.9% higher yield over better parent and top-parent. Hedau (2002) recorded F_1 hybrids DRG-1 x Pusa Nasdar, DRG-1 x PRG-7 and DRG-1 x AAUJ-3 are most promising combinations as they showed 93.09, 68.51 and 66.50 % heterosis, respectively over top-parent DRG-1.

Bulb Crops

Onion

Onion is one of the pioneer crop in which heterosis has been commercially exploited after the discovery of gene-cytoplasmic male sterility in 1944. Although, India is one of the leading onion producers, not much emphasis has been given to heterosis breeding in the past. One of the main reasons for not exploiting heterosis in onion is the availability of male sterile lines. In India, progress in the development of suitable male sterile and fertile inbred lines remained very slow past few decades. Sen and Srivastava (1957) attempted first time to develop F_1 hybrids in onion as early as in 1948 using exotic male sterile lines of Indian Local Stocks. The exotic male sterile lines have been found unsuitable in photoperiodically different environments in India. Later, very few workers attempted to test different hybrid combinations for heterosis and combining ability studies using male sterile lines (Joshi and Tandon, 1976 and Pathak *et al*., 1987). Recently, Lawande (2004) observed ample amount of heterosis with an increase of yield 19-60% in late *kharif* season and 16- 40% in *rabi* season over respective check of best open pollinated varieties. He developed R-1 hybrids under Hybrid Development Networking Programme at NRCOG by using two *cms* lines, indicating very high percentage of heterosis over standard checks, which ranged from 17 to 59% over N-2-4-1, 16 to 85% over B-780 and 11 to 77% over ALR.

Male sterility has been isolated from indigenous germplasm by several workers in India. Patil *et al.* (1973) in cultivar 'Niphad 2-4-1', Pathak *et al.* (1980) in 'Nasik White Globe' (IIHR-20) and Netra Pal *et al.* (1988 and 1999) in Pusa Red. Further, studies indicated strong cytoplasmic factor responsible for male sterility for Bombay 'White Globe' (Pathak *et al*., 1986). This male sterility has been transferred to several breeding methods. In hybrids of three male sterile testers and 20 inbred lines, various economic traits have been studies. Positive heterosis has been observed in nine hybrids over better parent for total bulb yield and it ranged from 47.9 to 89.5%, while, heterosis over best parent for marketable bulb yield recorded over 35% in three hybrids, MS_1 x NER-1, MS_1 x IIHR 21-1 and MS_8 x IIHR 21-1 and MS_8 x IIHR-52-1 (Pathak *et al*., 1987). Joshi and Tandon (1976) reported heterosis up to 72% over mid parent and up to 37% over better parent. In 1991, Aghora and coworkers observed that F_1 hybrids MS-1 x Arka Kalyan, MS-1 x Nasik Early Red, MS-8 x N-53 and MS-8 x Arka Niketan exhibited favourable heterosis over the best parent and they recommended for commercial cultivation. The gain from heterosis breeding in onion has remained controversial issue in India. Despite release of Arka Kirtiman and Arka Lalima hybrids from IIHR, Bangalore, and Hybrid-63 and Hybrid-35 from IARI, New Delhi, no hybrid has reached commercially to farmers fields.

Root Crops

Carrot

In carrot, heterosis has been reported for earliness, root length, root yield, carotene content, top weight, core diameter and root diameter. Poole (1937) was perhaps the first to report crossing of two inbred lines of carrot exhibited heterosis. Two inbred lines producing roots averaging 12.8 and 24.9 g, respectively when crossed, produced hybrids with roots weighing 80.5 g. The hybrid carrot has high yield (Bhagchandani and Choudhury, 1970) and better processing qualities (Michigan, 1978). Tokmakov and Rabunetz (1976) reported male sterility in carrot, the degree of sterility depends more on cultivar than on meteorological conditions. They also found that while primary umbels are male sterile, yielded 34.4-44.9% more and obtained higher amount of dry matter and sugars than those obtained from intervarietal pollination. Katsumata and Yasui (1965) reported that male sterility seems to be controlled by dominant genes. An inhibitor gene and specific cytoplasmic background also appeared to be involved. Male sterility forms of Nantes 4 and Moscow Winter A-515 has been found to segregate in I_1-I_3 generations of inbred lines (Timin, $1979_{a,b}$). Katsumata and Yasui (1965) reported that F_1 hybrids among the cultvars Yokono, male sterile Sansun and Kurola possessed good root form and high carotene quality, whereas, F_1 hybrid between Nakamura Sun and Kintoki (Golden Hran) has been found inferior.

Most of the F_1 hybrids exhibited heterosis for carotene content. Litvinova (1979) successfully used male sterile form of Nantes-14 and Chanteny-2461 in production of hybrids. He found heterosis 20-22% higher than in hybrids produced without using CMS (cytoplasmic male sterility) and F_1 hybrids showed better uniformity of morphological characters. On the contrary, Bonnet (1978) reported from France that root characters such as shape and colour are controlled by the genotype of the line not affected by the type of cytoplasmic male sterility. Development of high quality, high-yielding and disease resistant hybrids has also been reported by Baker (1978a) from the USA. Spartan Bonus is a three-way hybrid (Parent MS-41558, MS-45831B and MSU-9541B) with medium-long top that are resistant to *Cercospora carotae* and *Alternaria dauci*. In comparison with Danvers, Spartan Bonus has given higher yield and has less oily flavour and superior external and internal root colour (Baker, 1978_b). Report from Michigan State University indicated that release of three-way carrot hybrids. Spartan Classic from MS-45931A, MS-45931B and MS-45988C; Spartan Premium from MS-45931A, MS-45986B and MS-45988C and Spartan Winter from MS-45931A, MS-46000B and MS-45988C. All the three hybrids are early-maturing smooth rooted, moderately resistant or tolerant to *Cercospora carotae* and resistant to rusty root (*Pythium* spp.) and are suitable either fresh or processed. In comparison with open-pollinated. Nantes, Spartan Winter gives high yield (Baker and Valk, 1978). A late maturing hybrid Spartan North has also been developed from single cross between parental lines MS-45931A and MS-45986B (Baker, 1978c). This hybrid has long slim roots with good internal and external characters. Report from France indicates that male sterile lines show tolerance to *Erysiphe herachi*. Wild carrot species such as *Daucus siliculus*, *D. mamimus* and *D. dentatus* are also reported to be resistant. In crosses between 'Touchan' (susceptible) and *D. dentatus*, the resistance has been found monogenic and dominant (Anon, 1977-78). Very little systematic work has been done for the development of carrot hybrids in India. Recently, Verma *et al.* (2002) reported remarkable heterosis for length, girth and yield in carrot, in order of merit, the three best performing F_1 hybrids (240607 x N) x 1061, (28 x PY) x 1060 and (95 x N) x 1061 exhibited heterosis of 59.79, 44.52 and 27.55%, respectively for marketable yield over standard check. The first one also being the best, yielding 75.70% more than the top-parent 1061, has been found best to use as pollinator in the production of these F_1 hybrids. Heterosis on various attributes has also been reviewed by Bunin and Yoshikawa (1989) and Timin and Vasilevekii (1995). Some crosses of carrot showed high heterosis are given in Table 11.

Table 11: Crosses showing high geterosis in carrot

Cross	Attributes
Chantaney x Genanda	Increase in yield (25-33%).
Italian x Nantes	Increase in yield (12%).
Virginia Savoy x Old Domain	Resistant to cucumber mosaic virus.

Source: Peter (1998).

Radish

Frost (1923) studied the crosses between selfed lines of radish (*Raphanus sativus* L.). He observed that crosses showed very vigorous and usually exceeded the better parent for root size and all plant characters. Keppler (1941) also reported manifestation of hybrid vigour, especially in crosses within superior families. Incompatibility phenomenon existing in radish can be used for the development of F_1 hybrids for higher yield and early maturity. Pal and Sikka (1956) obtained high-yielding hybrids, which gave 30-60% more yield than better parent. Singh *et al.* (1971) obtained radish hybrids by crossing five cultivars, which gave significantly higher roots. However, no difference was observed for pithiness of hybrids and parents. Heterosis for root yield in radish has also been reported by Singh *et al.* (1971), Brar and Padda, (1972), Pandey *et al.* (1978), Singh *et al.* (1986) and Song and Wei (1996). Vinod *et al.* (2000) observed root yield increased up to 59.62 and 63.11% in crosses Mino Early White x Pusa Himani, Japanese White x DPR-1 (Palam Hirday), Chinese Pink x Nadauni, which manifested maximum heterosis for root yield and its traits. They also reported heterosis for yield, root weight and vitamin C content in promising F_1 hybrids like White Tipped x Icicle, Rapid Red x Sparkler, Desi x Japanese White, Icicle x Round Black Winter and Red White Tipped x Ruby. Kutty and Sirohi (2003) showed wide range of heterosis in radish over better parent for yield traits. They observed cross combination ACC-30205 x ACC-192 exhibited high heterosis for root yield and shoot weight, recorded maximum heterobeltiosis followed by yield/plant, root weight, leaf width, petiole length, root diameter, leaves/plant, leaf length, root length and root shoot ratio. The best performing hybrid ACC-30205 x ACC-192 is considered as a potential hybrid for commercial exploitation. There are some radish varieties which can be used as one of the parent for developing radish hybrids for resistant to diseases, are given in Table12.

Table 12: Resistant radish varieties to diseases

Disease	Resistant varieties
Wilt	Fix
Black rot	Summer Wonder, Regensberg Market
Downy mildew	China Rose White and Round Black Spanish

Legume Crops

Pea

The genetically diverse parents often produce maximum heterosis and offer a better chance of isolating transgressive segregants. But there are also reports that cross combinations showing maximum heterosis and divergence do not show any association with yield. Narsinghani *et al.* (1978a) reported that grouping pattern of varieties has quite random indicating that the geographical genetic diversity is not related. Subsequently, heterosis and inbreeding studies by Narsinghani and Singh, 1979 and Narsinghani and Rao, 1982 confirmed the non-significant relationship between genetic divergence and hybrid performance in peas (Rao and Narsinghani, 1987). Kalloo *et al.* (1980) highlighted that while selecting varieties just on the basis of geographical origin will not be useful. A high degree of heterosis has been noted in peas by Singh *et al.* (1975) and Singh *et al.(*1978a) for days to flowering, seeds/pod by Bhullar *et al.* (1978), 100-seed weight by Saxena (1977), but Singh and Singh (1970b) reported negative heterosis for 100 seed weight in pea. The hybrid vigour for seed yield in peas has also been observed by Srivastiva and Sachan (1975) and, Arvind and Dubey (1980), Parmar *et al.* (1990) and Tewatia *et al.* (1992). Kalia and Singh (1996) found that for yield, T-10 x Bonneville and P-185 x P-6583 cross-combination showed highest sca effect and heterosis, respectively. Kumar *et al.* (2000) observed appreciable heterosis over better parent and mid-parent for most of the yield and quality characters except TSS and protein content. In order to merit, F_1 hybrids JP-83 x EC-33866, JP-83 x Arkel and HUVP-2 x IP-3 for green pod as they exhibited significant heterosis of 59.14, 58.41, 73.18%, respectively over top parent PSR-18-6.

Cowpea

Hybrid vigour in cowpea is also dependent on specific parents used in hybrid combination. Some hybrids exhibit marked pod yield increased over parents (Mak and Yap, 1977 and Hazra, 1991). Among pod yield components, manifestation of heterosis for number of pods/plant, pod length and seed weight have been reported by Jones and Isbell (1956), Agble (1972) and Mak and Yap (1977). Hazra (1991) observed that when actual heterosis is considered against the corresponding genetic divergence between parents of crosses, one to one correspondence is not found owing to balancing or even cancellation of various components of heterosis. Tikka *et al.* (1977) reported that negative and moderate heterosis over better parent for days to 50% flowering and days to first pod picking. Bhaskaraih *et al.* (1980) studied the heterosis for yield component. They reported that heterosis over mid parents (MP) and better parent (BP), respectively, have been calculated for pod yield/plant (186.3 and 130.3%), number of pods/plant (166.4 and 152.8%), number of seeds/pod (36.1 and 19.5%), pod length (43.8 and 37.1%) and 100-seed weight (28.7 and 26.2%). Based on the heterotic studies, it inferred that number of pods/plant and 100-seed weight are most important yield-contributing characters in cowpea. Jain and Jain (1982) found a significant variation among parents and hybrids derived by line x tester analysis for fodder yield. They observed that seed yield and quality characters varied greatly in two environments and heterosis has been found higher in one environment than that of other for most of the characters in majority of crosses.

The non-additive gene effects predominated over additive for most of the characters. Zaveri *et al.* (1983) observed high degree of heterosis for grain yield, number of clusters / plant and number of pods/plant, but not for pod length, grains/pod and 100-seed weight, days to 50% flowering and maturity have been found positive heterosis over mid-parent and better parent. The heterosis for grain yield could be attributed to high heterosis in number of clusters and number of pods/plant. Selvaraj and Annappan (1983) observed heterosis for seed yield to the extent of 243% compared to the mid-parental values, (233%) over better parent and 64% over best parent. Patil and Shete (1987) recorded highest positive heterosis over mid-parent value which occurred for number of pod clusters/ plant (158%), followed by seed yield/plant (131%), whereas, over the better parent it has been observed highest pod clusters/plant (133%), followed by number of pods/plant (89%). Number of pods/plant and pod clusters/plant, supplemented by 100-seed weight, appeared to contribute to seed yield/plant. Therefore, it is suggested that segregating population from crosses in which heterosis for seed yield/plant is associated with heterosis for number of pod clusters and number of pods/plant provided useful material from which desirable line could be selected. Hazra *et al.* (1993) derived information for heterosis from pod yield and four important components like plant height, vine length, number of pods/plant, pod length and pod yield/plant. They observed that frequency and level of heterosis are related more to sca than to genetic divergence of parents as estimated by 'Mahalanobis', D^2 statistics. Sawant *et al.* (1994) recorded greatest positive heterosis over mid-parent (132.5%) for number of primary branches /plant (85.6%) and plant height (73.4%). Sangwan and Lodhi (1995) reported positive heterosis for yield ranging from 28.8 to 84.0% over better parent.

Heterosis up to 81.6, 35.6, 20.4 and 36.3% over better parent has been observed for number of pods/plant, pod length, number of seeds/pod and seed weight, respectively. Panmariammal *et al.* (1996) reported that majority of the hybrids have been found significantly superior over their parents except for days to 50% flowering. Bhore *et al.* (1997) recorded that significant heterosis over better parent as observed for days to 50% flowering (48.24%), days to maturity (91.52%), number of primary branches/plant (47.99%), plant height (9.33%), number of clusters/plant (37.54%), number of pods /cluster (24.91%), number of seeds/pod (17.84%), number of pods/plant (64.22%), green pod yield/plant (63.83%) and 100 grain weight (37.48%). Vishwanath *et al.* (1999) studied heterosis for yield and yield components in three inter-varietal crosses of cowpea over mid-parent and better parent has also been observed for most of the characters. Savithriamma and Lath (1999) recorded best hybrids combinations have been found for number of pods/plants, number of seeds/pod, seed yield, and 100-seed weight. Bhushana *et al.* (2000) found maximum heterosis over the mid-parent for secondary branches/plant (171.5%) followed by number of pods/plant (112.47%). Danam and Chaudhari (2000) observed maximum positive heterosis for seed yield/plant over MP (mid-parent), BP (better parent) and SC (standard check) to the extent of 64.47, 46.45 and 63.33%, respectively. They also reported that heterosis in seed yield/plant is due to the heterosis found in yield components, mainly, pods/plant, seeds/pod, clusters/plant, branches/plant and test weight. Pal *et al.* (2002) observed appreciable heterosis over better parent and mid parent in cowpea. They found that F_1 hybrids namely NDCP-13 x Arka Garima, Cowpea Local x Cowpea-263, Red Seeded x Pusa Komal, Cowpea Local x Cowpea-263 and Red Seeded x Pusa Komal showed high degree of heterosis.

French bean

Radkov (1979) found no reciprocal differences in hybrids of French bean. Singh and Saini (1983) studied heterosis in F_1, F_2 and inbreeding depression in F_2 generation. They reported that non-additive gene effects for most of the crosses. Guitierrez and Singh (1985) reported moderate to low heterosis for yield and yield traits. However, Savova (1985) observed significant importance of F_1 and F_2 hybrids of *Phaseolus vulgaris* L. (Savi) with stable lines of *P. vulgaris* x *P. coccineus*. Estimation of gene effects, genetic variance, heterosis and combining ability analysis in French bean have been studied by a number of workers (Vaid and Singh, 1985; Petova, 1985; Santos and Venocovsky, 1986 and Nienhuis and Singh (1988 a and b). Ram and Rajput (1998) observed that French bean varieties/lines VC-1, VL-6 and Pusa Parvati Loye have been found good general combiners and the crosses, *viz*. EC-43036 x HUR-30, EC-108101 x HUR-30 and VC-1 x Pusa Parvati exhibited high heterosis for plant height, pod length, number of pods/plant and total green pod yield / plant, respectively. In 1999, they also reported that cross VC-1 x PDR-14 showed maximum heterosis for total green pod yield/plant, number of pods/plant, pod length and earliness. They reported that heterosis for green pod yield is generally accompanied by heterosis for yield components.

4

REPRODUCTION, POLLEN BIOLOGY AND POLLEN BIOTECHNOLOGY

Reproduction

Since the life of an individual plant is limited in duration, it has developed certain mechanism by which it can reproduce itself inorder to continue the perpetuation of the species and also to multiply in number. Mode of reproduction determine the genetic constitution of crop plants, that is, whether the plants normally homozygous or heterozygous. A knowledge of the mode of reproduction of crop plants is also important for making artificial hybrids. The various modes of reproduction found in crop plants may be broadly grouped into the following two categories : (i) Asexual and (2) Sexual.

Asexual Reproduction

Heterosis breeding systems depend upon the reproduction system of plants, whether the plant is propagated asexually or sexually. Therefore, before development of hybrids, information regarding type of flower, flowering habits and floral biology is required in hybridization programme (emasculation and pollination). In a sexual reproduction, plants develop from vegetative parts of plants (vegetative propagation) or may develop from unfertilized gametes, ovular tissues, and modified cells (apomixes). Vegetative reproduction includes plant regeneration from underground stems such as tubers in potato, bulbs in onion and garlic, underground tubers such as sweet potato, corms in yams, and rhizomes in ginger and turmeric. In certain cases, aerial parts are used for propagation as stem cuttings in tapioca, sweet potato, watercress, and drumstick, root suckers in curry leaf and globe artichoke and turfts in chive. In apomixes, sexual reproduction or gamete fusion is replaced with other form of asexual reproduction. In this case, embryo develops without fertilization. Apomixis may be of different types: (1) Apogamy—where embryo develops from a gametic cell of embryo sac other than egg cell. Apogamy may be haploid or diploid. Diploid apogamy has been reported in onion. (2) Polyembryony—where more than one embryo develops from a seed, resulting in more than one seedling. In this case, embryos usually develop from nucellus integuments, or chalaza in addition to normal syngametic embryo. (3) Parthenogensis where embryo develops from an unfertilized egg. This may be at haploid or diploid level. Sometimes, embryo development take place from unfertilized egg of a diploid plant. This is called haploid parthenogenesis. It originates in plants due to environmental effects, especially due to effects of high or low temperature, use of foreign pollen, application of chemicals, and use of irradiated pollen. Haploids parthenogenesis produces haploids.

In cole crops, matromorphy has been observed. Matromorphy is synonymous with diploid parthenogenesis and the resultant plant produced is called a matromorphic plant. The phenomenon is found in Brassica. This is obtained after intergeneric crosses, interspecific crosses and use of foreign pollen. The main regions are the formation of unreduced embryo sac and parthenogenetic development of reduced egg cell and then its reduplication. Matromorphic plants developed from diploid embryos are largely heterozygous. The frequency of matromorphic seeds depends on temperature, pollen use, and application of growth regulators. Application of gibberellic acid increases the number of matromorphic seeds in *Brassica oleracea*. Patromorphy is synonymous with diploid anderogenesis, which is found in *Solanum* spp. In a few cases, male nucleus is unable to fertilize, but divides and gives rise to a haploid embryo. This is called androgenesis. In this case, egg cells degenerate. Apomixis is used in potato to produce homozygous autodiploid and tetraploid lines from haploids. It is also used to fix heterosis.

Sexual Reproduction

It involves fusion of male and female gametes to form a zygote, which develops into an embryo. In crop plants, male and female gametes produce a specialized structures known as flower. A complete flower consists of sepals (it is the lowermost leaf like organ, which protect the bud till it takes the form of flower), petals (present inside calyx, brightly coloured, produce nectar and odour of essential oils, which attracts insect for pollination), stamens (It is male part of flower consisting of filament, anther and connective long stalk which bears on sac and produces pollen-grains. The quantity and size of pollen-grain may differ) and pistils (Gynoecium- it is a female part of flower, which includes ovary style and stigma. The ovary contains ovules and stigma receives the pollen during pollination.). In vegetable crops, plants produce three types of flowers: staminate (androecious), pistillate (gynoecious or carpillate), and hermaphrodite. Majority of vegetable crops bear hermaphrodite flowers such as tomato, brinjal, pepper, okra, pea, beans, *etc*. Crops of family cucurbitaceae belong to the above-mentioned types or their combinations. Accordingly, there are the following types of flowering habits in cucurbits: (1) Trimonoecious plants bearing staminate, pistillate, and hermaphrodite flowers, (2) Monoecious plants bearing staminate and pistillate flowers, (3) Androecious plants bearing only staminate flowers, (4) Andromonoecious plants bearing staminate and hermaphrodite flowers, (5) Gynoecious-plants bearing only pistillate flowers, (6) Gynomonoecious plants bearing pistillate and hermaphrodite flowers, and (7) Hermaphrodite-plants bearing only hermaphrodite flowers.

The trimonoecious form is rare one and scarcely present in breeding lines of cucumber, muskmelon, and ridged gourd. Monoecious is the predominant one and found in ash gourd, bitter gourd, bottle gourd, cucumber, muskmelon, pumpkin, ridged gourd, round melon, smooth gourd, squash and watermelon. Generally, in monoecious plants, staminate flowers appear first and pistillate flowers in later stages. The probable sequence of flowers on plant is underdeveloped male, normal female, inhibited male, giant female, and parthenocarpic flowers. There may be a few deviations in sequence depending upon genotypes and environmental factors. In monoecious cucumber, staminate mixed, and pistillate phases are normally noticed. Andromonoecy is predominant in muskmelon and very rare in watermelon and ridged gourd, smooth gourd, and cucumber. Androecy and gynomonoecy are rarely seen in the breeding material- segregating population of ridged gourd, cucumber and muskmelon.

In crop plants, meiotic division of specific cells in stamen and pistil yields microspores and megaspores, respectively. Mitotic division of spore nuclei to produce gametes follows this; the male and female gametes are produced in microspores and megaspores, respectively. The production of microspores and megaspores is known as sporogenesis. Microspores are produced in anthers (microsporogenesis), while megaspores are produced in ovules (megasporogenensis). In microsporogenensis; each anther has four pollen sacs, which contain numerious pollen mother cells (PMCs). Each PMC undergoes meiosis to produce four haploid cells or microspores. This process is known as microsporogenesis (Fig.1). The microspore mature into pollen-grains mainly by a thickening of their walls. The megasporogenesis occurs in ovules, which are present inside the ovary. A single cell in each ovule differentiates into a megaspore mother cell. The megaspore mother cell undergoes meiosis to produce four haploid megaspores. Three of the megaspores degenerate, leaving one functional megaspore per ovule (Fig. 2). This completes megasporogenesis.

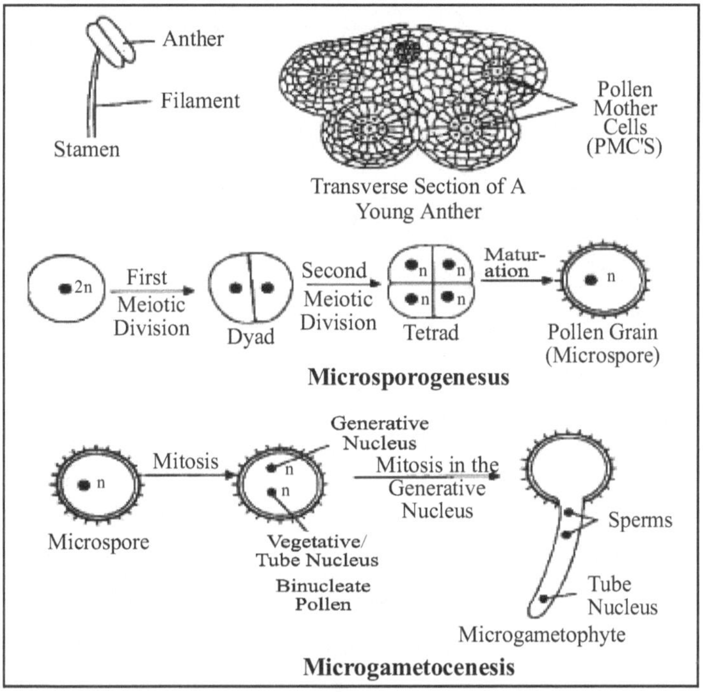

Figure 1: Microsporogenesis and microgametogenesis (*Source:* Singh, 2002, p. 51).

The production of male and female gametes in microspores and megaspores, respectively, is known as gametogenesis.This involves: (1) Microgametogenesis which refers to production of male gamete or sperm. During maturation of pollen, microspore nucleus divides mitotically to produce a generative or tube nucleus. The pollen is generally released in this binucleate stage. When pollen lands on to stigma of a flower, it is known as pollination. Shortly after pollination, pollen germinates. The pollen tube enters the stigma and grows through the style. The generative nucleus now undergoes a mitotic division to produce two male gametes or sperms. The pollen, although the pollen tubes, is known as microgametophyte (Fig.1). The pollen tube finally enters the ovule through a small pore, micropyle, and discharges two sperms into embryo sac. (2) Megagametogenesis in which the nucleus of a functional megaspore divides mitotically to produce four or more nuclei. The exact number of nuclei and their arrangement varies considerably from one species to another. In most crop plants, megaspore nucleus undergoes three mitotic divisions to produce eight nuclei. Three of these nuclei move to one pole and produce a central egg-cell and two synergid cells. One synergid is situated on either side of the egg cell. Another three nuclei migrate to opposite pole to give rise to antipodal cells. The two nuclei remaining in the center, the polar nuclei fuse to form a secondary nucleus. The megaspore, thus develops into a mature megagametophyte or embryo sac. The development of embryo sac from a megaspore is known as megasporogenesis. The embryo sac generally contains one egg cell, two synergids, three antipodal cells (all haploid, and one diploid secondary nucleus (Fig. 2).

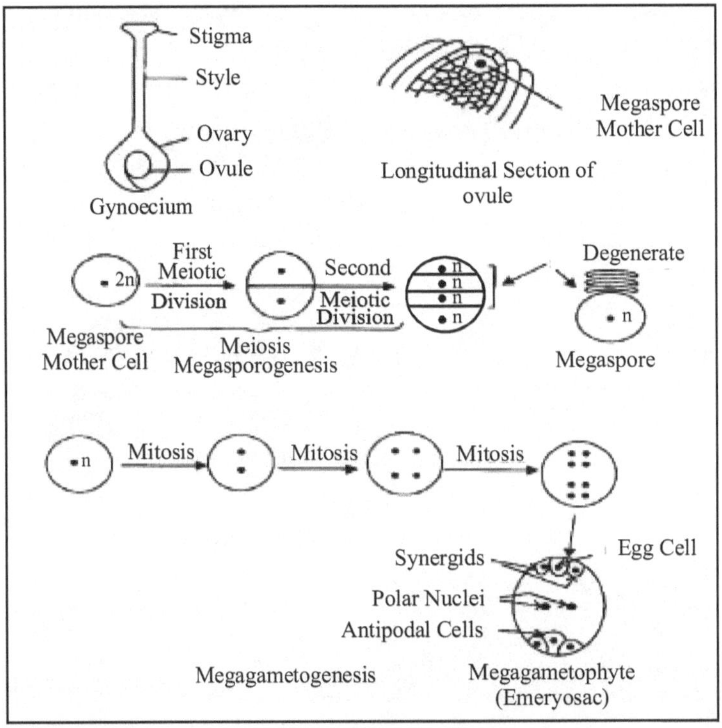

Figure 2: Megasporogenesis and megagametogenesis (Source: Singh, 2002, p.52).

Pollen and Pollen Biology

Pollen is the gold dust that carries the male germ of flowering plants and is the male partner in sexual reproduction and seed formation. They are generally shed in a desiccated condition and their moisture level is less than 20%. At the time of shedding, pollen-grains are either two celled—a large vegetative cell enclosing a generative cell; or three-celled—a vegetative cell and two sperm cells formed by the division of generative cell. There is a considerable variation in size and shape of pollen-grains. Although, majority of pollen-grains are spherical, in more marine angiosperms, such as *Amphibolis* and *Zostera*, they are filiform (up to 5 mm). The wall of pollen-grain is made up of two layers: an outer, acetolysisresistant exine composed of sporopollenin and an inner pectocellulosic intine. One of the conspicuous structural features of pollen-grains is ornamentation of wall formed by outer part of the exine. Followings are major events in pollen biology:

- Pollen development
- Pollen dispersion
- Pollination
- Pollen-pistil interaction
- Fertilization.

Pollen Development

Pollen-grains develop inside the anther and are dispersed by dehiscence of the anther. After dispersal, pollen-grains act as independent functional units and are exposed to prevailing environmental conditions for varying periods, depending on period and severity of environment, quality of pollen grains, particularly their viability and vigour, may be affected during pre-pollination phase. Eventually, pollen-grains are deposited on the stigma (pollination) through biotic or abiotic agents.

Pollen Dispersion

Pollens are dispersed by external agents such as insects (bees, flies, moths *etc.*), animals (birds, snails, *etc.*), wind and water. When insect carries pollination is known as entomophily (entomon, an insect; philein, to love). Pollination by insects is very general occurrence among flowers. Entomophilous or insect loving flowers have various adaptations by which they attract insects and use them as conveyors of pollen grains for the purpose of pollination. Principal adaptions are colour, nectar and scent. In some cases, pollination is brought about by wind is known as anemophily (*anemos*-wind). Anemophilous or wind-loving flowers are small and inconspicuous. They are never coloured or showy. They do not emit any smell nor do they secrete any nectar. The anthers produce immense quantity of pollen-grains, and its wastage during transit from one flower to another being considerable. They are also light and dry; sometimes, as in pine, they are provided with wings for facility of distribution by wind. Stigmas are comparatively large and protruding, sometimes branched and often feathery. Pollination may also be brought about in some aquatic plants particularly in submerged ones, though the agency of water is known as hydrophily (*hydro*-water). Those aquatic plants that lift their flowers above the water level normally achieve pollination through insects or wind or by water current. Hydrophilous flowers are as a rule small and inconspicuous.

Pollination

It refers to the transfer of pollen-grains from anthers to stigmas. Pollen from an anther may fall on to stigma of the same flower leading to self— pollination or autogamy. When pollen-grains from flowers of one plant are transmitted to stigmas of flowers of another plant, it is known as cross-pollination or allogamy. A third situation, geitonogamy, results when pollen from a flower of one plant falls on the stigmas of other flowers of the same plants, *e.g.* in maize. The genetic consequences of self-pollination are rapid increase in homozygosity. Therefore, populations of self- pollinated species are highly homozygous. Self-pollinated species do not show inbreeding depression, but may exhibit considerable heterosis. Cross-pollinated species preserve and promote heterozygosity in a population. Cross-pollinated species are highly heterozygous and show mild to severe inbreeding depression and a considerable amount of heterosis. The genetic consequences of geitonogamy are the same as those of autogamy.

Pollen Pistil Interaction

These interactions occur just after the pollen-grains reach the stigma and pollen-grains germinate on the stigma and the resulting pollen tubes grow through the tissues of stigma and style and ultimately enter the embryo sac. The process from pollination until the entry of pollen tubes into the embryo sac is referred to as pollen-pistil interaction or progamic phase. This phase plays an important role in determining the breeding system of the species/ population. During pollen pistil interaction, pollen is selected for quality and compatibility. Pollen-grains represent the gametophytes generation. A large number of genes, some unique to the pollen and others common to the sporophytic generation, are expressed during pollen development as well as during post-germination phase of pollen. Studies on expression of these genes are important for the application of pollen selection, for recombinant DNA technology, and for the induction of pollen embryos.

Fertilization

The fusion of one of the sperm with egg cell, producing a diploid zygote, is known as fertilization. The fusion of remaining sperm with secondary nucleus, leading to the formation of a triploid primary endosperm nucleus is termed as triple fusion. The process when both fusions (syngamy and triple fusion) occurs is known as double fertilization. The fertilized zygote develops into an embryo.

Pollen Biotechnology

Pollen biotechnology is the manipulation of pollen development and function, including alteration of its genome, for the production and improvement of crops and related products. Pollen biotechnology has direct application in hybrid seed production mainly by following pollination control mechanism.

- Geneic (nuclear) male sterility
- Gametocides or Chemical hybridizing agents
- Cytoplasmic male sterility
- Self- incompatibility

Geneic (Nuclear) Male Sterility

Male sterility is characterized by non-functional pollen-grains, while female gametes function normally. Male sterility can result from mutation in either the nuclear or cytoplasmic genes. Genic male sterility (GMS) is generally caused by mutation in genes that in recessive condition affect stamen and pollen development. However, GMS regulated by dominant genes is also known (Kaul, 1988). Geneic male sterility is reported in nearly every major crop and in a large number of dicotyledonous and monocotyledonous species. Generally, GMS is of spontaneous origin, but it can also be induced by chemical mutagens Fujimaki *et al;* 1977; Cross and Lady Man, 1991), by radiations (Driscoll and Barlow 1976), and by genetic engineering (Mariani *et al.,* 1990; Worrall *et al.,* 1992 and Goldberg *et al.,* 1993). Novel male sterile plants can also be generated by protoplast fusion (by cybrid formation, Kofer *et al.,* 1990 and Gourret *et al.,* 1992), by T-DNA transposon tagging (Aarts *et al.,* 1993), and by affecting the synthesis of flavonoids (Van Tunen *et al.,* 1994). Male sterility is classified into following groups:

(a) Genetic Male Sterility

- Temperature sensitive genetic male sterility (TGMS)
- Photoperiod sensitive genetic male sterility (PGMS)
- Transgenic genetic male sterility

(b) Cytoplasmic Male Sterility

(c) Cytoplasmic Genetic Male Sterility (CGMS)

(d) Chemically Induced Male Sterility

Genetics and morphology of GMS mutant

In nearly every crop species, number of male steriles (*ms*) loci have been identified, In tomato 12 *ms* gene has been reported (Stevens and Rick, 1986). For *ms* genes, a number of alleles exhibit varying degrees of male sterility. The linkage map of ms genes is available for most of the major crops. The morphology of male sterile mutants is variable. It ranges from complete absence of male reproductive organs to formation of normal stamens, with viable pollen, that fail to dehisce. In general, male-sterile plants are morphologically not distinguishable from the parent fertile plants with exception of flower structure. Male sterile flowers commonly smaller in size in comparison to fertile flowers. The size of stamen is generally reduced. The anther may or may not be collapsed and the filament length may be affected. In some cases, stamens may be modified as carpel, as in the stamenless –1 (sl-1) mutant of tomato (Fig.3). In others, structure with both anther and carpel features are produced, as in stamenless –2 (sl-2) mutant of tomato.

Effect of Environments on Expression of GMS

The expression of male sterility is known to be influenced by environmental factors. The two main physical factors that influence GMS are temperature and photoperiod.

Temperature

The effect of temperature on male sterility has been examined in the variable male — sterile (*vms*) mutant of tomato, *vms* plants exposed to temperature of 30° C and above are male sterile, but those grown in greenhouse are normal (Rick and Boynton, 1967). Experiments done in temperature controlled growth chambers on sl-2 mutants of tomato

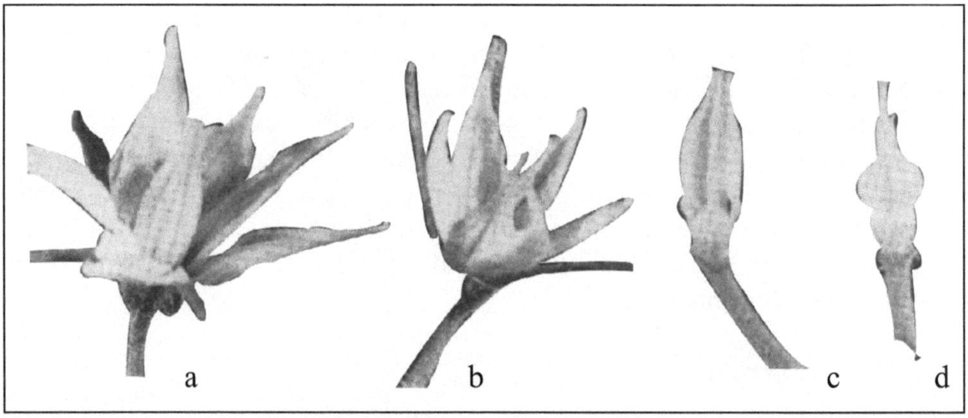

Figure 3: a-d. Lycopersicon esculentum: male fertile (a, c) and male sterile (b, d) flowers with calyx and corolla intact (a, b) and removed (c, d) showing developed (a, c) and underdeveloped (b, d) anthers. *(Source:* Kalloo, 1991, p. 46).

showed more accurately the effects of temperature on male sterility. When Sl-2 plants grown in high temperatures (28°C day / 23°C night) produce flowers with carpel-like structures instead of stamens, whereas those grown in low temperatures (18°C day / 15°C night) bear normal stamens with viable pollen (Sawhney, 1983). In intermediate temperatures (23°C day/ 18°C night), sl-2 flowers produce structure with half stamens and half carpel (Fig.3). Thus, a 5°C difference in temperature significantly modifies the expression of sl-2 allele. In a GMS mutant in *Brassica oleracea*, temperature sensitivity is even more precise at 10-11°C, the male-sterile plants can be converted into fertile in 30 days, but at 12° C, it takes over two months for partial fertility (Dickson, 1970). In many other GMS mutants, however, male sterility is not affected by a change in temperature. For example, a GMS line in *Brassica napus* is insensitive to temperature fluctuations.

Photoperiod

A change in photoperiod can have a strong influence on expression of male sterility. In tomato, a photoperiod-sensitive male sterile mutant 7B-1 is completely male sterile in summer field condition in Saskatoon. However at a day length of 8-10 hours, 7B-1 plants produce many flowers with normal anthers that contain normal and, viable pollen.

Cytological Changes

Pollen development is tightly programmed, sequential process that involves distinct changes in sporogenous tissue (gametophytic) and in tapetum and wall-layers (sporophytic tissues) of the anther. In GMS mutants, the breakdown in *microsporogenesis* can occur at a number of pre or post-meiotic stages. The abnormalities can involve, aberration in any one of the following stages: during the process of meiosis, in the formation of tetrads during the release of tetrads (that is, the dissolution of *callose*), at the vacuolated microspore stage, or at mature or near-mature pollen stage. Also, although, there is presumed specificity of action of ms genes, there is, in some cases, inter- and intra-plant variation. Abnormalities in pollen development in male sterile mutants of tomato are commonly associated with aberrations in tapetum development. In sl-2 mutant, for example, tapetal cells enlarge prematurely, are often highly vacuolated, and their degeneration is delayed. This mistiming of tepetum degeneration results in many microspore, being devoid of a wall. In the absence of wall, microspores enlarge and eventually burst, leaving a mass of debris in pollen sac. These observations on sl-2 mutant, and on ms mutants in different crops, demonstrate the importance of tapetum in microscope and pollen development. In male – sterile mutants induced by radiations and by chemical mutagens, pollen development can also be affected at pre-or post-meiotic stages. In these examples, the breakdown in microsporogenesis is also generally associated with abnormal tapetum development. The importance of tapetum in pollen development has been used to develop strategies at the molecular level that selectively destroy the tapetum and thereby induce male sterility in crop plants.

Biochemical Changes

Male sterility has been shown to be accompanied by qualitative and quantitative changes in amino acids, proteins, and enzymes in developing anther. The general approach in such studies has been to analyze the level of various amino acid, total proteins, and activity, and isozymes of specific enzymes believed to be associated with pollen development.

Amino Acid

Analysis of amino acid in male-sterile anthers of a number of species have revealed both reduced and increased levels of amino acid. In general, the level of proline, leucine, isoleucine, phenylamine and valine is reduced in male-sterile anthers, but that of asparagines, glycine, arginine, and aspartic acid increased. In GMS mutants, level of proline is particularly affected. In Japanese radish, for example, the mature male- sterile anthers contain one-eighth the amount of proline in comparison to the anthers. The level of other amino acids is also affected, but not to that extent. Kakihara *et al.* (1988) showed that filaments of male-sterile stamens accumulate proline, suggesting a problem in translocation of this amino acid to develop anthers. How proline is specifically required for pollen development is not understood. It is also not known whether the reduction in proline levels, or in other amino acids, is a primary or a secondary effect of mutations in MS genes.

Soluble Proteins

It has been observed that mutation in MS genes results in qualitative and / or quantitative changes in soluble proteins in male- sterile anthers. Differences in total protein content and a polypeptide profiles, obtained by 1-D and 2-D gel electrophoresis, have been shown between male-sterile and fertile anthers. In general, male - sterile anthers contain lower protein content and fewer polypeptide bands than fertile anthers. By feeding S-methionine to fertile and male- sterile floral buds of tomato at different developmental stages; Bhadula and Sawhney (1991) showed some polypeptides synthesized in normal stamens are absent in mutant stamens. As well, a 53-kd polypeptide is synthesized only in mutant stamens at the time when tapetum abnormalities appear in mutant anthers. Whether 53-kd polypeptide, or lack of a number of other polypeptides, is directly responsible for pollen degeneration in this tomato mutant is not known.

Enzymes

The sequential development of pollen-grains is involved specific enzyme- mediated reaction at different stages. Aberration in the activity of one or more of such enzymes can lead to abnormal pollen development and thus to male sterility. The enzyme callase is required for the breakdown of callose (β-1, 3-glucans) that surrounds PMC and, after meiosis, the tetrads. Microspores are released from tetrads by the activity of callase. In some GMS and CMS systems, mistiming of callase activity (that is, early or delayed) results in premature or delayed release of meiocytes and microspores, resulting in male sterility. Esterases have also been related to pollen development. They have been localized specifically in tapetum at early stages of microsporogenesis and later at the surface of pollen-grains. In both GMS and CMS systems, the activity of esterases is decreased. In sl-2 mutant of tomato, the number and intensity of isozymes of esterases are reduced at stages when abnormalities are observed. In tapetum and when pollen- wall development is affected, esterases are believed to have a role in hydrolysis of sporopollenin, the polymer required for pollen-wall formation. Aberration in the activity of esterase would, therefore, have an effect on pollen development.

The activities of other enzymes (for example, peroxidase, dehydrogenases, phosphatases and cytochrome oxidase) are also known to affect in male–sterile systems. Carbohydrates, particularly simple sugars, are essential components of pollen-wall development. It is, therefore, of interest to note that in a male-sterile tomato mutant, the activity of amylase is decreased, and it corresponds with high starch content and reduced levels of soluble sugars.

Gametocides or Chemical Hybridizing Agents

Chemicals capable of selectively inhibiting pollen development and thus blocking male fertility have been known for some time. Most are identified in screening programmes designed to discover chemical hybridizing substances (CHSs) for large-scale commercial production of hybrid seed, particularly in some grains. Driven by practical requirements this research has identified newer generation of CHSs with increasingly better selectivity and effectiveness. However, research on mode of action of this substance has lagged their practical application in breeding and hybrid production. Compounds are known that cause a range of effects, including feminization of male florets and inhibition of early anther development, inferences with tapetal functions and microspore development, and defective germination of mature pollen. Several CHSs also have demonstrated a stimulatory effect on the induction of androgenic plants from anther culture derived from treated plants. The strength of this response varies among genotypes. The molecular and physiological mechanisms of these effects are unknown.

Plant Hormones/hormone Antagonists

Many of the earliest known chemical-hybridizing agents are plant hormones and their antagonists. These include several auxins (IAA, 2, 4-D, NAA, TIBA *etc.*) and maleic hydrazide (MH), which may be either an anti-auxin or a growth inhibitor. Most of these substances have not been commercially successful as CHSs, because of their many inhibitory and phytotoxic effects unrelated to male fertility. These substances have not enjoyed commercial attention, recent research has been minimal. More promising results have been obtained with ethrel, gibberellins and abscisic acid. Except for ethrel , little is known about their mechanism of action.

Auxins and Auxin Antagonists

It has been observed that when lentil plants are treated with NAA, IBA, MH and coumarin, vegetative growth, flowering, pollen sterility, fruit setting, and seed yield are studied and found that there is general reduction of plant growth and seed yield. It has been noted enhancement of branching and floral abnormalities. Male sterility induced by the treatments has been found to express in several ways, including *in-situ* pollen germination, *in-situ* exudation of pollen cytoplasm, and modification of certain stamens into staminodes. With MH and IBA, the tapetum failed to enlarged and develop into the normal peripheral plasmodium, whereas with coumarin,the tapetal cells enlarged a typically and persistent.

Gibberellins and Antagonists

Gibberellins (GA) are well-known for their effects on sexual determination and floral development, which is the basis for chemical hybridization in sunflower. The response of plants to gibberellins and GA antagonists varies by species. In some species, GA interferes with developments of male floral organs or promotes feminization, whereas in others these effects are absent. Effects on pollen production in several crops have been demonstrated experimentally. For example, Mere and Bennekom (1982) reported reduced pollen shed in onion treated with a GA_{4+7} mixture. This growth regulator mixture did not affect the viability of the pollen that is shed. Although, used experimentally to produce hybrid onion for breeding purposes, the GA_{4+7} mixture is not commercially practical because of cost. In Solanaceae, the effects of GA applications vary by species. In tomato, normal plants treated with GA_{4+7} produce fertile flowers with separate stamens and split pistils. In pepper (Capsicum annum), however, GA_3 application produces feminization of flowers and inhibition of pollen development. In tomato, gibberellins synthesis inhibitor CCC (2-choloroethyl-trimethyl ammonium chloride) , at certain concentration, selectively inhibits the development of the stamen or otherwise suppresses pollen development. These effects are not sufficiently selective to serve as the basis for chemical hybridization in Solanaceae.

Abscisic Acid

In tomato, abscisic acid (ABA) caused effects on developing floral buds similar to those produced by CCC. In wheat, ABA caused male sterility if applied to plants just prior to or during meiosis of pollen mother cells. Thus, ABA may cause male sterility through more than one mechanism. It does not appear that ABA is currently under development as a CHA.

Other Substances that Disrupt Floral Development

Two synthetic growth regulators, LY1-95259 and TD-1123, are believed to interfere primarily with over-all male-flower development. The molecular action of these substances is not reported.

LY-195259

LY-1952-59 is reported as an effective chemical hybridizing agent for wheat. This compound is considered commercially promising at some times, but more recent publications are lacking. LY-195259 is applied when the flowering spikes are quite short and averaging 0.5 cm. High application rates at this, time prevented anther development, whereas lower dosages resulted in progressively reduced inhibition. Sterility at lower dosages is associated with smaller, abnormally twisted, and intensively pigmented locules. Rates are as low as 1.12 kg/ha produced greater than 95% male sterility. Seed setting as high as 80% of the male parent with treated cultivars. The hybrid seed appeared normal, and no other phytotoxic effects have been visually evident from rates as high as three times the mean effective dose. ^{14}C- uptake and translocation indicated that much of the compound remains at the site of application. Only a small percentage is translocated to plant organs above a treated leaf. It has been found that uptake of LY-195259 from soil is particularly effective in producing male sterility.

TD-1123

Undeveloped anthers that fail to dehisce are also produced by the cotton CHA, A variety of morphological effects have been observed at higher treatment levels.

Metabolic Inhibitors

Screening evaluation of off-the-shelf chemicals has identified several compounds that selectively depress the male fertility. These earlier findings are adequately reviewed by McRae (1985), and include the effect of halogenated aliphatic acids (for example, α, β-dichloroisobutyrate and 2, 2-dichloro-propinoate salts) and arsenicals (methanearsonate salts). Although, studies of these substances as hybridizing agents have been discontinued, investigation of their selectivity may help explain the physiological basis of cytoplasmic male-sterile mutants. Many of these mutants are associated with defects in mitochondrial proteins. Arsenates, in particular, would be expected to reduce the efficiency of normal metabolic processes.

Inhibitors of Microspore Development

Copper chelators

Studies of copper deficiency syndrome in small grains indicate that pollen development is irregular or absent. Abortive microspore development is associated with irregular or reduced formation of outer (exine) wall, which contains sporopollenin. The tapetal cells normally supplies of nutrients and sporopollenin precursors for developing microspores become disoriented and expanded. They contain irregular, and endopolyploid nuclei. It appears that their secretory functions are severely disrupted. It has been asserted that copper deficiency exerts these effects by inhibiting copper requiring oxidases that function in auxin metabolism. Graham (1976) proposed that copper deficiency could be manipulated for the production of hybrid wheat.Using copper chelators, particularly benzotriazole. Graham (1976) has been able to demonstrate selective induction of male sterility. However, practical application of these findings to hybridization has not been achieved.

Ethylene

Ethylene has been implicated as a natural regulator of development and maturation of several floral organs. Filament and corolla growth in flowers of *Ipomoea* spp. are inhibited by ethylene production. Anthers have been identified as a source of an early filament growth inhibitor *in vitro*. Anthers contained increased levels of ethylene precursor, aminocyclopropane carboxylate (ACC), from 15 to 6 hours before anthesis. Ethylene peaked to about 6 hours nearer anthesis than did ACC levels in this organ. ACC and ethylene levels in filaments fluctuated along with those in anthers.

Phenylcinnoline carboxylates (SC - 1058, SC - 1271 and SC - 2053)

Phenylcinnoline carboxylates are synthesized and developed by Sogetal Inc., a research laboratory belonging to Orsan, a French Biotechnology group. Hybrinova , a related company , has undertaken commercialization of hybrid wheat using Sogetal lead compound SC-2053. Hybrinova has taken advantage of earlier advanced research at sogetal and western plant breeders with this compound and predecessors, SC-1058, SC-1271 and SC-2053. Hybrinova is producing new hybrid combinations for testing in French. Published data have centered on three lead compounds:

- SC-1058, 1-(4'- trifluromethylphenyl)-4-oxo-5-fluorocinnoline-3-carboxylic acid.
- SC-1271, 1-(4'- chlorophenyl)-4-oxo-5-proprxycinnoline-3-carboxylic acid.
- SC-2053, 1-(4'- chlorophenyl)-4-oxo-5-methoxyethoxy) cinnoline-3-carboxylic acid.

The initial lead compound (SC-1058) has been identified from screening a series of 1,4 dihydro-4-oxo-1-proprxycinnoline- 3- carboxylic acid. The unsubstituted compound (1-phenyl-4- oxo cinnoline-3-carboxylic acid) has without pollen suppressant activity. Although, hybrid wheat could be made with SC-1058 compound, screening for agents with improved characteristics continued. A range of substitutions on cinnoline and phenyl rings have been prepared and evaluated. Significant pollen suppressant activity has been identified for compounds substituted at C-5 on cinnolin with fluorine or alkoxy groups. Compounds bearing an amino substituent showed little or no activity and compounds with methyl substitution at C-5 is inactive.

Phenyl ring substitutions are also explored systematically and found to modulate the activity of compounds. Symptoms associated with more phytotoxic compounds of this series, or with excessively high dosage levels, typically include chlorosis, plant height reduction, or necrosis. However no effect on seed quality is seen. The lead compound developed from this programme (SC-1271 and later SC-2053) for use as chemical hybridization agent are members of alkoxy series containing less than four carbon atoms, which have best balance of high male sterility and low phytotoxicity. The relationship between male sterility, seed yield, and hybrid purity has been reported to be complex. Male pollinator lines differed significantly in their effect on quantitative hybrid purity, even a constant high level of male sterility or seed yield data. It is necessary to develop direct analytical methods for determination of hybrid seed purity, such as analysis of seed storage proteins.

Genesis® (MON-21200)

Hybri Tech, a Monsanto affiliated company, has been developed a chemical hybridizing agent coded MON 21200 for commercial production of hybrid wheat. HybriTech has also developed elite germplasm for wheat hybridization using *Triticum timopheevi* cytoplasm and especially bred restorer lines. This group has also acquired the former wheat breeding and chemical hybridization programme of Rohm and Hass. Despite prior availability of CMS system originated from *Triticum timopheevi*, chemical has accelerated HybriTech's efforts to identified parental lines with superior combining ability of hybrid production. Monsanto earlier disclosed research on a series of pyrimidine monocarboxylates and benzoic acid analogues. These substances have been tested in growth chambers as potential chemical hybridizing agents for wheat. The 3-pyridinecarboxylic acid, 4-hydroxy-2,6-bis (trifluoromethyl) methyl ester, and 2,4- bis (trifluoromethyl) benzoic acid are only substances in each series that provide complete spike sterility. A total loss of activity accompanied minor changes in either molecule. With pyridine monocarboxylate, only substitutions at 4-position that contain acidic proton functionality, or that are subject to hydrolysis to 4-hydroxyl derivative, are active. The relationship of these chemical series to MON-21200, if any, has not been reported.

Inhibitors of Pollen Fertility

Azetidine-3-carboxylate (A_3 C, CHA)

It induces male sterility effectively in small grains particularly wheat. Anther development and pollen structure appear normal under light microscope and if applied after flag leaf has emerged from the boot, there is little general phytotoxicity. In treated rye-grass, A_3C also inhibits stem elongation substantially in parallel with pollen fertility. It also increases tillering but has no effect on accumulation of neutral-detergent-soluble substances.

Phloem Mobility and CHA Activity

Discovery of new chemical hybridizing agents have been through a process of screening, although good theoretical hunches have played a role in identification of starting points, particularly in the discovery of proline –like CHAs. Hsu and Kleier (1990) presented evidence that good mobility in phloem is a prerequisite for CHA activity. They postulate that phloem is necessary to overcome low transpiration rates and discontinuity of xylem in many reproductive tissues.

CHAs and Anther Culture

In 1972, Bennett and Hughes reported additional mitosis in microspores of plants that had been sprayed with ethrel. Work at that time was not successful in taking advantage of this effect to stimulate the frequency of androgenic callus formation. In the years that followed, the practice of conventional anther culture made strides, and few phytotoxic chemical hybridizing agents are discovered. Picard *et al.* (1987) took up the clue and reinvestigated the use of a chemical-hybridizing agent in wheat-anther-culture. Using fenridazon potassium (Hybrex), this group has been able to obtain a substantial improvement in the total yield of androgenetic green plants.

Cytoplasmic Male Sterility

Plants that do not produce viable and functional pollen-grains are male sterile. If such male sterility is exclusively maternally inherited, it is described as cytoplasmic male sterility (CMS). This type of male sterility is determined by cytoplasm (Fig. 4). Since, cytoplasm of a zygote comes primarily from egg cell, the progeny of such male sterile plants would always be male sterile. Cytoplasmic male sterility is known in many plant species, some of which are crop plants. Cytoplasmic male sterility may be transferred easily to a given strain by using that strain as a pollinator (recurrent parent) in successive generations of a backcross programme (Fig. 5). After 6-7 backcrosses, nuclear genotype of male sterile line would be almost identical to that of recurrent pollinator strain. The male sterile line is maintained by crossing it with pollinator strain used as recurrent parent in backcross programme, since nuclear genotype of the pollinator is identical with that of new male sterile line. Such a male fertile line is known as maintainer line or B line as it is used to maintain male sterile line. The male sterile line is also known as A line. There is considerable evidence that gene og conditioning cytoplasmic male sterility, particularly in maize, reside in mitochondria and may be located in a plasmid like element.

Although, CMS has been observed in more than 150 plant species, the mechanism by which CMS interrupts normal pollen development are not well-understood. Female fertility is generally not affected by CMS, so that male- sterile plants can and do set seed if viable pollen is provided. The affected organs and tissues in CMS plants are stamens (anther and filament) and pollen-grains (microspores). The stamen plays an integral role in crop production because it is responsible for carrying out male reproductive processes. Pollen-grains develop within the anther in a precisely timed, high-metabolic-demand process, with the tapetal layer of anther supplying enzymes, hormones, nutritive materials, nucleosides, and nucleotides for microsporogenesis to proceed. It appears that CMS results because of a multitude of effects, all more or less related to insufficient or mistimed supply of necessary resources for developing microspores. All CMS sources can theoretically be developed into functional CMS systems; only a few have actually been developed to the point where commercial quantities of hybrid seeds are being produced. There are several reasons for this, including insufficient of unstable male sterility, difficulties with restoration systems, difficulties with seed production, and undesirable pleiotropic effects of CMS pollination control system used. It is a long road from the discovery of CMS to commercial use of

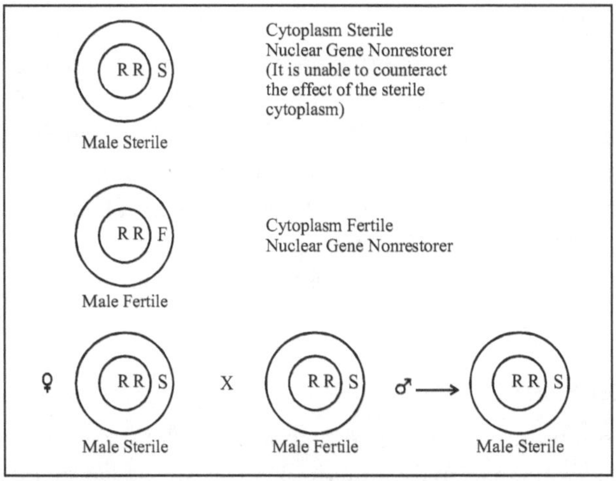

Figure 4: Cytoplasmic male sterility. Male sterility is caused by cytoplasm (depicted by S). The F cytoplasm is normal cytoplasm. If a nuclear restorer gene (the dominant allele of r) becomes available, male sterility will be included in cytoplasmic genetic group. This type of male sterility shows strict cytoplasm inheritance (*Source*: Singh 2002, p.70).

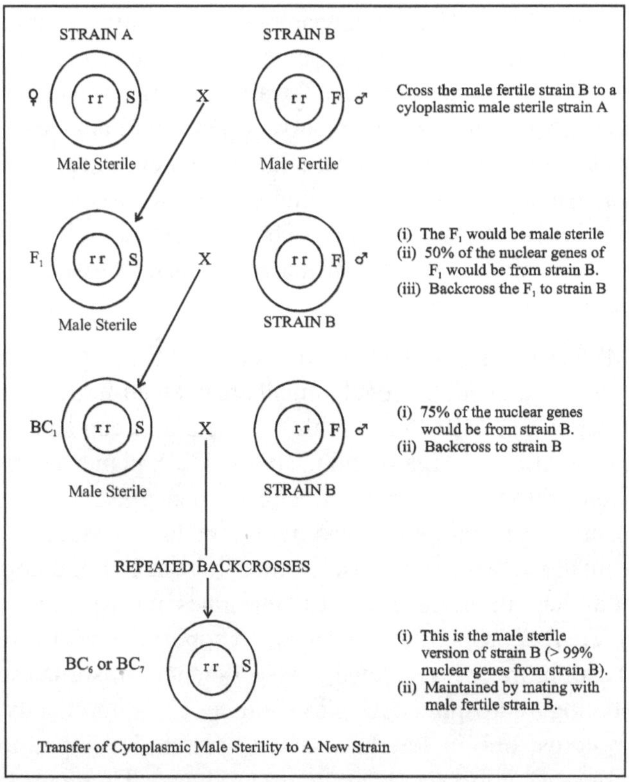

Figure 5: Transfer of male sterile cytoplasm from strain A to strain B. Strain A is used as non-recurrent parent in a backcross programme with strain B. After 6-7 backcrosses, the progeny will be similar to strain B in their genotype. The strain B serves as the maintainer of this new male sterile strain. (*Source*: Singh, 2002, p. 71).

CMS- based hybrids. Nevertheless, majority of hybrids grown today are based on CMS as pollination control mechanism.

Origin of Cytoplasmic Male Sterility

The CMS has several origins. These are:

From Interspecific Crosses

The CMS frequently arises from interspecific crosses. These cross types produce smaller differences between nucleus and cytoplasm, and therefore, somewhat decrease liklihood of obtaining a new CMS system. These crosses are easier to make intergeneric crosses, and many instances of these types of origins of CMS have been reported (Pearson, 1981). There are a few successful examples of CMS created by interspecific crosses, in general, either degree of male sterility has not economically satisfactory, or natural pollination mechanisms have not dependable, so that, except for the marginal use in Nicotiana and Gossypium, these CMS systems have not become a factor in the production of hybrid seed.

From Intraspecific Crosses

CMS also arises from intraspecific crosses. These are CMS types discovered in wild or segregated out of cultivars. CMS arising from intraspecific crosses is quite common and is indicative of species existing in two or more cytoplasms, at least one of which is a male-sterile cytoplasm. CMS systems derived from these sources are rapidly used in seed production programme in onion, corn, sorghum and rice (Pearson, 1981).

From Mutation

CMS may also arise apparently spontaneously. In this case, a male-sterile plant is discovered in a population of male- fertile plants. This is quite common in occurrence, and many successful CMS systems, for example, in onions, corn, and pearlmillet, originated spontaneously. CMS may be induced through the use of mutagens. A range of mutagens has been tried and occasional CMS has been uncovered. None of these sources of CMS have been found to be usable to date.

Characterization of CMS

The restorer genes required to overcome the CMS and to provide male-fertile progeny in male-sterile cytoplasm characterize CMS systems. CMS system is also characterized by DNA analysis technique applied to cytoplasmic organelles that are able to distinguish genetic differences between normal and male sterile cytoplasms. Current research in this area is primarily at the under laying CMS and relationship of nuclear- based restorer genes to cytoplasm-based male-sterility genes (Hanson, 1991). CMS restoration is by nuclear genes; frequently dominant in action, in many cases few in number (that is, 1-4 dominant genes, sometimes with minor male-fertility modifiers required to maintain complete restoration in some environments). The CMS restorer genes temporarily suppress the expression of CMS by mechanisms not well understood, permitting normal or near-normal pollen production.

DNA analysis techniques are uncovering restriction map differences between CMS and fertile cytoplasms in all crops so far studied.

Mechanism of CMS

Abnormal behaviour of tapetum in anthers has frequently been identified with CMS. Further, it has now been well-established that genetic determinant of CMS resides in the mitochondria and that nuclear genes control expression of CMS (Newton, 1988). The restorer genes are important components in the search for a CMS mechanism of action, since the restorer genes temporarily suppress male sterility even in the presence of CMS cytoplasm.

Limitations in Use of Cytoplasmic Male Sterility

There are several problems in utilization of CMS in vegetable hybrid seed production. These are:

- Pleiotropic negative effects of CMS on agronomic quality performance of plants in CMS cytoplasm.
- Enhanced disease susceptibility.
- Complex and environmentally unstable maintenance of male sterility and/ or male fertility restoration.
- Inability to produce commercial quantities of hybrid seed economically because of poor floral characteristics for cross-pollination.
- Danger of monoculture crop production.
- Continuous incorporation of small amount of male parent cytoplasm through each backcross during maintenance may lead to partial or complete break down of male sterility.

Future Developments in CMS

In addition to a better understanding of CMS and thus better control of CMS system, and extension of CMS to crops that presently have no CMS systems, other future uses of CMS can be suggested. New combinations of CMS mitochondria and normal or herbicide-resistant chloroplast by cybrid formation. Pelletier (1983) produced *brassica cybrids* with (1) ogu CMS mitochondria and nap cytoplasm and (2) ogu CMS mitochondria and atrazine-resistant chloroplasts using protoplast fusion and organelle sorting out. Similarly, Choung *et al.* (1988) were able to combine pole CMS mitochondria and atrazine-resistant chloroplasts to produce oilseed rape cybrids. In another approach, use of CMS-related mt DNA plasmids such as S_1 and S_2 (of corn from CMS-S) two plasmids that have many characteristics reminiscent of transposable elements of prokaryotes as potential vehicles for genetic engineering e.g, the mitochondria has been suggested. The S_1 and S_2 elements are linear duplexes with terminal repeat sequences. There is also good evidence that their disappearance as autonomous mt DNA species in association with CMS-S reversion is coupled with transposition of at least some of their sequences to mitochondrial genome. In light of this endeavour, vectoring of specific improved respiration-related genes or herbicide-resistance genes to mitochondria using S_1 and S_2 has been envisioned.

The CMS trait carried in mitochondria is used as a genetic marker to permit the identification of mitochondria in protoplast fusion-produced cybrids. Since, mitochondria genome recombination also frequently takes place in cybrid cells, this approach allows for recombination of mitochondrial genes to create new and improved mitochondria, with combinations of traits not formerly available. Nilkamp *et al.* (1983) developed genetic manipulation methods for plant-cell organelle genes. They studied molecular basis of CMS in *Petunia* and possibilities of transfer of the trait to other members of Solanaceae family by somatic cell cybridization. They used autonomous replicating sequences (ARSs) to try to vector mitochondria and chloroplast genes to their respective organelles. They propose the insertion of such ARSs in plasmids with an appropriate marker gene that can function as a selection marker (for example CMS) to produce a vector for plant organelles. The availability of such vectors and of techniques such as cybridization opens the way for introduction of organelle genes involved in economically important properties such as photosynthesis, CMS, resistant to herbicides, and resistance to plant pathogens into plant organelles.

Future molecular genetic work will help to elucidate molecular and functional basis of CMS. Techniques such as isolation and study of organelle DNA, development of restriction endonuclease, and cloning & sequencing techniques have permitted a closer look at CMS. In *corn* and *Petunia*, normal and CMS strains have been found to have significant mt DNA differences. In some crops, a specific mitochondrial gene has been directly implicated in CMS. For example, T-urf$_{13}$ gene and CMS-T, and Petunia pcf and CMS-S are functionally related in corn and Petunia, respectively. Despite, identification of CMS- related loci in some crops, molecular mechanism of CMS is still unknown. Thus, a cause-and-effect relationship for CMS has yet to be established for any crop. Once this relationship is established, possibility of improving or creating CMS in species desired, through the use of molecular genetic techniques, becomes closer to reality.

Cytoplasmic Genetic Male Sterility

This is a case of cytoplasmic male sterility where a nuclear gene for restoring fertility in male sterile line is known. The fertility restorer gene, R is dominant and is found in certain strains of species, or may be transferred from a related species. This gene restores male fertility in male sterile line, hence, it is known as restorer gene (Fig. 6). The cases of cytoplasmic male sterility would be included in cytoplasmic- genetic system as and when restorer genes for them would be discovered. It is likely that a restorer gene would be found for all the cases of cytoplasmic male sterility if a thorough search are made. In this case, plants would be male sterile in presence of male sterile cytoplasm if nuclear genotype is rr, but would be male fertile if nucleus is Rr or by RR (Fig. 6).

New male sterile line may be developed following the same procedure as in case of cytoplasmic system. But nuclear genotype of pollinator strain used in such a transfer must be rr, otherwise fertility would be restored. The development of new restorer strains is somewhat indirect (Fig. 7). First, a restorer strain (say R) is crossed with a cytoplasmic male sterile line A. The resulting male fertile plants are used as female parents in repeated backcrosses with strain C used as recurrent parent, into which the transfer of restorer gene(s)

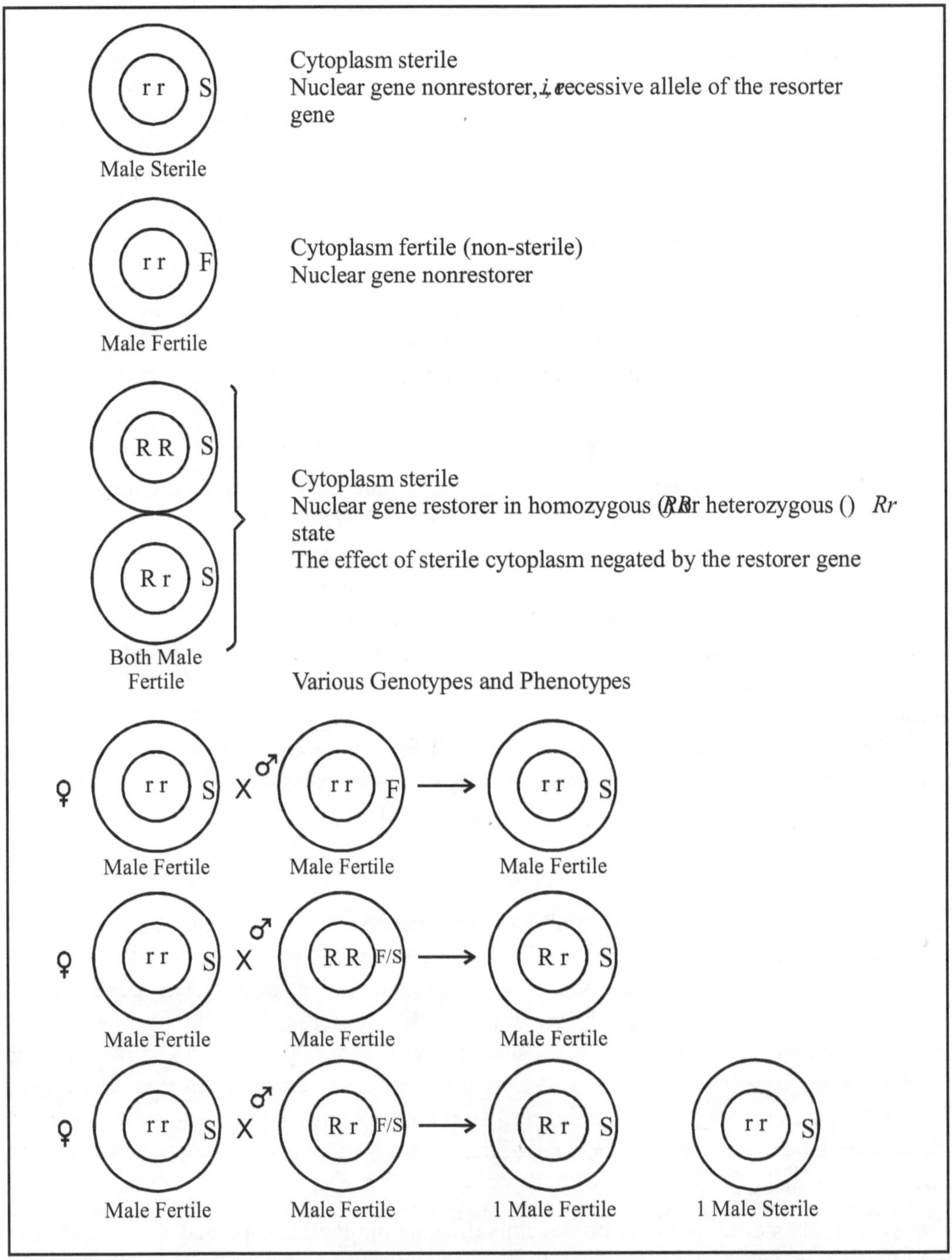

Figure 6: Cytoplasmic-genetic male sterility. Male sterility is due to a sterile cytoplasm; a restorer gene restores fertility, which is usually a dominant nuclear gene. F/S in cytoplasm indicate that cytoplasm may be either fertile (F) or sterile (S). (*Source*: Singh 2002, p.72).

is desired. In each generation, male sterile plants are discarded, and male fertile plants are used as females for backcrossing to strain C. This acts as a selection device for restorer gene R during backcross programme. At the end of backcross programme, a restorer line isogenic to strain C would be recovered.

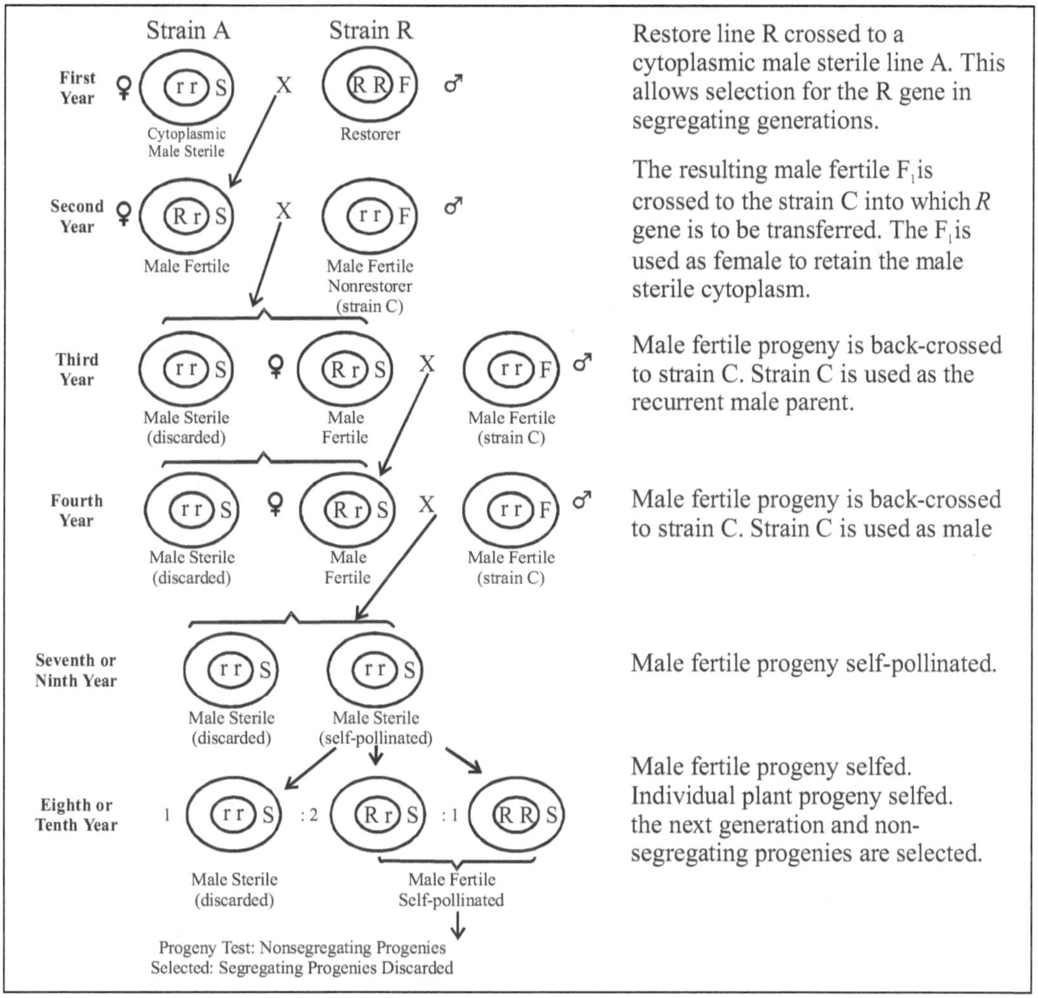

Figure 7: Transfer of a restorer gene R from a restorer strain (strain R) to a new strain (strain C) (*Source:* Singh 2002, p. 75).

Self-incompatibility

Self-incompatibility can be defined as an inability of viable pollen to fertilize an ovule carrying an identical *S-allele* and inhibitory self or sib-pollination. According to Brewbaker (1957) self-incompatibility occurs in more than 300 species belonging to 250 genera spread in about 70 families. In case of self-incompatibility, pollen-grains fail to germinate on stigma of flowers. If some pollen-grains do germinate, pollen tube fails to enter the stigma. In many species, pollen tubes enter the style, but they grow too slowly to effect fertilization before then flower drops. Sometimes, fertilization is affected, but embryos degenerate at every early stage. Self- incompatibility appears to be a biochemical reaction, but the precise nature of these reactions is not clearly understood. The genetic control of incompatibility reaction is relatively simpler. Lewis (1954) has suggested various classifications of self-incompatibility. A relatively simpler classification is as follows :

I. Heteromorphic system

II. Homomorphic system
   - Gametophytic control
   - Sporophytic control

Heteromorphic System

In this system, flower of different incompatibility groups are different in morphology. For example, in *Primula*, there are two types of flowers, *pin* and *thrum*. *Pin* flowers have long styles and short stamens, while *thrum* flowers have short styles and long stamens. This situation is referred to as distyly. *Pin* and *thrum* flowers are produced on different plants. The only compatible mating is between *pin* and *thrum* flowers. This characteristic is governed by a single locus; Ss produces *thrum*, while ss produces *pin* flowers. The incompatibility reaction of pollen is determined by the genotype of plant producing them. Allele *S* is dominant over *s*. The incompatibility system, therefore, is heteromorphic sporophytic. *Pin* flowers produce their pollen, would all be *s* in genotype as well as incompatibility reaction. The pollen produced by *thrum* flowers would be of two types genotypically, *S* and *s*, but all of them would be *S* phenotypically. The mating between *pin* and *thrum* plant would produce *Ss* and *ss* progeny in equal frequencies. (Fig. 8). This system is of little importance in crop plants. It occurs in sweet potato and buckwheat.

Homomorphic System

In homomorphic system, incompatibility is not associated with morphological difference among flowers. The incompatibility reaction of pollen may be controlled by genotype of plant on which it is produced (sporophytic control) or by this its own genotype (gametophytic control).

Gametophytic System

It is the primitive type of incompatibility from which other forms of incompatibility systems emerged (Arasu, 1968). East and Mangelsdorf first described it in 1925 in *Nicotiana sanderae*. The incompatibility reaction of pollen is determined by its own genotype, and not by the genotype of the plant on which it is produced (Fig. 9). This is because the biochemical substances involved in SI reaction of the pollen are produced after meiosis. In style, both S is determined by a single gene having multiple alleles, *e.g. Trifolium*, *Nicotiana, Lycopersicon, Solanum, Petunia etc.* The gametophyte system of self- incompatibility (GSI) is controlled by haploid pollen-grains and diploid pistil tissue. In most of the cases, this is controlled by single locus multiple alleles, but in some cases, this may be controlled by two

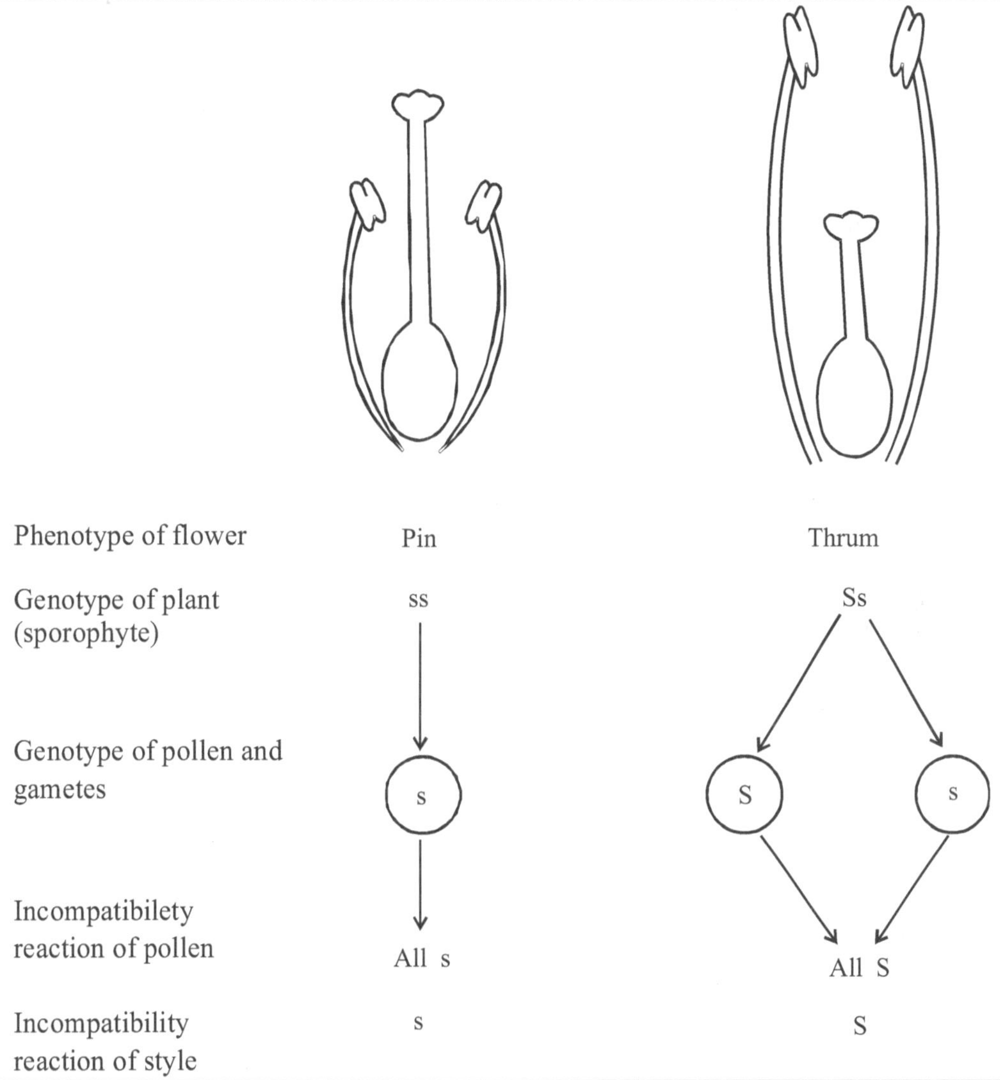

Mating		Progeny	
Phenotype	Genotype	Genotype	Phenotype
Pin x Pin	ss x ss	Incomplete mating	
Pin x Thrum	ss xSs	1 Ss : 1 ss	1 Thrum : 1 Pin
Thrum x Pin	Ss xss	1 Ss : 1 ss	1 Thrum : 1 Pin
Thrum x Thrum	Ss xSs	Incomplete mating	

Figure 8: Heteromorphic-sporophytic system of incompatibility. The relative position of stigma and anthers are governed by a single S/s. The incompatibility reaction of pollen is governed by genotype of sporophytic on which it is produced, while that of style is controlled by its own genotype. Allele S is dominant overs in determining incompatability as well as the length of style. Pollen of an incompatibility group would not be able to grow normally in style of the same incompatability group and would be unable to effect fertilization. Such mating is said to be incompitible (*Source*: Singh 2002, p.60).

multinucleate S loci and three or more S loci. In former case, pollen being haploid may have S_1, S_2, and S_3 - Sm, Sn alleles and pistil being diploid usually have S_{12} and S_{13} - Smn. There is fusion of pollen where unlike gametes and alike gametes never fuse (Fig. 9). With increase in number of SI alleles in population, there is an increase in the ratio of compatible to incompatible pollination. There is bifactorial control of GSI, which has been reported in *Solanum phureja* and *S. stenotomum*. Further more, S and Z alleles are also found to control this incompatibility reaction. Also, GSI is controlled by three or more loci. In β vulgaris, there is the involvement of four loci. It is possible to identify S chromosome having SI alleles.

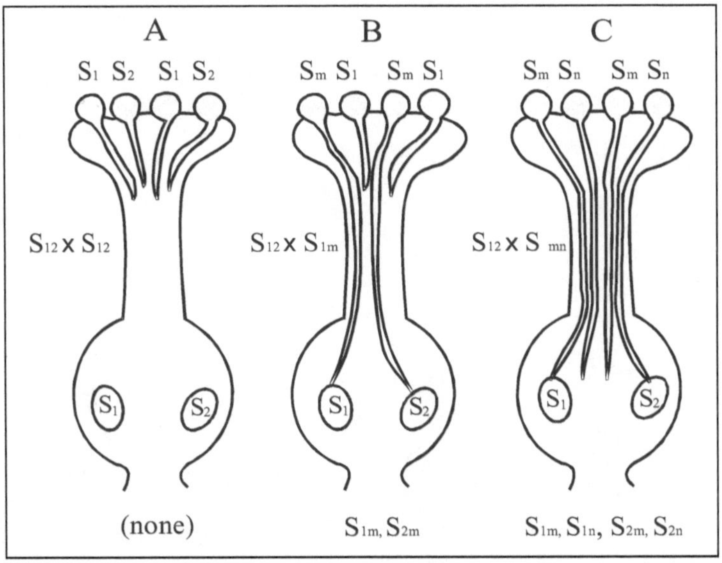

Figure 9: Pollen tube growth and fertilization in one S locus gametophytic incompatibility system. Three sources of pollen pollinating on S_{12} pistil. (A) pollen from an S_{12} ; all pollen tubes are inhibited, (B) pollen from an S_{1m} plant; S_1 pollen is inhibited and S_m is compatible, resulting in S_{1m} and S_{2m} zygotes and (C) pollen from an S_{nm} plant; both S_m and S_n pollen are compatable., resulting in S_{1m}, S_{2m} , S_{1n} , and S_{2n} zygotes. (*Source*: Kalloo, 1994, p.14).

Earlier, Martin (1961) in *L. chilense* observed that S_I allele located on chromosome 2, studied on the basis of a marker gene. In genus *Lycopersicon, L. peruvianum, L. hirsutum*, and *L. chilense* are self- incompatible. The pollen germination and growth in style have been studied and it has found that cessation of pollen tube growth took place in the upper part of style. Furthemore, inner wall of pollen tube of incompatible phenotype of *L. peruvianum* is temperature- dependent. The difference has been recorded in six lines of self- incompitable *L. peruvianum* at 25°C. But no such differences occurred at 10 - 20 °C. It indicates that there may be form of gene S that shows the effect of temperature, or there may be genetic modifiers of S gene action with temperature-dependent effect. Genetics of gametophytic self- incompatibility has been studied in *L. peruvianum* using enzyme loci. Such incompatibility has been found to control in *L. peruvianum* using enzyme loci. Such incompatibility is controlled by a single major locus on chromosome 1 close to *Prx* - 1 and *Idh*-1.

Biochemical basis of Gametophytic Self- incompatibility

On the basis of various researches it has been concluded that, within pistil extract, certain protein segregate with their respective S- alleles in genetic crosses. These S- proteins are present in high proportions in style tissues generally, 1- 10 % and are sufficiently diverse to be differentiated on SDS- polyacrylamide or isoelectric focusing gels. S- proteins have now been identified in the *Lycopersicon peruvianum* (Mau *et al.,* 1986), *Nicotiana alata* (Bredemeijer and Blass, 1981), *Petunia inflata* (Ai *et al.,* 1990), *Petunia hybrida* (Kamboj and Jackson, 1986), *Solanum chacoense* (Xu *et al.,* 1990a), and *Solanum tuberosum* (Krich *et al;* 1989). The molecular weights of these S-proteins range from 23 to 34kD; all possess high pI values- often higher than 8.0; and every S-protein of this type so far studied has beeen glycosylated. In addition to segregating with S-alleles, the temporal and spatial distribution of these proteins corelates with the onset of incompatibility and the site of pollen tube rejection.

Mechanism of Self-incompatibility Response

Since RNase activity is essential for S- protein function, therefore, RNase activity is an integral part of SI reaction, and the most obvious potential role would be in the degradation of pollen tube RNA. One hypothesis is that S-RNase enters incompatible pollen tubes as they grow through the extra cellular matrix and degrades RNA, resulting in the demises of pollen tubes. Since rRNA genes are not transcribed in pollen, this would provide an effective way of ensuring pollen tube death. This model necessitates a specificity in the interaction regulating the uptake of S-RNase into pollen tube. Another model for arrest of pollen tube growth envisages S-specific inhibitors that interferes with S-protein activity in an allele–specific manner. This model requires the male products of S-locus to be highly specific inhibitors that interact with S-proteins over a wide range of sequence diversity whilst remaining ineffective against one molecule, the self-S-protein. It is not as yet possible to rule out the inhibitor model entirely, as it is possible that a complex interaction, involving one or more pistil components and an S-specific inhibitor present in pollen tube cytoplasm, is required to inactivate S-RNases. RNase has been implicated in regulation of other key process in plant growth and development. Certain plant lectins may provide an interesting parallel to S- proteins. Ricin A and phytoalaxin, for example, are able to inactivate ribosome through an associated RNase activity in SI is therefore not an associated problem and is of potential significance to other areas of plant growth and development.

Bifactorial Gametophytic Self- incompatibility (Two-locus GSI System)

Lundquist first described it in the grasses in 1956. Most grasses investigated posses SI of this form, although all cereals except rye are self- compatible. The two-loci involved in the SI response are termed as S and Z and each are polyallelic. These appears to be cooperation between S and Z in pollen, but independent actions in the pistil. Each combination, therefore, gives rise to a distinct specificity in haploid pollen, and rejection occurs when this specificity is matched by one of the four possible combinations of S- and Z- alleles in diploid stigma. This results in two locus system's being more resistant to decay than that governed by a single locus, and two-locus system is less likely to suffer from employing two loci in an SI system is that is less likely to acquire self-compatible mutants, for S- and Z- loci act in both a complementary and an independent manner. Thus, if one locus mutates, other gives rise to incompatibility. The biofactorial system differs from other gametophytic systems not only in having two locus control, but also in exhibiting many cytological features that are much more akin to those of saprophytic systems (pollen is trinucleate, and inhibition occurs at the stigma surface). It therefore, has been suggested, that gametophytic SI in grasses has arisen independently from self- compatible plants. Thus, it is not surprizing that the operation of biofactorial SI differs fundamentally from that found in other gametophytic SI plants.

Molecular Cloning of S-gene Product and S-locus Structure in GSI

The molecular cloning techniques have considerably advanced our knowledge of S-gene and its expression in female tissues. So far, deduced amino acid sequence of 18 S-proteins from five Solanaceous species has been reported. These include seven S- alleles of *Nicotiana alata* (Anderson *et al.,* 1986 and 1989 and Kheyr- Pour *et al.,* 1990), three of *Petunia inflata* (Ai *et al.,* 1990), three of Petunia hybrid (Clarke *et al.,* 1990), two of *Solanum chacoense* (Xu *et al.,* 1990 b), and three of S. tuberosum (Kaufmann *et al.,* 1991). Two independent experiments have unequivocally established S- proteins as the pistillar component of SI response. Both studies involved transformation with S- protein constructs to alter pistil phenotype. Lee *et al.* (1994) demonstrated both loss and gain of female S-function of S_3 allele in *Petunia inflata* through transformation with antisense and sense constructs of S_3 allele driven by its own promoter. Murfett *et al.* (1994) also demonstrated gain of female S- function in a *Nicotiana alata* x langsdorffii hybrid with a sense construct of *N. alata* Sa_2 allele driven by a chittanase promoter (ChiP). The advent of these transformation experiments heralds a new era of opportunity in Si research, opening up the possibility of studying S- gene structure- function relationship. Analysis of aligned sequences revels 34 perfect conserved residues and 27 sites that are only conservatively replaced. Of the 34 conserved residues, 16 are also conserved in RNase T2 of *Aspergillus oryzae* and RNase Rh of Rhizopus niveus. These short stretches of homology would not normally be considered significant. However, because these stretches include the sequences surrounding the catalytic domains of fungal ribonuclease- the histidine residues essential for catalysis (Kawaka *et al.,* 1990) and four-cystine members of two-disulfide bonds- this observation assumed the greater significance. These findings led to the discovery that S-proteins do indeed exhibit ribonuclease (Rnase) activity (McClure *et al.,* 1989).

Amino acid sequence identity between S- proteins so far published varies considerably, from 38.74 to 93.47% (Singh and Kao, 1992). This tremendous diversity between alleles of the same species contrasts sharply to other multiallelic genes, which differ only in a few amino acids.Conserved residues within S-proteins essentially fall into five regions, designated C1-C5. A window, averaged hydrophobicity plot shows that C1, C4, and C5 are hydrophobic, suggested that they might from the core structure of S- proteins. C2 and C3 are hydrophilic and are also conserved in the fungal ribonucleases; hence they are probably involved in catalytic activity. Two regions of extreme variability have also been identified—hypervariable (Hv) regions a and b (Ioerger *et al.,* 1991). These are consistently the most hydrophilic regions in each S-protein and may constitute the domains involved in encoding S-allele specificity. The role of carbohydrate side chains of S-proteins in S-allele specificity has recently been investigated. This involved transformation of *Petunia inflata* with an S_3 construct in which single asparagines glycoslation site is converted to asparatic acid. It has been observed that transformants expressing this protein levels comparable to wild type S-proteins gained S_3 function. Though some S-proteins possess more than one side chain, this result strongly suggests that glycosyl groups of S-proteins play no role in S-specificity and thus this function must be encoded by amino acid backbone.

Sporophytic Self–incompatibility

It was first reported by Hughes and Babcock in 1950 in *Crepis foetida*, and by Gerstel in *Parthenium argentarum* (in the same year). It is governed by a single gene, S, with multiple alleles; more than 30 alleles are known in *Brassica oleracea*. In general, the number of S-alleles is considerably larger in gametophytes than in sporophytic system. The incompatibility reaction of pollen is governed by genotype of the plant on which pollen are produced, and not by the genotype of the pollen. (Fig. 10). In sporophytic system, S alleles may exhibit dominance, individual action (co-dominance) or competition. Consequently, there may be many complex incompatibility relationships. Lewis has summarized following characteristics of this system.

- There are frequent reciprocal differences.
- Incompatibility can occur with the female parent.
- A family can consisted of three incompatibility groups.
- Homozygotes are consisting of three incompatibility groups.
- An incompatibility group may contain two genotypes.

The best characterized sporophytic system that operate in membrane of Brassicaceae (Cruciferae). The crucifer stigma is capped by a layer of papillate cells and, in a compatible pollination, there is a rapid series of events, including adhesion to the stigma, hydration and germination of the pollen-grains, and subsequent invasion of the papillar cell wall by emerging pollen tube (Elleman *et al.,* 1992). The phenomenon associated with incompatible pollination varies according to the strength of alleles involved. In strict form of incompatible pollination, adhesion to dry stigma surface is poor and even hydration is absent, but with weaker alleles, hydration and germination may occur. Resultant tubes succeed in penetrating the stigmatic cuticle, but they fail to invade the stigma cell-wall (Kanno and Hinata, 1969). Generally, there is deposition of β- 1, 3-glucan callose within stigmatic papillae at the site of contact with incompatible pollen. This can be useful diagnostic phenomenon when determining the cross-compatibility. However, recent data suggest that this is related more to the leakage of gametophytic proteins from necrotic pollen-grains or tubes than to playing a mechanistic role in incompatibility reaction itself (Elleman and Dickinson, 1994). The SSI is also developmentally regulated and comes into operation 1-2 days pre-anthesis. Because stigmas are receptive prior to this point, there is a window in development when it is possible in generation and maintenance of S-locus- homozygous lines for both research and breeding purposes. Cabbage has a sporophytic self-incompatibility system. The sporophytic system of self-incompatibility can be either heteromorphic (heterostyly) or homomorphic. Heterostylic sporophytic incompatibility is controlled by one gene with two alleles in distyly and by two genes, each with two alleles in tristyly (Kalloo, 1988). Homomorphic sporophytic incompatibility is controlled by one locus with multiple alleles. In cabbage modifier genes influence the self-incompatibility S alleles for sensitivity to environmental factors (Nasrallah and Wallace, 1968).

Frankel and Galun (1977) have given following characteristics of sporophytic system of self-incompatibility.

- Incompatibility is controlled by one S locus having several alleles. Over 50-70 alleles have been found in *B.oleracea*. (Hodgkin *et al.,* 1988 and Kalloo, 1988).
- The reaction of pollen is characterized by the genotype of sporophytic tissue in which the pollen are formed, and therefore, controlled by two S alleles.
- All the pollen of a plant have the same incompatibility reaction.
- The two S alleles may react as co-dominance (independent) or they may interact by one being dominant over the other.
- Active allele identity in both pollen and pistil leads to incompatibility.
- The dominance/independence relationship of *S alleles* in pollen and in pistil may differ.
- The reaction of S_1S_3 x S_1S_2 (male) in sporophytic system in cabbage has been given in Table 11 (Frankel and Galun, 1977).

Table 11: Sporophytic self-incompatibility system (S_1S_3 x S_1S_2 (male)

Pollination	Pistil reaction	Compatibility
Independent	Independent	Incompatible
S_1 dominant to S_2	Independent	Incompatible
S_2 dominant to S_1	Independent	Compatible
Independent	S_1 dominant to S_3	Incompatible
Independent	S_3 dominant to S_1	Compatible
S_1 dominant to S_2	S_1 dominant to S_3	Incompatible
S_1 dominant to S_2	S_3 dominant to S_1	Compatible
S_2 dominant to S_1	S_1 dominant to *S3*	Compatible
S_2 dominant to S_1	S_3 dominant to S_1	Compatible

The *S- allele* relationships in some cabbage cultivars studied by Haruta (1966) are given in Table 12.

Table 12: Types of incompatibility relationship of S-alleles in pollen and stigma found in cabbage cultivars.

Type of incompatibility	Dominance relationship among S-alleles		
	In stigma	**In pollen**	**Cultivars**
I	S_b dominant to S_a	S_b dominant to S_a	Alde, Copenhagen Market, Slow Bolting, Flat Dutch and Naketo Kyoku Sosci
II	S_b independent of S_a	S_b dominant to S_a	Surehead, Danish Ball Head, and Copenhagen Market.
III	S_b dominant to S_a	S_b dominant of S_a	Maeda Banso

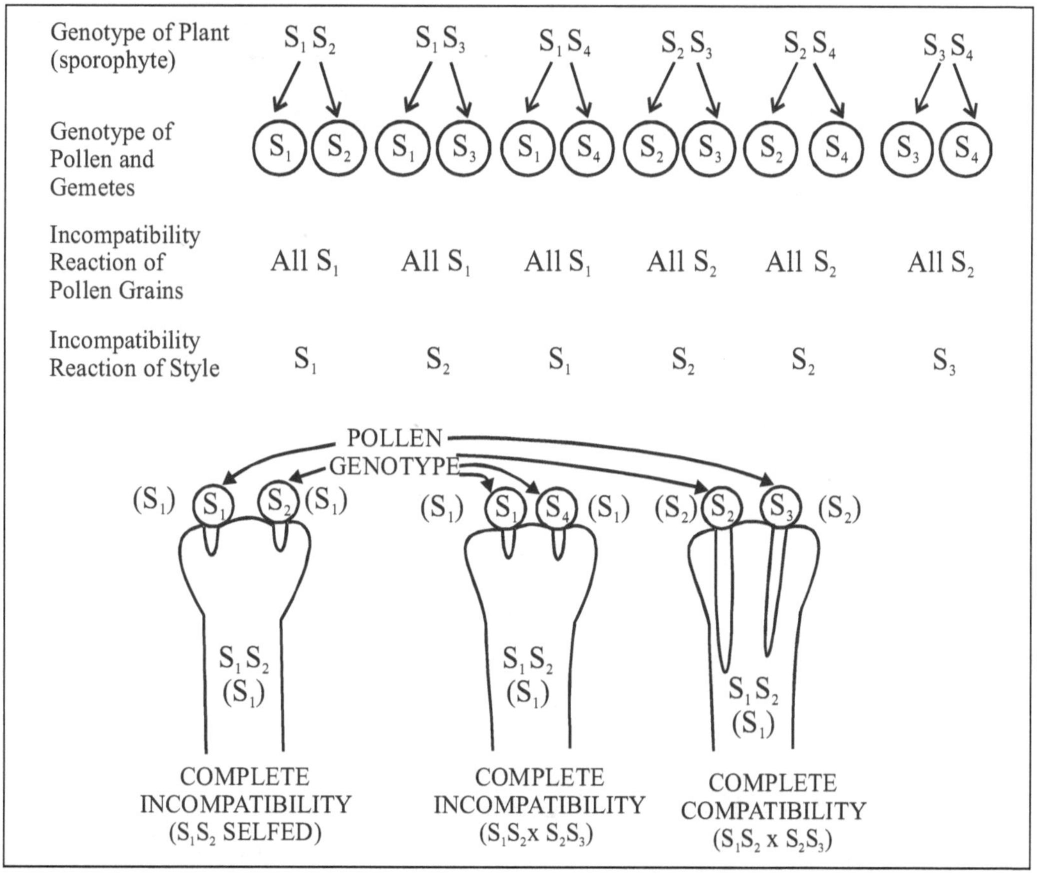

Figure 10: Sporophytic system of self- incompatibility. The incompatibility reaction of pollen-grains is controlled by the genotype of plant (sporophytic) on which they are produced, while that of style is governed by its own genotype. For simplicity, it is assumed that the incompatibility alleles (S-alleles) show complete dominance in the manner $S_1>S_2>S_3>S_4$ - *etc.* The actual situation is much more complicated and several different types of allelic relationship are encountered (*Source*: Singh, 2002, p.63).

Comparison between Gametophytic and Sporophytic Self- incompatibility

A comparison between gametophytic and sporophytic systems are listed in Table 13.

Table 13: Comparison between gametophytic and soaprophytic system (Lewis, 1949).

Character	Gametophytic System	Sporophytic System
Number of loci	One	One of more
Number of alleles	Multiple	Two alleles when one or two genes distyly or tristyly. Multiple alleles for a single locus -homomorphic.
Reaction of self-incompatibility factor	Pollen itself determines independent of other S alleles reaction. Either 100 % pollen grains function or 50 % or more may not function at all	May show independent action or dominance or complex interaction. The reaction of pollen is determined by both alleles present in the sporophytic. Pollen-grains produced by plant either do not function at all or function hundred per cent.
Reciprocal differences	No reciprocal differences	Reciprocal differences do occur.
Population structure	Homozygoten are not normal part of the system. The parental genotypes do not occur in the progeny	Homozygoten are a normal part of the system. Parental types may occur in the progeny.
Pollen morphology	Binucleate except grasses, which have trinucleate pollen-grains.	Generally trinucleate except *Lythrum primula,* which has binucleate pollen grains.
Site of action	Pollen tube inhibited in style and ovary	Inhibition of pollen germination or pollen tube growth on stigma except *Linum kelianthus*, where incompatibility reaction takes place in stylar tissues.
Effect of polyploidy	May result into self-compatibility	No effect.
Time of S gene action	Post - meiotic	Premeiotic or meiotic.

Diallelic Sporophytic SI and Heteromorphy

Unlike gametophytic SI systems, sporophytic systems can exist with only two alleles, and in such cases two alleles have been termed S (dominant) and s (recessive). This is possible because the diploid heterozygous male parent produces pollen of a single S-phenotype though the pollen genotypes are both S and s. The most studied genus possessing this type of SI is *Primula.* Almost all diallelic SI systems display floralmorphism, usually in the form of di- or heterostyly. In distyly, one form has a short style and long anthers (the *thrum* form), and the other has opposite (the pin form). Various other floral characters often vary between morphs, including pollen size and sculpturing, cell shape in stylar conducting tissue, and stigma morphology. Generally, there is tight correspondence between phenotype and heteromorphy, and it might be reasonably suggested that incompatibility phenotype and heteromorphy are pleiotropic expression of the same gene. However, recombinants between incompatibility phenotypes and heteromorphy, and between character states within the heteromorphy, are known, the best studied being the primula homostyles (Mather and Winton, 1941). It is likely, therefore, that in most cases, heteromorphy is controlled by two linkage groups - one comprising genes encoding morph-associated characters, and other S/ s- incompatibility- locus - with two groups themselves closely linked in a super gene. The rarity of recombinants within linkage groups, either between incompatibility phenotype and heteromorphy or between various characters involved in heteromorphy, strongly suggests that various features controlled by loci of super gene are co- adapted.

Mechanism of Self-incompatibility

The mechanism of self-incompatibility is quite complex and poorly understood. The various phenomena observed in self-incompatibility mating are grouped into three broad categories: (1) pollen-stigma interaction, (2) pollen tube-style interaction, and (3) pollen tube-ovule interaction. All these categories are discussed below:

Pollen-Stigma Interaction

These interactions occur just after the pollen-grains reach the stigma and generally prevent pollen germination. At the time they reach stigma, pollen-grains generally have two nuclei in gametophytic system, while they have three nuclei in sporophytic system. This was once considered to be basis for the two incompatibility systems, but available evidence indicates otherwise. However, the structure of stigmatic surface appears to be definitely involved in the differences between the two systems. In gametophytic system, the stigma surface is plumose having elongated receptive cells and is commonly known as 'wet stigma'. Incompatible pollen-grains generally germinate on reaching the stigma; the incompatibility reaction occurs at a later stage. There are clearcut serological differences among pollen-grains with different S genotypes; such differences have not been observed in sporophytic system.

In sporophytic system, stigma is papillate and dry. It is covered with a hydrated layer of proteins known as 'pellicle'. There is evidence that pellicle is involved in incompatibility reaction. There are striking differences in stigma antigens related to S- allele composition. Within few minutes of reaching the stigmatic surface, pollen releases exine exudates, which is either protein or glycoprotein in nature. These exudates induce immediate callose formation in papillae (which are in direct contact with pollen) of incompatible stigma. Often callose is also deposited on young protruding pollen tubes preventing any further germination of pollen. Thus, in sporophytic system, stigma is the site of incompatibility reaction; once the pollen tube crosses stigmatic barrier, there is no further inhibition of pollen tube growth. In homomorphic sporophytic system, incompatibility reaction of pollen is probably due to deposition of some compounds from anther tapetum onto pollen exine.

Pollen Tube-Style Interaction

In most the cases of gametophytic system, pollen-grains germinate and pollen tubes penetrate the stigmatic surface. But in incompatibility combinations, growth of pollen tubes is retarded within stigma, *e.g.* in Oenothera, or a little later in style, *e.g.* in *Petunia, Lycopersicon, Lilium etc*. In latter case, there is a cessation of protein and polysaccharide synthesis in pollen tubes, which leads to degeneration of tube wall and bursting of pollen tube.

Pollen Tube-Ovule Interaction

In some cases, *e.g.* Theobroma cacao, pollen tubes reach the ovule and effect fertilization. However, in incompatible combinations, embryos degenerate at an early stage of development. Generally, such species have hollow styles, *e.g. Lilium, Ribes, Narcissus, etc.*

Importance of Self-incompatability

Self-incompatibility effectively prevents self-pollination. As a result, it has a profound effect on breeding approaches and objectives. These are given below:

- In self-incompatible fruit trees, it is necessary to plant two cross-compatible varieties to ensure fruitfulness. Further, cross-pollination may be poor in adverse weather conditions reducing fruit setting. Therefore, it would be desirable to develop self-fertile forms in such cases.
- Some breeding schemes, *e.g.* development of hybrid varieties *etc.* initially require some degree of inbreeding. Although, sibmating leads to inbreeding, but for the same degree of inbreeding, it takes twice as much time as selfing. Further, for the maintenance of inbred lines, selfing would be necessary.
- Self-incompatibility may be used in hybrid seed production. For this purpose: (1) two self-incompatible, but cross-compatible; lines are interplanted; seed obtained from both the lines would be hybrid seed, (2) alternatively, a self-incompatible line may be interplanted with a self-compatible line. From this scheme, seed from only the self-incompatible line would be hybrid, (3) schemes for the production of double cross and triple cross hybrids have also been proposed and their feasibility has been demonstrated in brassicas.
- The gametophytic system has been used to a limited extent for hybrid seed production in clover and *Trifolium (*Leguminosae*)*. In Solanaceae, cultivated species are generally self-fertile, and self-incompatibility is confined to wild species. Primarily, the Japanese seed companies have exploited the sporophytic system for hybrid seed production in brassicas. In compositae, another economically important family showing sporophytic self-incompatibility, the cultivated varieties are generally self-fertile.

Problems in Exploiting Self-incompatibility

These are:

- In species with self-incompatibility continuous inbreeding results in the appearance of new incompatibility relationships, thus causing self-compatibility. Such mutated forms cannot be used as parents for producing hybrids.

- Continuous inbreeding in many brassica crops may lead to complete loss of inbred lines. It will, therefore, by more desirable if tissue culture techniques for propagation of parental lines are followed. Sib-mating for maintenance of these lines may also prove useful.
- Pseudo-incompatibility may lead to pure seed in an otherwise hybrid seed.
- Reduction of strength of incompatibility by environmental factor, *e.g.* elevated temperatures may weaken incompatibility or may even break it down. Similar may be the situation under high-humidity conditions. Thus, hybrid seed production under unfavourable conditions may result in high proportion of self or sib-seed, *i.e.* up to 30% or even higher. Such seed is of no commercial value. On the other hand, high temperature and high humidity are ideal conditions for propagation of inbred lines used as parents.
    - Hybrid seed would be expensive if self-incompatible lines are difficult to maintain, produce low-volume and weak seed (strength of incompatibility).
    - *Restriction of pollination within parental lines*. It is found that bees do not visit the plants randomly. Each bee prefers to stay within one parental line. This preference is stronger when two parental lines differ morphologically. This results in more of sib-pollination and less of outcrossing. This may be overcome by increasing the morphological similarities in parental lines and using some other pollen vectors like flies.
    - *Isolation of s-alleles*. This can be only known from the breeding behaviour of strains. Self-incompatibility status has been found to be high among Indian cauliflowers, while it is low among snowball types. It is therefore, essential that for harvesting of F_1 hybrid seed without contamination of selfed seed, strongly self-incompatible strains have to be used.

Elimination of Self- incompatibility

In order to break self-incompatibility, some techniques and measures could be adopted. These are :

- Pollen irradicated at low doses of 0.1-2 krd is applied on stigma. Similarly, application of NAA or IAA on flower produces a few self- seeds.
- Heat treatment of pistil up to 55-60°C reduced the self-incompatibility in L. peruvianum (Hogenboom, 1972). In cole crops, at high temperature incompatibility is decreased (Johnson, 1971). High temperature has been used to break self-incompatibility in *Brassica oleracea, B.oleracea L. var. gemmifera* (Visser, 1977) and Raphanus sativus (Matsubara, 1980 and 1984).
- Allopolyploidy or doubling the chromosome number of diploids to tetraploids may break self-incompatibility. The B.*oleracea*, B.*campestris*, and B.*nigra* are self-sterile but their allopolyploids (B.*napus*, B.*juncea*, and B.*carinata*) are self-fertile.

Autopolyploidy has broken self-incompatibility in *L .peruvianum* (Nettancourt *et al;* 1974). Doubling the chromosome number of diploids to tetraploids in *B. campestris*, *B.rapa*, and Raphanus sativus, however, has no effect on incompatibility (Howard, 1942).

- Self-incompatibility alleles can be mutated to self-fertile alleles using physical and chemical mutagens.
- In crucifers, bud pollination (application of pollen on stigma of flower bud of the same plant) may produce the self-seeds (Pearson, 1931).
- Self-compatible alleles can be transferred in self-incompatible plants.
- In *B.oleracea*, application of electric potential differential of 100V between pollen and stigma at the time of pollination may break self-incompatibility (Roggen, 1972). Electric-aided pollination has been found effective to break S.I. in Brussels' sprouts and savoy, white cabbage (Roggen, 1973). Application of glutamic acid, folic acid and nicotinic acid (50 mg/litre) broke self-incompatibility and thus resulted in seed setting in radish (Matsubara, 1984). By repeated pollination in self-incompatible varieties, self-seeds are produced in chicory (Nieuwhof, 1961). The CO_2 treatment has been found effective to break self-incompatibility in brassica (Nikanishi, 1969 and 1975). The CO_2 concentration of 4-6% applied for 8,16,or 24 hr at 100% relative humidity proved to be the most effective treatment to reduce the self-incompatibility effect. Such treatment has been found a favourable effect on pollen adhesion, pollen germination, and increased the number of pollen tube in style (Palloix, 1985).

Barriers to Hybridization

Hybridization is one of the most effective methods of crop improvement programmes. Most of the hybridization work carried out so far has used genetic variability within the species, and thus crossability barriers are not the main constrain in breeding programmes. However, extensive breeding for yield and genetic uniformity over the years has led to reduce genetic variability and consequently loss of many useful genes in crop species. In recent years, breeding programmes have been increasingly dependent on interspecific and intergeneric crosses, often covering wild and weedy species. Wide hybridization refers to the crosses made between distantly related species or genera. Crossability barriers are very frequent in interspecific and intergeneric hybridization programmes.

Classification of Barriers

Crossability barriers prevent the fusion of male and female gametes originating from individuals of different species/genera and or the development of a fertilized ovule into viable seed. The barriers also include those that limit effective utilization of hybrids for gene introgression. Stebbins has presented the first comprehensive discussion on crossability barriers in 1958. The crossability barriers may be grouped into three major categories. The first is physical and the other two are physiological barriers. These barriers are discussed as below :

Temporal and Spatial Isolation of Parental Species

Non-synchronous flowering of parental species is quite common, particularly in wide hybridization programmes. Unless an effective method, that is, early or staggered sowing or suitable photoperiodic treatment, is available to achieve synchronous flowering of the parents, non-synchronous flowering is a major barrier. Similarly, hybridization between parental species growing in different geographical areas is sometimes necessary. When species that normally are geographically isolated from one another are grown together, their flowering response may be affected, resulting in non-synchronous flowering, particularly when parental species grow in different ecological conditions. In tree species, this approach is not convenient as the plants raised from seeds could take many years to flower.

Pre-fertilization Barriers

Fertilization is the result of successful completion of a series of sequential events following pollination. Prefertilization barriers prevent fertilization by arresting post-pollination events at one or many levels. These barriers can be conveniently described under three categories, depending of levels of barrier of the pistil. Cytological details of prefertilization barriers have been investigated only a limited number of crosses. Until 1960s, this was probably due to the lack of suitable techniques to study pollen germination and pollen tube growth in the. The methods of clearing and/or staining used by early investigators to study post-pollination events in pistil are also not very effective. The aniline blue fluorescence method developed to study pollen tube growth during 1950s, is the most effective technique. The method is based on the affinity of the fluorochrome, water-soluble aniline blue, for callose. Because, pollen tubes invariably contain callose deposition along the wall, they can be readily observed under the fluorescence microscope following staining with aniline blue.Pre-fertilization barriers in crop plants may be due to following reasons.

Unilateral Incompatibility

In many interspecific and intergeneric crosses, incompatibility operates in one direction, whereas the reciprocal cross is successful. Such incompatibility is generally referred to as unilateral incompatibility. Most of the authors restrict the term UI to the barriers operating before fertilization. UI has been reported in a large number of crosses. UI is more common when the cross includes a self-compatible (SC) and self-incompatible (SI) species. The crosses, in general, show incompatibility when SI species is used as the female parent (SC x SC). However, in many crosses, UI also operates in other combinations (SC x SI, SC x SC, and SI x SI). Many recent studies have reported UI in crosses between wild and cultivated species. Irrespective of presence or absence of SI in one of the species, pistil of cultivated species generally inhibits pollen of the wild species, whereas, the pistil of the latter permits good pollen germination and satisfactory tube growth of cultivated species.

Active versus Passive Inhibition

Self-incompatibility inhibition is the result of active recognition of the pollen. Self-pollen is positively recognized as a result of the interaction of *S-allele* products in pollen and pistil. Positive recognition results in the activation of metabolic processes in pollen and/or the pistil to bring about pollen inhibition. Many investigators have also implicated active recognition mediated through the *S-allele* in interspecific incompatibility. As mentioned in the previous section, the SI genes have been implicated in UI. Pandey (1968 and 1969) suggested that the *S-gene* imparts two types of specificity: (A) primary specificity, which controls interspecific incompatibility, and (B) secondary specificity, which controls self-incompatibility. According to de Nettencourt *et al.* (1974), the loci inhibiting pollen tube growth in the cross *Lycopersicon peruvianum* x *L. esculentum* are either closely linked to, or allelic to, the *S-locus*.

Inhibition on Stigma Surface

Pre-fertilization barriers on stigma surface result in the arrest of pollen germination or pollen-tube entry into the stigma. This is one of the frequent barrier, particularly in distantly related species. The causative factors for the failure of pollen germination are: (a) lack of effective adhesion, (b) lack of full hydration, and (c) absence of pollen germination factors of stigma. Pollen adhesion and hydration are prerequisites for germination. Pollen adhesion largely depends on nature and extent of surface components of pollen and stigma. Pollen adhesion is not a constraint in species having wet stigma. In species with dry stigma, pollen adhesion may be a problem, particularly when pollen-grains do not possess sufficient pollen coat materials. Failure of pollen germination may be the result of insufficient or uncontrolled hydration. Pollen hydration is the result of transfer of water from the stigma to pollen through an osmotic gradient. Insufficient hydration may result in crosses in which the osmotic potential of pollen does not match with that of stigma. Because pollen-grains of many species gradually hydrated for germination, the rapid hydration that occurs on a wet stigma covered with aqueous exudates may lead to the failure of pollen germination.

Inhibition in Stigma and Style

Failure of the pollen tube to reach the ovary is perhaps the most common interspecific pre-fertilization barrier. This is due to the arrest of pollen tubes in stigma, or just below the stigma, or further down the style. Such arrested pollen tubes often show abnormalities in the form of thicker tubes, excessive deposition of callose, swollen tips, and branching of tubes. It is now well-established that growing pollen tubes utilize stylar nutrients. One of the possible reasons for the arrest of pollen tube growth is an inability of pollen tubes to utilize stylar nutrients. This may be due to lack of suitable nutrients in transmitting tissue or lack of pollen tubes.

Inhibition of the Ovary

Unlike the inhibition of pollen tubes in the style, inhibition of pollen tubes in the ovary is not well investigated. However, there are many examples of the phenomenon, particularly in crosses involving in graminaceous species in which pollen tubes grow normally and reach the ovary. In the cross-maize x sorghum, no inherent barriers exist for the growth of pollen tubes through the style, but pollen tube entry into the ovule seems to be a problem. Because pollen tube entry into the ovule seems to be dependent on the presence of a chemotropic substance in the micropyle, failure of pollen tubes to enter the micropyle may be due to the absence of such a substance. Another abnormality reported in maize x sorghum cross is the hypertrophy and lysis of nucellar tissue after the entry of pollen tubes into the upper part of the ovary. Even after the entry of pollen tubes into the embryo sac, there may be disturbances in double fertilization. In wheat x maize, 80% of the ovules showed development of the embryo but not of endosperm; 8% of the ovules showed development of endosperm but not of embryo; only 12% showed development of both embryo and endosperm.

Post-fertilization Barriers

Post-fertilization barriers result in the failure of fertilized ovules to develop into mature seeds. Post-fertilization barriers are more prevalent than pre-fertilization barriers in interspecific crosses. These barriers may operate at different stages of embryo development or during germination and subsequent growth of F_1 hybrids. Many causative factors have been identified for the breakdown of embryo development at different stages. These include the presence of lethal genes, genic disharmony in the embryo, and failure or early breakdown of endosperm. Arrest of normal development of embryo and endosperm may also result from unfavourable interactions between embryo sac and surrounding ovular tissues. Often, embryo sac is associated with the proliferation of the nucellar or integumentary cells.

Methods for Overcoming Interspecific Crossing Barriers

The phenomenon underlying crossing barriers are incompatibility and incongruity. Incompatibility operates in interspecific crosses and is the result of the activity of *S-alleles*. Incongruity occurs in interspecific crosses as a result of lack of genetic information in one partner necessary to complete pre-and-post pollination processes in other. There are several methods used to overcome incongruity.

Techniques for Overcoming Stigmatal and Stylar Barriers

Genetic Variation in Interspecific Crossability

The statement that "two species are not crossable, is controversial unless a broad genetic variation of the parental species has been used and the cross-combinations have been carried out on a large scale under a wide range of environmental conditions." This is because of crossability, is determined by both genetic and environmental conditions. It is, therefore, necessary to test different accessions of both the parents for hybridization programmes. Unilateral incongruity is the phenomenon in which a cross is successful only in one direction, whereas the reciprocal cross fails. In lily, crossing barriers can be overcome by using cut-style pollination, but mostly in one direction. In interspecific crosses between sorghum and maize, Laurie and Bennett (1989) found the variation of pollen tube inhibition to be due to a single gene.

Use of Mixed and Mentor Pollen

The use of mixed pollen (a mixture of compatible and incompatible pollen) and mentor pollen (compatible pollen genetically inactivated by irradiation but still capable of germination) mixed with incompatible pollen is reported to overcome inhibition of stigma and in the style in many plant species. For lily, mentor pollen have been found effective in overcoming self-incompatibility but not in interspecific crosses.

Influence of Environmental Conditions

The presence of an optimal level of receptivity of stigma can vary from several hours (in mango) to more than one week (in lily). It determines the optimal time of pollination. A positive effect of high temperature in overcoming self-incompatibility and incongruity has been detected and been applied in breeding programmes of lily by heating the style or by pollinating at high temperatures. Heat-sensitive inhibitors of pollen tube growth are probably inactivated at high temperature. Comparable effects of floral aging on pollen tube growth have been reported by Ascher and Peloquin (1966).

Style and Ovary Manipulations

As it was first demonstrated more than 50 years ago in *Datura*, pollen-tube growth inhibition in style can be overcome using different pollination techniques in which style and ovary are manipulated. One of these manipulations involves removal of the stigma and a part or whole of the style and pollinating the cut end. This is referred to 'stump pollination,' or 'cut style,' or 'intrastylar,' or ' amputated-style' pollination. In lily, two types of incongruity are distinguished, based on the arrest of pollen tube growth at different levels of style. In the first type, the site of arrest of pollen tube growth is just below the stigma. This 'upper inhibition' occurs 12-24 hours after pollination and results in short pollen tubes. In other type, pollen tube growth ceases halfway in the style 3-4 days after pollination. This 'lower inhibition' results in medium - sized pollen tubes. In a comparison of several pollination methods, it has been shown that using the cut-style technique could circumvent pre-fertilization barriers in lily.

Following stump pollination, many pollen tubes of lily, *Lathyrus* and *Fritillaria*, for example, grow normally in the ovary. In this way pollen circumvents stylar and stigmata barriers that inhibit pollen tube growth. However, a complication associated with this method in lily is the low seed setting, probably caused by premature arrival of pollen-tube in the ovary. A majority of pollen tubes either grow past the inner integument or grow along, but not into, the micropyle after cut-style pollination. Activation of the ovary by stigmatic pollination preceding inter-stylar pollination did not result in an increase in the percentage of ovule penetration. The percentage of ovule penetration after cut-style pollination did increase when a longer part of the style is retained on the ovary. Despite the low seed setting, a large number of unique interspecific lily hybrids are obtained by using this method. Also in crosses between Fritillaria imoerialis and *F. raddeana*, Wietsma *et al.* (1994) had able to obtain interspecific hybrid using the cut-style technique.

The stylar grant technique has been applied successfully to improve the results of the cut-style technique. In this method, pollen-grains are deposited on a compatible stigma. After one day, the style of the pollen donor is cut 1-2 mm above the ovary and grafted on to the ovary of another plant. Style and stigma are joined *in vivo* using a piece of a straw filled with *L. longoflorum* stigmatic exudates or stuck together with only the exudates. *In vitro*, a piece of 'water agar' has been placed on the style, shows the effect of different pollination method on average seed setting. It can be observed that cut-style pollinations result in a low seed setting, whereas grafted-style method gives a better seed setting but is not successful in a high percentage of the pollinated ovaries.

Chemical Treatments

Application of growth regulators, such as auxin, cytokinins and gibberellins, to the pedicel or the ovary at the time of, or soon after, pollination may improve fruiting and seed-setting after interspecific pollination. Application of growth regulators to delay abscission of the style shows positive effects on the development of young fruits. In many crosses, application of growth substances promotes post-pollination development up to a stage when hybrid embryos can be excised and cultured. In interspecific crosses, treatment of stigma with organic solvents such as hexane and ethyl acetate before pollination has been reported to be effective in overcoming pre-fertilization barriers. Immunosuppressors such as amino-n-caproic acid, salicylic acid, and acriflavin have been used to produce wide hybrids in many crops. The female parent is treated with immunosuppressor before and/ or after pollination of many days, and the resulting embryo is cultured on a suitable medium. It has been suggested that crossability barriers between distantly related taxa are mediated through a specific inhibition reaction analogous to immunochemical mechanism of animals. Treatment with immunosuppressors inactivates these immunochemical reactions. However, there is so far no evidence to indicate the involvement of immunochemical reactions in interspecific crosses.

Techniques for Overcoming Post-fertilization Barriers

A range of *in vitro* methods have been developed to overcome post-fertilization barriers in a number of plant species. When abortion occurs at a very early stage and maternal tissue has no negative influence on development of seeds, ovary culture can be applied. Young fruits can be grown *in vitro* to a stage at which dissection of embryos is possible. In some crops, ovary is large, and slicing the ovary into small parts and then culturing them is a better option for rescuing hybrids *in vitro*. This technique is referred to as 'ovary-slice culture.' When the mismatch between embryo and endosperm development starts very early and ovary culture and / or ovary-slice culture fails, ovules can be dissected out of the ovaries and cultured *in vitro*. Since, Hannig (1904) employed embryo culture for the first time 100 years ago, this technique is applied in numerous crops.

Ovary Culture and Ovary-slice Culture

Culture has been applied in many species: *Brassica, Lilium, Nerine*, *Tulipa*, and *Phaseolus*. Ovary-slice culture has been applied by Kanoh *et al.* (1988), Straathof *et al.* (1987), and Van Tuyl *et al.* (1991) for the production of interspecific *Lilium* hybrids. Ovaries are harvested 7-40 days after intrastylar pollination and, after surface sterilizing, sliced into 2-mm-thick disks. Seed germination occurred 30-150 days after pollination (DAP). By this method, plantlets are obtained from very small embryos.

Ovule Culture

In those crops in which the fruit is aborted before embryo culture can be applied. Ovule culture is an easy and fast method. This technique is applied in *Alstroemeria, Cyclamen, Lycopersicon, Nicotiana* and *Vitis*. In *Alstroemeria* for example, fertilization occurs 24 hours after pollination. Ovaries are harvested 2 DAP; the ovules are dissected and placed on a Murashige and Skoog (MS) medium containing 9% sucrose. Six weeks later the ovules are transferred to an MS medium with 4%-5% sucrose. Germination of ovules starts 1-2 weeks later. Histological studies of the *in vitro* cultured ovules revealed that the ovules within the first two weeks of culture enlarged to twice their size due to proliferation of inner and outer integuments. No development of cellular endosperm *in vitro* is found, whereas, embryo cells divided almost normally. By 2 DAP, the first plantlets arose from the cultured ovules, whereas normal seed development in *Alstroemeria* takes 2.5-3 months. Depending on the genotypic combination of interspecific crossing, percentage of seedlings obtained from ovule culture varied from 0.5 to 22.5%, whereas in *in vivo* situation on the plant, no seed could be harvested.

Embryo Culture

Embryo culture can be applied successfully in crosses in which pollinated flowers stay on the plant for a considerable time before natural abscission occurs. This method has been applied in a large number of crops. Some of the recent examples are: *Allium, Alstroemeria, Howea, Lilium, Lycopersicon,* and *Solanum*. To overcome the problems involved in isolating the young embryos from ovules and providing suitable conditions for their growth, embryo culture has been modified in some systems. The ovule is cut in half and the cut halves, or only the halves containing the embryo, are cultured in a liquid medium. Out of these half ovules, germinating embryos emerged, which could be raised to plantlets on a solid medium.

A large number of interspecific and intergeneric hybrids have been produced in *Brassica* through sequential culture of ovary, ovules, and often embryos. In sequential culture, ovaries are initially cultured for 6-10 days. Enlarged ovules are excised from cultured ovaries and recultured on a fresh medium. Hybrids are realized either directly from cultured ovules or after excising and culturing the embryos. Przywara *et al.* (1989) followed sequential culturing of ovules and embryos to raised hybrids between *Trifolium repens* and *T. hybridum*.

Techniques for Overcoming F_1 Sterility

Barriers occurring after a successful embryo rescue is hybrid breakdown and F_1 sterility. Hybrid, breakdown results in the loss of hybrids before flowering and is a result of the unbalanced new gene combinations. F_1 sterility of interspecific hybrids is very common, and may be the consequence of reduced chromosome pairing during meiosis. Breeding at polyploid levels is widely used in interspecific hybridization programme of many ornamental crops, such as *Alstroemeria, Freesia, Chrysanthemum, Gladiolus* and *Lilium*.

Chromosome Doubling

Interspecific hybrids may display sterility owing to lack of chromosome pairing during meiosis. Thus sterility hampers further breeding. Somatic chromosome doubling may induce homologous pairing of chromosomes and restores fertility. Colchicine has been used successfully to produce fertile allotetraploids in many crops, such as *Anigozanthos, Arachis, Brassica, Lilium, Phaseolus* and *Tagetes*. As an alternative for colchicine, oryzalin-a herbicide with antimitotic activity is being recently used successfully. Cross ability between parents may be improved in plant breeding by equalizing their functional ploidy level. Allopolyploids may function as fertile bridges for gene introgression into the cultivar assortment.

Application of 2n-gametes

Application of meiotic polyploidization in interspecific breeding programmes may be of great importance for introgression of characters from diploids to tetraploids. In many species, polyploidization results in functional 2n-gamates in one or both parents. Such gametes may result from mechanisms of meiotic restitution. Normally, the frequency of 2n-gametes is low. Wide interspecific lily hybrids are usually completely male and female sterile. In rare cases, however, some fertile pollen is detected. In a group of more than 50 hybrids from the cross *L. longiflorum* x *L. candidum*, raised through embryo culture, only one hybrid showed pollen fertility of 25%. Meiosis in this hybrid has been found highly irregular and all pollen contained 2n- gametes. Comparable cases are found in lily by back crossing oriental hybrids with '*Shikayama* x *L. henryi* and *L. auratum* x *L. henry*, resulting in triploid progenies. Back-crossing these triploids with *L. auratum* x *L. henryi* resulted in *aneuploids* with chromosome numbers ranging between 36 and 48. In contrast to wide interspecific crosses, the cross between Asiatic hybrid' Enchantment and related *Lilium pumilum* produced fertile pollen. Meiotic studies of most of these hybrids showed the formation of not only haploid pollen but also relatively high percentages of 2n-pollen. In *Alstroemeria*, interspecific hybrids produce 2n-pollen up to 60-70% in some of the genotypes of crosses between Chhilean and Brazilian species. In this crop, sexual polyploidization appears to be a convenient system for breeders to produce triploid and tetraploid interspecific hybrids. For example, if a trispecific triploid generates 2n-gametes, then in one step of unilateral sexual polyploidization, a tetraploid with four different genomes can be produced.

Storage of Pollen

The pollen longivity of different species varies between minutes and years depending primarily on taxonomic status of the plant and on abiotic environmental conditions. For a number of agronomically important taxa, including the short-lived graminaceous pollen, special storage conditions are needed to preserve the viability and fertilizing ability of pollen for a long period. Recent, sophisticated methods such as Nuclear Magnetic Resonance (NMR) Spectrometry, Fourier Transform Infrared Spectroscopy (FTIR), and different ultra microtechniques for electron microscopy have helped to carry out precise studies on water regime in pollen, parallel to molecular changes occurring in membranes during pollen dehydration and rehydration. Cryopreservation seems to be the most efficient method for the long-term preservation (now up to 10 years) of partly dehydrated pollen-grains. Beyond the classical role of pollen banks the promising application of modern *in vitro* techniques of plant biotechnology (isolation and fusion of reproductive cells, and DNA transformation of artificially produced zygotes and embryos) has opened new challenging opportunities for germplasm cryopreservation in the new future. The pollen storage has been used to satisfy following practical needs:

- To hybridize plants that flower at different times and locations or show non-synchronous flowering
- To provide a constant supply of short-lived (recalcitrant) pollen.
- To facilitate supplementary pollination for improving yields.
- To eliminate the need to grow male lines continuously in breeding programmes.
- To obviate the variability incidental to daily collection of pollen samples.
- To study pollen allergens and the mechanism of self-incompatibility.
- To provide materials for international germplasm exchange.
- To ensure the availability of pollen throughout the year without using nurseries or artificial climate for plant growth.

Causes in Losses of Pollen Viability

There are two causes of loss of pollen viability:

Biochemical Changes in Pollen

The major cause of loss of viability during storage appears due to deficiency of metabolites which is due to continued metabolic activity of pollen, even though this activity is going on at a very reduced rate, Linskens and Pfahler (1973) reported considerable changes in carbohydrate and organic acid content of pollen of various species during long-term storage. Deficiency of respiratory substrates can strongly contribute to the loss of pollen viability, especially in the case of three-celled pollen, where respiration rate is much higher that in the two-celled type. Pollen-grains of grasses, however, contain a satisfactory amount of starch grains even after losing viability. Consequently, viability decrease might be associated with the inactivation of enzymes, such as the amylases and phosphatases involved in the degradation of reserves stored in pollen-grains.

Desiccation and Cellular Changes of Pollen

The water content of living pollen-grains vary considerably between different families, with most recorded values between 15 and 35% of fresh weight at the time of shedding. The water content of gramineae pollen, however, is generally high (35-60%) and the metabolism is not retarded greatly. To some extent, the original pollen moisture content depends upon the environmental conditions, particularly temperature, air humidity, and water supply to pollen donor plant. The initial pollen moisture content is generally expressed as a percentage of the fresh weight. Recently, nuclear magnetic resonance (NMR) spectrometry, which is more accurate and nondestructive than the conventional method, has been used to measure the water content of pollen. The regulation of pollen water content might be an important adaptive mechanism for survival after pollen dispersal. The pollen of cross-pollinating species can survive longer in nature than the pollen of self-pollinating species. Pollen-grains that remain viable after hydration are termed desiccation tolerant; those that loss viability parallel to dehydration are termed desiccation sensitive. In contrast to the research on the molecular mechanism of membrane deterioration during the dehydration and storage of seeds, little is published in this field for pollen.

The correlation between the condition of vegetative cell membrane and pollen viability was first discussed by Heslop-Harrison (1979) and by Shivanna and Heslop-Harrison (1981). Studies on the membrane state using the fluorescein diacetate test in the pollen of several taxa indicated that most of the pollen samples exposed to dry conditions lost their membrane integrity, and invariably they failed to germinate. Barnabas (1985) showed that pollen-grain viability in maize has been found closely and positively correlated with desiccation tolerance. In fact, maize-pollen could sustain up to a limit of 80% water loss without detrimental changes in normal pollen functions. If molecular re-organization taking place in the pollen membranes as a result of dehydration does not lead to an irreversible alteration in the structure, the membranes of desiccated grains can be restored by the uptake of water while on the stigma. The period and extent of desiccation after which pollen membranes can recover their structure vary from species to species. The way in which desiccated pollen is rehydrated prior to viability assessment is important for the restoration of membrane integrity. For most of the desiccated pollen systems, gradual hydration in humid air is favourable for restoration of membrane integrity. Hoekstra and Van der Val (1988) and Crowe *et al.* (1989) recommended that the gradual rehydration of dry pollen causes a shift in membrane lipids from the gel phase to the liquid crystalline phase and that imbibitional damage end leakage could be reduced or eliminated by adequate hydration.

Methods of Storage of Pollen

There are different methods of pollen storage. These are:

Short-Term Storage of Pollen

Effect of Temperature and Humidity

Studies on water regime in pollen indicates that the loss of pollen viability in short-term storage is related to irreversible structural changes in plasma membrane and cytoskeletal elements of pollen rather than to a deficiency of essential metabolites. It has also been observed that pollen of a large number of taxa can be successfully stored for a limited period of time through the manipulation of storage temperature and humidity. Pollen can conveniently be stored in glass or polyethylene vials or in other suitable small container. The unsealed containers are frequently put into desiccator containing suitable dehydrating agents (dried silica gel, saturated solutions of different salts, or various concentrations of sulfuric acid) to maintain the desired relative air humidity. To prevent sticking, powdered diluents, such as *Lycopodium* powder, *talc*, or corn or wheat flour, are mixed with the pollen. In general, low temperature and relative humidities are favourable for most taxa. Tricellular pollen, especially that of graminae, requires sophisticated environmental conditions to preserve viability and fertility even for a short period of storage.

Pollen Storage in Organic Solvents

It was first demonstrated by Iwanami (1972) and Iwanami and Nakamura (1972 a and b) observed that pollen-grains of various taxa could be successfully preserved in different organic solvents. The pollen of *Camellia japonica, Impatiens balsamina* and *Lilium spp.,* has been found able to survive storage in benzene, petroleum, diethyl ether, acetone, and chloroform. Pollen of *Citrus* spp. maintained viability in various organic solvents for three months. Liu *et al.* (1985) found that of the 11 species examined (*Camelia japonica, Armenica vulgaris , Prunus salicina, Pyrus ussuriensis, Prunus triloba, Malus pumila, Prunus persica, Zea mays, Juglans regia, Salix babylonica*, and *Gingko biloba*), the pollen of all the insect-pollinated species showed viability after storage at 4°C in a suitable organic solvent for 35-40 days. A correlation between polarity of organic solvent and its efficacy for pollen storage has been established. Pollen-grains stored in nonpolar organic solvents such as hexane, cyclohexane, and diethylether retained viability and showed very little leaching of phospholipids, sugar, and aminoacid into the solvents. Pollen-grains stored in polar solvents, on the contrary, lost viability and showed extensive leaching of these substances. The efficiency of individual organic solvents varies greatly for different plant species; however, solvent has proved to be better for storage of any pollen than have low temperature and humidity. Also, whereas, the technique of storing pollen in organic solvents seems to be fairly simple, this method can be used with advantage only for short-term storage. Serious attempts should be made to extend this pollen storage method to a large number of taxa.

Long-Term Storage of Pollen

During storage at temperature above 0°C, the metabolic activity in pollen, although slow, will nevertheless continue, resulting in a gradual decrease in, and finally total loss of pollen viability. Thus, cryogenic conditions seem to show promise for the really long-term preservation of pollen viability.

Storage at sub-zero temperatures

The longevity of bicellular pollen that is considered to be desiccation tolerant or that has low original water content has successfully been prolonged using storage temperatures between -10°C to -34°C. All the data listed in the review by Towill (1985) come from the direct exposure of pollen to the storage temperatures. Pollen survival varied with the species examined, but ranged from 1 to 3 years.

Freez or vacuum-drying (lyophilization)

The technique of freez-drying involves the rapid freezing of pollen to a subzero temperature (frequently to -60°C or-80°C) and then the gradual removal of water under vacuum sublimation. Vacuum drying does not include the initial freezing step; the pollen is directly exposed to a vacuum and simultaneous cooling, and the moisture is withdrawn by evaporative cooling. To develop useful procedures, the optimum pollen water content, cooling method, and duration and extent of drying and rehydration must be determined. Pollen storage at sub-zero temperatures, especially if combined with lyophilization, can be very effective for a number of taxa. Pollen samples may respond even better if they are air-dried before freez-drying. The storage atmosphere may influence the longivity of freez-dried or vacuum-dried pollen. Both vacuum and an inert gas, such as helium or nitrogen, have been used and compared so far.

Cryopreservation by Deep Freezing

Ultra low temperatures (from -70°C to -196°C) can be used for preserving pollen in an unaltered condition with great potential, especially where other storage methods are not successfully applicable, or where long-term preservation is desired. Pollen of most species does not require any specific rate of cooling or warming when the moisture content is satisfactory for freez storage. A reduction in pollen water to below a threshold level prior to low temperature exposure seems to be crucial for achieving survival. These observations suggest that partially dehydrated pollen possesses less freezable water and can survive deep-freezing. Further experiments are necessary to determine the critical moisture level (pollen water content below which the fertilizing ability of pollen is retained after low temperature exposure) for successful, long-term cryostore, retained age of pollen of a wide range of species. Recent investigations on the molecular structures of membranes using Fourier Transform Infrared Spectroscopy (FTIR) indicate that lipid transitions in membranes may cause major damage during freezing or warming after freez-thaw has been completed. Membrane phase transitions seem to be responsible for imbibitional damage in dry pollen. Damage can, therefore, be avoided by gradual hydration or by raising the temperature true to effect transition to the liquid crystalline phase. FTIR spectroscopy might be a useful tool in examining the biochemical properties of differently treated pollen, and it has a potential for evaluating optimum cryopreservation protocols. Such special pollen handling methods are needed for cryostorage of desiccation-sensitive type of pollen.

Estimation of Pollen Viability After Storage

A number of cytological and cytochemical methods have developed to determine the viability of preserved pollen. These methods are described below:

Quick Methods based on Enzyme Assay

Several tests have been standardized to demonstrate cytochemically the activity of certain enzymes in living pollen, such as dehydrogenases by triphenyl tetrazolium chloride (TTC) test (Stanley and Linskens, 1974), catalase, by the benzidine test (King, 1960), or esterases by the fluorescein diacetate (FDA) test (Heslop- Harrison and Heslop- Harrison, 1970). Of these, FDA test has shown positive correlation with in vitro germination as well as seed setting. Viability tests based on the determination of enzyme activity may often lead to false estimations, because the presence of a functional enzyme in the cell itself does not guarantee that the pollen, though recovered, is still viable and functionable. Enzyme assay are better used only for a rough estimation of pollen survival.

In vitro Pollen Germination

Germination in *in vitro* of pollen is the most frequently used method for checking pollen viability. It is relatively simple and quantitative, especially when applied to bicellular pollen. In numerous texa, the percentage of *in vitro* germination of stored pollen is correlated with its fertilizing ability. The BK medium is suitable for *in vitro* germination in more than 86 species. Gradual, controlled dehydration before *in vitro* germination is recommended to avoid imbibitional damage to the pollen. However, application of this test is limited because of difficulties in achieving *in vitro* germination in three- celled taxa.

Pollen Germination on Stigma

Intact or excised stigmas pollinated with stored pollen are able to provide a true picture of pollen function. A 'semi *in vitro* germination test' is devised for the efficient examination of viability in dehydrated and deep-frozen stored graminae pollen. The percentage of germinating pollen and the length of pollen tubes developed can be measured after staining with cotton blue or aniline blue. The semi *in vitro* germination of graminaceous pollen (stored for a long period) after three hours incubation at room temperature and stained with cotton blue.

Fruit and Seed-setting

Fruit and seed setting give the most exact estimation of pollen viability and fertilizing capacity. Seed setting is determined after the artificial pollination of well-protected stigmas in previously emasculated flowers. Because the environmental conditions existing at the time of pollination may have a strong influence on the success of fertilization, a complex comparative analysis of the data from *in vitro* and semi *in vitro* pollen germination and from seed setting is recommended. Especially in case of stress-sensitive pollen, only a full examination can provide a reliable idea of changes in pollen functions occurring during dehydration and storage. The ultimate aim of storage has been establishment of pollen banks through which the pollen of any species can be obtained at any time of the year and at any place in the world. Until now, assessment of quality of stored pollen is based exclusively on viability (*in vitro* germination/ seed setting). Extensive studies carried out on stored seeds have clearly shown that the assessment of vigour (speed of germination) is a better indication of seed quality than germinability. A study by Shivanna and associates have shown that in pollen, as in seeds, storage and other stresses affect vigour before affecting viability. Loss of vigour may affect the competitive ability of pollen and the quality of subsequent progeny. Thus, it is important to study the vigour of stored pollen in addition to its viability.

Pollination Constraints and Management of Pollination Insects for Crop Production

Insect pollination is a crucial component in the production of many commercially important crops. Enhancing the amount or quality of insect pollination can lead to increase in crop value by decreasing the time to crop maturity and by increasing crop uniformity, quantity, and quality. The diverse group of insect species that play a role in crop pollination range from those that are incidental flower visitors to those that are superbly adapted for pollinating flowers on which they forage exclusively for their food. Each species of insect has specific morphological, physiological, behavioral, and life history constraints that affect its abundance, effectiveness, and distribution on pollinated crops. These constraints have a direct bearing on the ways in which insects can be most effectively and economically manipulated for enhancing crop pollination. The species most suited to crop pollination forage under a wide range of environmental conditions, feed exclusively on flowers for pollen and nectar, visit flowers of the same species in succession, move frequently between flowers, and have morphological adaptation that facilitate pollen pick-up and its transfer to flower's stigma surface. Problems in crop pollination arise when pollination and plant are not perfectly matched. This happens frequently because the few species of pollinators that are easily managed must be utilized on a diverse range of crops. There are two basic strategies to improve pollination: increase insect visitation rate or increase the rate and quality of pollen transfer to flowers. Each of these strategies must be considered within the context of their relative impacts on yield and input costs. Following stratigies that can be utilized to enhance and maintain populations of insects pollinators on agricultural crops including vegetables and to enhance their rate and quality of pollen transfer.

Increasing Pollinator Visitation Rates

When insect pollinated crops are limited, increasing the pollinator's visitation rate increases seed or fruit setting up to the point when plant resources become limiting. Beyond this point, there will be no further positive impact on crop yield. Management to increase pollinator visitation is warranted when crops are pollen limited and costs are associated with increasing pollinator visitation rates fall below the associated increase in crop value. The degree of impact that the increases in pollinator visitation rates will have on yield depends on the crop, its degree of self-incompatibility, and the need for outcrossing. Pollinator visitation rates can be increased by either enhancing local populations of native pollinators or supplementing existing populations with commercially managed species.

Enhancing Populations of Native Pollinators

One approach to increase overall pollinator visitation on pollen-limited crops is to enhance by local populations of native species through habitat management. There is considerable regional, annual, and seasonal variation in population abundance of diverse pollinator guilds foraging on crops that reflect inherent differences in their morphology, sociality, foraging preference, number of generations per year, and adult phonologies. Habitat management strategies are likely to be more effective if directed toward the members of a pollinator guild that are the most efficient but that also have low annual variation in abundance on target crop. The guilds of pollinators associated with a particular crop require diverse habitats to 'fit' their behavioral and life history constraints. In addition to having nesting sites provided, a succession of forage sources must be an available before and after the target crop blooms to maintain populations of long-lived solitary and social species. Pollinators must also be provided with a supply of appropriate nesting materials, and sites for overwintering and mating. In some cases, it may be desirable to alter cropping practices to more closely suit the characteristics of pollinator. Timing of planting dates, so that flowering periods of red clover coincide with peak populations of bumblebees can be increased by capturing queens in artificial nests and moving them to a target crop. Such introductions can triple local populations on red clover. However, without further management, populations must be reestablished each year.

Importing Commercially Managed Pollinators

For commercial crop pollination, we currently can effectively and economically manage by a number of social species of bees, including honeybee, *Apis mellifera* L., and several species of bumblebees. The honeybees large foraging populations, transportability, year-round availability, and broad dietary preference at the colony level, coupled with floral constancy of individual foragers, make it the pollinator of choice for a wide range of crops. Honeybees are suitable for use in diverse cropping systems and are commonly used to pollinate oilseeds, legumes, fruit, vegetables and nuts. Bumblebees have commercially viable in applications where their large size, long tongue-length, ability to vibrate flowers requiring buzz-pollination, and ability to fly at relatively low temperatures or ability to fly in enclosed spaces, make them more efficient pollinators than honeybees.

To date, in part because of the technical difficulties and high cost associated with bumblebee management, their use has been restricted chiefly to use in high value greenhouse crops or in crops such as red clover, where their population can be enhanced with little difficulty. Several solitary species of bees have also been exploited successfully for crop pollination, including the alkali bee (*Nomia melanderi* Ckll.), leaf -cutter bee (*Megachile rotundata* F.) and orchard bee (*Osmia* spp.) These species all lived in closed proximity to one another reproduce in artificial nesting sites. These two aspects facilitate the manipulation and management of their population. Alkali and leaf cutter bees have both proven extremely valuable in commercial alfalfa seed pollination. Leaf-cutter bee management system for alfalfa can also be easily adapted to increase in seed production in other crops, such as clover, *Trifolium* spp., polish rape, *Brassica campestris* and *Brassica napus*. The commercial use of orchard bee has been more limited, but there is good potential to further develop these bees for use in pollinating a number of fruit and nut crops. Before bees are moved to a crop, they should be manage to maximize the number of foragers that are available when the target crop comes in to bloom.

Maintaining Pollinator Populations on Target Crops

Pollination of the target crop can be reduced if neighbouring crops, native plants, or weeds provide greater nectar or pollen rewards for pollinators. The negative impacts of floral competition on pollinator abundance on a target crop can be reduced by the following methods: altering the timing when bees moved to the crops; replacing or rotating colonies on the crop; spatially or temporally isolating the crop; enhancing crop attractiveness through crop breeding and agronomic practices; using attractant sprays; training bees to crop odours; and reducing the attractiveness of competing plants. These management techniques are described as below :

Timing of Pollinator Introduction

Floral visits are usually maximized when managed pollinators are moved to crops after the initiation of flowering. Importing bees too late in the bloom period can be detrimental because flowers produced early in the season are not pollinated, which can reduce yields. Importing bees too early can cause other problems. For example, sources of nectar and pollen are required to support populations of species with limited flight ranges, such as the leaf-cutter bee, and to prevent their emigration from the field. Their vegetation rate is maximized when incubation is timed carefully, so that the adult emergence is synchronized with floral density. In contrast, emigration from the field is not a problem with bumble-and honeybees, as these social species are capable of surviving on non-blooming crops for extended periods. They maintain a store of food and, with larger flight ranges, are capable of exploiting alternative sources of forage beyond the confines of the crop border. Nevertheless, timing of colony introduction has a major effect on the proportion of bees foraging on the targeted crop. Floral visitation rates are higher when bees are moved to crops after the initiation of bloom, because bees moved to crops before bloom quickly become trained to alternative forage sources. The exact timing of colony placement is not strictly controlled by crop phenology because environmental conditions can alter bee foraging patterns and delay the ' optimal' time to introduced colonies and maximize fruit set.

Colony Replacement and Rotation

In crops that bloom over long periods, sequential introduction of pollinators can be used to keep place with expanding bloom or to compensate for bee losses. When honeybee colonies are moved to a new location, workers tend to visit the closest crop available and slowly radiate outward from their colony to forage. Thus, even if competing crops are nearby, many workers will be retained on the target crop for a few days until they discover more attractive sources of forage. Foragers can be maintained on crops by replacing colonies or sequentially moving in additional colonies throughout the blooming period, even though the bees from colonies introduced early in the blooming period, having switched to competing crops. This technique has variable success, however, when honeybee colonies are sequentially increased on alfalfa fields. The number of nectar foragers increases but foragers trip fewer the flowers. In this case, the practice is not economically feasible, especially once the high of colony rental and transport are considered.

Spatial and Temporal Isolation

Spatial isolation can be used to maintain foragers on the target crop by increasing the distance to competing crops to the point where it is beyond the economical flight range of the pollinator. Isolating crops pollinated by species such as the leaf-cutter bee, which typically forages within several hundred meters of its nest, is much easier than isolating crops pollinated by honey- and bumblebees, which forage several kilometers from their colony. Increasing the isolation distance from other crops can increase bee foraging but can also alter the type of forage collected and the quality of pollinator visits. Honeybees are ineffective pollinators of alfalfa in many geographical regions due to their tendency to avoid pollen collection and to ' thieve' nectar from flowers. However, bees do trip flowers regularly in dry areas where alfalfa is grown under irrigation and little floral competition exists.

Temporal isolation may be effective in areas where, due to the acreage grown, spatial isolation is difficult to achieve. Plants that are highly attractive to pollinators can be grown in close proximity if overlap in bloom is avoided. In regions with long growing seasons or in crops with short flowering periods, it may be feasible to avoid overlaping in bloom by altering planting dates to avoid competition with neighbouring crops. In orchards where dandelion and white clover often attracts pollinators away from the target crop, spatial isolation to reduce competition can be achieved through mowing or herbicide applications. Competing crops should not be remove from orchards, however, if they retain nectar-collecting bees that subsequently switch to foraging for pollen on the target crop.

Plant Breeding and Agronomic Practices

Polytrophic bees tend to forage on the most attractive source available. Long-term selective breeding programmes can increase the attractiveness of crops to pollinators. In cases, where pollinators are poorly adapted to crops, it may be feasible for plant breeders to select for lines that contain one or more attractive characteristics. Most insect pollinators forage on flowers to collect nectar, pollen or both. Thus, selection of plants to enhance the quality, quantity, or accessibility of nectar or pollen rewards should improve the ability of plant to attract pollinators. If rewards produced by flowers are too high, however, the number of flowers or plants visited by each pollinator may decrease, reducing the amount of cross-pollination. Pollinators preferentially visit plants based on no rewarding characters such as floral colour, the numbers of flower per inflorescence, and odour. Thus, breeding to enhance visual or olfactory cues may also be effective. Attractiveness of alfalfa to honeybees depends, in part, upon the relative amounts of attractive and repellent components in volatile floral odours. Selection of plants to optimize quantities of these compounds may also lead to increased pollinator visitation.

Nectar production is influenced by external factors affecting photosynthesis and plant growth. Thus, implementing cultural practices to alter one or more aspects of soil moisture or fertility can increase crop attractiveness. Soil moisture can be regulated by irrigation. Watering increases nectar production in crops such as cotton, *(Gossypium* spp.), but attempts to use irrigation to regulate nectar secretion have mixed success. Experiments on snapdragons (*Antirrhinum majus)* indicate that deviations from optimal soil moisture levels in either direction can lead to reductions in nectar yield. Irrigation can also negatively affect bee visitation. Applications of water during evening periods significantly reduced nectar production in hybrid canola, *Brassica napus* L., field, resulting in a 1.7-fold reduction in the abundance of foraging bees. Bee activity has been found to reduce for up to a day after irrigation occurred.

Nectar production is enhanced through fertilization where soils are deficient in one or more nutrients or trace elements. In addition to enhancing floral nectar secretion, fertilizer applications can increase the attractiveness to pollinators by increasing the number of flowers/ plant. Fertilizing with nitrogen, phosphorous, potassium, and various trace elements such as boron, alone or in different combination, increased nectar secretion and flower production. In some cases, higher nectar production resulting from fertilizer treatment increase bee visitation and crop yields. The effects of nutrients on plant growth and nectar secretion are independent and can not be studied in isolation. In general, combinations of soil conditions that place no restrictions on growth but promote a balance between vegetative and reproductive growth are most likely to produce optimal conditions for nectar production.

The approach of influencing nectar production through breeding or manipulation of soil fertility may have limited application, since climatic factors often have an overriding influence on nectar production.

Attractant Sprays

Various substances have been tested as attractant sprays, some of which show promise. Although, some attractants increase honeybee abundance on a crop, they do not always increase crop yield. Sugar syrup sprays and food supplement sprays often increase the abundance of foragers but rarely increase crop yields, because the foragers tend to collect the sugar syrup rather than visit the flowers for nectar or pollen. Pheromones- chemicals used for communication between members of the same species can increase activity of bees foraging on flowers when applied to crops as dilute sprays. Most of the research on directing pollinators to crops has focused on the honeybee because of its wide spread use as a pollinator on many crops and our knowledge of its pheromonal systems. Because these chemical messages are species specific, pheromone sprays directed at one species do not enhance activity of alternative pollinators.

Crop Odours

Social insect pollinators such as honeybees and bumblebees, can be directed to forage on target crops by feeding colonies with sugar or protein supplements laced with floral odours. Increased pollinator visitation has been achieved by feeding colonies of honeybees from containers of sugar syrup covered with flowers of the target crop or by feeding sugar syrup containing distillates of floral odours. In some crops, the increased recruitment of foragers increases. Workers also become somewhat conditioned to their food sources and, when moved to a new site, tend to visit the same type of floral sources, if available. Thus, rotating colonies between two locations several kilometers apart may help to maintain bees on crops, where bees are moved from areas with abundant bloom to smaller plots near competing forage sources. Further, research is required before techniques to condition foragers to specific crops can be recommended to commercial growers.

Reducing the Attractiveness of Competing Crops

Flooding an area with large numbers of pollinators can reduce the relative attractiveness of competing crops. Pollinators are unable to assess a plant's capacity to produce nectar but they do assess nectar availability. When pollinators forage on the most attractive population of flowers in an area, they changed the reward spectrum for other pollinators. Thus, as the pollinator's population increases, the available foraging niches 'fill up' and, over time, the available nectar rewards of all the plants in an area tend to become equivalent. In cropping system, increasing pollinator's abundance on more attracting competing crops decreases the value of their nectar and pollen rewards relative to the target crop. Introducing a large number of honeybee colonies near a target crop has been used successfully to ameliorate the impact of floral competition. It is suggested that repellent compounds could be sprayed on competing crops to reduce their attractiveness to bees. Many compounds have been identified that could be used to reduced pollinator visitation when sprayed on competing crops. However, use of repellent sprays to prevent floral competition is likely to be limited only to specialized situations involving highly valuable crops, such as breeders seed plots, because application of repellent sprays to other flowering crops might reduce their yields, permission to spray would have to be obtained from neighbouring land owners, and costs in most situation would be prohibitive.

Pollen Efficiency of Insects

Insect is an important vector of pollen for many vegetable crops, especially in cross-pollinated vegetable crops. Insects visiting flowers to collect nectar, pollen, and other reward often serve as important incidental agents of pollination. The efficiency of insect species as pollinators varies greatly like: (1) the interaction between foraging insects, (2) comparison of pollinating abilities of insect sexes and castes, (3) comparison of nectar and pollen forages, (4) characteristics of pollen-grains and insect body hairs and their influence on pollen movement, and (5) pollen removal, and various other pollen losses, during insect visits. The descriptions about these factors are described below:

Interaction Between Foraging Insects

Interaction exists between foraging insects and have direct impact on pollination. For insistences, encounters causing disturbance, both between and within insect species, can change the duration of flower visits and alter foraging rate. Interactions range from aggressive to subtle. Insects foraging at the same floral resource may become physically competitive. Brain (1957) found that competitive between *Bombus* spp. Sometimes involved deliberate displacement of a floral occupant by a new comer. It has been observed that *Rosa carolina*, contact with other bees averaged once every 14-15 flowers, and smaller bees (B. *ternaries* and B. *vagans*). Kikuchi's (1965) observed that there is a strong positive correlation between dominance rank and insect-head width. Dominant species of hover flies forced subordinates to plant foliage. Size of foraging force, however, can overcome differences in body size and governed successfully displacement of larger *Apis mellifera* by aggressive *Trigona* spp. A study of honeybees and stingless bees, involving an artificial feeder, interspecific aggression is directed mainly to the next largest species. The smallest and most aggressive, bees (*Trigona iridipensis*) have over twice number of pollen species stored in their colonies than the largest bee (*Apis dorsata*), yet their respective flight ranges have 100 m and >5 Km. Aggression is more common on artificial than on natural flowers and on smaller than larger inflorescence and can affect related species very differently. It is also seen that all competitive interactions between species take the form of over hostility. Individuals may avoid confrontation by changing their foraging patterns. The final interaction considered is also the consequence of activity of previous floral visitors. Certain 'primary robbers' chew through the base of sympetalous corollae to steal nectar without contacting the anthers or stigmata at the flower front. 'Secondary robbers', are opportunists (usually species lacking the mandibular dentition necessary to be primary robbers themselves), take advantage of these perforations at the flower base by collecting nectar from there. Most reports are from legumes, where honeybees utilize holes chewed by *Bombus* spp; such diversionary foraging by the reduced their pollinating value (Kendall and Smith, 1976).

Comparative Pollinating Abilities of Insect Sexes and Castes

The pollination capabilities of male versus female individuals of the same insect species have been studied. In bees, it is significant that males do not provision ? brood with pollen; instead, when foraging they seek only nectar and do not actively gather pollen, nor groom it from their bodies into pellets for return to the nest. Indeed, male bees often live outside the nesting area. For balance, it has been observed that marginally favours females as superior pollinators. The forging behaviour of males and females can vary. Male bees worked faster, but made less extensive contact with the reproductive parts of *Clatonia* flowers (Motten *et al.,* 1981), visited pistillate flowers of *Cucurbita pepo* only half as often (Tepedino, 1981) and visited greater numbers of flowers/plant (Jennersten *et al;* 1991) than their female counter parts. Both male and female bees of the same species robbed nectar without causing pollination. Female bees exhibited higher fidelity than male to a particular plant species. Non- oviposting, female cabbage - white butterflies are as constant to plant species as foraging males, although egg-laying females switched plants regularly. Relatively less information is available on the importance of castes (queen and workers) as pollinators. The evidence todate comes from bumblebees, and on a per bee basis, favours the queens. Although, abilities of queen at pollen removal are similar, flower- handling times are also same or faster. Differences in tounge length, a significant factor of pollinator foraging behaviour, partially explain this variation in foraging speed.

Influence of Insect Tongue Length on Flower Visitation Pattern

Numerous studies have focused on the question of the length of the proboscis of flower's visitors and its influence on pollinator foraging behavior. It has been observed that tongue length is always longer in queen bumblebees than in workers. And in male, it is usually intermediate. Thus, queen of *'shorttongued'* bumblebees forged on flowers with deeper corolla tubes than workers. Workers may assort to flower species possessing corolla depths in accordance to their tongue lengths; longer tongues, allowing exploitation of both short and long - tubed flowers. In the same manner, larger workers within a single *Bombus* spp. may attend different flowering species than smaller individuals. Relatively short tongue limits accessibility to nectar at the base of a tubular corolla. Bumblebees do not always yield to inaccessible flowers; instead some chew through the perianth base, thereby robbing nectar without necessarily pollinating the flowers or dispersing its pollen. Disparity in tongue length has been found apparently in sufficient to always give queens the advantages, because nectar robbing by both queens and workers of the same species can occur. In B. vagans, workers with longer tongues robbed nectar less often. Not only do insect tongues play a role indicating foraging behaviour, they are also important vehicles of pollen transfer on bees, butterflies and flies. On flies, pollen-grains have been found on tongues almost exclusively.

Influence on Pollen Movement of Pollen Grain Traits and Insect Body Hairs

Size of pollen-grains is variable and influences pollen removal and transfer. Even with in a single species, pollen size depends on ploidy, flower position within the plant, heteroanthery, and environmental conditions.Pollen -grains of species pollinated by wind are intermediate in size (17-58 µm), whereas those adhering to insects are much more variable, ranging from 4.5 µm (*Myosotis*) to > 200 µm (*Cucurbita*). Pollen-grain size has been found negatively correlated with number of grains adhering to body hairs of honey and bumble bees (Lukoschus, 1957 and Thomson, 1986) and butterflies (Levin and Berube, 1972). Besides size, adherence of pollen to insects also has been attributed to pollen morphology. Murphy (1984) found that grains of *Lasthenia* adhered to checkerspot butterflies better than to those of three other species visited more frequently, and he suggested that small grains with rough surfaces may be key parameters for pollen transfer by Lepidoptera. Pollen-grains that are spherical but strongly echinate (for example, Malvaceae- cotton and okra) adhered poorly and are rarely carried as pollen loads by honeybees. A form of pollen 'packaging,' attributed to threads of viscin, may account for the clumping of pollen on visitors to *Epilobium* and *Oenothera*.

Adherence is not only a function of pollen characteristics, but also of the insect. The number, length, and complexity of body hairs differ markedly between insect species and are inferred to have profound influence on an insect's carrying capacity. Hair density is also important. Kearns (1992) suggested that small quantities of *Linum* pollen on muscoid flies occurred because the grains (56 µm) did not lodge between their body bristles. In contrast, body loads of Cardamine and Chrysopsis pollen (23-30 µm) have been found much larger. Electrostatic charges on bees and flower parts may also cause pollen to adhere to bees or to jump from bees to stigma. In special case of social insects, microgametophytes might also be transferred among nest mates inside the colony.

Pollination by Bees Foraging for Nectar or Pollen

Unlike many other flower-visiting insects, adult bees are involved with the nutrition of their immatures. Although, there are several types of ' floral rewards' sought by insects, it is chiefly the nectar and pollen that are collected as primary food sources by bees for return to their nests. The pertinent floral sources-nectaries and anthers often are situated near opposite extremities of the floral structure, and different floral postures are taken and different behavior (associated with specific body movements) are invoked during a bee's pro-curation of each food type. Also, easily recognizable are foragers with or without loads of pollen in the corbiculae of their hind legs (for example, honey and bumble bees) or in the scopa on the underside of their abdomen (for example, leaf-cutter bees). Consequently, many researchers have been able to compare nectar collectors with pollen gatherers which often take a small amount of sticky nectar as a medium to consolidate their loads of pollen) for various aspects of pollinator effectiveness.

For bees to collect loads of nectar and pollen, the number of flowers visited, and the time required, are highly variable. Honeybees required much more time to collect nectar loads and spent seven times longer on inflorescence of Banksia when foraging for nectar. Bumblebees stayed almost three times longer on flowers of Pedicularies for nectar. However, the collection of pollen from Kalmia took longer than that of nectar of honeybees, bumblebees and andrena bees. Furthermore, honeybees attending Helianthus visited more florets/capitula as nectar than as pollen gatherers; the reverse has been found true for red clover. This lack of agreement between studies, on the effort required to collect either pollen or nectar, is not surprising and indicates inherent interspecific differences in nectar and pollen production per flower, and variability caused by prior depletion of these floral rewards.

Pollen Removal and Pollen Losses

Several studies have quantitatively investigated various aspects of pollen removal by insects. The bulk of these studies concern Hymenoptera, likely a reflection of the vital pollen-collecting activities of bees. Bee hair acquire pollen readily and bees usually bear higher quantities of body pollen than do other insect taxa, such as beetles, butterflies and flies. Investigations have dealt with the effects on pollen removal of floral-visit duration, number of visits a flower receives, and various pollen losses from flowers. As bee visits to previously unvisited flowers lengthen, consistently more pollen is removed. The effect of duration on pollen removal is not linear across all insects. For example, workers of *Bombus flavifrons* spent shorter times during initial visits to virgin flowers of *Aconium delphonifolium* than did workers of *B.mixtus* yet removed significantly more of a flower's total pollen.

The quantity of pollen removed varies during the first visit to virgin flowers. Despite these plant species possessing a wide range of floral morphology, bees usually removed large quantities (30-80%) of microgametophytes. Interspecific variation in pollen removal is related to contact of the pollinators with the anthers. Visits to the legumes Lupinus and Vicia, whose flowers dispense pollen in small quantities per offering are exceptional. The trend that pollen removal increases with increased duration of floral visits also pertains to the number of visits received by the flower or inflorescence. Pollen removed from flowers during insect visits has a range of destinies. Grains removed, but not carried by the insect, represent a passive loss. The simple recoiling of the proboscis post feeding resulted in 15% (*Phlox pilosa*) and 52% (*P. glaberrima*) loss of pollen from the tongues of Colias butterflies. On an average, one-seventh of *Erythronium* pollen fell to the ground during bee's visits, a loss correlated positively with the visit duration. Furthermore, several reports of pollen rejection exist whereby some individuals deliberately groomed off much of the pollen that attached to their body hairs during nectar feeding, for example, honeybees and bumblebees. Pollen discarded is dedicated to that behaviour. When pollen does not leave the flower on the body of the departing insect, it can be lost to another plant species, that is, 'misplaced pollen'. A lack of floral constancy by the vector can result in heterospecific pollination.

Pollen carried on the body surface of certain insects can also lose its viability. Viability of *Brassica* pollen is reduced to 20% when placed on Atta texana and other ant species; this loss is attributed to antibiotics on the ant's bodies. In contrast, viability averaged over 98% for the thorax of bees and wasps. Perker and Hatley (1978) found that viability of onion pollen brushed experimentally from the body hairs of *Apis melifera* was only 11%, compared to 24-40% for flies, wasps, and solitary bees. This loss is related to inhibitory enzymes in the liquid utilized by honeybees to pack pollen in their corbiculae, that is, 'active pollen loss-germination inhibition'. On the contrary, apple pollen collected directly from flowers or brushed from the bodies of a variety of insect visitors showed a consistent reduction in pollen viability for only one hover fly species. Pollen residing on the hair coat of honeybees confined overnight to their hive (where the brood-nest temperature is maintained around 34ºC) lost its viability by next morning. However, in one test, apple pollen has been found still viable after a 21 hours exposure inside the hive.

The loss of pollen from circulation as a result of a bee's grooming actions, performed in order to gather microgametophytes into loads for transport back to its home, has been termed 'active loss'.Various factors are known to influence grooming behaviour. The frequency or vigour of grooming movements by bumble bees rose with increasing amounts of pollen removed per flower and also varied between different bee species visiting bees generally occur in flight, between flowers, such that pollen grooming is reduced when bees simply have to walk (instead of fly) to nearby florets on an inflorescence. Grooming actions disturb the original location of pollen-grains on the insect's body; such rearrangement can have consequences for pollination. There is a relationship between weight of pollen loads carried and bee size. Clements and Long (1923) provided data for several bee species that gathered pollen from raspberry; collected quantities of pollen varied widely. Only the two bumble bee species carried loads per foraging trip that approached or exceeded half their body weight in pollen, with average loads well below 10 mg. Despite these various types of pollen losses, much of a flower's pollen may never leave the anthers. Two studies involving relatively large, echinate pollen-grains, those of cantaloupe and *Malvaviscus arboreus*, showed that even after 40-50 flower visits, up to 25% of the pollen remained in microsporangia. And although most radish flowers received numerous visits, at least 20% of presented pollen is not withdrawn. In contrast, the combined visits of hover flies and halictid bees removed 99.7% of pollen produced by *Jepsonia*.

5

POPULATION IMPROVEMENT OF PARENTS/HYBRID VARIETIES

Concept

Basic concept of this type of improvement is to accumulate, infuse and improve the frequency of favourable genes in established population. This includes the removal or substitution of an undesirable gene with a counterpart favourable one with the superior performance in a given environment, incorporation of a balanced gene combination for better performance - *per se* or may be even bringing in a better plant type congenial for better physiological activity and biochemical versatility, resulting in better performance of a population. As per theory of Hayes (1963), most of the commercial varieties/hybrids can be improved in one way or the other for one or more of the agronomic attributes. This is primarily because some of the lines in each hybrid generally excel in comparison with other in their ability to transmit desirable characters to their hybrid progeny and thus provide scope for the improvement of the weak inbred lines of the hybrid to further better its performance. The choice of the particular population improvement technique mainly depends upon the genetic information. Since by and large, population improvement techniques envisage controlled crossing followed by selection and recrossing by further selection. This information is vital. Therefore, before deciding upon the choice of a particular methodology, such information as the type of gene action involved in the expression of attribute under improvement, nature of genetic associations, extent of heritability, various possibilities of genetic manipulations should be available. Some of the procedures, which could be successfully utilized in population improvement of parents used in released hybrid populations, are backcross method of breeding, convergent improvement, gamete selection, technique for the removal of complementary genes and step ladder method of breeding.

Backcross method of breeding can be conveniently utilized for the incorporation of a specific gene or genes systems governing the characters like resistance to diseases and insects, physical and chemical quality characteristics, male sterility and fertility restorer genes *etc*. in componential inbred lines of an established hybrid with a view to improve its performance and make it better fitting for the production purposes. Suppose if a particular single cross, three-way cross or double-cross hybrid suddenly becomes susceptible to a particular pathogen or pest, then to improve the hybrid for further commercial production, resistance to disease may be incorporated through a standard backcross-breeding programme. Various steps involved in the incorporation of resistance to a disease say for example leaf blight in a double cross (A x B) X (C x D), would be to grow the double cross hybrid and its inbreds under artificial epiphytic of leaf blight and then to identify the componential inbred line of double cross which contributes to susceptibility to disease (say line A). Then the next step would be to cross this source of resistance to leaf blight is dominant, then select the resistant plants in backcrossed progenies and again backcross with A. If the gene governing resistance is recessive in nature, then self the backcross progeny to identify resistant plant and again backcross it with the recurrent line A. Repeat the backcrossing for 4-6 generations, depending upon the non-adaptability attribute of donor source F, and finally recover most of the genotype of line A, designate it as A. Reconstitute the double cross hybrid as (A x B) x (C x D) and compare its performance with the original double cross hybrid (A x B) X (C x D). Backcrossing technique can also be used for the incorporation of genes for certain quality characteristics, various plant type attributes *etc.* in population improvement programmes of hybrid varieties.

Convergent Improvement

Richey (1927) first suggested this procedure. Convergent improvement is a system of double backcrossing for improving each of the two inbred lines of a cross without much interfering, changing, or modifying the combining ability of the cross. Basically in this system, backcrossing is utilized. It basically improves the inbred lines and betters the performance of the cross by accumulating favourable gene (s) obtained from counterpart lines by careful selection following each backcross and by discarding less favourable genes. Various steps involved in convergent improvement as given by Hayes *et al.* (1955) are as follows:

Step 1: Select a high-yielding desirable single cross.

Step 2: Backcross the F_1 to both of the parents and then further backcross for 4-6 generations simultaneously in two series to respective inbred parents. Select only vigorous plants, which have desirable characteristics, needed to be improved during the period of backcrossing and use only these in making the backcrossing.

Step 3: Self the backcrossed lines after 4-6 generations of backcrossing and select within the selfed lines.

Step 4: In order to obtain further improvement, repeat the above steps (2-3) with the recovered inbred lines.

Gamete selection

Stadler (1944) suggested this method for the first time. Gamete selection is a method of selecting and combining desirable gametes from a genetically variable heterozygous population into the background of an inbred line of known performance and combining ability. In the improvement of hybrid populations, this is done primarily to improve the desirability of a particular weak inbred line, which is otherwise good. The basic procedure of this type of selection involves the crossing of a good inbred line of known performance with a random sample of pollen from an open-pollinated variety or any other known heterozygous source, self-pollinating the individual F_1 plant obtained from the variety x inbred cross, simultaneously topcrossing them to an appropriate tester stock. This is followed by conducting a yield trial of the inbred x variety cross progenies, selecting the better performing progenies of the crosses and then finally, continuing selection with the selfed seed of the original F_1 seed.

The following steps are involved for this method.

Step 1: Pollinate A with mixed pollen obtained from the desirable plants of variety (V) selected as a source of superior gametes.

Step 2: Self pollinate selected individual plants from these crosses and use the pollen of each individually to pollinate plants of (C x D) to obtain crosses of A (C x D).

Step 3: Test the crosses produced by step 2 in elaborate yield trials and study all the characters of importance including yield, in comparison with crosses of A (C x D).

Thereafter, either of the two methods of further selection could be followed. Firstly, selection in selfed lines obtained from the original selfed plant from the step 2, which has performed very well in topcrosses with (C x D) single crosses. The other alternative selection may be that about 10 or so best performing original selfed plants from step 2 may be mixed together to constitute and produce a new population. From this population, a second cycle of gamete selection may be started after one generation of open pollination to ensure recombination of genetic factors. Selected plants from this population may be handled as in step 2 and 3. Finally, after two or three cycles of selection for superior plants through recurrent selection from this population, superior lines may be isolated to take the place of A in the double cross (A x B) x (C x D). Finally, this may be compared with original double cross hybrid (A x B) x (C x D). Gamete selection is essentially a method of early testing. It has been indicated that it may be useful for selection and isolation of inbred material from the open-pollinated varieties.

Removal of Undesirable Complementary Genes

Complementary genes which individually are grown with expression of the same character but when they come together in a hybrid, they interact to produce a distinct and different expression. A typical example of this has been the genes governing white flower colour in peas. When the two white colour flowered varieties with $A_1A_1a_1a_1$ and a_1a_1 A_2a_2 genotypes are crossed together they produce a hybrid ($A_1a_1A_2a_2$) with pink colour. The breeding scheme, which could be effectively followed for the removal of such complementary factors from a double top cross hybrid, has been given as follows:

Step 1: Self-pollinate a number of plants in the male parent of double topcross, *i.e.* in open-pollinated variety (V). Number them serially and simultaneously cross them individually with single cross female parent (A x B).

Step 2: Grow these crosses individually the following crop season. Self them. Examine the fruit setting on individual plants.

Step 3: Go back to the self-seed obtain in step 1 of the selected plants and reconstitute the variety from the selected plants. Compare the performance of original (V) and the reconstituted (V') variety.

Step 4: Reconstitute the double topcross as (A x B) x V if their yields are equal and compare its performance with the original double top cross hybrid (A x B) x V.

Step Ladder Method of Breeding

Earlier, Dhawan (1965) has suggested a step ladder method of breeding for a phased elimination of inferior inbred lines and by substitution of an altogether new inbred lines for further genetic improvement of an established double cross hybrids in maize, say for example (A x B) X (C x D). This method involves by making three way crosses, *i.e.* (A x B) x E, F, G, H *etc.* by using the better parental single cross (A x B) or say (SC1), as a top cross tester to screen the new inbred lines (E, F, G, H *etc.*) emerging in a breeding programme. This is followed by the identification of superior specific combining inbred lines from three-way cross test for evaluating them in a replicated yield trial. The further step, which follows, is the prediction of performance of a new double cross. In this, the tester SC1 forms one parent and a new single cross SC2 A (E x G) is developed from two of the superior lines E and G. The superiority of lines E and G has already been established by three way-cross test. It forms as one of the parent in new double cross. In this way, inferior single cross (C x D) is replaced by a new and superior single cross (E x G) and the new double cross hybrid (A x B) x (E x G) is developed. This new hybrid not only gives the better yield than original double cross but also usually agronomically superior.

6

HYBRID VARIETIES TESTING AND RELEASE PROCEDURES IN INDIA

Hybrids are released for commercial cultivation either by the Central or State Varieties Release Committee. The release of a hybrid for commercial cultivation is based on conclusive demonstration of its superiority over the best existing hybrids (included as check in evaluation trials) in yielding ability or in some other features of economic importance, *e.g.* abiotic and biotic stresses *etc*. Due to above reason, hybrids are extensively evaluated for their performance, resistance to diseases and insects, and quality in multilocational trials. The multilocational trials are conducted under All India Coordinated Vegetable Improvement Projects, which play a key role in testing, identification and release of new hybrids. Before release of hybrids, three activities and operations are involved. These are:

- Evaluation
- Identification
- Release and Notification

Evaluation

It consists of following seven different types of trials / tests conducted during evaluation. In all these trials, best existing hybrid is included as a check for comparison. In general, more than one (often two or three) hybrid is included as check.

- Station trials
- Multilocation trials
- National trials
- Adoptive trials
- Minikit trials
- Trials on disease and insects
- Trials for quality test

Station Trials

The breeder who has developed the hybrids, conducts it. It may be conducted for one or two years and is often called as preliminary yield trial. The objective of station trial is to make sure that hybrid/ hybrids developed by breeders, are superior in performance (at that location) to the best available hybrid for the region before they are included in trials of All India Coordinated Vegetable Improvement Project. This is of great importance because in the trials of coordinated project, hybrid / hybrids would not only have to compete with the best existing hybrid, but also with the hybrids developed by other breeders. The salient points in station trial are as:

- To avoid disappointment and rejection in early stages of testing, the breeder should himself make sure to superiority of hybrids, he has isolated.
- The best existing hybrids should be used as check.
- Standard package and practices should be adopted for cultivation.
- Disease and insect relation with hybrid should also be seen.
- The station trial may be repeated once make sure that new hybrids are superior to existing hybrids.

Multilocational Trials

These trials are conducted under the All India Coordinated Vegetable Improvement Project. The objective of these trials is to evaluate the performance of newly-developed hybrids at several locations, distributed over the region. Since, soil and climatic conditions show a large variation from one region of the country to the other, the country has been divided into several agroclimatic zones (Table 1). Each zone consists of areas having similar agroclimatic conditions. It should be noted that zones have not been created on the basis of political boundaries, but on the basis of agroclimatic conditions, prevailing in the area. Thus, it may be expected that agroclimatic conditions at any two locations within a zone would be more comparable than those at two locations belonging to any two different zones.

Table 1: Eight agroclimatic zones and states under each zone

Agroclimatic zone	States
Humid Western Himalayan region	Jammu & Kashmir, Himachal Pradesh and Uttranchal
Humid Bengal-Assam Basin	West Bengal and Assam
Humid Eastern Himalayan and Bay Islands	Sikkim, Meghalaya, Manipur, Nagaland, Mizoram, Tripura, Arunachal Pradesh and Andaman and Nicobar Islands
Sub-Humid Sutlej Ganga Alluvial Plain	Punjab, Uttra Pradesh and Bihar
Sub-Humid to Humid Eastern South Eastern Uplands	Eastern Madhya Pradesh, Orissa and Andhra Pradesh
Arid Western Plains	Rajasthan, Gujarat, Haryana and Delhi
Semi Arid lava Plateau and Central High Lands	Madhya Pradesh (excluding eastern area) and Maharashtra
Humid to Semi Arid Western Ghats and Karnataka Plateau	Karnataka, Tamil Nadu and Kerala

The trials conducted under coordinated projects are grouped in two categories:

- Initial evaluation trials
- Advanced hybrid trials

Initial Evaluation Trials and Advanced Hybrid Trials

The initial evaluation trials are conducted at least 10-12 locations within a zone and it is conducted under irrigated conduction. The hybrids are tested in IET is for two years only. These are promoted for advanced hybrid trials (AHT) if their performance is outstanding, otherwise, they are rejected. The objective of IET is to eliminate relatively inferior entries, so that their number is reduced to a manageable size for AHT evaluations. Both in IET and IHT, the crops are raised under good environmental conditions. The plot size and spacing in both IET and AHT trials varies from crop-to-crop as given in Table 2.

Table 2: Plot size and spacings for IET and AHT trials

Crop	Plot size (m)	Spacing (cm)	Design	Replication
Brinjal	4 x 3.6	60 x 45	RBD	3
Tomato	3.6 x 3	60 x 45	RBD	3
Chilli	4.3 x 3	50 x 45	RBD	3
Capsicum	4.3 x 3	50 x 45	RBD	3
Okra	3 x 2.7	60 x 30	RBD	3
Cucumber	4 x 1.8	100 x 30	RBD	3
Bottle gourd	10 x 3	300 x 50	RBD	3
Bitter gourd	5.5 x 2.5	150 x 300	RBD	3
Watermelon	2 x 5.4	150 x 90	RBD	3
Ridged gourd	6 x 2.5	300 x 50	RBD	3
Cauliflower	4.5 x 3.2	45 x 40	RBD	3
Cabbage	4.5 x 3	45 x 30	RBD	3
Broccoli	4.5 x 3	45 x 30	RBD	3

National Trials

The national trials are conducted throughout the country in all the zones. The entries in national trials consist of one entry from each zone that ranked first in that zone in uniformity. The national trials serve as IET for an entry in zones than that in which it was developed and tested in Uniformity Regional Trials. The purpose of national trials is to evaluate outstanding entries of one zone in other agoclimatic zones to see if they perform well in other zones as well. It has been suggested that national trials should include IET entries of entire zone so as to evaluate all the entries on a national level.

Adoptive Research Trials

These trials are conducted on research stations or farms of state governments. The hybrids identified by the workshop in All India Coordinated Vegetable Improvement are included in these trials. Trials are conducted in without replication in very large area with agronomic practices as same for which the concerned entry has been identified. The data from these trials are considered for release of the identified entry as a new hybrid.

Minikit Trials

The minikit trials are conducted farmers' fields. These trials are conducted along with the adoptive trials in following crop season after the entry has been identified for release.

The objective of minikit trials is to popularize new hybrid among farmers of the zone. It also serves another purposes; the seed of a good new hybrid reaches the farmers.

Trials for Diseases and Insect-Test

Entries are evaluated for diseases and insects resistance throughout the period of testing, that is, during IET as well as AHT. Diseases and insects resistances are tested both under natural epidemic as well as artificial epiphytotic (epidemic) conditions. The disease reaction tests for different diseases of various crops are carried out at different places where epidemics of concerned disease occur regularly. The places where a disease occurs at a high intensity, that is, the disease occurs in an epidemic form, every year are known as hot spots for that disease. Serious diseases and insects occurs in vegetable crops are given in Table 3.

Table 3: Major diseases and insects of vegetable crops

Vegetable Crops	Diseases	Insects
Tomato	Fusarium wilt, verticilium wilt, late blight, septoria leaf spot, anthracnose, bacterial wilt, bacterial canker, leaf curl virus, tomato mosaic and root-knot nematodes	Fruit-borer and white fly
Brinjal	Bacterial wilt, phomopsis blight, little leaf and root-knot nematodes	Shoot- and fruit borer
Bell pepper and chilli	Fruit rot, cercospora leaf spot, powdery mildew, bacterial leaf spot, phytophthora root rot, root -knot nematodes and common TMV	Thrips, mite, aphid, and fruit-borer
Okra	Yellow vein mosaic virus, fusarium wilt, cercospora leaf spot and fruit rot.	Fruit- and shoot borer, jassids and whitefly
Cauliflower and cabbage	Black rot, sclerotinia rot, alternaria blight, irwinia rot and cabbage yellows.	Cabbage butterfly caterpillar, aphids and diamond back moth
Carrot	Alternaria blight and cercospora leaf spot.	
Onion	Purple blotch, basal rot, stemphyllium blight and bacterial storage rot.	Thrips
Cucurbits	Powdery mildew, downey mildew, anthracnose and cucumber mosaic virus.	Red pumpkin beetle, fruit fly and aphids

Trials for Quality Test

Quality tests are conducted to determine the suitability of an entry for various uses of its produce. The quality in vegetable crops is not a simple character, and is often not easy to determine. Therefore, a number of different tests have to be conducted for which sophisticated equipments may be needed for speedy and reliable results. In each crop, a different set of quality tests has been developed in order to determine the suitability of new hybrids for various uses. The quality parameters, which are required in vegetable hybrids are given in Table 4.

Table 4: Quality parameters in vegetable crops

Vegetable Crop	Quality Parameters
Tomato	Large round fruit with adequate firmness and shelf-life, uniform fruit size, shape, red colour and freedom from external blemishes or abnormalities for fresh market. Large fruit size, high fruit quality and continuous production for home garden tomatoes. Deep uniformally red coloured tomatoes, pH below 4.4, high total soluble solids (4.5-7%) and high alcohol insoluble solids (AIS) in processing tomatoes.
Brinjal	Fruit shape, size and colour as per consumer's preference with soft flesh, low proportion of seed and lower olenine content.
Bell pepper and chilli	Oblete or round fruit in bell pepper and long fruit in chilli, pleasing flavour, high sugar/acid ratio, high pigment content and vitamin C in bell pepper and high capsaicin ($C_{18}H_{27}NO_3$) and high oleoresin in chilli
Okra	Dark green, tender, thin, medium long, smooth 4-5 ridged pods at marketable stage. Pods should be free from conspicuous hairs and optimum seed setting ability.
Pea	Long, attractive green pods with more seeds/pod. Sweetness in pods with high shelling percentage.
Cauliflower	Non-ricey, compact bract free protected curds with retentive cream/white colour and bettter seeding quality.
Cabbage	Longer staying capacity in field after formation, narrow, short and soft core with short stem.
Knol-khol	Spherical swollen bulb like stem and it should be free from bolting.
Radish	White, long/stump roots with thin tap root and non-branching habit, non-pithy roots with slow bolting habit. Pungency of roots should be as per consumer's preference.
Carrot	Scarlet/orange colour roots with uniformity in shape and size. Smooth root surface, thick flesh roots, and thin and self-coloured core in roots, broad shoulder red, cylindrical, uniformly tapering or stump rooted with non-branching habit. Early rooting with free from cracking. High sugar- and dry-matter content in roots.
Turnip	Root colour as per consumer's preference, white, purple types more liked in India that Golden Ball types. Stump-rooted hybrids with thin tap root and non-branching habit. Slow bolting habit and appropriate dry matter (8-9%) in roots.
Beet root	Dark red & uniformly coloured roots along with uniform root shape. Absence of internal white rings in roots. Slow bolting with manogerm seeds.

Vegetable Crop	Quality Parameters
Onion	High TSS is important for dehydration industry producing onion chips and powder.
Cucumber	Attractive green or dark green fruits with smooth surface and without prominent spines or prickles. Uniform long cylindrical shape without crook-neck. Fruits should be free from carpel separation without hollow spots with free from bitterness along with less seeds at edible maturity.
Muskmelon	Fruits should be attractive round/spherical in shape with thick flesh and attractive in colour. Small seed cavity, sweet, juicy musky flavoursome fruits. TSS should not less than 10% with tough netted skin of fruit.
Watermelon	Tough-skinned fruits for long distance transportation. TSS content not less than 10%. Fruits with smaller and fewer seeds with attractive deep red flesh and firm flesh.
Squash and Pumpkin	Thick fruit flesh and small seed cavity. Round/oblong/flat round fruit shape. Orange flesh colour, rich in β-carotene, a precursor of vitamin A.
Bottle gourd	Attractive green fruit of round/long/club-shaped with sparse hairs persisting on skin and non-fibrous flesh at edible stage. Fruits should be non-bitter.
Bitter gourd	Less ridged fruit surface with thick fruits particularly suitable for stuffing. Fruit size variation as per consumer's preference (small 7.5-10 cm long, medium-long 10-15 cm, long 15-20 cm and extra long 20-40 cm). Immature seeds for longer period during green edible stage.

Hybrid Testing Centres In India

Following organizations in various states of India are actively engaged in testing of vegetable hybrids under All India Coordinated Vegetable Improvement Project (Table 5).

Table : 5. List of organization and their respective crops

Organisation	Crops
Indian Institute of Vegetable Research, 1, Gandhi Nagar (Naria), Varanasi 221 005 (Uttar Pradesh)	Brinjal, tomato, chilli, capsicum, okra, cucumber, bottle gourd, bitter gourd, cauliflower, watermelon, ridged gourd, broccoli, cabbage and onion.
Department of Vegatable Crops, Indian Agricultural Research Institute, New Delhi 110 012	Brinjal, tomato, chilli, okra, cucumber, bottle gourd, bitter gourd, cauliflower, ridged gourd, broccoli, cabbage and onion.
Division of Vegetable Crops, Indian Institute of Horticultural Research, Hessaraghatta, Bangalore 560 089. (Karnataka)	Tomato, chilli and okra.

Organisation	Crops
IARI Research Station, Katrain 175 129, Kullu Valley (Himachal Pradesh)	Capsicum, cucumber, broccoli and cabbage.
Vivekananda Parvitya Krishi Anusandhan Shala, Almora 263 601 (Uttaranchal)	Tomato, okra and broccoli.
Horticulture and Agroforestry Research Programme, Namkum, Ranchi 834 010 (Jharkhand)	Brinjal, tomato, onion and ridged gourd.
Department of Horticulture, Horticulture College and Research Institute, Tamil Nadu Agricultural University, Coimbatore 641 003 (Tamil Nadu)	Brinjal, chilli, okra and onion.
Department of Vegetable Crops, Punjab Agricultural University, Ludhiana 141 004 (Punjab)	Brinjal, tomato, chilli, okra, bitter gourd, watermelon, broccoli, cabbage and onion.
Department of Horticulture, Mahatma Phule Krishi Vishawavidyalaya, Rahuri 413 722 (Maharashtra)	Brinjal, chilli, capsicum, cucumber, bitter gourd, cauliflower, watermelon, ridged gourd and onion.
Department of Horticulture, College of Agriculture, Orissa University of Agriculture and Technology, Bhubaneswar 751 003 (Orissa)	Okra.
Department of Horticultre, University of Agricultural Sciences, Dharwad 580 005 (Karnataka)	Chilli, capsicum, watermelon, ridged gourd and onion.
Department of Horticulture, Agricultral Research Station, Durgapura, Jaipur 302 018 (Rajasthan)	Tomato, watermelon, ridge gourd and broccoli.
Department of Vegetable Crops, Narendra Deva University of Agriculture and Technology, Narendra Nagar (Kumarganj), Faizabad 224 229 (Uttar Pradesh)	Tomato and chilli.
Department of Vegetable Crops, Choudhury Charan Singh Haryana Agricultural University, Hisar 125 004 (Haryana)	Chilli, bitter gourd and onion.
Agricultural Research Institute, Andhra Pradesh Agriculture University, Rajendra Nagar, Hyderabad 500 030 (Andhra Pradesh)	Okra
Department of Vegetable Crops, College of Agriculture, Jawaharlal Nehru Krishi Vishawavidyalaya, Jabalpur 482 004 (Madhya Pradesh)	Tomato and okra.
Department of Horticulture, Assam Agricultural University, Jorhat 785 013 (Assam)	Okra
Vegetable Research Station, Gujarat Agricultural University, Junagadh 362 001 (Gujarat)	Okra and ridged gourd.
Department of Horticultre, Vegetable Research Station, Chandra Sekhar University of Agricultre and Technology, Kalyanpur, Kanpur 208 024 (Uttar Pradesh)	Chilli, okra, bottle gourd, bitter gourd, cauliflower, ridged gourd and onion.

Organisation	Crops
Department of Horticulture, Bidhan Chandra Krishi Vishawavidyalaya, Distt. Nadia, Kalyani 741 235 (West Bengal)	Tomato, chilli, watermelon and onion.
Department of Vegetable Crops, Govind Ballabh Pant University of Agricultre and Technology, Udham Singh Nagar, Pantnagar 263 145 (Uttar Pradesh)	Tomato, cauliflower, bottle gourd, bitter gourd, cauliflower and ridged gourd.
Department of Vegetable Crops, College of Horticulture, Dr. Y.S. Parmar University of Horticulture and Forestry, Solan 173 230 (Himachal Pradesh)	Tomato, capsicum, cucumber, broccoli, and cabbage.
Division of Olericulture, Sher-E-Kashmir University of Agriculture and Technology, Shalimar, Srinagar 191 121 (Jammu & Kashmir)	Brinjal and capsicum.
Department of Olericulture, College of Horticultre, Kerala Agriculture University, Vellanikkara 680 654 (Kerala).	Bitter gourd.
Department of Botany and Plant Breeding, B.A. College of Agricultre, Gujarat Agriculture University, Anand 388 110 (Gujarat)	Brinjal, tomato, chilli, capsicum, okra and cauliflower.
Department of Vegetable Crops, Himachal Pradesh Krishi Vishawavidyalaya, Palampur, Distt. Kangra, (Himachal Pradesh)	Cabbage.

Identification of Hybrids for Release

Outstanding hybrids are identified for specific zone at the annual workshops of All India Coordinated Vegetable Improvement Projects. The proposal for identification of hybrids is an open forum by constituting a committee in the workshop. The respective breeder who has developed the concerned hybrid prepares a proposal for identification of entries. They prepared according to a prescribed proforma that includes performance of hybrid in AHT for at least two years' data from pathological, entomological, quality tests *etc*, and those from agronomic trials. The criteria for identification of entries vary considerably from one crop to other. An entry may be identified if it falls into one of the following two categories:

- An entry significantly superior to the check in yield and comparable to the check in disease resistance.
- An entry comparable to the check in yield but constantly and markedly superior to it in resistance to diseases.

These criteria are fixed in the workshop of the concerned coordinated projects and reflect the experiences of workers with concerned crops. As a result, they are not fixed entities, and are likely to vary as per the needs of the nation as well as the experiences of concerned workers. Tremendous progress has been made by public and private sectors in the development of hybrids in several vegetables including tomato, brinjal, chilli, bell pepper, cabbage, cauliflower, cucumber, muskmelon, watermelon, carrot, onion *etc*., and till date 84 vegetable hybrid varieties have been recommended for cultivation in various parts of country based on multilocational trials conducted under All India Coordinated Vegetable Improvement Project (AICVIP). These hybrids along with parental organization and recommended zones are given in Table 6.

Table 6: Public and private sectors bred hybrids identified/released for cultivation

Crop/hybrid	Developing Centre	Recommended Zone(s)	Year of Identification
Brinjal			
Arka Kusumkar	IIHR, Banglore	VIII	1981
KT-4	IARI, Katrain	VIII	1987
Pusa Hybrid-6	IARI, New Delhi	IV	1990
Arka Navneet	IIHR, Banglore	-	1991
Pusa Hybrid-5	IARI, New Delhi	IV, VII, VIII	1992
ARBH-201	Ankur Seeds, Jalana	IV, V, VI, VII	1993
NDBH-1	NDUA & T, Faizabad	IV, VII, VII	1993
MHB-10	MAHYCO Seeds, Jalana	IV, VI, VII	1993
NDBH-6	NDUA & T, Faizabad	IV	1995
Pusa Hybrid-9	IARI, New Delhi	VI	1997
ARBH-541	Ankur Seeds, Jalana	All	2001
PBH-6	Pandey Seeds, Faizabad	All	2001
JBH-1	GAU, Junagadh	All	2001
BH-1	PAU, Ludhiana	IV	2002
BH-2	PAU, Ludhiana	IV, V,	2003
VRBHR-1	IIVR, Varanasi	IV, VI	2003
NDBHL-54	IIVR, Varanasi	IV	2004
ARBH-786	Ankur Seeds, Jalana	IV	2004
VNR-51	VNR Seeds, Chattisgarh	IV, VI	2005
BWBH-3	IIHR, Banglore	IV, VI	2005
COBH-3	TNAU, Coimbatore	VIII	2005
HABH-3	HARP, Ranchi	IV	2005
BCTH-17	BCKV, Kalyani	II, V	2005
Tomato			
ARTH-4	Ankur Seeds, Jalana	IV, VII	1992
MTH-6	MAHYCO Seeds, Jalana	VII, VIII	1992
Pusa Hybrid-2	IARI, New Delhi	I, IV, VI, VII	1993
NA-501	Nath Seeds, Aurnagabad	VI, VII	1995

Crop/hybrid	Developing Centre	Recommended Zone(s)	Year of Identification
DTH-4	IARI, New Delhi	VII	1995
Avinash -2	Novartis Seeds, Pune	VI	1998
HOE-303	Novartis Seeds, Pune	IV	1998
Sun-496	Sungrow Seeds, New Delhi	II, IV, VIII	1999
BSS-20	Beejo Sheetal Seeds, Jalana	All	2001
DTH-8	IARI, New Delhi	IV	2002
CHTH-1	HARP, Ranchi	IV	2002
ARTH-128	Ankur seeds, Jalana	VIII	2002
KTH-2	CSAUT, Kanpur	IV, V	2003
JKTH-3055	J K Seeds, Secunderabad	I, IV	2004
KTH-1	CSAUAT, Kanpur	IV	2004
Nun-7730	Nunhems Seeds, Hyderabad	I, IV	2004
TH-01462	Syngenta Seeds, Pune	I, IV, VI, VI	2005
NTH-6	IIVR, Varanasi	IV, VI	2005
TLBRH-9	IIHR, Bangalore	IV, VI, VII	2005
HATH-3	HARP, Ranchi	IV	2005
BCTH-4	BCKV, Kalyani	II, V	2005
Chilli			
HOE-888	Novartis Seeds, Pune	IV, VIII	1997
ARCH-236	Ankur Seeds, Jalana	IV	1997
Sungro-86-235	Sungro Seeds, New Delhi	IV, VIII	2002
ARCH-228	Ankur Seeds, Jalana	IV, V, VI	2003
CCH-3	IIVR,Varanasi	II, IV, V, VI	2005
MSH-149	IIHR, Bangalore	IV, VI, VIII	2005
KCH-3	CSAUAT, Kanpur	IV	2005
MSH-172	IIHR, Bangalore	IV, V, VI, VIII	2005
Bell Pepper			
KT-1	IARI, Katrain	I	1990
Lario	Syngenta Seeds, Pune	I	2002
DARL-202	DARI, Almora	I, IV	2003
KTCPH-3	IARI, Katrain	I, VI, VII	2005
Cabbage			
Pusa Synthetic	IARI, Katrain	I, II, IV	1992
Sri Ganesh Gol	MAHYCO Seeds, Jalana	V	1992
Nath-401	Nath Seeds, Aurangabad	I, IV, V, VI, VII	1993
BSS-32	Bejo Sheetal, Jalana	VII	1995
Quisto	Novartis Seeds, Pune	IV	1998
SYCFH-202	Syngenta Seeds Pune	IV, VIII	2004
Summer King	Sungrow Seeds, New Delhi	I, IV	2004
KGMR-1	IARI, Katrain	I, IV	2005

Crop/hybrid	Developing Centre	Recommended Zone(s)	Year of Identification
Cauliflower			
Pusa Hybrid-1	IARI, New Delhi	II, IV	1992
DCH-541	IARI, Ned Delhi	II, IV	2003
SYCFH-2003	Syngenta Seeds, Pune	IV, V, VII	2005
Muskmelon			
M-3 Hybrid	IARI, New Delhi	IV	1990
MHY-5	RAU, Durgapura	VII	2002
Watermelon			
Arka Jyoti	IIHR, Bangalore		1981
MHW-6	MAHYCO Seeds, Jalana	IV, V, VIII	1999
Bottle Gourd			
PBOG-2	GBPUA & T, Pantnagar	VII	1998
PBOG-1	GBPUA & T, Pantnagar	IV, VI	1999
NDBH-4	NDUAT, Faizabad	All	2001
Okra			
DVR-1	IIVR, Varanasi	IV, VII	1998
DVR-2	IIVR, Varanasi	VI	1998
DVR-3	IIVR, Varanasi	All	2001
DVR-4	IIVR, Varanasi	IV, V, VII	2002
HBH-142	HAU, Hisar	IV, V, VI, VI	2005
SOH-152	Syngenta Seeds, Pune	IV, VII, VIII	2005
Bitter Gourd			
Pusa Hybrid-2	IARI, New Delhi	IV, V, VI	2003
Cucumber			
NBGH-167	Nirmal Seeds, Maharashtra	IV	2004
Hybrid-1	Century Seeds, New Delhi	IV, VII	2004
PCUCH-3	GBPUAT, Pantnagar	I, IV	2005

Besides identified hybrids, several vegetable hybrids are popular among farmers. Seeds of these hybrids are also being marketed by private organizations (Table 7). Realizing potential of vegetable hybrid technology in India, ICAR, New Delhi, initiated a network project entitled "Promotion of Hybrid Research in Vegetable Crops" during 1995-96. This project was initiated nine vegetables, *viz.* tomato, brinjal, chilli, capsicum (bell pepper), okra, onion, cabbage, cucumber and bitter gourd with following objectives: (i) to promote hybrid research to increase productivity *per se* the country, (ii) to incorporate biotic stress resistance in hybrids, (iii) to strengthen hybrid seed research and hybrid seed production technology, (iv) to test the proven hybrids for their potentiality and (v) to establish their production technology. This project has completed with considerable achievements. The hybrids promising at various locations are given in Table 8.

Table 7: Popular hybrids developed and marketed by private sector

Crop	Hybrid Variety
Tomato	Rupali, Naveen, Rashmi, Tolstoi, Minakshi, Vaishali, Karnataka Hybrid, Lerica, NS-815, NS-2530, TH-2312 and TH-802.
Brinjal	Suphal, Manju, Kalptaru and HOE-404.
Chilli	Agni, NS-1101, CH-1, CH-3, NS-1101, Tejashwani and HOE-808.
Capsicum	Indira, Bharat, Lerio, Early Bonty, NS-33 and Hira.
Cabbage	Cabbage-8, Questo, NS-22, Hari Rani Gol, Krishna and Vishesh.
Bottle gourd	Barad and Kaveri.
Watermelon	Madhur, Milan, Nath-101 and MHW-4.
Muskmelon	Abhjeet, MCH-5 and MHW-4.
Okra	Varsha, Vijay and Vishal.

Table 8: Performance of promising hybrids at various locations

Crop	Locations	Promising hybrids
Tomato	Varanasi	Avinash-2, Hemlata, TH-2312 and Ratna
	Bangalore	TH-2312, ARTH-13, Avinash-2 and DTH-6
	Hisar	ARTH-3
	Coimbatore	Avinash-2
Brinjal	Varanasi	NDBH-8, ARBH-201 and PDBH-13
	Bangalore	IIHR-500 A x Arka Keshav and IIHR-104 x IIHR-491
	Coimbatore	190 x Annamalai, 190 x Co-2 and Aruna x SM-8
Chilli	Varanasi	BSS-188, BSS-141 and HOE-818
	Bangalore	HOE-888, MHP-59 and ARCH-236
	Coimbatore	Kiran
	Anand	Kiran, Tejaswini and BSS-138.
Cabbage	Katrain	KT CBH-127, KT CBH-95 and KT CBH-139
Cucumber	Varanasi	CH-13,
	Bangalore	PCUCH-1
	Rahuri	DH-9, 18 and 17 (developed form gynoecious line)
Onion	Bangalore	MS-65 x PBR-243 and MS-65 x PBR-245
	New Delhi	Hybrid-63 and Hybrid-35

Besides public sectors developing high-yielding hybrids in vegetable crops, some private seed sectors have also been playing a pivotal role in developing disease resistant hybrids (Table 9).

Table 9: Disease resistant hybrids developed by private sector seed companies

Seed Company / Crop hybrids		Characters
Mahyco, Jalna		
Cabbage	Harirani Gol & Kranti	Resistant to yellows and black rot
	SriGanesh Gol	Resistant to yellows
Ridged gourd	Surekha	Tolerance to mildews
Watermelon	Amrut, MHW-6	Field tolerance to fusarium wilt
Capsicum	Green Gold & MCPH-11	Resistant to TMV
Okra	No.7, No.8, No.10	Tolerant to yellow vein mosaic virus
Century Seeds, New Delhi		
Tomato	Rishi, Maitri and Manmohan	Marked resistant to fusarium wilt and verticellium wilt
Cabbage	Geetanjali	Tolerance to black rot and fusarium wilt
Cauliflower	Early Himlata	Tolerance to black rot
Muskmelon	Madhumati	Tolerance to downey mildew
Chilli	Hybrid-1	Tolerance to mosaic, leaf curl, wet rot and dieback
Okra	Panchalic and Adhunik	Tolerant to yellow vein mosaic virus
Beejo-Sheetal Seeds, Jalna		
Cabbage	Swati, Sarita, Bajrang, Shanti, BSS-44, Swarna and Sonali	Resistant to fusarium
Tomato	Meenakshi and Menka	Resistant to fusarium
Okra	Tara	Resistant to yellow vein mosaic virus
Sungro Seeds, New Delhi		
Tomato	Sungro-501	Tolerant to leaf curl
	SC-1, SC-3	Tolerant to wilts
	Sungro-830	Tolerant to alternaria
Okra	Supriya	Tolerant to yellow vein mosaic virus
Namdhari Seeds, Bangalore		
Chilli	NS-2302	Tolerant to virus
	NS-1820	Tolerant to virus
Tomato	Suraksha	Tolerant to bacterial wilt
Pro-Agro Seeds, New Delhi		
Tomato	Abhiman	Tolerant to early blight
Cabbage	Bahar	Tolerant to black rot

Seed Company / Crop hybrids		Characters
	Pragati	Tolerant to black rot
Chilli	Vardan	Tolerant to anthracnose
	Kranti and Krishna	Tolerant to TMV
Okra	Uphar	Tolerant to yellow vein mosaic virus
Sandoz Seeds, Pune		
Tomato	Rasika and Ratna	Tolerant to fusarium and verticillium
	Lerica	wilt.
	Avinash-II	Tolerant to tomato leaf curl virus
Chilli	Bijli and Agni	Tolerant to TMV
Capsicum	Indra	Tolerant to TMV and PVY
Cabbage	Quisto and Vishwas	Tolerant to fusarium yellows
Cauliflower	Pawas	Tolerant to xanthomonas, pernospora, pseudomonas
Indo-American Hybrid Seeds (I) Pvt. Ltd, Bangalore		
Okra	Varsha, Vijay and Vishal	Highly tolerant to yellow vein mosaic virus.
Tomato	9501, 9502	Highly tolerant to leaf curl virus.
	Akash, Utpan and Ruchi	Tolerant to foliar diseases.
Chilli	No.5, 6 and 9	Tolerant to leaf curl and dieback.

Source: Seshadri (1998).

Release of Hybrids by Central Sub-committee on Crop Standards and Release of Varieties/Hybrids

After identification, a hybrid is tested at least for one year on adoptive research trials. During this period, disease tests and quality tests are also conducted. Based on data from adoptive trials, and disease and quality tests, the breeder who has developed the concerned strain, submits a proposal for its release as new hybrid for consideration by the Central Sub-committee on Crop Standards and Release of Varieties. The proposal is prepared according to a prescribed proforma as given below. The breeder usually gives a name or he may give it a number, which is included in the proposal. After a hybrid has been released for a zone by the Central Sub-committee on Crop Standards and Release of Varieties, the Director, High Yielding Varieties/Hybrids, Ministry of Agriculture and Irrigation, Government of India, notifies the concerned authorities and the states within that zone for seed development and distribution of varieties/hybrids. This is known as notification of varieties/hybrids.

Proforma for Submission of Proposal of Release and Notification of Vegetable Varieties/ Hybrids to Central Sub-committee on Crop Standards

Notification and Release of Varieties/Hybrids of Vegetable Crops

1. Name of the crop with latin name:
2. a) Name/code/designation under which tested :
   b) Proposed name of the hybrid:
3. Sponsored by
   a) Institution or agency responsible for developing hybrid (with address) :
   b) Name of person (s) / scientist responsible for the development of the variety :
4. a) Breeding objectives :
   b) Breeding method :
   c) Parentage with details of its pedigree :
5. State the hybrid which most closely resemble the proposed hybrid in general characteristics :
6. a) Whether recommended by workshop/ group meeting :
   b) If so, its recommendation with specific justification for the release of proposed hybrid :
   c) Specific areas/region for which recommended :
7. Recommended ecology including suitability under controlled/field conditions :
8. Description of Hybrid
   a) Plant height :
   b) Distinguishing morphological characteristics :
   c) Characteristics different from those of the parents :
   d) Maturity (range in number of days) seeding/transplanting to flowering/fruiting) :
   e) Maturity on ground (early, medium and late) wherever such classification exists :
   f) Reaction to major diseases under field and controlled conditions :
   g) Reaction to major pests (under field and controlled including storage pests) :
   h) Agronomic features (e.g. resistance to lodging, shattering, fertilizer) :
   i) Quality attributes purpose for which recommended :

(i) Table/proceeding/report or any other :
(ii) Table quality attributes/nutritional quality :
j) Reaction to environmental stress :
k) Kind of root stock to be used, if applicable :

9. Description of parents :
a) Plant height :
b) Distinguishing morphological characteristics :
c) Days to 50% harvest :
d) Maturity/ripening (range in number of days seed to be used) :
e) Is there any problem of synchronization? If yes, methods of over it :
f) Reaction to major disease (under field and controlled conditions (reactions to physiological strains/races bio types to be indicated wherever possible) :
g) Reaction to major pets (under field and controlled conditions including storage pests)
(h) Agronomic features (*e.g.* fertilizer responsiveness, seed rate *etc.*) :
(i) Reaction to environmental stresses :

10. Yield data in AICRP trial/regional trial on farm trials (levels of fertilizer application, density of plant population and superiority over local control standard variety to be indicated :

11. Premium attributes based on table 1 :

12. a) Agency responsible for maintaining breeder seed :
b) Quantity of breeder seed for 5 ha for low volume seed rate and 10 ha for high volume seed rate. :

13. Information on the acceptability of the variety by farmers/consumers/industry attach data generated on these aspects :

14. Specific recommendations, if any, for seed production/seed technology :

15. Any other pertinent information :

Signature Director
Indian Institute of Vegetable Research
1, Gandhi Nagar (Naria)
Varanasi 221 005 (UP)

Signature of Breeders

Release of a Hybrid by State Variety Release Committee

In addition to the Central Sub-committee on Crop Standards and Release of Varieties, each state has its own variety release committee. The State Variety Release Committees usually require a separate set of trials conducted at the Regional Research Station of the state, in addition to the data from coordinated multiplicational trials. A hybrid released by a State Variety Release Committee must be notified by Ministry of Agriculture, Government of India, before its certified seed is produced. Further, only the ministry notifies those varieties/ hybrids, which have been identified by the workshop of All India Coordinated Vegetable Improvement Project. The State Variety Release Committee consists of the following: (1) Director of Agriculture of the state (Chairman), (2) Director State Seeds Corporation; (3) Director, State Seed Certification Agency; (4) Additional Directors of Agriculture of the state; (5) Joint Directors of Agriculture of the state; and (6) Directors of Research of all agricultural universities of the state. In addition, (7) Deputy Directors of Agriculture of the state; (8) Officers-in-charge of all Regional Agricultural Testing and Demonstration Centres; (9) Breeders, agronomists, pathologists and entomologists of all agricultural universities and agricultural institutes in the state are also invited. It may be noted that the persons listed under items 7, 8 and 9, infact, are not the members of the committee, but they are invited to assist the committee in its deliberations. The prescribed proforma for the proposal of releasing hybrid by State Variety Release Committee is similar to proforma as given above in Central Sub-committee on Crop Standards and Release of Varieties/ Hybrids.

7

GENETIC PURITY TESTING OF HYBRID SEEDS

Meaning and Importance

Genetic purity denotes the percentage of seeds (by weight) belonging to the variety/hybrid under certification. It is one of the primary criteria of seed quality. The efforts of plant breeders become fruitful only when sufficient quantities of good quality seeds of improved varieties/hybrids reach farmers / growers. This is necessary to take all round care at every step of seed multiplication before the seed is delivered for marketing. Hence, genetic purity testing is of critical importance not only to farmers but also to plant breeders, seed producers as well as quality control enforcing agencies. Increasing emphasis on hybrid cultivars due to their better performance and quality produce, more money and expertise have been put in further research of hybrid development. In this context, protection of parental lines is assuming critical importance and for that purpose, several breeding institutes do adopt various strategies for identification, characterization and protection of their parental lines as to avoid pilferage and misuse of their proprietary lines by other organizations. Use of random amplified polymorphic DNA (RAPD) and other molecular markers for evaluating seed purity of F_1 hybrids and their parental lines are closely related, thus, helps in assessing contamination of F_1 seeds even with the seeds of sib (inbred seed and seed from self-pollination of parental lines). For genetic purity testing of hybrids seeds, it is imperative that parental lines of hybrid be adequately characterized on the basis of maximum number of distinguishable descriptors (morphological, biochemical or molecular markers).

Advantages of Genetic Purity

The advantages of genetic purity testing are :

- For long-term research investment in genetic resources conservation and utilization over greater time frames, for facing the newer challenges.
- Plant variety protection, registration, certification and patents, wherein the whole phenomenon is on the basis of distinctness, uniformity and stability (DUS) testing of released/notified cultivars in changing paradigms of Plant Variety Protection and Farmers Right laws as promulgated in Indian parliament.
- To detect even minute genetic differences between cultivars *vis-à-vis* for the existence of novelty among essentially derived varieties (EDVs).
- Measurement of genetic diversity, which at times offer efficient opportunities to hedge genetically and unforeseen maladies.
- Assurances of genetic purity for ensuring better agronomic performance and predicted expectations.

- Improvement of germplasm in term of agronomic performance.
- Prevention of misappropriations and willful admixture of seeds/cultivars at commercial or farmers' level.

Methods of Testing

Conventionally, conducting grow-out tests of the seed lots and comparing the morphological characteristics of hybrids to assess their purity. The seed lot of hybrid to be tested is grown in field along with the check (authentic/reference sample of the same hybrid) lot and various morphological characteristics are compared from seedling stage to maturity (McDonald, 1988). Since, expression of morphological character is liable to be influenced by environment and grow-out test is also time and labour-intensive, more precise and rapid alternative techniques are required. In the past two decades, an intensive effort has been made to develop appropriate techniques for assessing genetic purity of seed under laboratory/ controlled conditions. These are discussed below:

Laboratory Examinations

The seeds are analyzed using one or more of the following tests.

Morphological Test

Examination of morphological features of seeds, such as, length, width, thickness, shape, weight, colour, seed coat colour, *etc*., and comparing them with those of authentic sample. The seeds may be examined with naked eye or magnified with a hand lens. The seeds may be examined with rapidly comparing length, width, thickness *etc*; of seed samples at a very reasonable cost. The structure of seed coat surface may be examined and compared with that of authentic sample with the help of a scanning electron microscope. But the equipment is expensive and not suited for routine analysis.

KOH-Bleach Test

Seeds are immersed in a 1:5 (wt/vol.) solution of potassium hydroxide (KOH) and fresh household bleach (5.25% NaOCl) for 10 (in case of seeds with brown coats) or 5 minutes (in case of seeds with white seed coat); the mixture is stirred gently periodically. Seeds are rinsed with tap water, dried on paper towel and observed for the presence of dark pigmented tannic acid and light pigmented testa.

NaOH Test

This test is used to distinguish red and white coloured seeds that have been treated with a fungicide or weather damaged. The seeds are immersed in a 5% sodium hydroxide (NaOH) solution and observed for 60 minutes. Red seeds become bright yellow (just after immersion), then turn to dark orange (after 15 minutes) and finally become bright yellow on immersion; the colour fades to straw colour (after 10 minutes) and remain as such.

Peroxidase Test

Seed coats of individual seeds are removed and placed in separate test tubes. In each test tube, following reagents are added: 10 drops of 0.5% guaiacol; after 10 minutes, one drop 0.1% H_2O_2; 1 minute later, reddish brown colour of solution indicates the seed coat to be peroxidase positive, while a colourless solution reveals it to be peroxidase negative.

Phenol Test

Seeds are soaked in water for 16 hr at 20 ± 1°C, placed in a Petri dish lined with filter paper soaked with 1% phenol solution and kept at 30 ± 1°C, and the reaction noted after 4 hrs phenol vapour reacts chemically (not enzymatically) with seeds to produce black, dark-brown, brown, light brown colours or there may be no reaction. For a better resolution, the colour development may be observed every 30 minutes.

Grow Out Test

The seed sample is sown in a controlled condition with authentic sample. Genetic purity is determined on the basis of observations made on plant morphological characters with reference to authentic sample. If genetic purity is worked out both at seed and plant level, then the size of submitted sample is to be doubled and genetic purity is reported in percentage. The procedures for grow out test are as given below:

- While raising the desired population, standard and recommended agronomic/cultural practices (*e.g.* field preparation, size of plot, row length, distance between plants, irrigation, fertilization *etc.*) in respect of individual crops are followed both for the unknown sample and its control.
- The possibility to prove the genuineness of a cultivars by grow out test is based on hereditary characteristics of plants. Usually, the cultivars differences are more distinct if growth conditions are favourable. Crop should be grown that the genetically differences express themselves as clearly as possible. In self-fertilizing species, the individual of cultivars may be theoretically identical, whereas individual of cultivars in cross-pollinated fertilizing species may not be genetically similar, but comprises a number of types. Therefore, it is easier to determine the cultivars purity in self-fertilizing species than in cross-fertilizing species where examination for greater part is based on mutual comparison between samples to be tested and standard sample. Hence, it is essential to sow various samples of the same cultivars in succession and standard samples are sown at suitable intervals (for example one standard sample for every 10 samples to be tested). The size of plots, row length *etc.* will differ crop-to-crop.
- The seed rate is adjusted depending on germination-percentage of individual samples and sowing may be done by dibbling. Subsequent, thinning is not recommended.
- The test crop may be raised along with the control either in the areas recommended for the variety or in off-season nurseries. The authentic control sample from the

originating plant breeder/breeding institute is to be maintained by testing station / agency following standard procedures. A minimum of 200 plants from the control sample will be raised along with the test crop.

- All plants are to be studied keeping in view the distinguishing characters described for the cultivars both in the test crop as well as the control. Necessary corrections are incorporated if the control is found to be heterogeneous.
- The specification of the field plot and row length are determined. The number of plants required for taking observation is dependent on maximum permissible off types which are given below:

Maximum Permissible off types (%)	Minimum Genetic Purity (%)
0.10	99.9
0.20	99.8
0.30	99.7
0.50	99.5
1.00 and above	99.0 and below

- The analysis employed for conduction 'grow out test' should possess the basic qualification as identified under Seeds Rules, 1968. These are given below:

Standard	Reject Numbers for Sample Sizes of	
	800	400
99.5 (1 in 200)	8	*
99.0 (1 in 100)	16	8
95.0 (5 in 100)	48	24
90.0 (10 in 100)	88	44
85.0 (15 in 100)	128	64

* Indicates that the sample size is too small for a valid test.

Greenhouse/Growth Chamber Test

The variety/crop is grown under controlled environmental condition in a specified growing medium, which helps in better development of distinct morphological features at a particular stage of development. Though, this method is less time-consuming, it is based on limited number of plants and also expensive (Payne, 1993).

Application of Molecular Markers—Proteins/Isozymes/DNA Based Markers

There are several markers/ descriptors being used for identification of varieties and parental lines of hybrids on the basis of morphology, proteins and isozymes and recent techniques based on DNA polymorphism. As to overcome the innate constraints associated with certain technologies and to meet the day- to- day challenge of identifying and cataloguing newer and essentially derived varieties (EDVS), a research for better marker systems to establish the uniqueness of a particular variety is on. A suitable alternative for assessing genetic purity (Smith and Register, 1998) could be, PAGE/SDS PAGE of protein/isozymes, IEF/Ultra thin layer IEF of proteins and Electrophoresis of DNA based markers-RFLP/ RAPD/AFLP/Microsatellites/minisatellite *etc.* Molecular markers posses several distinct advantages over morphological and other conventional classical markers are:

- While phenotypes of most the morphological markers can only be recognized at the plant level, genotypes of molecular loci can be established at cellular, tissue and plant levels.
- The number of distinguishable alleles at morphological or biochemical loci is rather limited and often dependent on application of exogenous mutagens. RFLPs however, provide a virtually unlimited number of markers, as the polymorphism does not need to alter the phenotype or the change of a protein to become detectable. With markers, for all chromosomes at intervals of 10-15 map units, the flow of any gene of interest through segregating generations in a breeding programme can be monitored, using a linked (anonymous) RFLPs markers as tag.
- Morphological markers are usually associated with unwanted phenotypic effects, which is not the case with molecular markers.
- As off types (%) alleles of most of the molecular markers are codominant, all possible genotypes can be distinguished in any segregating generation.
- Only a few epistatic or pleiotropic effects occur with molecular markers, allowing a virtually unlimited number of segregating markers to be monitored in a single population.

Protein and Isozyme Based Molecular Markers

Electrophoresis of Proteins and Isoenzymes

Proteins are the product of one / multiple loci. Their difference is based on number of amino acids, constituting a protein and number of peptide chain and also type of peptide chain present. Depending upon the nature of amino acid, may be acidic or basic. Isoenzymes are product of two or more loci. These are different molecular forms of an enzyme with same function and substrate specificity. But they show age, tissue and developmental specificity (Markert and Mollar, 1959). Electrophoresis offers powerful tool to help in characterization and identification of varieties or parental lines of a hybrid. One variety differs from other varieties in one or more protein or enzyme constituents. In this technique, an electric current is applied across a support media, which may be polyacrylamide gel, starch gel paper or special plastic material. The electric current accelerates differential mobility and consequent separation of compounds such as mixtures of proteins, which are spotted at one side or end of the supported media. Variant of electrophoretic techniques is used to analyze for isoenzymes, which are different molecular forms of protein (enzyme) that exhibit the same basic enzyme activity. In this technique, plant juices or extracts are electrophoresed, *i.e.* subjected to an electric field to separate the isoenzymes, which are followed to act on various substrate materials before the latter are stained. The following properties of isozymes make particularly useful for testing genetic purity of seed (Arus, 1983).

- The expression of isozyme loci is codominant and generally not altered by the environment. Thus, it is convenient to distinguish true hybrids from the contaminants.
- Many isozymes are expressed in seeds/seedlings; hence the test is generally completed within a short time. On the other hand, seed/seedling-morphological markers are very few. Particularly, the outer seed coverings being maternally inherited, it is difficult to distinguish hybrid seed from its female parent.
- More precise than chemical test and less expensive and simpler than molecular markers.
- Verification of correct variety leveling and compliance.
- Electrophoresis of proteins / isozymes is a powerful yet cost-effective tool, it can accurately detect the presence of unwanted plant types in hybrid and inbred seed. Seed sellers and buyers can use this information to verify the seed products for deviation from the labeled / prescribed purity standards and expected performance levels in customers fields. Isozymes loci are usually the markers of choice for identifying and quantifying off types in inbred lines and selfs and off types in single cross hybrid population of tomatoes, melons, *etc.*
- Molecular or biochemical markers can be of great value to seed purity maintenance in larger interest provided the techniques/ facilities are available for: (i) highly accurate and cost-effective determination of levels of inbred selfing and hybrid out - crossing, unbiased by environmental stress factors or lack of phenotypic expression, (ii) obtain accurate genetic purity of information before seed conditioning beings, (iii) quick confirmation of variety leveling, (iv) verification of plant variety in field environment to resolve customer complaint issues and (v) use of Breeders Seed Purification Programme to eliminate impurities in foundation lines. Biochemical markers used in identification of parental lines and genetic purity testing of some vegetables crops are given in Table 1.

Table 1: Biochemical markers used in identification of parental lines in vegetable crops

Crop	Marker Systems and Purpose
Tomato	EST and ADH (alcohol dehydrogenese)
Pepper	Total seed protein polymorphism (12 bands with 22-98 Kda and POX)
Radish and cabbage	EST, GOT (glutamate oxalo-acetate transaminase), PGM
Pea	Protein poly morphism gave 70-96% similarity among cultivars
Beans	EST, GOT, 6-PGD
Chinese cabbage	PGM, POX, EST, 6-PGD
Muskmelon	Peroxidase, shikimate dehydrogenase, acid phosphatase
Lettuce	Protein polymorphism

Source: Rao *et al.* (2004).

Polyacrylamide Gel Electrophoresis (PAGE)

In this case, support matrix used for carrying out electrophoresis is polyacrylamide gel, which is a polymer of acryl amide monomer. The polymerization takes place in presence of an initiator TEMED, catalyst-ammonium persulphate and cross linker methylene bis-acrylamide. PAGE can be performed either in a denatured system (*i.e.* SDS-PAGE) or in a native system (*e.g.* alkaline or acid PAGE), depending upon the kind of proteins used. PAGE of salt soluble proteins has been used for genetic purity assessment in maize. Isozymes analysis has been used for testing genetic purity in white cabbage (Baranski, 2000).

Isoelectric Focusing (IEF)

IEF is separation of amphoteric substances in aqueous solution, based on their isocentric nature. The separation of proteins mixed with carrier ampholytes takes place on a matrix between two electrodes. As a result of electrolysis a pH gradient is created between the two electrodes, which is stabilized by the buffering of ampholytes. Proteins along with ampholytes get fixed when it reaches the isoelectric point, where the net charge becomes zero. IEF of storage proteins has been used for testing hybrid purity in tomato and varietal purity in potato & peas.

DNA-based Markers

DNA-based markers such as RFLP, RAPD, AFLP and SSR are used in genetic purity assessment (McDonald, 1988). There are following advantages of DNA-based markers over proteins/isozymes.

- *Direct Analysis of Genome:* Inspite of great interest in protein/isozyme marker systems, the major limitations is non-detection of mutation, may occur in non-coded part of genome and will not be translated into protein. But DNA marker could overcome this problem.
- *No Tissue and Age Specificity:* Proteins/isozyme profile varies if taken from different tissue and at different age, which may be due to post translational changes. But this is not the case of DNA markers.
- *Independent of Environment:* Proteins/isozyme profile vary if taken on normal and stress condition. But this is not so in case of DNA-based marker.
- *Unlimited Polymorphism:* Proteins/isozyme as there are functional in nature so polymorphism obtained is less compared DNA-based marker.

Restriction Fragment Length Polymorphism (RFLP) Marker

It denotes that a single restriction enzyme produces fragments of different lengths from the same stretch of genomic DNAs of different strains or related species. RFLPs are detected as follows: (i) large molecular weight genomic DNAs are isolated from several strains or related species, (ii) these DNAs are then digested with the same selected restriction enzyme, (iii) fragments present in these digests are separated through electrophoresis, (iv) the resulting gel are transferred to a suitable solid support and exposed to a suitably labeled appropriate DNA probe under conditions favouring DNA: DNA-hybridization (southern hybridization), and (vi) the bands to which the probe has hybridized are detected by filming them on a suitable photo film through radioautography.

A restriction fragment length polymorphism (RFLP) is detected when autoradiographed band migrates to two more different positions (in different lanes of the gel). The pattern of RFLPs generated will depend mainly on the following: (i) the differences in DNAs of selected strains/species, (ii) the restriction enzyme used and (iii) DNA probe employed in southern hybridization. Detectable RFLPs are generated due to following changes in the DNA of an organism: (i) changes in the base sequence of recognition sites of the restriction enzymes, (ii) relatively large deletions and (iii) relatively large additions in the concerned region of the genomic DNA. A very large number of restriction enzymes are now available permitting the selection of such enzymes that would generate RFLPs. The DNA probes may be obtained from (i) genomic libraries, (ii) cDNA libraries obtained from addition/ substitution lines, flow stored chromosomes, chromosome-specific repeated sequences or micro-dissected chromosomes. Single-copy sequences (representing, most likely, structural genes) are the best probes, but low copy and even multiple copy sequences are also used. Generally, probes prepared from the same species are used, but those from other species may also be employed.

The RFLP is detected as a differential movement of a band on the gel lanes from different species/strains; each such band is regarded as a single RFLP locus. It may be noted that an RFLP locus is definable only by the combination of a specific restriction enzyme with a specific DNA probe. The linkage among different RFLP loci and that between RFLP loci and oligogenes/polygenes is readily determined by studying (i) a set of recombinant inbred lines derived from a suitable cross, or (ii) F_2 or backcross progeny from such a cross. (a suitable cross means the cross between two strains differing for concerned RFLP loci and /or oliogenes/polygenes). RFLP maps similar to the conventional linkage maps can be readily prepared, and these can be effectively integrated with the genetic maps prepared conventionally. Efficient protocols for such mappings have been developed. The RFLP maps can be assigned to specific chromosomes/chromosome arms based on: (i) linkage with genetic markers already assigned to specific chromosomes, (ii) use of addition/ substitution lines, (iii) study of suitable translocation stocks, (iv) employing monosomic/ trisomic lines and (v) *in situ* hybridization to polytene chromosomes.

Advantages

- The number of RFLP loci is very large so that even very small segments of chromosomes can be mapped; ideally every cistron could be mapped.
- Mapping does not necessarily depend on gene function.
- Even quantitative trait loci can be mapped, which is virtually impossible through conventional techniques.

- It is astoundingly rapid as compared to conventional linkage mapping.
- Fewer individuals (25-50 individuals/F_2 generations) need to be studied. However the technique is 100 to 1000 times as costlier as the conventional linkage mapping.

Applications

- Identification and isolation of a gene known to be tightly linked with an RFLP locus.
- Fingerprinting of strains/varieties for their unequivocal identification.
- Linkage mapping of quantitative trait loci.
- Identification of the most important loci affecting a quantitative trait.
- Highly efficient indirect selection for tightly linked quantitative trait loci and even for those oligogenses direct selection, which may either be difficult or costly, give best results.
- Determination of chromosome segments alteration for which is likely to yield the best results.
- Establishing the relationship among various strains/species.
- Understanding the identity and function of thus for 'mysterious' polygenes *etc.* RFLP maps are being generated several crop species/ genotypes including analysis of hybrid pepper (Livneh *et al.,* 1990).

RFLP Linked Mapping in Vegetable Crops

A high degree of unique identification has been made among cultivated varieties and inbred lines of many species. These are possible by means of one or more profiling methodologies. For those several crop species for which sufficient inherent variability already exists, genetic profiling for many breeding application, measurement of genetic diversity, minimum distance assessment and variety identification are not only an economic and practical feasibility but also a necessary component. A list of vegetable crops where, RFLP maps have been prepared and RFLP linked mapping genes of economic value in vegetable crops is given in Tables 2 and 3, respectively.

Table 2: A list of vegetable crops where RFLP map have been prepared

Crop	Gametic chromosome (n)	Number of RFLP loci	Map length (cM)	Genome size (bp)	Resolution bp/cM
Tomato	12	>1000	1300	715×10^6	0.550×10^6
Potato	12	350	690	$750\text{-}1000 \times 10^6$	$>1.0 \times 10^6$

Table 3: Some examples of RFLP linked mapping of major genes of economic value in some crops

Crop	Gene Located on RFLP Map	Lines Used
Tomato	*tm-2a locus* for resistance to TMV	Near isogenic lines (NILs)
	Gene 12 and II for resistance against *Fusarium oxysporium*	Near isogenic lines and BC population.
	Gene *Mi* for resistance against root- knot nematode (*Meloidoyne* spp.)	Near isogenic lines (NILs)
	Gene sm for stemphylium resistance	Near isogenic lines (NILs)
Potato	Gene for resistance against cyst nematode *(Globodera rostochiensis)*	Two diploid clones
	Two gene (R x 1, R x 2) for resistance against potato virus x (PVX)	Two F_1 populations

Random Amplified Polymorphic DNAs

Random amplified polymorphic DNAs (RAPDs) are detected by repeated cycles of DNA denaturation-renaturation-DNA replication in a PCR (polymerase chain reaction) equipment. The genomic DNA of selected pair of oligonucleotides (a short nucleotide), and this mixture is then used for DNA amplification in a PCR. The segment of genomic DNA amplified (replicated repeatedly) will be flanked by the sequences that are complementary to the two primers used in the mixture; one primer will be complementary to one end of this segment, while the other will be complementary to other. Those strains that have a segment of DNA, the ends of which are homologous to oligonucleotides used as primers, a distinct band will be seen on the agarose gel, while this band will be absent in other strains. Thus, RAPDs bands are either 'present' or 'absent' in a strain and the F_1s will always show the band (dominant expression). In contrast, RFLP bands are detected by their differential migration in parents (codominant expression). RAPDs analysis if facilitated by the availability of base sequence of a short segment of genomic DNA, variation in this segment can now be detected by generating an oligonucleotide that is homologous to it. But oligonucleotides having any base sequence can be used in these studies. The oligonucleotides can be prepared using an oligonucleotides synthesizer.

The information obtained from RAPDs can be used for linkage mapping much in the same manner as that from RFLPs. It is important that RAPDs approach does not utilize restriction enzymes (which are very costly), and probes of any kind (this saves the effort needed for developing suitable probes, and the risks of handling radio labelled materials). In the PCR utilizes it DNA polymerase, which is resistant to high temperatures needed for DNA denaturation, so that, it is reused repeatedly during DNA amplification. Thus, RAPDs represent savings in both cost and effort, and their use is expanding rather rapidly. The RAPDs have the same applications as the RFLPs. However, doubts have been expressed about the reproducibility of RAPDs. In view of this, alternative PCR-based markers have been and are being developed. These markers are superior to RAPDs in one or more aspects. A polymorphism in a restriction pattern occurs when the mutation of a single base pair results in the loss or creation of a new restriction site or when by insertion/deletion, the size of a restriction fragment is altered. These alterations are detected on an autoradiography, when these fragments bind the hybridization probe. The polymorphism between individuals result from sequence differences in primer binding sites and are visible as presence or absence of a particular RADP band. (Williams *et al.,* 1990). Based on this principle RAPD marker has been used for assessing purity of F_1 hybrid seed of tomato by Rom *et al.* (1995), broccoli by Crocett *et al.* (2002) and Chinese cabbage by Meng *et al.* (1998).

Amplified Fragment Length Polymorphism (AFLP) Markers

Though RFLP markers are reliable, time consuming, technically demanding and not amendable to automation and random amplified fragment length polymorphism (RAPD) markers based upon PCR are rapid, less expensive and highly sensitive to reaction conditions, but Vos *et al.* (1995) and Antonius, (1999) have developed a new set of markers termed Amplified Fragment Length Polymorphism (AFLP) that combines the advantages of both RFLP or even RAPD as well as PCR based markers. It has the reliability of RFLP and efficiency of PCR. In addition, its ability to detect genetic variations is much more than RFLP or even RAPD. The technique is based upon PCR amplification of subsets of genomic DNA, which have been digested with restriction endonucleases. Double stranded adapters are ligated to the ends of the digested fragments. These adapters and ends of restriction fragments together provide annealing sites for the first amplification known as preselective amplification. This is followed by selective amplification wherein a different set of primer having additional selective nucleotides at the three ends are used to initiate DNA synthesis. This step results in the amplification of only a subset of restricted fragments because the selective nucleotides extend into the restriction fragments and only those fragments having complementary with the selective ends serve as templates. Each amplification reactions results in generation of large number of fragments. These fragments are resolved on denaturing polyacrylamide gels and 50-100 bands per individual samples are produced. AFLPs can be conveniently visualized with the help or radio active or fluorescence levels, or alternatively by silver staining (Hartl and Seefelder, 1998). AFLP marker has been used for genetic purity testing of *Brassica* spp. Following techniques are used:

Restriction Digestion of DNA and Ligation of Oligonucleotide Adapters

A pure, high quality and native genomic DNA preparation is used for generating restriction fragments, one being a rare cutter and other frequent cutter. The restriction fragments obtained have a rare cutter sequence on one end and a frequent cutter sequence on the other end. These fragments are ligated with double stranded adapters. These adapters consist of a core sequence and an enzyme specific sequence. Adapter ligated restriction fragments provide annealing site to AFLP primers. Therefore, this strategy permits the amplification of a large number of fragments without having the knowledge of their sequence.

Pre-selective Amplification

The pre-selective amplification involves amplification of the adapter ligates restricted fragments using low primers, one corresponding the frequent cutter and other corresponding to the rare cutter restriction endonuclease. These primers are not radioactively labeled. The products of this amplification serve as templates for the subsequent amplification after dilution.

Selective Amplification of Restriction Fragments

Selective amplification is performed using different sets of primers that extend into the restricted fragments. These primers contain 0-3 selective nucleotides at the three ends. One of the two primers used in this step is labeled. These primers initiate DNA synthesis only form a subset of restriction fragments, which have matching nucleotides. The number as well as composition of selective nucleotides in the primers used for selective amplification is important factors in determining the number of restricted fragments getting amplified.

Gel Analysis of Amplified Fragments

After selective amplification, the reaction products are separated using denaturing polyacrylamide gel electrophoresis and the final banding pattern is obtained by autoradiography by non-radioactive method.

Simple Sequence Repeats (SSR) Marker

The genomes of higher organisms contain arrays of repeated sequences that display length variations. These repeat sequences are Simple Sequence Reports (SSR). The core sequences and SSRs are di, tri and tetra nucleotide repeats. For example AT, AC, AG, CG, ATT, AAC, AGG, AATC, AAGG *etc*. SSRs are also referred as micro satellites (STMS).

PCR Analysis of SSR Loci

PCR analysis on SSR loci requires identification of DNA sequences at flanking regions and desirable suitable primers to specifically amplify the repeat region of interest. According to Smith and Register (1998), to make any profiling technology suitable for use in genetic purity assessment, the following conditions must be fulfilled.

- The degree of discrimination among varieties must be at least 75% and preferable higher.
- Unlinked loci, preferable on different chromosomes should be assayed.
- At least 10 unlinked loci may be preferred. In case of EDVs at least 60-100 loci are to be assayed.
- The data must be reproducible and easy to generate, score and store.
- Sample throughout and cost must be 2-5 times better than the conventional approaches.
- Methods must be standardized across laboratories on an international basis.
- It would be desirable for technology to be both cost-effective and capable of providing more detailed genetic data.

Microsatellite or Minisatellite Markers

These are multi-locus probes belonging to repetative DNA group and are usually non-species specific, occurring ubiquitously (Joshi *et al;* 1999). Their analysis procedure has been derived from RFLP analysis. For visualization, specific fragments are first hybridized with a labelled micro or minisatellite probe and fingerprints generated by these probe are termed as oligonucleotide fingerprints. Microsatellites are short tadem repeats (STRs) or simple sequence repeats (SSRs), comprising about 1-6 bp long monomer sequences which are repeated several times. As the tandem repeats, present on these loci, vary in number of repeat units between genotypes/ cultivars, these are termed as variable number of tandem repeats (VNTRs) which can be defined as single locus containing variable number of tadem repeats between units. Similarly, numerous loci containing tadem repeat within a genome, as hypervariable regions (HVRs). Complex banding pattern exhibited by micro and minisatellites and simultaneously detection of multiple DNA loci makes them an ideal marker system for cultivars identification. Besides this, these probe act as dominant finger printing markers as well as codominant sequence microsatellite site (STMS) markers.

Comparative Advantages of Common Molecular Techniques

Among molecular markers, isozyme markers have proven their usefulness already in genetic studies and breeding programme of various crops (Tanksley and Rick, 1980; Tanksley and Orton, 1983). For tomato, an isozyme linkage has been constructed contains 41 isozymes genes corresponding to 15 unique enzyme reactions. Thirty-six of these genes have been mapped to their respective chromosomes (Tanksley and Bernatzky, 1987). Thus, tomato isozymes have served many purposes, *e.g.* verifying the purity of hybrid seed (Tanksley and Jones, 1979), tagging genes of economic importance (Rick and Fobes, 1977; Tanksley *et al.,* 1984), detecting genes and chromosomes from wild species following introgression (Rick *et al.,* 1986), identifying somatic hybrids and mapping genes underlying quantitative variation (Vallejos and Tanksley, 1983). Nevertheless, isozyme loci suffer several limitations, which are not inherent to the use of DNA markers. All the modern molecular techniques belong to the basic approaches: southern hybridization or polymerase chain reaction (PCR). The southern hybridization based approach involves restriction digestion of genomic DNA; size fractionation of digested DNA through gel electrophoresis; transfer of fractionated DNA fragments on nylon membranes; and finally hybridization with labeled probe sequences to visualize DNA polymorphisms. RFLP (Restriction Fragment Length Polymorphism) is the most widely used techniques under this approach. PCR based approaches involve, amplification of specific/random region's of genomic DNA using primers of different specificities, through polymerase chain reaction followed by visualization of amplified products on gel after staining with ethidium bromide (EtBr) or fluorescence. The differences between various molecular techniques used for testing genetic purity of hybrids are given in Table 3.

Sample Size for Genetic Purity Testing

A key step in the seed production process is the design of sampling and testing procedure used to evaluate purity of seed lot. However, due to uncertainties in such methods, there is always a risk of incorrectly rejecting or accepting a seed lot. To manage this misclassification risk, especially with regard to the presence or absence of transgenic traits, certain sampling plan and formulae for determining the sample sizes necessary to reduce the size to the

Table 3: Differences in various molecular techniques

CHARACTERS	RFLP	RAPD	Microsatellites	AFLP
Principle	Endonuclease restriction, southern blotting hybridization	DNA amplification with random primers	PCR of simple sequence repeats	Endonuclease restriction and PCR amplification
Type of polymorphism	Single base changes, insertions and deletions	Single base changes, insertions and deletions	Changes in length of repeats	Single base changes, insertions and deletions
Genomic abundance	High	Very high	Medium	High
Level of polymorphism	Medium	Medium	High	High
Dominance	Co-dominant	dominant	Co-dominant	Dominant
Amount of DNA required	2-10 μg	10-25 μg	50-100 μg	2-10 μg
Sequence information required	No	No	Yes	No
Radioactive detection required	Yes/no	No	No	Yes
Development costs	Medium	Low	High	Medium
Start-up costs	Medium/high	Low	High	Medium

Source : Singh *et al.* (2001).

lowest possible and to control these misclassification errors when accepting or rejecting seed lots are required to be worked out. In this regard, sequential sampling procedures would offer satisfactory solutions. Conventionally, for testing genetic purity of hybrid seed lot, for assessing one percent off type plants, maximum permissible limit, is at least 400 plants needed to be tested under field plot grow out test (GOT). That case, an equal number of seeds (*i.e.* 400) therefore, need to be tested individually in electrophoratic analysis. However, for total replacement or supplementing fields grow-out test by electrophoretic analysis, such huge sample size will be a major constraint. The cost involved in electrophoretic analysis of 400 individual seed would be a limiting factor. However, sequential analysis technique may be followed for reducing the sample size to a batch of 20 or more, which would mean up to 95 % saving in work load, if specification is predetermined. Various research workers have worked out on the smallest individual seed samples for assessing genetic purity in different crops, *viz* 10 - 20 seeds in pepper, 25-30 in Chinese cabbage, 40 in tomato and 400 seeds in chicory have been reported to provide reliable assessment of genetic purity (Rao *et al;* 2004).

8

BIOTECHNOLOGY AND VEGETABLE IMPROVEMENT

Origin, History and Definition

The term biotechnology was coined by Karl Ereky, a Hungarain engineer, in 1919. The origin of biotechnology can be traced back to prehistoric time when farmers and gardeners have been making use of hybridization and grafting among plants with a view to bring about an improvement in existing attributes. Using these approaches, several hundred varieties in ornamental plants were produced. Microorganisms were already used for processes like fermentation, formation of yoghurt and cheese from milk, vinegar from molasses, production of butanol and acetone from starch by *Clostridium acetobutylicum* or the production of antibiotics like penicillin from *Penicillium notatum*. Four defferent phases of biotechnological research were witnessed during the last century. In 1920, Chain Weizmann used *Clostridium acetobutylicum* to convert starch into butanol and acetone (during world war I, acetone was used as an essential component of explosives). Alexander Fleming in 1929 discovered penicillin from *Penicillium notatum* and wounded soldiers were given treatment using this antibiotic during world was II. In 1970s with the discovery of restriction enzymes, which led to the development of recombinant DNA, technology brought a revolution in the field of biology. The latest phase in the history of biotechnology is the study of structure and function of whole genome (DNA) and whole proteome (protein). This area of research is described as genomics and proteomics, respectively. Some more important activities in the field of biotechnology are given in Table1

In fact, biotechnology is the product of interaction between science of biology and technology. It is the technological exploitation and control of biological system. Biotechnology has wide range of its uses and accordingly Office of Technology Assessment described biotechnology as: (i) old biotechnology that includes fermentation, *i.e.* formation of vinegar from molasses and that of cheese and yoghurt from milk, production of antibiotics from certain fungi and process of baking and brewing and (ii) new biotechnology that includes the techniques of recombinant DNA and PCR polymerase chain reaction, cell culture and fusion and bioprocessing which became possible only through the researches in molecular biology. However, no efforts have been made to distinguish between old and new biotechnology. Keeping in view the importance of biotechnology, attempts have been made to define it. Following are some of the definitions of biotechnology, however, none of the definitions of biotechnology seems to be complete. The first definition seems to be comprehensive and wide.

- Biotechnology is any technique that uses living organisms or substances from these organisms, to make or modify a product to improve plants or animals or to develop microorganisms for specific uses (Office of Technology Assessment of United States Congress).

- Biotechnology is the controlled use of biological agents, such as microorganisms or cellular components, for beneficial use. (US National Science Foundation).
- Biotechnology is a technology using biological phenomena for copying and manufacturing various kinds of useful substances (Japanese Biotechnologist).
- Biotechnology is the application of biological organisms, system or process to manufacturing and service industries (British Biotechnologist).
- Biotehnology is the application of microorganisms and biological systems to the production of goods and services that are beneficial to human welfare (Malik, 1989).

Table 1: Biotechnological activities

Activity	Year
Animal biotechnology	
Yeast used to make wine and beer	Before 6000 BC
Yeast used for making leavened bread	About 4000 BC
Copper mined with the aid of microbes (Rio Tinto. Spain)	Before 1670 AD
Sewage treatment systems using microbes developed / established	About 1910 AD
Large-scale production of acetone, butanol and glycerol using bacteria	1912-1914
Large-scale production of penicillin	1944
Mining of uranium with the aid of microbes (Canada)	1962
First successful genetic engineering experiments	1973
Marketing of human food of fungal origin (UK)	1980
The use of monoclonal antibodies of diagnosis approved in USA	1981
Approval for the use of insulin produced by genetically-engineered microbes (GEMs) (USA and UK)	1983
Animal interferons, produced by GEMs, approved for the protection of cattle against diseases	1984
Conceptualization of human genome project	1986
Launch of human genome project	1990
Development of human genetic map for all 23 chromosome	1992
Dolly, a cloned sheep produced	1997
Drosophilla genome sequence published	2000
Work draft of human genome sequence published	2001
Rat genome sequence published	2004

Activity	Year
Plant biotechonology	
First attempt of plant tissue culture	1902
Embryo culture of selected crucifers attempted	1904
Asymbiotic germination of orchid seeds *in Vitro*	1922
In vitro culture of root tips	1922
Use of embryo culture technique in interspecific crosses of *Linum*	1925
Successful culture of tomato roots	1934
In vitro culture of cambial tissues of ulmus to study adventitious shoot formation	1940
Use of coconut milk containing a cell division factor for the first time in *Datura*	1941
In vitro culture of crown gall tissues	1941
In vitro adventitious shoot formation in tobacco	1944
Raising whole plants of *Lupinus* and *Tropaeolum* by shoot tip culture	1946
Use of meristem culture to obtain virus-free Dhalias	1952
First application of micrografting	1952
Production of haploid callus of the gymnosperm *Ginkgo biloba* from pollen	1953
First plant from a single cell	1954
Discovery of kinetin a cell division hormone	1955
Discovery of regulation of organ formation by changing the ratio to auxin: cytokinin	1957
Regeneration of somatic embryos *in-vitro* from the nucellus of *Citrus ovules*	1958
Regeneration of embryos from callus clumps and cell suspensions of *Daucus carota*	1959
Publication of first handbook on plant tissue culture	1959
First successful test tube fertilization in *Papaver rhoeas*	1960
Use of microculture method for growing single cells in hanging drops in a conditioned medium	1960
Enzymatic degradation of cell walls to obtain large number of protoplasts	1960
Filtration of cell suspensions and isolation of single cells by plating	1960
Development of Murashige and Skoog nutrition medium	1962

Activity	Year
Production of first haploid plants from pollen grains of *Datura*	1964
Restriction endonuclease term coined to a class of enzymes involved in cleaving DNA	1968
Selection of biochemical mutants *in-vitro* by the use of tissue culture derived variation	1970
First achievement of protoplast fusion	1970
Discovery of first restriction endonuclease from *Haemophillus influenzae* Rd: It was later purified and named *Hind II*	1970
Preparation of first restriction map using *Hind II* enzyme to cut circular DNA of SV 40 into 11 specific fragments was prepared	1971
Regeneration of first plants from protoplasts	1971
First report of interspecific hybridization through protoplast fusion in two species of *Nicotiana*	1972
First recombinant DNA molecular produced using restriction enzymes	1972
Joining of two restriction fragments regardless of their origin produced by the same restriction enzyme by the action of DNA ligase	1972
Development of a procedure in which the appropriate enzyme can be added to fill in any single-stranded gap and the use of DNA ligase to join two fragments thus giving rise to recombinant DNA	1972
Discovery of reverse transcriptase: In certain cancer- causing animal virus, genetic information flows in the reverse form	1972
Use of Lobban and Kaiser technique to develop hybrid plasmid-insertion of EcoRi fragment of DNA molecular into circular plasmid DNA of bacteria using DNA ligase. Gene from African clawed toad inserted into plasmid DNA of bacteria	1973
Cytokinin found capable of breaking dormancy in excised capitulum explants of Gerbera	1973
Regeneration of haploid *Petunia hybrida* plants from protoplasts	1974
Biotransformation in plant tissue cultures	1974
Discovery that the Ti plasmid is the tumor inducing principle of *Agrobacterium tumefaciens*	1974
Positive *in vitro* selection of maize callus cultures resistant to toxin of *Helminthosporium maydis*	1975
Development of the high resolution two dimensional gel electrophoresis procedure, which led to the development of proteomics	1975
Shoot initiation from cryopreserved shoot apices of carnation	1976

Activity	Year
Octopine and nopaline synthesis and breakdown found to be genetically controlled by the plasmid of *A. tumefaciens*	1976
Successful integration of the Ti plasmid DNA from *A. tumefaciens* in plants	1977
A method of gene sequencing based on degradation of DNA chain.	1977
Discovery of split genes	1977
Somatic hybridization of tomato and potato resulting in pomato	1978
Cocultivation procedure developed for transformation of plant protoplasts with *Agrobacterium*	1979
Use of immobilized whole cells for biotransformation of digitoxin into digoxin	1980
Commercial production of human insulin through genetic engineering in bacterial cells	1980
Restriction fragment length polymorphism (RFLP) technique developed	1980
Studies on the structure of T-DNA by cloning the complete EcoRI digest of T-37 tobacco crown gall DNA into a phage vector, thus allowing the isolation and detailed study of T-DNA border sequences	1980
Introduction of the term somaclonal variation	1981
Incorporation of naked DNA by protoplasm resulting in the transformation with isolated DNA	1982
Polymerase chain reaction (PCR), a chemical DNA amplification process idea conceived	1983
A Murray, J Szostak constructed the first artificial chromosome	1983
Transformation of tobacco with *Agrobacterium*; transgenic plants developed	1984
Alec Jeffreys developed the technique of DNA fingerprinting	1984
Development of the genetic fingerprinting technique for identifying individuals by analyzing polymorphism at DNA sequence level	1984
K Mullis and co-workers invented polymerase chain reaction (PCR)	1985
TMV virus- resistant tobacco and tomato of coat protein gene of TMV	1986
Development of biolistic gene transfer method for plant transformation	1987
Isolation of Bt gene from bacterium (*Bacillus thuringensis*)	1987
KA Barton isolated Bt gene from Bacillus thuringensis	1987
Formal launch of the Human Genome project	1990

Activity	Year
Development of the random amplified polymorphic DNA (RAPD) technique	1990
JGK Williams *et al.* Developed the technique of RAPD	1990
Development of DNA microarray system using light directed chemical synthesis system	1991
Reporting by the Institute for Genomic Research of the complete DNA sequence of *Haemophilus influenzae*	1995
Development of DNA fingerprinting by amplified fragment length polymorphism (AFLP) technique	1995
M Schena developed DNA chip "microarray" system	1995
P Vos developed AFLP technique	1995
Calgene released first transgenic crop (flavr savr tomato) for commercial cultivation	1996
Sequencing of *E. coli* genome	1997
Sequencing of the genome of a multicellular organism (*Caenorthabditis elegans*)	1998
'International Rice Genome Sequencing Project' (IRGSP) launched	1998
Sequenceing of human genome successfully completed	2001
Syngenta and BGI published draft rice genome sequences	2002

Applications

In coming years, biotechnological tools have to play a very crucial role in various ways including development of specific parental lines for hybrids in vegetables. Some of the potential thrust areas where biotechnology can have additional aid to conventional hybrid breeding are (Kalloo *et al.*, 2001):

- Molecular markers can be used to identify genic male sterile lines at seedling stage, identified male sterile plants can serve the purpose of female parent in F_1 seed production field. For this purpose, polymerase chain reaction (PCR) based markers seem feasible particularly in vegetables having less plant population per hectare (*e.g.* cucurbits). However before utilization, cost-effectiveness of such techniques needs to be worked out.
- For developing multiple biotic stress resistant lines, already available molecular markers having establish linkage with stresses may be used to enhance breeding programme for development of parental lines (Appendixes 1-10).
- In vegetable crops, where male sterility system is not available, *e.g.* okra, male sterility system can be developed through genetic transformation. There are a number of

molecular strategies, which can be used to develop male sterility system depending upon the availability of gene constructs and plant materials (Kumar *et al.*, 2000).

- Anther culture technique can be utilized for development and maintenance of self-incompatible lines in cole vegetables and also to develop inbred lines in other cross-pollinated vegetables.
- Protoplast culture and fusion technique may be used to transfer traits of interest, *e.g.* cytoplasmic male sterlity cytoplasm, resistant gene (s) for biotic stresses inside the parental lines from cross-incompatible genus/texa. Embryo culture technique can also be used for the same purpose.
- Most of the genic male sterility is controlled by single recessive gene, however, dominant gene controlling male sterility are also available in some vegetables, *e.g.* cabbage. Advancement of biotechnology has made possible to utilize such dominant male sterility for F_1 seed production through micropropagation in those vegetable crops where vegetative part is consumed. In the first step, male sterile line with homozygous dominant gene can be multiplied by clones and then crossed as female sterile line. In second step, seeds of male sterile lines can be used to raise female parent seedlings for hybrid seed production programme (Fang *et al.,* 1997).
- Genetic transformation technique can be used to develop trait specific such as to prevent and minimize post-harvest losses due to over-ripening, physical damage, chemical injury and pathological decay.
- Transgenic parental lines could be developed for enhanced nutritional value of vegetable hybrids.
- Senescence free transgenic parental lines of hybrids could be developed.
- Biotic stress, *viz.* salt, drought, cold and heat tolerance transgenic parental lines of hybrids could be developed.
- Biotechnology can play a role in the development of apomictic hybrids of vegetables.

Approaches

There are two basic technique used in plant biotechnology. These are: (1) plant tissue culture and (2) genetic engineering.

Plant Tissue Culture

In broad sense, *it is the cultivation of plant organs, tissues or cells in test tubes on artificial media: often the techniques of plant cell and tissue culture are called in-vitro techniques*. The *in-vitro* techniques were developed initially to demonstrate the totipotency led to the development of techniques for cultivation of plant cells under defined conditions, that is, *tissue culture*. This was made possible by the brilliant contributions from R.J. Gautheret in France and P.R. White in USA during the third and the fourth decades of 20th century. White reported shoot bud regeneration from tobacco suspension cultures in 1939, but the first plant from a mature plant cell was regenerated in 1959 by Brawn. Most of the modern tissue culture media have been derived from the work of Skoog and co-workers during 1950s and 1960s. Hanning did the first embryo culture work in 1904 that cultured mature embryos of some crucifers; Laibach soon applied the technique in 1925 to recover hybrid progeny from an interspecific cross in *Linum*. Subsequently, contribution from several workers led to the refinement of this technique. Maheshwari and Guha first produced haploid plants from pollen-grains in 1964 by culturing anthers of Datura. This marked the beginning of anther culture or pollen culture for the production of haploid plants. Many workers, more notably by J.P. Nitch, C. Nitch and coworkers, have further developed the technique. These workers have been shown that isolated microspores of tobacco produce complete plants. Plant protoplasts are naked cells from which cell-wall has been removed. In 1960, Cocking produced large quantities of protoplasts by using cell-wall degrading enzymes. The techniques of protoplasts production have now been considerably refined. It is now possible to regenerate whole plants from protoplasts and also to fuse protoplasts of different plant species. There are several techniques which are used in tissue culture are described in below:

Plant Regeneration (Mass Propagation of Inbreds/hybrids)

Some breeding materials, *i.e.* male sterile lines, self-incompatible lines, inbred lines and haploids *etc.* are not easily propagated by seed, therefore, tissue culture method is employed for their propagation. The tissue culture method for regeneration, supplements the conventional breeding methods. Plants can be propagated asexually on a large-scale under *in-vitro* conditions either directly from apical, auxillary and adventitious buds or indirectly through somatic embryogenesis. Micropropagation is very useful especially to propagate superior manually produced hybrids for several generations without any genetic change. So for 200 crop species have been micropropagated. It might be suitable for clonal propagation of F_1 hybrids in crops like cabbage and their parental *SI* or *CMS* lines on a large-scale. Until recently, auxillary shoot production has been found the most advanced method of micropropagation but since the mid 80's interest in somatic embryogenesis has increased tremendously with the success achieved in many crops. Now production of somatic embryos and their conversion into artificial or synthetic seed appears as an efficient system of propagation of true-to-type hybrids. By propagating male sterile and male fertile or self-sterile parental lines through synthetic seeds, a pollination control system can be introduced in the plant population. A small number of hybrids can be produced by hand-pollination and subsequently mass multiplied by artificial seed technology. Thus, making it possible to exploit hybrid vigour in crops like beans where the production of hybrid seed is very difficult and reduces the cost of hybrid seed. Information is available regarding use of micropropogation in vegetable crops which can also be used in hybrid seed production which are summarized below:

Solanaceous Vegetables

Tomato

Geetha *et al.* (1998) used seedling explants and multiplication was achieved on MS medium with 20 μM NAA. hypocotyl and cotyledonary explants are the most preferred explants in tomato micropropogation. A shoot regeneration protocap has been designed by Gamburg *et al.* (1997) where nodal portion and shoot explants were used, giving rise true-to-type tomato plants bearing normal flowers. The variation occurs where callus culture takes place but Mirghis (1995) produced true-to-type, normal plantlets using culture with apical and axillary buds. Other reports on use of micropropagation in tomato are given in Table 2.

Table 2: Micropropagation in tomato (*Lycopersicon esculentum*)

Explant	Medium	Response	Reference
Cotyledon	2.5µM NAA+ 1.5 µM Kin	Normal plantlet regeneration	Jawahar *et al.*, 1998
Seedling leaf	20µM BA+ 1µM NAA	True to type	Geetha *et al.*, 1998
Node / shoot/ shoot tips	MS major salt + B5	Flowering plantlets	Gamburg *et al.*, 1997
Hypocotyl	BA	Shoot proliferation	Newman *et al.*, 1996
Apical shoot	Various regulators	True to type	Mirghis *et al.*, 1995
Cotyledon	0.1 µM Zeatin	Shoot proliferation	Ichimura and Oda, 1995
Hypocotyl/ cotyledon	1µ M NAA+ 1µM Seating	Normal plantlet regeneration	Ye *et al.*, 1994
Cotyledon	6 µM IAA+ 5µM kin	High yielding rapid multiplication	Farash *et al.*, 1993
Shoot / shoot tips	4 µM BA+ 4 µM IAA	Normal plantlet regeneration	DeVilliers *et al.*, 1993

Source: Kumar *et al.* (2003).

Brinjal

It is very good material for tissue culture studies as it can be cultured in semi-solid as well as liquid medium without loss of regeneration ability. Like other Solanaceous vegetables in brinjal also hypocotyls and cotyledon are most utilized explants. Sharma *et al.* (1995) succeeded in regenerating eggplant on MS medium with 11.1 µ M BA and 2.9 µM IAA. Bhansali and Ramavat (1993) culture plants of *S. melongena* and shoot proliferation were obtained by placing axillary buds on woody plant medium with 0.1 mg/l BA. Other informations regarding use of micropropogation in brinjal are given in Table 3.

Table 3: Micropropagation in brinjal (*Solanum melongena*)

Explant	Medium	Response	Reference
Leaf explant	2.9 µM Kin	High frequency of plantlet regeneration	Raghuveer *et al.*, 1994
Cotyledon	3.5µM IAA + 0.7 µM Zeatin	Shoot formation	Brsancha *et al.*, 1990
Cotyledon, hypocotyl leaf	11.1µM BA+ 2.9 µM IAA	Plantlet regeneration	Sharma *et al.*, 1995

Source: Kumar *et al.* (2003).

Chilli

Efficient micropropagation system in chilli serves for propagation of elite plants like male sterile plant, F_1 plant displaying heterosis etc. In recent past, numerous, studies have been done on propagation via shoot organogensis induced on explant prepared from seedling at different stages. In a few cases shoot generation on a medium devoid of any growth regulators have observed (Ezura *et al.*, 1993 and Binzel *et al.*, 1996). Usually in all papers cited there after micropropogation has been induced and accomplished by exogenous growth regulators supplied to medium (Table 4). The maximum utilized explant consists of half seed explant (Rogoniska and Drozdowska, 1996), cotyledon and hypocotyls (Ramirez and Ochoa, 1996). Commonly explants were placed onto an agarfied shoot induction medium supplemented with cytokinin (BA or TDZ) and often an auxin (IAA, IBA) (Arec and Lopez, 1994). Many times shoot generation was achieved by transferring bud cluster to an elongation medium *in vitro* because, shoot elongation is a problem in chilli cultures (Steinitz *et al.*, 1999). Gibberelic acid is usually added to the elongation medium. Zhou *et al.* (1994) used PD medium with MS inorganic salts and silver nitrate to achieve rapid regeneration of *Capsicum* cultures.

Table 4: Micropropagation in chilli (*Capsicum annuum*)

Explant	Medium	Response	Reference
Apical / axillary buds	3μM BA + 0.9μM IAA/ 5μM BA+ 1.5μM IAA	High frequency of plantlet generation	Mohamd *et al.*, 1998
Apical / axillary buds	3μM BA+ 0.5μM IAA	Rooted plantlets	Gupta *et al.*, 1998
Hypocotyle	MS medium	Shoot proliferation	Rermirez and Ocha, 1996
Immature zygotic embryo	9μM 2,4-D+ 10% CW+ 8% sucrose	Normal plantlet generation	Binzel *et al.*, 1996
Cotyledon	PD+ MS inorganic salt+ ZH organic+ 5 BA+ 0.2 μ M IAA+ 2 $AgNO_3$	High frequency of plantlet generation	Zhou *et al.*, 1994
Shoot / shoot tips	B5+ 300Mg/l chloram ate or B5+ 50μM thiophnate methyl	Rooted plantlets	Ma *et al.*, 1991
Node / axillary bud	Half strength MS medium	Rooted plantlets	Yamamoto, 1994
Shoot/ shoot tips	MS basal medium	Normal plantlet generation	Ezura *et al.*, 1993
Cotyledon	2.5μM TDZ+0.1μM IAA+ 2μM GA_3	Plantlet generation	Arce and Lopez, 1994
Shoot and shoot tips	5μ M BA + 0.5μM IAA	Rooted plant lets	Mirza and Narkhede, 1996

Source: Kumar *et al.* (2003).

Cucurbitaceous Vegetables

In most of the studies with cucurbits species, MS medium has been widely used. Earlier, Kathal (1988) described adventitious shoot formulation from cotyledon explant on a medium containing 1 µM BA and 2ip each. *In-vitro* multiplication of cucumber (*Cucumis sativus*) is well advanced. Navratilova (1987) reported microprogation through leaf, cotyledon, apical meristem and nodal portion (Table 5). Szymanska and Molas (1995) observed direct regeneration of plantlets from leaf explants on MS medium supplemented with 0.5 mg/litre IAA, 2mg/litre BA and 1 mg/litre $AlCl_3$. Burza and Malepszy (1995) used young leaves to develop rapid and efficient regeneration in cucumber. Kumar *et al* (2002) reported micropropagation using apical and axillary bud explant in *Coccinia grandis* and *Trichosanthus dioica*. The micropropagation in squash has been reported by Cassells (1991) through shoot apices. Sheo *et al.* (1991) and Zhao *et al.* (1998) reported micropropagation using shoot tip/ axillary buds with MS medium (1.0 mg/litre BA) in pumpkin. Mythili and Thomas (1999) observed that cucurbitaceous species regenerate directly or indirectly from explant rooted easily when cultured *in-vitro* on MS basal or ½ strength MS basal medium, with 3% sucrose. In most of the species, there is no need of growth regulators but in some cases, a low level of auxin (0.01 NAA/ IAA) is required.

Table 5: Micropropagation in cucurbits

Explant	Medium	Response	Reference
Cucumber (*Cucumis sativus*)			
Leaf explant	0.5 µM IAA+ 2.0 µM BA+ 1.0µM $AlCl_3$	Direct organogenesis	Szymanska and Molas, 1995
Cotyledon/ leaf/ nodal portion	3.0 µM BA	Field transfer of plants	Mishra and Bhatnagar, 1995
Cotyledon/ leaf/ nodal portion	9.0 % sucrose	Somatic embryo	Lou and Kako, 1994
Apical meristem	0.01 µM IBA + 0.1 µM BA+ 10.0 unit Ascorbic acid	Regenerated plants	Kathal *et al*., 1994
Pumpkin (*Cucurbita moschata*)			
Shoot tip/ axillary bud	1 µM BA+ 0.5µM IAA	Direct organogenesis	Zhao *et al*., 1998
Shoot tip/ nodal portion	0.01M BA + 0.1µM NAA	Field transfer of plants	Islam *et al*., 1992
Shoot tip/ nodal portion	1µM BA	Field transfer of plants	Seo *et al*., 1991
Watermelon (*Citrullus lanatus*)			
Axillary buds	2.0 µM BA+ 0.1 µM IAA	Multiple shoot	Spestidis, 1996
Root explant	1.0 µM BA	Field transfer of plants	Kathal *et al*., 1994
Cotyledon	2.0 µM BA	Plants with quality fruits	Song *et al*., 1988

Source: Kumar *et al.* (2003).

Cole Crops

In recent past, many workers have achieved direct regeneration from meristematic tissues and curd pieces. In cauliflower, most responsive explant reported is curd pieces. Vandermoortele *et al.* (1991) achieved multiple shoot from curd pieces on MS medium containing 2.5 µM BA, 5µM NAA and 168µM sucrose (Table 6). Another very successful report is by Borovec (1988) where he achieved 95 % survival of cauliflower plants after transfer in soil. In Indian cauliflower, there are efficient regeneration systems available to multiply plants using different explants like hypocotyls, cotyledon (Singh *et al.*, 2002). The cabbage has been mostly utilized for genetic transformation but not for large-scale multiplication. Souza *et al.* (1998) successfully regenerated cabbage with plants from axiallary buds on MS medium supplemented with 28.5 µM IAA, 118.58µM kinetin and 0.87 µM TDZ, and transferred the platelets to soil with high survival rate.

Table 6: Micropropagation in cole vegetables.

Explant	Medium	Response	Reference
Cauliflower (*Brassica oleracea var. botrytis*)			
Hypocotyl, curd pieces/ branches	2.44µM BA+ 5.0µM NAA+ 168µM sucrose	Multiple shoot	Vandermoortele, *et al.*, 1999
Curd meristem	2.0 Kin liquid	Field transfer of plants	Keiffer *et al.*, 1995
Head explant	1.0 NAA+ 0.1 IAA (mg/l)	Plants with high survival rate	Borovec, 1988
Internode and infloresence axis	5.0µM BA	Regenerated plants	Singh, 1988
Leaf and stem	8.88 µM BA	Regenerated plants	Singh *et al.*, 2002
Cabbage (*B. oleracea var. capitata*)			
Axillary buds	28.5µm IAA+ 18.58µM Kin + 0.87µM TDZ	Regenerated plants field transfer	Souza *et al.*, 1998
Shoots	0.5 µM BA+ 0.1µM IBA	Regenerated plants field transfer	Sretenovic *et al.*, 1997

Source: Kumar *et al.* (2003).

Bulb Crops

In bulb crops, micropropagation is being utilized for producing virus-free stocks. The most utilized explants in garlic are cloves and immature bulbs. It has been reported up to 60,000 plantlets can be obtained from single bulb in a year (Yaseen *et al.*, 1994). Various reports on use of micropropagation in onion and garlic have been given in Table 7.

Table 7: Micropropagation in bulb crops

Explant	Medium	Response	Reference
Onion *(Allium cepa)*			
Immature flower / ovary	1.0μM NAA+ 2.0 μM BA+ 0.02 μM TDZ	A large number of plantlets through direct organogensis	Germs and Martinovitch, 1998
Immature flower bud	BDS+ 2.0 μM BA+ 2.0 2, 4-D+ 100 mg/litre myoinositol + 10% Sucrose	High frequency plantlets	Jeong and Park, 1997
Shoot primordial	1.0 μM 2,4-D+ 0.5 μM picloram	High frequency plantlets	Jeong and Park, 1997
Sliced basal plate	4.44 μM BA	Shoot proliferation with no variation in regenerated plants	Barringer *et al*., 1996
Garlic *(Allium sativum)*			
Shoot tip	0.3 μM IAA + 2.0μM 2IP	Plantlet formation	Yun *et al*., 1998
Leaf explant	2 Kin+ 0.5 IAA μM	High frequency of plantlets through globular bodies	Zheng *et al*., 1998
Meristems	2 μM IP+ 5μM Jasmonic acid	High frequency plantlet regeneration	Uhcman *et al*., 1998
Root and leaves	5μM IAA	Somatic embryogensis	Bockish *et al*., 1997
Root explant	1μM NAA+ 6μM BA	High frequency plantlets regeneration	Haque *et al*., 1997
Bud and basal plate	B5, pH 7.5 + 15^{0}C	Bulblet formation	Liu *et al*., 1996
Meristem	MS+ B5 basal	Microbulbils	Hernandez *et al*., 1994
Shoot tips	MS + 0.5 % ascorbic acid + 12% sucrose	Commercial low cost bulbils	Yaseen *et al*., 1994
Clove shoot	0. 5 μM NAA	Virus-free plantlets	Masuda *et al*., 1994

Source: Kumar *et al.* (2003).

Somaclonal Variation

Micropropagation or mass propagation as discussed above, shoot tips or axillary buds are used for direct propagation on culture medium without intervening callus phase. The generation of considerable variation during tissue culture has been described as somaclonal variation. *The cells and plants regenerated from cell cultures showing heritable variation is known as somatic variation*. These variations are detected in plants derived in any form of cell culture. However, somaclonals are particularly used to refer plants regenerated from cell cultures originated from somatic tissues in contrast to gametoclones to refer to plants regenerated from cell cultures originated from gametic tissues. Plants regenerated from cell culture irrespective of tissue of origin are referred to as R or R_0 plants. The self-fertilized progeny of R plants are referred to as R_1 plants. Subsequent generations produced by self-fertilization are termed R_2, R_3, R_4 *etc*. Somaclonal variations could be utilized in breeding programme. The breeding strategy for the use of gametoclonal variation and somaclonal variation for the development of varieties / hybrids have been given (Figs. 1 and 2). Somaclonal variations most likely arise as a result of:

- Chromosome structural changes (*e.g.* small deletion and duplications. These are particularly important in polyploid species).
- Gene mutations (both dominant and recessive mutations have been recovered, the latter being more common).
- Plasmid gene mutations (*e.g.* streptomycin resistant mutants due to mutational changes in cpDNA.
- Gene amplification (*e.g.* in some cases of herbicide resistance).
- Small chromosome re-arrangement.
- Transposable elements (these may be induced to transpose due to stress during *in vitro* culture.
- Mitotic crossing over.

It may be necessary to remember that above variation may be transient (epigenetic) or genetic; only the latter is transmitted to next generation and is thus important for crop improvement. The selection of useful genetic somaclonal variation, two approaches have been followed:

In the first approach selection is exercised in cells cultured for different periods and screened for derived traits (*e.g.* for resistance to specific herbicides, fungal toxin, pollutants, extremes of temperatures and salinity), and from these selected cells or cultures, plants are regenerated. In the second approach, selection is exercised at the phenotypic level in generated plants. The former approach of selection at cellular level has the advantage of screening millions of cells with a relatively little effort and resources. But this method can not be used for traits like flower characters or grain yield. Further, phenotype expressed by selected cells may not be expressed by regenerated plants. Therefore, characters for which selection is exercised at single cell level need to be checked in regenerated plants.

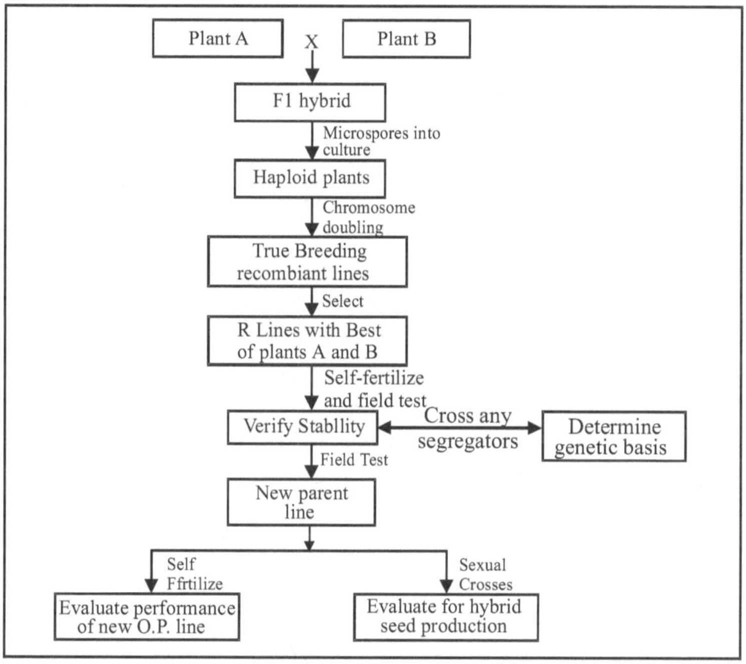

Figure 1: Breeding strategy for the use of gametoclonal variation, using a hybrid sorting technique, for the development of new plant varieties (*Source:* Ram, 1997, p. 126).

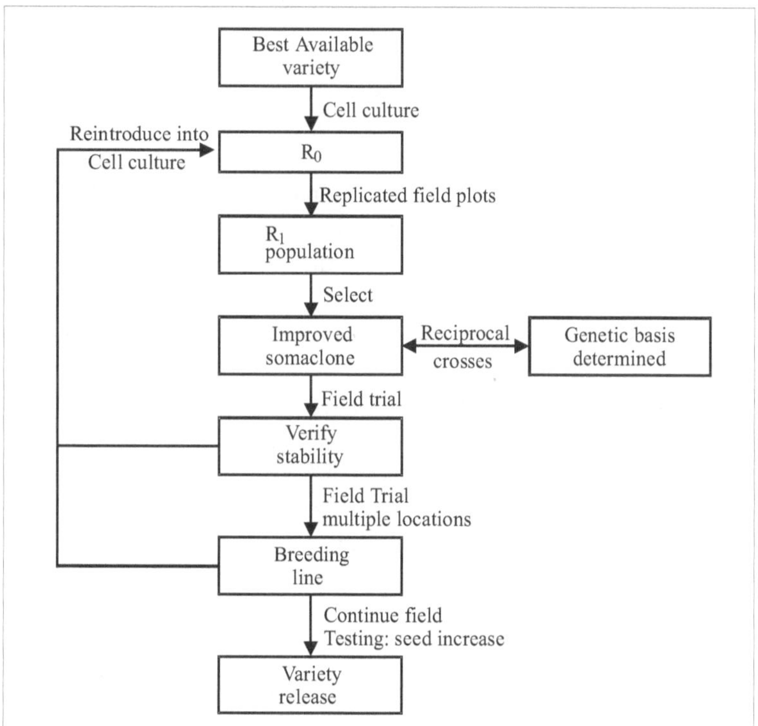

Figure 2: Breeding stragey for the use of somaclonal variation for the development of new plant varieties (*Source:* Ram, 1997, p. 127).

Distant Hybridization

Concept and History

A large numer of allied species of different vegetable crops-wild, semi-wild, and semi-cultivated-are being explored continuously as genetic material for breeding purposes. Some of these carry resistance to diseases, insect-pests, stress environments, and are rich in quality attributes. The importance of allied texa of vegetable crops to combine resistant genes in commercial cultivates was released long ago and practically, interspecific gene transfer has been demonstrated in few vegetable crops. Crosses between varieties of the same species represent vast majority of crosses used in crop improvement. But in many cases, it may be desirable, or even necessary to cross individuals belonging two different species or genera. When individuals from two distinct species of the same genus are crossed, it is known as *interspecific hybridization.*When individuals being crossed belonging to species from two different genera, it is referred to as *intergeneric hybridization*. Hybridization between individuals from different species, belonging to the same genus or to different genera, is termed as *distant hybridization*, and such crosses known as *distant crosses* or *wide crosses*. This is because individuals used for hybridization in such cases are taxonomically more distantly related than different varieties from the same species.

The first authentic record of a distant hybridization for crop improvement is the production of a hybrid between carnation (*Dianthus caryophyllus*) and sweetwillian (*Dianthus barbatus*) by Thomas Fairchild in 1717. Subsequently, a large number of interspecific hybrids were produced. Most of the interspecific hybrids were of no agricultural value, and were of academic interest only. But many interspecific hybrids, particularly in case of ornamentals, served as commercial varieties. An interesting intergeneric hybrid, *Raphanobrassica*, was produced by a Russian scientist named Karpechenko in 1928. *Raphanobrassica* is amphidiploid from a cross between radish (*R. sativus*) and cabbage (*B. oleracea*). *Raphanobrassica* was produced with a view to combine the root of radish with the leaves of cabbage, but it had roots like cabbage and leaves like radish, a useless combination of characters. But the first intergeneric hybrid, and with a greater potential than *Raphanobrassica*, was *Triticale*, which was first produced by Rimpaue in about 1890.

Triticale is the amphidiploid obtained from crosses between wheat and rye. About 50 years of intensive research characteristics of *triticales* have been improved considerably. It is beginning to become an accepted crop in certain areas, *e.g.* Canada. At present, distant hybridization has a definite role to play in crop improvement programmes: improvement of a few crop species is primarily based on interspecific hybridization, *e.g.* sugarcane (*S. officinarum*), but in most of the cases, it is used for the transfer of a few desirable genes, particularly those for disease resistance. In many crop species, the genes for disease resistance transferred from the species are extremely useful. In 1976, Jack Harlan concluded that 'there are several examples in which genes from wild relatives stand between and starvation or economic ruin.'

Crossability

The main objective of interspecific hybridization is to transfer resistance characteristic(s) of wild species to commercial varieties. Therefore, knowledge of crossability of important species is essential. The term crossability refers the percent seed setting in crosses per percent seed setting in selfs x 100. The crossability behaviours of wild species with cultivated species in vegetable crops are described below:

Tomato (*Lycopersicon* spp)

Important allied species of the genus *Lycopersicon* are *Lycopersicon pimpinellifolium*, *L. peruvianum*, *L. hirsutum*, *L. hirsutim* f. *glabratum*, *L. cheesmani* and *L. chilense,* as they have a lage number of genes for resistance to insect-pest, nematodes and environmental stresses (Table 8). Morphologically, wild species are very much different than cultivated varieties. It has been observed that *L. esculentum* is easily crossable with *L. pimpinellifolium* and there is no barrier of any consequence in such hybridization. The crossability of *L. esculentum* and *L. hirsutum* under adverse conditions requires application of auxin. The *L. hirsutum* f. *glabaratum* is not crossable with *L. pennellii* as *L. esculentum* is crossable with *L. peruvianum*, *L. hirsutum*, *L. chilense* and *L. pennellii* as female parent. On the basis of crossability, *Lycopersicon* species are classified into two groups: one *L. peruvianum* group consisting of *L. chilense* and *L. peruvianum* and the other *L. esculentum* group including *L. pimpinellifolium*, *L. hirsutum*, *L. cheesmanii*, *L. parviflorum*, *L. chmielewskii* and *L. penellii.* The *L. esculentum* hybridizes freely with *L. pimpinellifolium* and *L. cheesmaniii*, giving coloure fruited tomato. Also, it is compatable with *L. hirsutum* f. *glabaratum*, *L. hirsutum* f. *typicum*, *L. parviflorum*, and *L. chmielewskii* only as female parent. The success of *Esculentum peruvianum* and *Esculentum chilense* hybrids needs special aids, *i.e.* embryo culture. Earlier, Rick emphasized the close affinity of *Solanum pennelli* (now *L. pennelli*) with *Lycopersicon* species. Interestingly, *S. pennellii* hybridizes unilaterally with *L. esculentum*, *L. cheesmannii*, *L. pimpinellifolium*, *L. parviflorum*, *L. hirsutum*, and *S. lycopersicodes*. However, it does not cross with other species of *Lycopersicon or Solanum.* *S. lycopersicoides* hybridizes unilaterly with *L. esculentum*, *L. pimpinellifolium*, *L. cheesmanii*, and *L. pennellii*; *S. rickii* hybridized only with *S. lycopersicodes*. However, *S. ochranthum* failed to hybridized in every combination (Fig. 3)

Table 8: Tomato species and their resistant characteristics

Species	Resistance
Lycopersicon pimpinellifolium	Fusaruim and verticillium wilt, septoria leaf spot, leaf mould root rot, blight, bacterial wilt, bacterial canker and tobacco mosaic virus.
L. peruvianum	Septoria leaf spot, leaf mould root rot, tobacco mosaic virus, leaf curl virus, tomato etch virus, spotted wilt virus and rot knot nematode.
L. hirsutum	Fusaruim wilt, septoria leaf spot, leaf mould, early blight, bacterial canker, tomato etch virus, pin worm, colorado potato beetle, spider mite, aphids, and leaf minor.
L. hirsutum var. *glabaratum*	Root rot, tabaco mosaic virus, leaf curl virus, fruit worm, pin worm, horn worm, spider mites, leaf minor, cold and frost.
L. glandulosum	Septoria leaf spot and fruit rot.
L. cheesmanii	Alkalanity and salanity.
Solanum pennellii	Spider mites, aphids, and drought conditions.
L. chilense	Curly top virus.

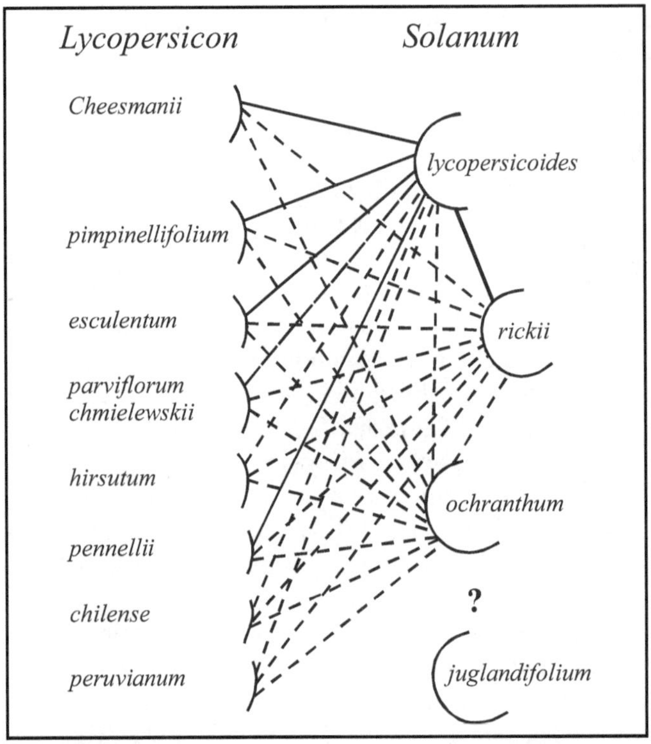

Figure 3: Crossability relationship between *Lycopersicon* and *Solanum*. The width of line indicates the degree of compatability. Dashed lines cross failure (*Source*: Kalloo, 1994, p.143).

Brinjal (*Solanum* spp)

It is widely cultivated species. Its wild allied species have good promise with respect to transferring hardiness and resistance to pest and diseases. Important allied species, which can be involved in breeding programme of *S. melongena* are *S. sisymbrifolium*, *S. khasianum*, *S. xanthocarpum*, *S. indicum*, *S. gilo*, *S. integrifolium*, and *S. incanum*. Nishio *et al.* (1984) classified 11 *Solanum* species into 3 groups on the basis of their interspecific compatability: (1) *S. melongena*, *S. incanum* and *S. macrocarpon*, (2) *S. integrifoluim*, *S. gilo* and *S. nodiflorum*, and (3) *S. indicum*, *S. mamosum*, *S. torvum*, *S. sisymbrifolium*, and *S. toxicarium*. Very limited efforts have been made in the utilization of wild species for the improvement of brinjal. However, earlier Khan and Rao (1976) observed that when *S. melongena* as male parent hybridized with *S. indicum,* produced a number of hybrids and its reciprocal crosses. Rao (1979) also studied crossability among 10 species involving mostly the aforementioned species. Important parameters in studying crossability are: normal seed set on selfing, normal seeds which grow well, normal seed but seedling dead, normal seed which did not germinate, fruit set but seeds shrunken, fruit set but parthenocarpic, and unsuccessful cross (Fig. 4). *S. melongena* cultivars as female parent hybridized with *S. melongena* var. *incanum*, *S. indicum*, *S. intergrifolium*, and *S. gilo* as resulted in functional seeds. It produced only shrunken seeds with *S. khasianum*, which probably may need special aid for production of a functional hybrid plant. *S. khasianum* as female did not hybridize with *S.melongena*. *S.melongena* as female did not hybridize with *S.indicum*, S. *xanthocarpum*, and *S.sisymbrifolium*. Rai (2005) also observed that when *Solanum melongena variety* Arka Kashev was crossed with *S.gilo* as male parent, causing parthenocarpic fruits (Fig. 5).

Chilli and Bell pepper (*Capsicum* spp).

The important domisticated taxa and associated wild of *Capsicum* are *Capsicum annuum, C. chinese,C. pubescens, C. eximium,* and *C. cardenasii.* Crossability in various species were studied by Eshbaugh (1957), and this has been described in Fig. 3. This indicates similarity between *C. cardenasii* and *C. eximium* and *C. eximium* and *C. pubescens*, whereas *C. cardenasii* and *C. pubescens* are distantly related. *Capsicum annuum, C. chinese, C. frutescens* hybridize to each other but produce a partially fertile F_1. The F_1s of *C. annuum* with *C. microcarpum* and *C. chacoense* produce a highly sterile F_1. The *C. pendulum* hybridized with *C. annuum* with difficulty and produced several sterile plants. *C. annuum* X *C. pendulum* produced sterile sesquidiploid.

♀ \ ♂	*S. melongena cultivar*	*S. melongena var. insanum*	*S. incanum*	*S. integrifolium*	*S. gilo*	*S. inicum*	*S. zuccagnianum*	*S. xanthocarpum*	*S. khasianum*	*S. sisymbrifolium*
S. melongena cultivar	■	●	●	●	●	▲	▲	▲	⊖	▲
S. melongena var. *insanum*	●	■	●	●	●	▲	▲	▲	▲	▲
S. incanum	●	●	■	●	●	●	▲	▲	▲	▲
S. integrifolium	●	●	●	■	⊕	⊕	▲	▲	▲	▲
S. gilo	⊕	⊕	⊕	⊕	■	▲	●	▲	⊕	⊕
S. inicum	●	●	●	●	●	■	◐	⊕	▲	▲
S. zuccagnianum	▲	●	▲	▲	▲	▲	■	⊕	▲	▲
S. xanthocarpum	⊕	●	⊕	▲	▲	●	▲	■	▲	▲
S. khasianum	▲	▲	○	▲	▲	▲	▲	▲	■	▲
S. sisymbriifolium	▲	▲	○	▲	○	▲	▲	▲	○	■

Figure 4: Diagrammatic representation of the results of the crosses attempted between Solanum species: Normal seeds set on selfing; ■ Normal seeds which grew well; ● Normal seeds, but seedlings died; ◐ Normal seeds, but did not grow; ⊕ Fruits set, but seeds shrunken; ⊖ Fruits set, but parthenocarpic; ▲ Cross unsuccessful.

Source: Kalloo, 1994, p. 138.

Figure 5: *Solanum melongena* X *S. gilo* crosses causing parthenocarpic fruits. *Source* : Rai, 2005

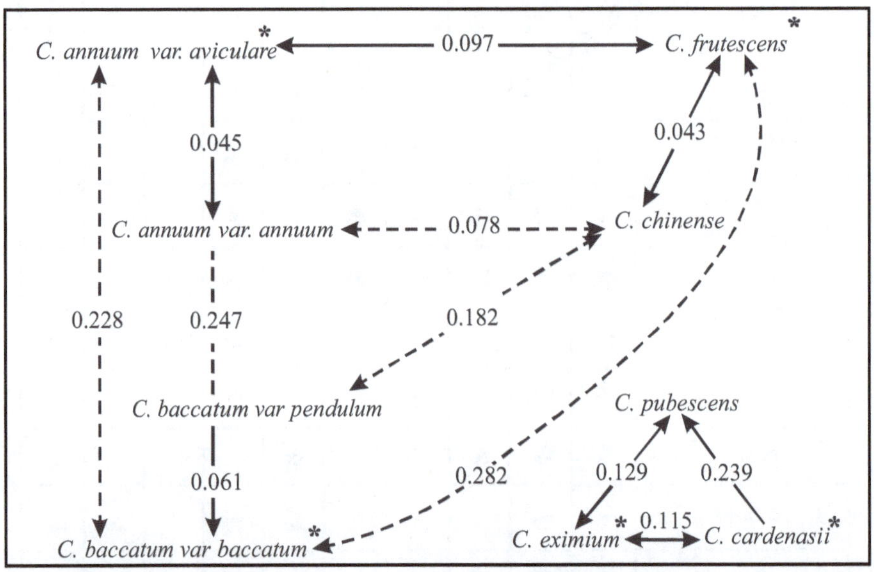

Figure 6: Summary crossing diagram from Eshbaugh (1975) with *Net's* standard genetic distance coefficients, based on starch gel electrophoresis data, superimposed to show the correlation of these two independent analyses in *Capsicum* spp. (*Source*: Kalloo, 1994, p. 144).

↔ Pollen stainability = 99 to 100%

— Pollen stainability = 50 to 89%

--→ Pollen stainability = 15 to 40%

--→ F_1 seed produced but inviable

* Wild progenitor species.

Okra (*Abelmoschus* spp)

It is one of the most important vegetable crops of the tropics and subtropics. A fairly large number of species have been collected from African and Asiatic countries. There is enormous variation in chromosome number of various species of *Abelmoschus*, therefore, species hybridization in *Abelmoschus* is complex. *Abelmoschus manihot* (2n = 68), *A. moschatus* (2n = 66), *A. ficulneus* (2n = 72), *A. angulosus* (2n = 58), *A. tuberculatus* (2n = 58), *A.pungens* (2n = 138), *A.tetraphyllus* (2n = 138), and *A. crinitus* (2n = 138) are probable allied species of *Abelmoschus*. *A. manihot* is being extensively involved in breeding programme of okra for yellow vein mosaic virus disease resistance. Quite a few studies have been carried out in the interspecific hybridization of genus *Abelmoschus* with varying degrees of success. *A. esculentus* (2n = 124) is crossable with *A. manihot* is crossable with *A.moschatus* and *A. ficulneus* and produces amphidiploids (2n = 192) with the help of colchicine. *A. esculentus* (2n = 68) with special aids such as embryo and ovule cultures. *A. esculentus* (2n = 130) is crossable in both direction with *A. tuberculatus* (2n = 58).

Cucurbits

This is a very wide group in which a large number of genera and species are included. But practical importance of wide crosses has been realized only in genera *Cucumis* and *Cucurbita*. In *Cucumis*, muskmelon (*C. melo*) and cucumber (*C. sativus*) are of great commercial importance. The West Indian gherkin (*C. anguria*) and African horned cucumber (*C.metuliferus*) are grown on a small scale. The resistant characteristics are available in wild species. The allied species of genus *Cucumis*, namely *africanus*, *anguria*, *ficifolius*, *myriocarpus*, *metuliferus* and *angolensis*, posses breeding potential but limited efforts with success have been made for interspecific hybridization. Therefore, there is ample opportunity to explore the possibilities of transfer of desirable genes from these species to commercial varieties. On the basis of crossability, incongruity has been observed between Asian and African groups of *Cucumis* species, *i.e. C. sativus* on one side and *C. melo* and African species on the other side. *C. sativus* as female parent is crossable with *C. hardwickii,* indigenous to India, but it is not compatible with other 10 species. *C. melo* also did not hybridized with most of the species. However, with the aid of hormones, fruit development has been noticed in crosses with *C. ficifolius*, a tetraploid. *C. ficifolius* resulted in a self-sterile hybrid when crossed with *C. africanus, C. anguria* and *C. dipsaceus*.

The comprehensive interspecific hybridization have been derived the interrelationship of *Cucumis* species and accordingly this can be organized into four cross-sterile groups (Fig. 7). *Cucurbita* is one of the most important genera of the family Cucubitaceae. *Cucurbita pepo, C. maxima, C. moschata* and *C. mixta* are the species cultivated in various parts of the world, and *C. ficifolia* and *C. lundelliana* have some desirable characteristics which can be utilized for the improvement of cultivated *Cucurbita*. The important objectives of interspecific hybridization in *Cucurbita* are to combine plant growth habit and resistance to diseases and insect-pests. *C. pepo* is crossable with *C. maxima* and *C. mixta; C. pepo* and *C. moschata* are crossable with the aid of embryo cultures. *C. moschata* is compatible with *C. maxima, C. lundelliana* is tolerant to powdery mildew and possesses perennial habit and is crossable with *C. maxima, moschata, ficifolia, pepo* and *mixta*.

There are possibilities for developing an interspecific gene pool. An attempt has been made to combine the germplasm of five species of *Cucurbita* into a common gene pool and to use this gene pool for interspecific gene transfer. Such fertile interspecific hybrids combining the characteristics of 2-4 species can be maintained.

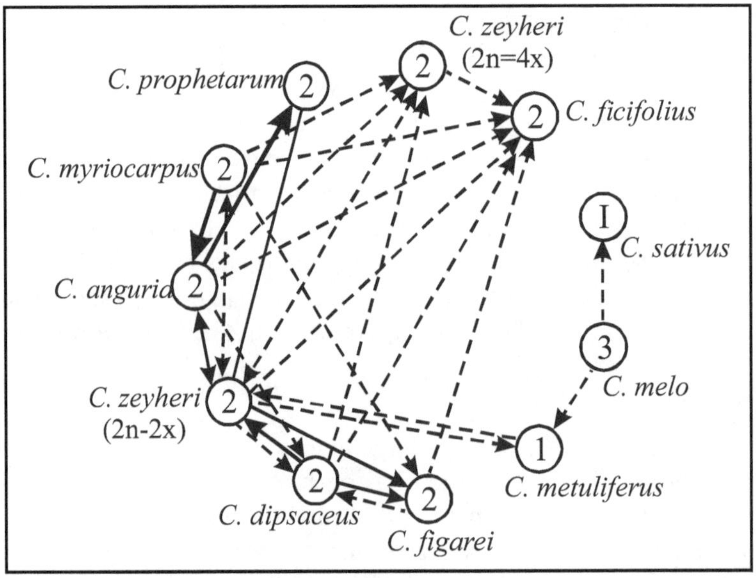

Figure 7: Crossability polygon of *cucumis* species. (*Source*: Kalloo, 1994, p.147).

←→ good to moderately self-fertile hybrids

←--- sparingly self-fertile hybrids

------→ self-sterile, sparingly cross-fertile hybrids.

------→ inviable seeds of

Arrow point in the direction of the female parent

Allium species

Allied species of *Allium* can be used to a great extent to develop disease resistance, winter hardiness, and earliness. In *Allium, Allium fistulosum* as a female parent crosses easiely with *Allium cepa. Allium fistulosum* as a male parent is not easiely crossable with *Allium cepa. Allium cepa* is compatible with *A. galanthem*, *A. drobovii* and *A. pskemense* and the F_1's are found fertile. F_1 can be backcrossed to either of the parents. However, *A. cepa* x *A. galanthem* and *A. cepa* x *A. pskemense* crosses are not successful and a certain degree of sterility has been observed. It has been observed that hybridization of *A. cepa* with *A. altaicum* and *A. fistulosum* with *A. vavilovii* produced 7.4 - 36 % seed setting, respectively.

Cole Crops

Within *Brassica oleracea*, there is a wide variation in morphological types, therefore, these are classified as botanical varieties. These are Brussels' sprouts (*B. oleracea, gemmifera* group). sprouting broccoli (Italica group), curly kale (Acephala group), cabbage (Capitata group), khol-rabi (Gongylodes group), wild cabbage, kale and cauliflower. These botanical varieties have been crossed in all possible combinations excluding reciprocals. It has been found that crosses marrow-stem kale x sprouting broccoli and marrow-stem kale x cauliflower are promosing. It is also seen that *brassica oleracea* and *B. campestris* are hybridized at a temperature 25°C with specific aid.

Barriers in Production of Distant Hybrids

The difficulties encountered in production of interspecific hybrids may be grouped into three broad classes:

- Failure of zygote formation
- Failure of zygote development
- Failure of seedling development

Failure of Zygote Formation

In many cases, distant hybrids are not produced because the zygote is not formed due to failure of fertilization. Fertilization often fails because the pollen tube is unable to reach the embryo sac and consequently the sperms are not available for fertilization. The pollen tube may burst in style of another species and thus fail to reach the embryo sac. This occurs in some species combinations like in *Datura*. In some cases, style of species used as female is longer than pollen tube. Therefore, pollen tube does not reach the embryo sac. In such cases, cross may be made by using the species with shorter style as female or by cutting off a portion of style of the parent with longer style. Pollen tube of polyploid species are usually thicker than those of diploid species. When a diploid species is used as female and polyploid one as male, the polyploid tube grows at slower rate in diploid style than in a polyploid style. This difficulty may be overcome by using the diploid species as the male parent or by shortening the style of diploid species being used as a female. These are some of more common barriers to zygote formation; other barriers may be encountered in specific cases.

Failure of Zygote Development

In many cases, fertilization does take place and zygote is produced. But the development of zygote is blocked at various stages. Failure of embryo development may take place due to:

- Lethal genes.
- Genotypic disharmony between the genomes of two parental species.
- Chromosome elimination.
- Cytoplasmic incompatability.
- Endosperm abortion.

Lethal Genes

Some species carry a lethal gene, which causes death of zygote during early embryonic development. Such lethal gene generally do not have any effect in species carrying them but affects only interspecific hybrid embryos. Interspecific hybrids in such cases may be obtained by using a (of the species that carries lethal gene) carrying the non-lethal allele for making interspecific crosses.

Genotypic Disharmony Between Two Parental Species

In some cases, death of embryo may occur due to genetic imbalance rather than the action of specific lethal gene. There is a possibility that cases of genotypic disharmony may, infact, be cause of lethal gene action but due to insufficient genetic analysis, the lethal genes have not been detected so far. The hybrid from these crosses die at various stages of development ranging from early-embryonic development to reproductive phase of F_1 plants. When failure of zygote development is due to genetic imbalance in hybrid, interspecific hybrids can not be obtained since they would not survive owing to genetic disturbances. But if such cases are complicated cases of lethal gene action, interspecific hybrids may be obtained by using strains that do not have lethal allele. Therefore, in case where genetic imbalance seems to be the reason for the failure of zygote development, several strains of each species should be crossed with each other. If some combinations produce viable progeny, while others do not, the case is most likely one of lethal gene action. Different genotypes / strains of a species show consisderable variation in case with which they produce interspecific hybrids. The genetic basis and the mechanism of this variation is not precisely known.

Chromosome Elimination

In some case of distant hybridzation, chromosomes are gradually eliminated from zygote. This generally does not prevent embryo development, but the resulting embryo and F_1 plants obtained from such crosses are not true interspecific hybrids since they do not have two parental genomes in full. Generally, chromosomes from one genome are successively eliminated due to mitotic irregularties, and in extreme cases, chromosomes of only one genome may remain in embryo; such embryos are, consequently, haploid.

Cytoplasmic Incompatability

The embryo development may be blocked by an incompatability between cytoplasm of species used as female and genome of species used as male. Such an interaction, more generally, lead to hybrid weakness and male sterility in hybrids. But in a few cases, embryo development may also be prevented.

Endosperm Abortion

Seeds from a large number of distant crosses are not fully developed and are shrunken due to poorly developed endosperm. Such seeds show poor germination, and may often fail to germinate. When endosperm development is poor or is blocked, the condition is generally known as endosperm abortion. The endosperm may abort during early embryo development, preventing further development of embryo, or it may abort at a later stage allowing an almost complete development of embryo. In case of endosperm abortion, hybrid plants may be raised by culturing hybrid embryos on a suitable culture medium. The embryos in such cases, are removed from young seeds before endosperm abortion takes place. This is termed as embryo rescue.

Failure of Seedling Development

Some distant hybrids die during seedling development or even after initiation of flowering. Some interspecific hybrids are chlorophyll deficient and fails to survive. Growing them *in vitro* may grow such plants to maturity by grafting them onto normal plants. The mechanism involved in failure of seedling development most likely involves complementary lethal genes. Interspecific and intergeneric F_1 hybrids show both chlorosis and necrosis, which is often lethal. Chlorosis refers to improper chlorophyll development, resulting in variable degree of chlorophyll deficiency. In an extreme case, very little chlorophyll may be produced.

Unilateral Incompatability

Unilateral interspecific incompatability is mainly conditioned by the genes of pistil or style, *S-alleles* of plants, and cytoplasmic factors. The dominant gene (s) present in pistil may decide to accept foreign pollen. Pollen tube inhibition in style is the main reason for unilateral incompatability. Self- incompatability, in many cases, also decide the acceptance or rejection of foreign pollen. This observation has been observed in *Solanum* and *Lycopersicon.* Usually style of self- incompatable plants does not accept the pollen of self-compatable plants. In potato, style of several self-incompatible plants failed to accept the pollen of self- compatible species. In tomato, it has been observed that *L. peruvianum*, *L. hrsutum*, and *L. chilense* did not accept the pollen *of L. esculentum* and *L. pimpinellifolium*. The inhibition of *esculentum* pollen tube in style of *peruvianum* is due to dominant gene. Interspecific incompatability and self- incompatability are closely related phenomena. But Hogenboom (1972) described interspecific incompatability as incongruity, which is independent of self- incompatability. Incompatibility is the inhibition of fusion of normal gametes, whereas, incongruity refers to all barriers to crossing. In tomato, unilateral incongruity is observed between *Eulycopersicon* and *Eriopersicon* species. However, this does not truly exhist in *L. hirsutum* f. *glabaratum*. Unilateral incongruity of *L. peruvianum* x *L. esculentum* can be broken with selection and breeding. Cytoplasmic factor may be strong barriers for the failure of reciprocal crosses. It has been seen in cross *S. indicum* x *S. melongena*. In this case, while using *S. indicum* as the female parent, the crosses are successful. But as male is not; this may be due to obvious incompatability reaction shown by pollen of *S. indicum*. Further more, cytoplasm of *S. melongena* may not be able to accommodate the gene or gene complexes of *S. indicum.* The cytoplasmic can be altered by changing the ploidy level of one of the parents. *S. melongena* as tetraploid has been found crossable with *S. indicum* in both directions.

Breaking Hybridization Barriers

The success of utilization of a genus or species depends upon its hybridization with cultivated varieties. Various measures like embryo- and callus- culture rechnique, bridge crosses, early pollination, repeated pollination, regulating the ploidy levels, widening genetic variability in species, biosystematic studies, and use of growth substances are being employed to improve the success of distant hybridization. Most of the incompatability barriers have been discussesed in *Chapter 4*, Reproduction, Pollen and Pollen Biotechnology. However, some breaking barrier are summarized below:

- Biosystematic studies are effective to select the strains in which the cross-incomparability is broken. There is enormous variation in species of *Lycopersicon peruvianum.* Some of the stains of *L. peruvianum*, in which unilateral incompatibility is broken, can be used and interspecific hybrids are produced by taking *L. esculentum* as male parent. Similarly, in *Cucumis*, genetic variation and biosystematic relationship has been studied with the species to explore the possibility of breaking the crossability barriers. There is ample variation within the species of *C. anguria* and *C. zeyheri* which carry cucumber green mottle mosaic virus resistance. This variability can be utilized to overcome hybridization barriers.
- Bridge cross can also employed to overcome the hybridization barriers. The glaring example of involvement of bridge cross for achieving successful hybridization is available in potato. *S. bulbocastanum* has some valuable characteristics but not crossable with *S. tuberosum*. *S. pinnatiseanum* has been found crossable both with *S. bulbocastanum* and *S. tuberosum*, exhibiting the possibility to be used as a bridging species to make hybridization successful between *S. bulbocastanum* and *S. tuberosum.* Triploid *S. acaule* x *S. bulbocastanum* is doubled with colchicine and resulting fertile haploid F_1 is pollinated with *S. phureja* to produce a triple hybrid which is pollinated with *S. tuberosum*, as a result, a quadruplex hybrid is produced. Such hybrids are vigorous, having a resistance gene to *Phytophthora infestans*. Corresponding, *S. acuale* can be used as a bridging species to make hybridization successful between *S. pinnatisectum* and *S. tuberosum.* Besides, high resistance to *P. infestans* has been transferred from *S. bulbocastanum* by using *S. acule* and *S. phureja* as bridging species. Resistance to virus Y has been transferred from non-tuberous *S. etuberous* using *S. verrucosum* and *S. pinnatisectum* as bridging species.
- Hybridization barriers can also be overcome by changing the ploidy level. In potato, there is a wide range of variation in ploidy levels, therefore attempts have been made to regulate the ploidy levels to overcome the barriers. In *Solanum*, chromosome doubling of diploid *S. chacoense* has led to successful hybridization with *S. tuberosum.* Similarly, by doubling the chrosome number, *S. polyadenium*, *S. chacoense* and *S. ribinii* can be hybridized with *S. tuberosum*. Tetraploid *L. peruvianum* as pistillate parent exhibited better crossability with *L. glandulosum*. In *Cucumis* also, difficult, weak, or poor crosses can be made successfully at various ploidy levels because endosperm may act as a source of nutrition for young embryos. *Brassica oleracea* (*Capitata* group) and *B. campestris* (turnip rape) do not with each other as diploid

but hybridize as tetraploids. However, an embryo culture has to be employed for the recovery of embryos.

- Application of some growth substances such as IAA, IBA, NAA and PCPA on flowers or flower pedicels enhances fruit setting and seed formation. In *Solanum* and *Lycopersicon*, application of auxin prevented flower assist pollen hydration and prevented desiccation. Application of NAA and potassium gibberelate applied to pedicel of pollinated flowers in the cross *Phaseolus vulgaris* x *P. acutifolius* improved pod setting.

Limitations of Distant Hybridization

An extensive use of distant hybridization in crop improvement has been limited by several problems that are summarized below:

- Although many distant crosses can be made with the aid of various techniques, several distant combinations can not be crossed. When a cross fails, it is generally difficult and time-consuming to establish the exact cause of failure. When the reason for failure of a cross is not known, one can attempt to overcome the barrier only by trial and error. Distant hybridization is the most common, and perhaps the most useful, at the interspecific level (within the same genus), but several intergeneric crosses have also been of great promise, *e.g.* Triticale, Raphanobrassica *etc.* Interfamily crosses are neither possible, and in all probability, nor profitable. The technique of somatic hybridization is being developed to produce incompatible distant crosses with some notable successes.
- F_1 hybrids from distant crosses generally exhibit variable sterility; rarely F_1 is fully fertile. Partial sterility of F_1 does not present a serious problem, and the F_1s are difficult to maintain, and in some cases doubling of the chromosome number of F_1 may not be feasible or desirable. In any case, F_1 sterility confronts the breeder with a problem to overcome which he has to resort to specialized techniques demanding additional resources and time.
- The production of new crop species through allopolyploidy following distant hybridization suffers from several problems, *e.g.* lower economic yields, poor agronomic characteristics, sterility *etc.* But these characters can be improved through breeding and selection. However, this does require considerable efforts, resources and time.
- Transfer of oligogenes from related species is mainly limited by reduced pairing of the alien chromosome with that of the recipient species. Reduced recombination may be overcome by chromosome manipulation, but this is limited to polyploid species, and even their considerable cytogenetic analysis is required to make sure that only a small segment of alien chromosome is transferred. In a vast majority of cases, however, gene transfer has to depend on spontaneous pairing and recombination. Elaborate techniques are being developed by C.J. Diriscoll in wheat to promote crossing over between the alien chromosome segment carrying the desirable gene and chromosome of recipient species.

- Linkage of the gene (s) under transfer with some undesirable gene (s), called linkage drag, presents a major problem in using distant hybridization in crop improvement. In fact, undesirable linkages have prevented commercial exploitation of many alien gene transfers. In few cases, continued breeding has been able to break the undesirable linkages, but this demands considerable time, efforts and resources.
- In many distant hybridization programmes, transfer of recessive traits and of quantitative characters is not feasible. In fact, monogenic dominant characters are most easily transferred, and such transfers constitute the bulk of interspecific gene transfer.
- F_{1s} from some interspecific hybrids fail to produce flowers. Hybrids between species belonging to the Rhizomatosae and those from section *Arachis* of peanuts do not flower; all the techniques to induce flowering in these hybrids have failed. Similarly, some interspecific hybrids in the genus *Glycine* fails to flower, but when they are grafted onto *Glycine max* they are readily induced to flower.
- In distant hybridization programmes, it is important to use the best variety of a crop species as one of the parents. Transfer of genes from a related species into an agronomically poor type will necessitate further breeding efforts to use that gene. In many cases, the best varieties of a crop species cannot be crossed with wild species. The wild relatives generally cross more easily with land races than with highly improved varieties.
- In some cases, F_1 seeds of interspecific hybrids exhibit dormancy. For example, some interspecific hybrids in *Arachis* remain dormant for 5-10 years regardless of the treatments applied for inducing germination. Thus, seed dormancy may render the use of some interspecific F_1hybrids difficult and in some cases, it may effectively prevent their utilization in crop improvement.

Somatic Hybridization

The results of the fusion of two somatic cells of different species, genera or family are considered in somatic hybridization. This is also known as protoplast fusion or para sexual hybridization. Protoplast fusion or somatic hybridization is one of the most important uses of protoplast culture. This is particularly significant for hybridization between species or genera, which can not made to across by conventional method of sexual hybridization. Although, somatic hybridization was successfully achieved first in animals and later in plants, its significance has been realized first fully in plants because hybrid cells can be induced to regenerate into whole plants. Somatic hybridization has a great advantages over sexual methods for interspecific and intergeneric hybridization.

- A number of species and genera are incompatible to each other due to barriers in sexual hybridization in vegetable crops. The common pre-syngamic and post-syngamic sexual hybridization barriers are overcome by this technique.
- There is the possibility to transfer some very valuable resistant genes from wild species to cultivated varieties.
- Transfer of organelles such as nucleus, mitochondria and chloroplast from species to another.

Steps Involved in Somatic Hybridization

The technique of somatic hybridization involves following four steps.

Isolation of Protoplast

It is readily achieved by treating the cells / tissues with a suitable mixture of cell- wall degrading; usually a mixture of pectinase or macerozyme (0.1- 1.0 %) and cellulase (1-2 %) is appropriate for most plant materials. Osmotic concentration of enzyme mixture and of subsequent media is elevated (usually by adding 500-800 m mol/litre) sorbitol or mannitol to stabilize the protoplasts and to prevent them from bursting. The cells and tissues are incubated in enzyme mixture for several hours. Naked protoplast virtually devoid of cell-wall is gradually released in the enzyme mixture. For isolation of protoplast, virtually all plant parts but leaf measophyll is the most preferred tissue. In general, fully expanded leaves are surface sterilized, their lower epidermis is peeled off with a pair of forceps. The peeled areas are cut out with a scalpel and suspended in the enzyme mixture. After incubation, protoplast is washed with a suitable washing medium to remove the enzymes and debris. The protoplast may be cultured on a suitable medium in a variety of ways. They readily regenerate cell-wall; undergo mitosis to form macroscopic colonies, which can be induced to regenerate whole plants.

Fusion of Protoplasts

A number of strategies have been used to induce fusion between protoplasts of different strains / species. Of these, relatively more successful strategies are described below:

$NaNO_3$ Treatment

This method was successfully utilized for fusion of protoplast from root tips of oat and maize seedling but is not preferred due to low frequency of fusion, particularly when highly vacuolated mesophyll protoplasts are used. In this method, protoplast is suspended in an aggregation mixture (5.5% sodium nitrate in 10 % sucrose solution). This mixture works as a fusion inducing mixture and causes fusion on incubation (water-bath maintained at 35°C). In order to obtain a high frequency of fused protoplasts, the mixture may be centrifuged and the pellet resuspended and incubated for one or more additional cycles. Finally, the mixture is placed by a liquid medium and the protoplast in this mixture is incubated again. The cycle may be repeated once or twice before plating the protoplasts on solid medium. The fusion of protoplasts may be monitored at different steps through examination under an inverted microscope (Fig. 8).

Treatment with Calcium ions (Ca++) at High pH

This method involves spinning (centrifugation) of protoplasts in a fusion inducing solution (0.05M $CaCl_2$ $2H_2O$ in 0.4M mannitol at pH 10.5) for 30 minutes at 50 g, after which the tubes are placed in water bath (37°C) for 40-50 minutes. This leads to fusion of 20-50 % of protoplasts. This method has been found superior to other methods in some cases, but high pH may be toxic in other cases.

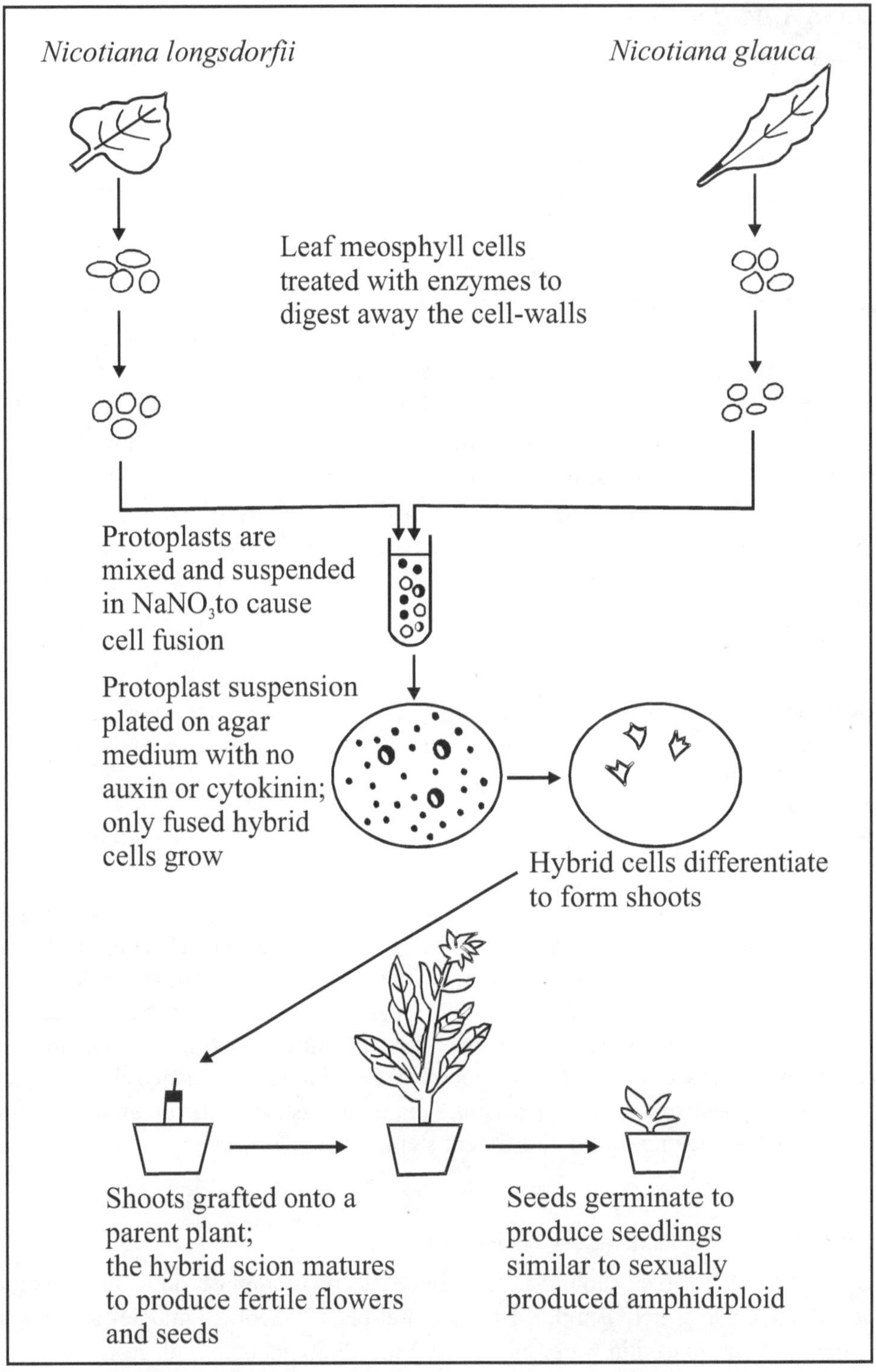

Figure 8: Procedure of $NaNO_3$ treatment for production of interspecific somatic hybrids in genus Nicotiana. *(Source*: Gupta 2001, p. 332).

Polyethylene Glycol (PEG) Treatment

Since 1974, protoplast fusion has been successfully achieved in several crops, using polyethylene glycol (PEG) as an *fusagen*. This technique gives high frequency of fusion with reproducible results and involves low cytotoxicity. The technique can be used for fusion of protoplasts from unrelated plant taxa. The agglutination of protoplasts during PEG – treatment, can be brought about by the following two different methods: (i) when protoplasts are available in sufficient quantities, 1 ml of culture medium with suspended protoplast is added to 1ml of 56 % solution of PEG and the tube taken for 5 seconds. The protoplasts are allowed to sediment for 10 minutes, washed with growth medium and examined for successful agglutination and fusion, (ii) if protoplast is available in micro quantities, drop culture can be used. Two types of protoplasts are mixed in equal quantities and 4-6 micro drops (100µl each) are placed in a small Petri plate and allowed to settle for 5-10 minutes at room temperature. Two to three micro drops (50µl each) PEG are added from the periphery in each drops, which is incubated for 30 minutes at room temperature (24^{0}C). This leads to agglutination of protoplasts. Sometimes, a cover glass is placed in the middle of Petri plate before protoplast suspension is poured. This avoids sticking of protoplasts to the floor of Petri plate and also makes it convenient to handle the protoplasts including their fixation, attaining and examination. After PEG treatment, protoplasts are gradually washed and during this process, most of the fusion is achieved. PEG is then replaced by culture medium to allow growth of fused protoplast.

Electrical Fusion

If protoplasts are placed into a small culture cell containing electrodes and a potential difference is applied, then the protoplasts will line up between the electrodes. If now an extremely short, square wave electric shock is applied, protoplast can be induced to fuse (Fig. 9).

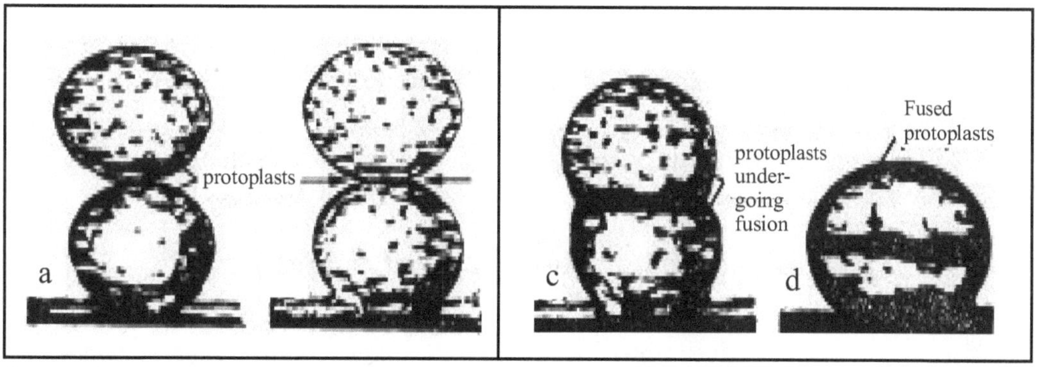

Figure 9: Protoplast fusion induced by the application of electric shock.
(*Source*: Gupta, 2001, p. 334).

Selection of Somatic Hybrid Cells

The protoplast suspension recovered after treatment with a fusion inducing agent (fusogen) consists of following cell types: (i) unfused protoplasts of two species / strains,

(ii) products of fusion between two or more protoplast of the same species (homokaryons) and (iii) hybrid protoplasts produced by fusion between one (or more) protoplasts of each of the two species (heterokaryons). In somatic hybridization, only heterokaryotic or hybrid protoplasts, particularly those resulting from fusion between one protoplast of each of the two species, are of interest. However, they form only a small production of population (at 0.5 –10%, Fig. 10). Therefore, an effective strategy has to be still an active area of investigation. A number of strategies have been used for the selection of hybrid protoplasts. Some visual markers, *e.g.* pigmentation of parental protoplast may be used for identification of hybrid cells under a microscope. Hybrid cells are then mechanically isolated and cultured. For example, the protoplasts of one species may be green and vacuolated (from mesophyll cells), while those of the other may be non-vacuolated and non-green (from cell cultures). Where such features are not available, the protoplasts of the two parental species may be separately labelled with different fluorescent agents, This approach is time-consuming, and requires considerable skill and effort. Several workers have attempted to devise systems, which especially select hybrid cells. In simple words, these systems exploit some properties (usually, deficiencies) of parental species, which are not expressed in hybrid cells due to complementation between their genetic systems. These properties may be sensitivity to culture medium constituents, antimetabolites, temperature *etc*; inability to produce an essential biochemical (auxotrophic mutants), chlorophyll or some other pigmentation etc. These properties may be naturally present in parental species or may be artificially induced through mutagenesis. Efforts have been made to select heterokaryons or calli. Following methods have been used for selection:

- The hybrid *Nicotiana glauca* x *N. longsdorfii* could grow on a medium on which none of the parents could grow (Fig.11)
- Hybrids of *Petunia parodii* (green) with *P.hybrida*, *P. inflorata* or *P. parviflora* (albino), could be selected in the form of green callus, which represented only hybrid cells; similar approaches also have been followed in hybrid of *Datura innoxia* (albino) with other species of *Datura* or with *Atropa belladonna* (wild type-green).
- The protoplasts of two parents may be labelled by different fluroesent agents, which will then enable the selection of hybrids, as done in the genus *Nicotiana* by fluorescent activated cells sorter (FACS) (Fig. 12).

Culture of Hybrid Cells and Regeneration of Hybrid Plants

Once the hybrid calli are obtained, plants are induced to regenerate from them since this is a prerequisite for their exploitation in crop improvement. Further, hybrid plants must be at least partially fertile in addition to having some useful property, to be of any use in breeding schemes. The culture techniques have been refined to a state where plant regeneration has been obtained in a number of somatic hybrids (Table 9). But even today, it has not been possible to recover hybrid plants and/ or calli from a number of somatic combinations. This phenomenon is known as *somatic incompatibility*. The reasons for somatic incompatibility is not clearly understood.

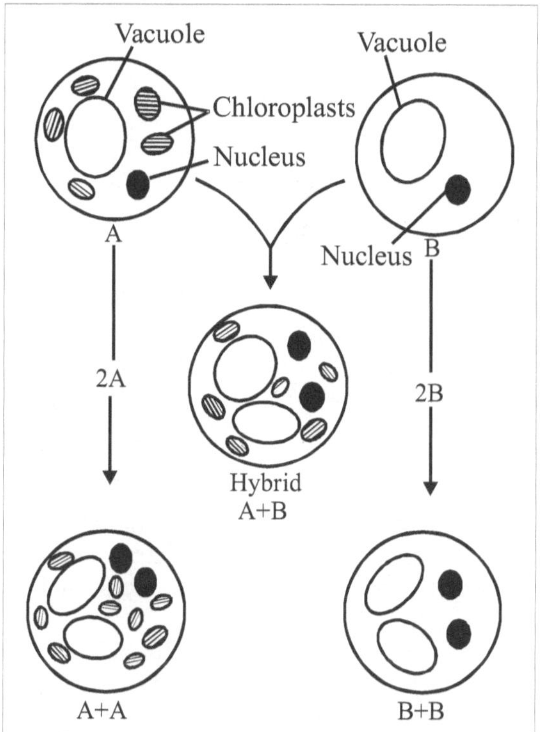

Figure 10: Different possible product obtained due to fusing of two types (A and B) of protoplasts. *(Source*: Gupta, 2001, p. 335).

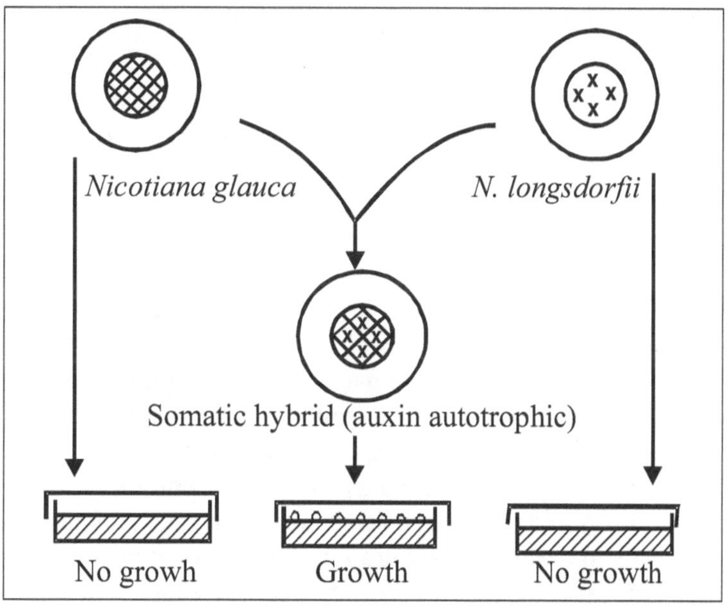

Figure 11: Auxin independent growth of hybrids of *Nicotiana glauca* and *N. longsdorffii*; the two parental lines cannot produce auxin and thus do not grow on auxin-free medium; hybrid cells produce auxin and are thus able to grow and form callus. *(Source*: Gupta, 2001, p. 335).

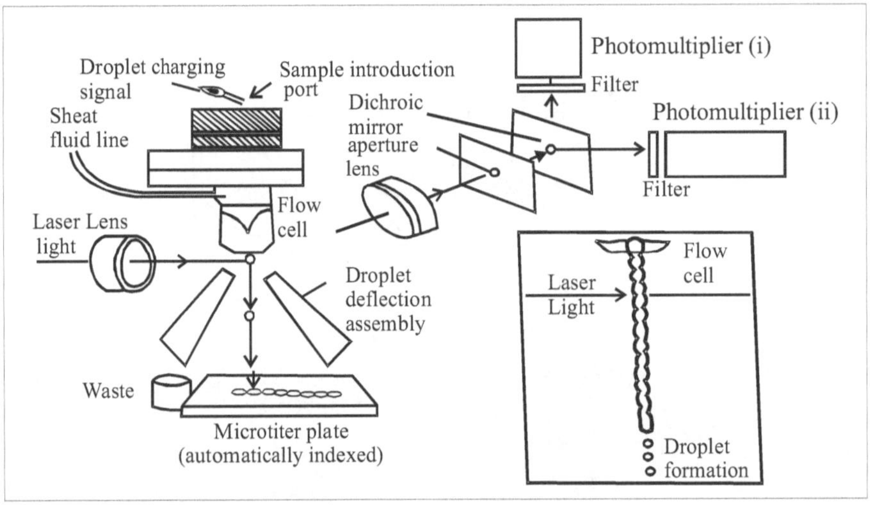

Figure 12: Simplified diagram of principle of a fluorescent activated cell sorter; the droplet containing a single cell is electrically tested for its fluorescence property; by applying a change to plates, through which the droplet falls, it is possible to separate the parental protoplast and hybrids. *(Source :* Gupta 2001, p. 336).

Table 9: A list of some distant somatic hybrid plants

Symmetric or near – symmetric hybrids	Asymmetric hybrids
*Solanum tuberosum + Lycopersicon esculentum**	*Daucus carota + Aegopodium podagraria*
Datura innoxia + Atropa belladona	*Daucus carota + Petroselinum hortense*
Arabidopsis thaliana+ Brassica campestris	*Hyoscyamus muticus+ Nicotiana tabacum*
Atropa belladonna+ Nicotiana chinensis	*Datura innoxia+ Physalis tabacum*
	Datura innoxia + Physalis minima
	Nicotiana tabacum + Daucus carota

* The + symbol indicates that hybrid was obtained through protoplast fusion. (*Source:* Singh, 2002).

Anther Culture

Guha and Maheshwari proposed the first haploid plant through anther culture in 1964. In anther and pollen culture, haploid plants may be obtained from pollen-grains by placing anthers or isolated pollen-grains on a suitable culture medium. Anthers may be taken from plants grown in the field or in pots, but ideally, these plants should be grown under controlled temperature, light and humidity. The optimum condition may differ from species to species. Often, the capacity for haploid production declines with the age of donor plants. Exposure of excised flower buds to a low temperature for some times, *e.g.* at 5ºC for 72 hours for tobacco, prior to removal of anthers for culture may markedly enhance the recovery of haploid plants. In some species, however, a brief exposure of anthers to a high temperature is reported to have a promotory effect, *e.g.* at 35ºC for 24 hours in *Brassica campestris*. The medium requirements may vary with vegetable crops. Anther cultures are generally maintained in altering period of light (12-18 hours; 5000- 10,000 lux m^2) at 28°C and darkness (12-6 hours) at 22°C, but the optimum conditions vary with the species. The wall of responsive anthers turn brown and after 3-8 weeks they burst open due to the developing callus or embryos. After seedling (from embryos) or shoots (from callus) become 3-5 cm long, they are transferred to soil in the same way as other *in-vitro* regenerated plantlets. Pollen-grains of cultured anthers show remarkable cytological changes during first 6-12 days, called the *inductive period.* The early divisions in responding pollen-grains may occur in one of following four ways:

- The uninucleate pollen grain may divided symmetrically to equal daughter cells both of which under go further divisions, *e.g* in *Datura innoxia* (Pathway I).
- In some cases like chilli, uninucleate pollen divides unequally (as one or two divisions. The callus/ embryo originates due to successive divisions of vegetative cell (Pathway II).
- But in a few species, *e.g.* in *Hyoscimus niger*, pollen embryos originate from the generative cell alone. The vegetative cell either does not divide or divides only to a limited extent forming suspensor like structure (Pathway III).
- Finally in some species, *e.g.* in *Datura innoxia,* uninucleate pollen-grains divide unequally producing generative and vegetative cells. But both these cells divide repeatedly to contribute to the developing embryo (Pathway IV).

The responsive pollen-grains become multicellular and ultimately burst open to release the cell mass. This cell mass may either assume the shape of a globular embryo and under go the development stages of embryogeny, or its may develop into callus depending on plant species. The production of haploid plants by anther culture has been presented in Fig. 13. Earlier, anther culture has been attempted by several workers in vegetable crops (Table 10).

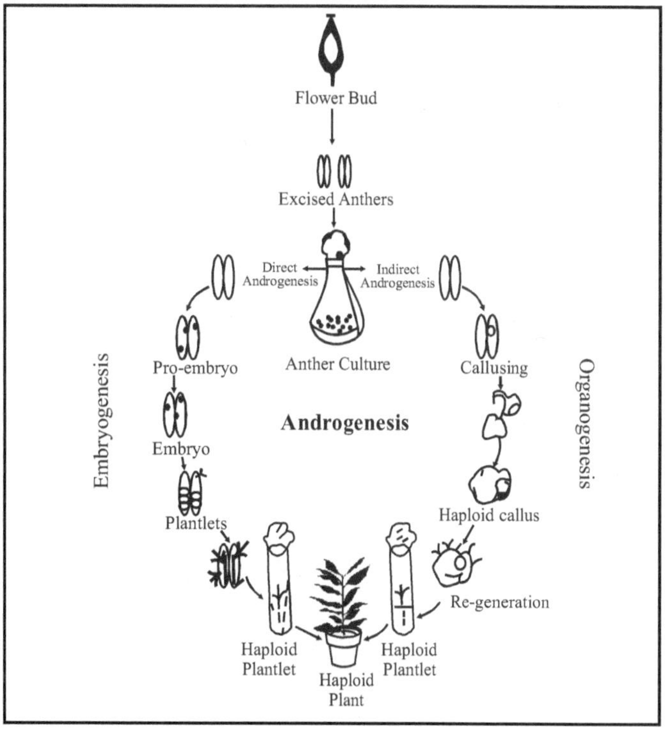

Figure 13: Schematic representation of culture of excised anthers and development of haploid plants directly by embryo formation, or through haploid callus.
Source: Kalloo, 1994, p. 138.

Table 10: Vegetable crops in which haploid tissue, embryos calluses, or plants have been obtained by culture of excised anther/pollen

Crop	Reference
Solanum tuberosum	Sopory and Rogan, 1976 and Spory *et al.,* 1978
S. verrucosum	Weatherhead and Henshaw, 1979
S. melongena	Raina and Iyer, 1973 and Isouard *et al.,* 1979
Lycopersicon esculentum	Gresshoff and Doy 1972; Sharp *et al.,* 1971 and Sharp *et al.,* 1972
Capsicum annuum	Wang *et al* 1973., Sibi *et al* 1979; Sibi *et al.,* 1980; Vagera and Hairane, 1985 and Dumas *et al;* 1981
Capsicum spp.	Novak, 1974
Asparagus officinalis	Pelletier *et al.*, 1972; Dore, 1973; Choudhury, 1985 and Kruger, 1985
Phaseolus vulgaris	Peter *et al.,* 1977
P. aureus	Bajaj and Singh, 1980
Vicia faba	Hesemann, 1977
Vigna unguiculata	Ladeinde and Bliss, 1977
Glycine max	Gaungchu *et al.,* 1980
Cucumis melo	Dryanoska and Ilieva, 1983
Brassica oleracea (Italica group)	Ockendon, 1984
B. oleracea (Gemmifera group)	Ockendon, 1985

Source: Kalloo (1994).

Achievements of Tissue Culture

Plant tissue culture techniques, in combination with recombinant-DNA technology, are essential requirements for development of transgenic plants. However, culture techniques like anther/pollen/ovule/meristem can themselves be used for improvement of different traits or as an aid to conventional plant breeding. In addition, through the use of protoplast fusion and hybrid embryo rescue, hybridization between distant taxa/species has been successfully achieved. Somaclonal variation generated during culture can also be utilized for recovery of variants showing disease resistance or other desirable traits. The use of such techniques for improvement of vegetable crops is given in Table 11.

Table 11: Achievements of tissue culture for improvement of vegetable crops.

Crop	Technique applied for	Reference
Anther/ microspore/ ovule culture		
Brassica oleracea	Haploid production	Lilo and Hansen. 1987; Phippen and Ockendon, 1990 and Farnham, 1998
Capsicum annum	Haploid production	Wang *et al;* 1973; Harn *et al.,* 1975
Lycopersicon esculentum	Haploid production	Sharp *et al.,* 1971; Greshoff and Doy, 1972
Solanum melongena	Haploid production	Raina and Iyer, 1973
Cucurbita pepo	Haploid production	Metwally *et al.,* 1998 a
Brassica rapa	Haploid production	Kuginuki *et al.,* 1997
Embryo rescue		
L. esculentum x L. peruvianum	To rescue hybrid embryo	Thomas and Pratt, 1981
P. vulgaris x P. angustissimus	To rescue hybrid embryo	Belvanis and Dore, 1986
Vigna pubescens x V. unguiculata	To rescue hybrid embryo	Fatokun and Singh, 1987
S. melongena x S. torvum	To rescue hybrid embryo	Blestsos *et al.,* 1998
Meristem culture		
Allium sativum	Elimination of onion yellow dwarf virus	Walkey *et al.,* 1987
A. ascalonicum	Elimination of leek yellow dwarf virus	Walkey *et al.,* 1987
Pisum sativum	Elimination of pea seed borne mosaic virus	Kartha and Gamborg, 1978
Solanum melongena	EMCV	Raj *et al.,* 1991
L. esculentum	Cryopreservation	Grout *et al.,* 1978
P. sativum	Cryopreservation	Kartha *et al.,* 1979
Protoplast fusion		
S. melongena + S. integerefolium	Resistance to Pseudomonas	Kameya *et al.,* 1990
S. melongena + S. saintwongesi	Resistance to Pseudomonas	Asao *et al.,* 1994
B. oliracea + B. compestris	Disease resistance	Itoh *et al.,* 1991
B. oleracea + B. napus	Atrazine resistance	Jourdan *et al.,* 1989
B. oleracea + Armoraia	Asymmetric hybridization	Navratilova *et al.,* 1997
L. esculentum X L. peruvianum + L. lycopersicoides	Fertile somatic hybrids	Matsumoto *et al.,* 1997
L. esculentum + L. peruvianum	Somatic hybrids with intermediate morphology	Scala *et al.,* 1984
Somaclonal variation		
L. esculentum	Tolerance to Fusarium oxysporum culture filtrate	Scala *et al.,* 1984
	Resistance to *F. oxysporum*	Shahin and Spivey, 1986
	Tolerant to TMV	Cassells *et al.,* 1986
	Resistance to bacterial wilt	Toyoda *et al.,* 1989

Source: Singh *et al.* (2001).

Genetic Engineering

Concept

The necessary techniques for genetic engineering in plants are being constantly refined, but the available techniques have already demonstrated the power and usefulness of approaches. The general approach for genetic engineering in plants may be outlined as (i) introduction of gene of interest into the cells of concerned plant species, (ii) integration of this gene into nuclear/organellar genome of plant cells, (iii) expression of transferred gene in new genetic background, (iv) regeneration of whole plants from genetically-modified cells, and finally (v) transmission of transferred gene to sexual progeny of these plants. It may be kept in mind that any gene that is to be transferred into a plant has to be first obtained to multiple copies. This is ordinarily achieved either by first cloning the gene in *E. coli* or by amplifying the gene using PCR (polymerase chain reaction). It may further be noted that gene transfers in plants are primarily based on tissue culture (therefore, a reliable protocol for the regeneration of whole plants from tissue cultures of concerned plant species is a pre-requisite for such work) and that integration and expression of introduced genes must be stable to be transmitted through sexual process. *A plant in which a foreign gene has been transferred through genetic engineering is called a transgenic plant and the gene so transfer is called transgenic*. Progress in this exciting area of research for production of transgenic plants has been so spectacular that by the turn of the century, we hope to be growing crops, which have been tailored to market specifications by the addition, subtraction or modification of genes. Transgenes will also be important in increasing the efficiency of crop production systems. For instance, transgenic plants resistant to herbicides, insects, viruses and hosts of other stresses have already been produced. Transgenic plants have also been produced, which are suitable for food processing (*e.g.* bruise resistance and delayed ripening in tomato). Another exciting example is the production of male sterile (due to *barnase* gene) and fertility restorer (due to *barstar* gene) plants in *Brassica napus*, so that hybrid seed in future will be conveniently produced without manual emasculation and controlled pollination. This has also eliminated the need for a search of cytoplasmic male sterility (*cms*) and fertility restoration system in crops, where hybrids are intended to be produced for higher yields. Another major goal for production of transgenic plants is their use as bioreactors or factories for production of especially chemicals and pharmaceuticals. This area is described as 'molecular farming' or 'molecular pharming'. The transgenic plants have also been produced for identification of regulatory sequences for many genes, using gene constructs with overlapping deletion.

Methods of Gene Transfer

Following methods are used for transferring the genes for development of varieties/ hybrids having resistance to biotic and abiotic stresses.

Ti Plasmid

This plasmid is found in *Agrobacterium tumefaciens*, which produces crown gall (tumour) in a large number of dicot species. Similarly, Ri plasmid occurs in *A. rhizogenes*, which causes the hairy root disease of several dicots and monocots. The Ti (tumour inducing) and Ri (root inducing) plasmids are essentially similar; they both carry genes for the production of tumours and roots, respectively. Both Ti and Ri plasmids have a *t- region* (the region transferred into host genome; generally called *T- DNA*) and a *vir* (= virulence) region (Fig.14). The T-DNA is flanked on both the sides by a border (B) of direct repeat sequences each of 25bp (base pairs). The T-DNA has four important genes; two genes, tms 1 (now called *iaaM*) and *tms 2* (now called *iaaH*), govern the synthesis of IAA; one gene, *tmr* (now known as *ipt*) determines the production of a cytokinin (isopentenyl adenosine); while the remaining gene *os,* is concerned with opine (a nitrogenous compound synthesis). The genes *iaaM, iaaH* and *ipt* are involved in tumour production and are usually called oncogenes; these genes have to be removed from the *T-DNA* of a Ti plasmid before it can be used as a vector. The process of removal of three tumor-producing genes from T-DNA is called *disarming* of *Ti* plasmid, and the resulting plasmid is referred *disarmed.*

The border sequences delimiting T-DNA are the sites where an endonuclease, encoded by *virD* operon of vir region, nick (produces a single – strand break) the Ti plasmid. Thus, these sequences are essential for the excision of T- DNA and its subsequent integration into plant genome must be integrated in T- DNA between these two border sequences. The *vir* region is essential for T-DNA transfer, although Vir region itself is not transferred into plant cells. The *vir* region contains 6 or 8 operon depending on type of *Ti* plasmid. Operon virA of vir region encodes the protein *VirA*; this protein is activated (by autophosphorylation) by acetosyringone. Acetosyringone is a phenolic compound produced by dicot plant cells located at wound sites. Activated *VirA* now phosphorylates *VirG* (encoded by *virG* operon), which forms dimers. *VirG* dimers bind to promoters of all *vir* operon leading to their transcription. One product (*VirD2*) of the *VirD* operon is an endonuclease, which nicks the *T- DNA* at right border; ultimately a single-stranded copy of *T-DNA* is produced. *VirD2* remains covalently bound to the 5'-end of this single-stranded *T- DNA* copy and guides it into plant cell nucleus. Transfer of *T- DNA* occurs through a tube like structure formed between the bacterial and plant cells by certain ultimately integrates into plant genome.

There are two types of Ti vectors, cointegrate and binary. The gene to be transferred is usually first cloned in an *E. coli* vector. This *E. coli* vector is ultimately integrated within the *T- DNA* of a disarmed *Ti plasmid*; the Ti vector so produced is called cointegrate vector. Such a vector transfers it into the plant cells the sequences of *E. coli* vector along with the transgene. In case of binary vectors, a pair of vectors is essential for gene transfer. One vector of this pair (called *mini – Ti*) contains disarmed *T-DNA*, but it does not have the *vir* region. The other vector of the pair (called helper *Ti*) has *vir* region but does not have *T-DNA*. Mini-*Ti* can propogates both *E. coli* and *Agrobacterium*. The transgene is integrated within *T- DNA* of mini *Ti*, which is then cloned in *E. coli*. The recombinant mini-*Ti* is introduced into such a *A. tumefaciens* cells, which contain the helper *Ti*. The *vir* products of helper *Ti* bring about transfer of mini- *Ti T- DNA* containing the transgene.

The reporter gene, *e.g,* for antibiotic resistance, is placed in the disarmed *T-* region; this reporter gene permits the direct selection of transformed (containing the *T- DNA*) plant cells. The gene is DNA segment being transferred is also integrated into *T-DNA*. This especially constructed plasmid is introduced into *A. tumefaciens*, which is then co- cultivated with leaf discs or some other suitable explant, or with protoplasts. After a period of co-culture, the leaf discs etc. are transferred to a selective medium, which (i) selectively prevents the multiplication of *A. tumefaciens* cells and, at the same time (ii) allows a selective proliferation of only those plant cells into which the T-DNA from this bacterium has been integrated. The same medium may also induce plantlet regeneration; if not, the resulting cell mass (of transformed cells) may be transferred to a regeneration medium to obtain transformed plants. This approach is highly successful in dicot species. It has also been applied to monocots by adding exogenous acetosyringone during co-culture. The integration of T-DNA into the host genome is random (in terms of the site of insertion), leads to stable transformation and transgenes are transmitted to sexual progeny.

Plant Viruses

Techniques are being developed to use certain DNA and RNA plant viruses, *e.g,* cauimo viruses, gemini viruses *etc.* as vectors. The cauliflower mosaic virus (CaMV; a caulimo virus) is a DNA virus. A DNA segment may be integrated into the viral DNA, which is suitably modified to serve as a vector, and the host plants may then be infected with this especially constructed virus. Following infection, the virus spreads systematically into the host plant; plant cells usually have a high copy number of virus (upto 10^5 virions/cell). Therefore, each cell will also have a very high copy number of the transferred gene. Similarly, wheat dwarf virus (a gemini virus) has been cloned and reintroduced into wheat plants. The wheat dwarf virus infects several members of Graminae, while caulimo viruses infect many species of the family Cruciferae. Viral vectors usually produce only transient transformation, confined to that generation only; they do not lead to stable transformation.

Direct DNA Uptake

DNA is taken up directly by plant cells, plant protoplasts, pollen-grains, mature embryos and even whole seeds; in many cases a transient expression of foreign genes has also been reported. Genetic transformation through direct DNA uptake by protoplasts was first reported in case of tobacco. Polyethylene glycol (PEG) promotes DNA uptake by protoplasts (by changing the membrane properties) as well as the frequency of genetic transformation (upto 10^{-3}). PEG-induced DNA uptake by protoplasts and genetic transformation.

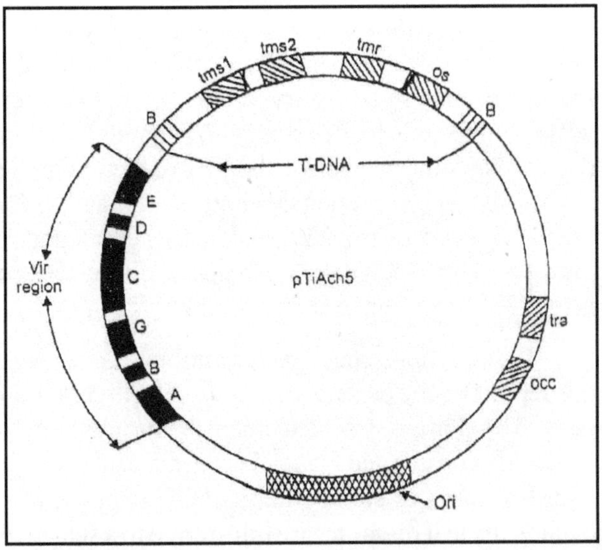

Figure 14: A schematic representation of Ti plasmid of *A. tumefaciens*. The T- region (*T- DNA*) is bordered on both sides by a direct repeat sequence of 25 bp (B) ; T-DNA contains genes tms1 (now *iaaH*), *tmr* (now *ipt*) and *os*. The *vir* region has atleast six operons; *vir D* operon codes for an endonuclease responsible for excision of a single stranded copy *of T- DNA* from *Ti* plasmid and then guiding this DNA into plant cell nucleus.

Source: Singh, 2002, p. 724.

Microinjection and Macroinjection

Plant regeneration from transformed protoplasts, still remains a problem. Therefore, cultured tissues that encourage the continued development of immature structures, provide alternate cellular targets for transformation. This immature structures may include immature embryos, meristems, immature pollen, germinating pollen, isolated ovules, embryogenic suspension cultured cells, *etc*. The main disadvantage of this technique is the production of chimeric plants with only a part of plant transformed. However, from this chimeric plants, transformed plants of single cell origin can be subsequently obtained (Fig. 15). Utilizing this approach, transgenic chimeras have actually been obtained in oilseed rape (*Brassica napus*). When cells or protoplasts are used as targets in the technique of microinjection, glass micropipettes with 0.5-10 μm diameter tips are used for transfer of macromolecules into the cytoplasm or nucleus of a recipient cell or protoplast. The recipient cells are immobilized on a solid support (cover slip or slide, *etc.*) or artificially bound to a substrate or help by a pipette under suction (as done in animal systems). Often a especially designed micromanipulator is employed for microinjecting the DNA. Although, this technique gives high rate of success, the process is slow, expensive and requires highly skilled and experienced personnel.

The DNA microinjection employing needles with diameters greater than cell diameter has also been tried. In rye (*Secale cereale*), a marker gene has been macroinjected into the stem below the immature floral meristem, so as to reach the sporogenous tissue (De la Pena *et al.*, 1987) leading to unsuccessful production of transgenic plants. Unfortunately, this technique could not be successfully repeated with any other cereal, when tried in several laboratories. Therefore, doubt is expressed about the validity of earlier experiments conducted with rye (Potrykus, 1991).

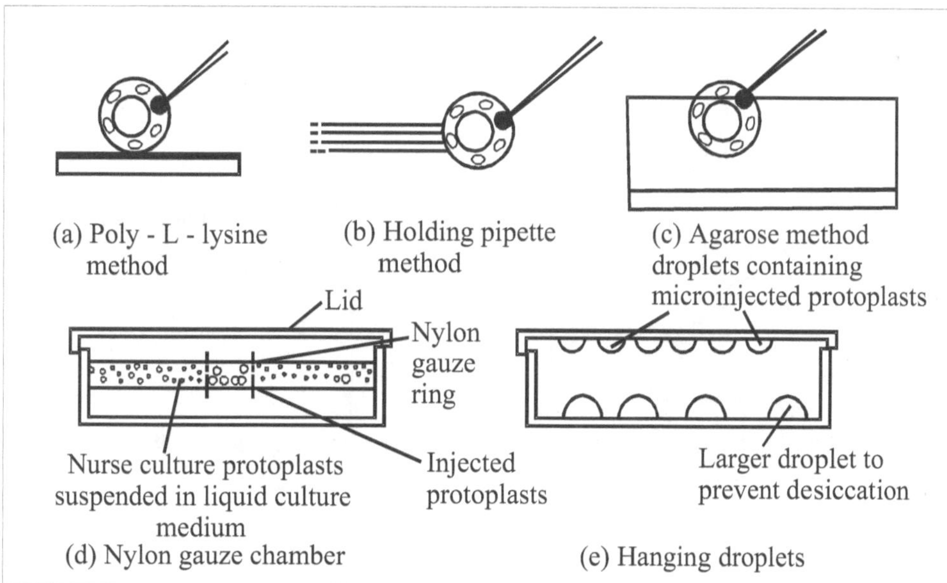

Figure 15: Methods for microinjection and culture of plants protoplasts.
(*Source* Gupta, 2001, p. 362).

Microjectiles (biolistics or particle gun)

In recent years, it has been shown that DNA delivery to plant cells is also possible, when microparticles (tungsten or gold) coated with DNA of interest are accelerated to a very high initial velocity (1,400 ft per sec). These microprojectiles, normally 1-3 nm in diameter, are carried by a 'macroprojectile' or the bullet and are accelerated into living plant cells (target cells can be pollen, cultured cells, cells in differentiated tissues and meristems) so that they can penetrate cell- walls of intact tissue (Fig. 16). The acceleration is achieved either by an explosive charge (cordite explision) or by using shock waves initiated by a high voltage electric discharge. The design of two particle guns used for acceleration of microprojectiles is shown in (Fig. 16). Transformed plants using the above technique have been obtained in many cases including soybean, tobacco, maize, rice, wheat, etc. Transient expression of genes transferred in cells by this method has also been observed in onion, maize, rice and wheat. There is no other gene transfer approach, which has met with so much of enthusiasm. Consequently considerable investment has been made in experimentation and manpower for development of this technique. The advantages of this method over microinjection include the following: (i) thousands of particles are accelerated at the same time, causing multiple hits resulting in transfer of genes into many cells simultaneously, (ii) since intact cells can be used, some of the difficulties encountered with the use of protoplasts are automatically circumvented; (iii) the method is universal in its application, so that cell type, size and shape or the presence/absence of cell-walls do not significantly alter its effectiveness. In view of this, particle bombardment method using microprojectiles has a great promise in a variety of plant species.

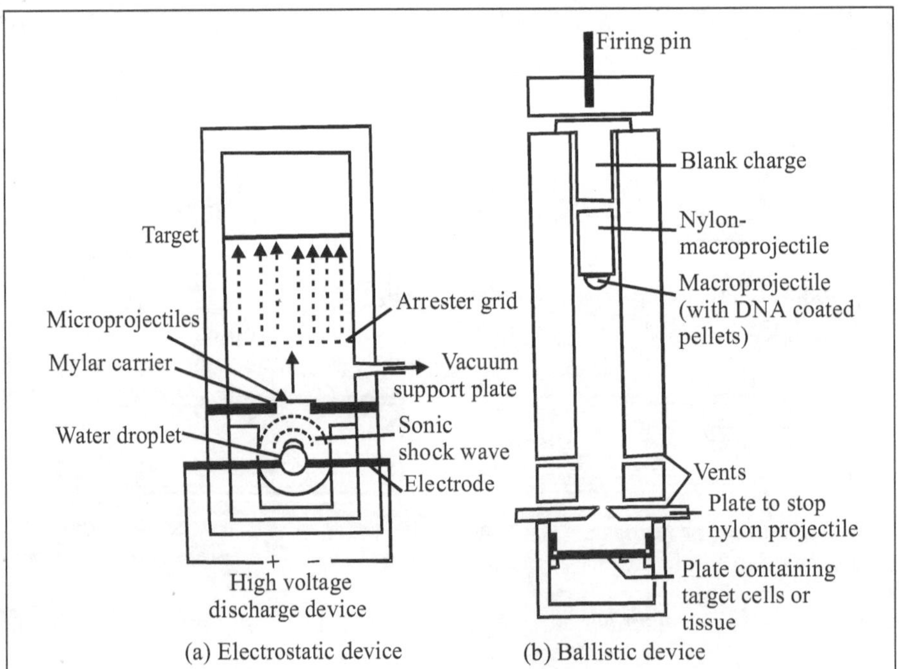

Figure 16: Microprojectile acceleration devices. *(Source :* Gupta 2001 p. 364).

Electroporation Method

This method is based on the use of short electrical impulses of high field strength. These impulses increase the permeability of protoplast membrane and facilitate the entry of DNA molecules into the cells, if the DNA is in direct contact with the membrane. In view of this, for delivery of DNA to protoplasts, electroporation is one of the several routine techniques for efficient transformation. However, since regeneration from protoplasts is not always possible, cultured cells or tissue explants are often used. Consequently, it is important to test whether elctroporation could transfer genes into walled cells. Discharging a capacitor across the electrodes in a specially designed electroporation chamber generates the electroporation pulse. Either, a high voltage (1.5 kV) rectangular wave pulse of short duration or a low voltage (350 V) pulse of long duration is used. The latter can be generated by a home-made machine (Fig. 17).

Protoplast in ionic solution containing the vector DNA, is suspended between the electrodes, electroporated and then plated as usual. Transformed colonies are selected as described earlier. In this case, cat gene associated with a suitable promoter sequence is transferred. Transformation frequencies can be further improved by: (i) using field strength of 1.25 kV/cm, (ii) adding PEG after adding DNA, (iii) heat shocking protoplasts at 45 ^{0}C for 5 minutes before DNA adding and (iv) by using linear instead of circular DNA (Fig. 18).

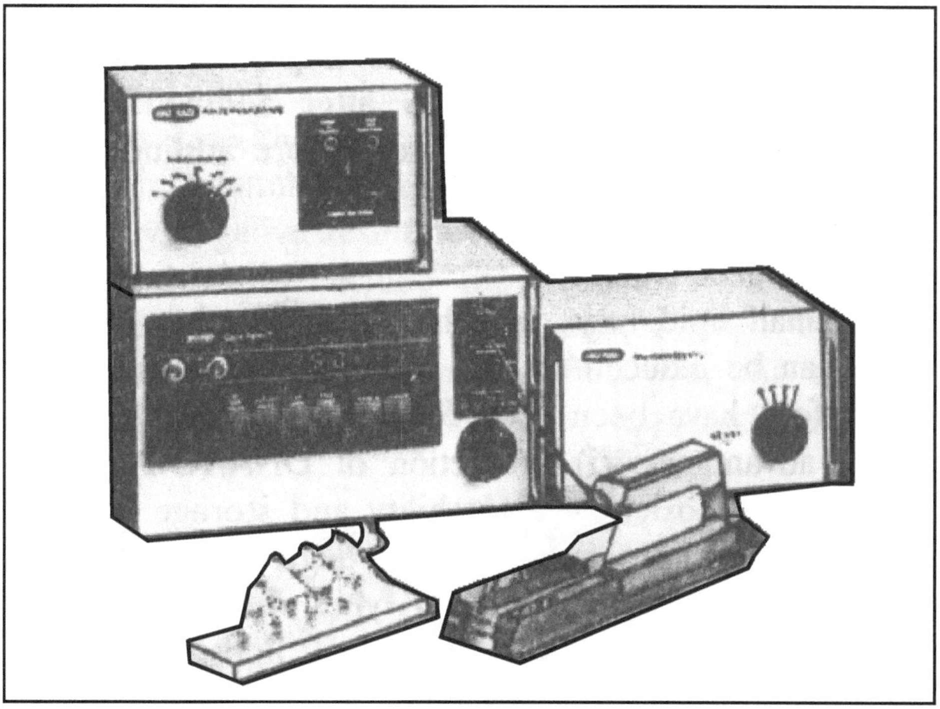

Figure 17: An eectroporation system, suitable for transformation of bacteria, mammalian cells and plant cells. *(Source* : Gupta, 2001, p. 365).

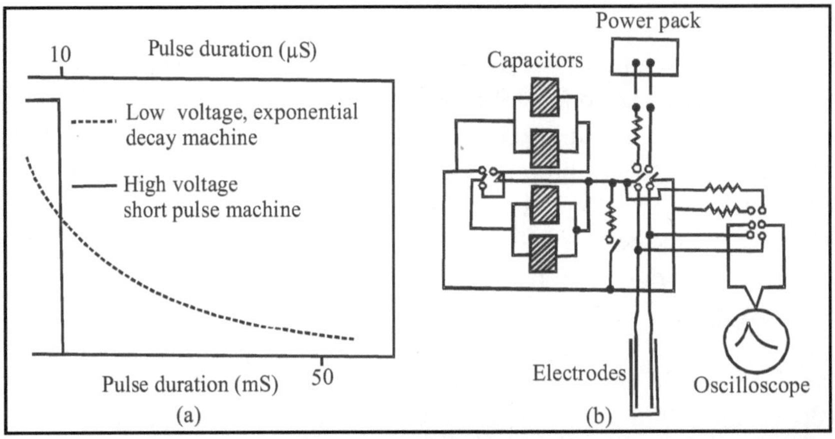

Figure 18: (a) Details of output characteristics of two principal types of electroporation devices in use: (b) circuit diagram of low voltage, exponential decay machine. (*Source*: Gupta 2001, p. 365).

Liposome Mediated DNA Uptake

Liposomes are small lipid bags, in which a large number of plasmids are enclosed. They can be induced to fuse with protoplasts using devices like PEG, and therefore have been used for gene transfer. This technique offers following advantages: (i) protection of DNA/RNA from nuclease digestion, (ii) low cell toxicity, (iii) stability and storage of nucleic acids due to encapsulation in liposomes, (iv) high degree of reproducibility and (v) applicability to a wide range of cell types. In this technique, DNA enters the protoplasts due to endocytosis of liposomes, involving following steps: (i) adhesion of liposomes to protoplast surface, (ii) fusion of liposomes at site of adhesion and (iii) release of plasmids inside the cell. The technique has been successfully used to deliver DNA into the protoplasts of a number of plant species (*e.g.* tobacco, petunia, carrot, *etc.*). A illustration of this technique is given in Fig. 19.

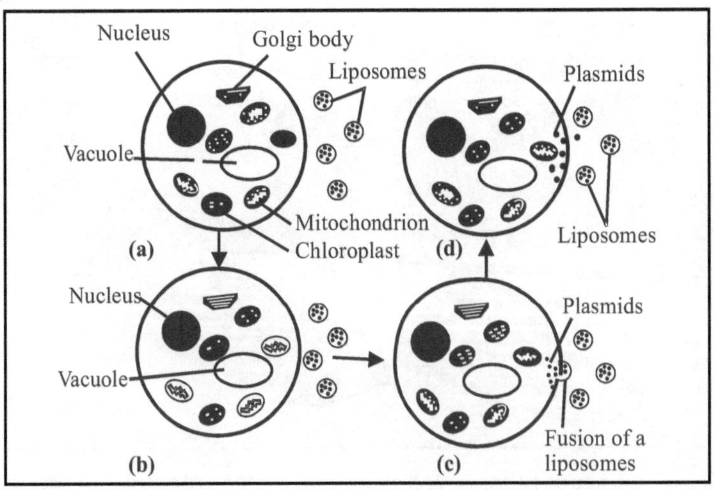

Figure 19: Fusion of plasmid-filled liposome with isolated protoplast as a method of plasmid uptake. *(Source* : Gupta 2001, p. 366).

Calcium Phosphate Precipitation

Foreign DNA can also be carried with the Ca^{++} ions, to be released inside the cell due to precipitation of calcium in the form of calcium phosphate. In the past, this method was considered to be very important for gene transfer in plants.

Transformation Using Pollen or Pollen Tube

There has been a hope that DNA can be taken up by the germinating pollen and can either integrate into sperm nuclei or reach the zygote through the pollen tube pathway. Both these approaches have been tried and interesting phenotype alterations suggesting gene transfer have been obtained. In no case, however, unequivocal proof of gene transfer has been available. In a number of experiments, when marker genes are used for transfer, only negative results are obtained. Several problems exist in this method and these include the presence of cell wall, nucleases, heterochromatic state of acceptor DNA, callose plug in pollen tube, *etc.* transgenic plants have never been recovered using this approach and this method, though very attractive, seems to have little potential for gene transfer.

Incubation of Dry seeds, Embryos, Tissues or Cells in DNA

Incubation of dry seeds, embryos, tissues or cells in known DNA (viral or non-viral having defined marker genes) has been tried in many cases and expression of defined genes has been witnessed. However, in no case proof of integrative transformation could be available. In all these cases, plant cell walls not only work as efficient barriers, but are also efficient traps for DNA molecules. It would be very surprising if DNA can cross cell-walls efficiently without permeabilizing them either by PEG, or by electroporation or by any other device.

Transformation by Ultrasonication

In wheat, tobacco and sugarbeet, explants after being cultured for a few days, are sonicated with plasmid DNA (carrying marker genes like cat, npt II and gus). When sonicated calli are transferred to selective medium, shots are obtained, although all control calli (not sonicated with plasmid DNA) died. In tobacco transgenic plants are obtained at a frequency of 22%. This work has been conducted at the Biotechnology Research Centre at Beijing (China).

Applications

Dispite the totipotency of plant cells, thus obviating the need of transforming specially the germ line cells as required in animals, production of a large number of transformed plants was, till recently, resticted mainly in tobacco, petunia and tomato. The efforts made use of genetic engineering for developing transgenic plants for various attributes are discussed below:

Engineering for Disease Resistance

Virus Resistance

There are several approaches, which have been used to engineer plants resistant to viruses. The first approach aims at introducing genes encoding viral coat protein to produce transgenic plants resistant to plant virus infection. The viral coat protein produced by the transgenic plants appear to block a receptor in the plant cell that is required for uncoating the viral nucleic acid, or to interfere with viral replication or expression of viral genes. Resistance is manifested by delay in symptom development or escape from infection altogether. A second approach to engineering virus resistance in plants has centered on satellite RNA's. It has been observed that expression of satellite RNA's in transgenic plants can reduce infection with cucumber mosaic virus. The third approach aims at using virus RNA to govern transgenic resistance. Following the successful field-testing of coat protein mediated resistance against TMV in tomato plants (Nelson *et al.*, 1988), additional testing of a number of plant lines has occurred. A better understanding of the cellular and molecular mechanisms that confer transgene mediated resistance may lead to second generation viral genes that improve resistance beyond current levels and extend its application to still more viruses and crop plants. The transgene (s) in vegetable crops against viruses is given in Table 12.

Table 12: Vegetables engineered against viruses

Purpose of genetic manipulation	Transgenic Product	Transformed Plant	Origin of Transgene	Reference
Potato virus X (PVX)	Viral coat protein	Potato	PVX	Homenway *et al.*, 1988
Potato virus Y (PVY)	Viral coat protein	Potato	PVY	Stark and Beachy, 1989
Tomato yellow leaf curl virus	Coat protein	Tomato	TYLCV	Kunik *et al;* 1994
Tomato spotted wilt virus	N-gene	Tomato	TSWV	Pang *et al*., 1992
Tomato yellow leaf curl virus	Truncated version of CI gene of TYLCV	Tomato	TYLCV	Brunetti *et al.*, 1997
Tomato mosaic virus	Antisens RNA	Tomato	TOMV	Beachy *et al.*, 1987
Cucumber mosaic virus	Satellite RNA	Tomato	CMV	Stommel *et al.*, 1998
Cucumber mosaic virus	Viral coat protein	Cucumber	CMV	Gonsalves *et al.*, 1992

Source: Singh *et al.* (2001).

Fungal Resistance

Several examples are now available, where transgenic plants resistant to fungal pathogens have been produced. One major approach for engineering plants resistant to fungal diseases includes use of *chitinase* genes of plant and/or bacterial origin. *Chitinase* enzyme degrades chitin rich walls of fungi and delay symptom development. Other antifungal protein genes include osmotin, *glucanase thionin, peroxidase etc*. Recently, an effective antifungal protein from barley seeds called *ribosome-inactivating protein* (RIP) has been isolated. RIP acts by inhibiting protein synthesis *via* N-glycosidase modification of fungal 28 S rDNA. Transgenic crop plants expressing RIP proteins are now being tested against many fungi (Sharma and Kumar, 1995).

Bacterial Resistance

Most of the crop plants are susceptible to bacterial diseases, which have been difficult to control. Bactericides are not a complete solution because bacteria quickly evolve resistance to them. Using molecular biological tool, a novel approach for engineering resistance against bacteria has been undertaken to isolate the *tabtoxin* resistance genes (*ttr*) from bacteria *Pseudomonas syringae* pv. *tabaci* and subsequently to introduce into susceptible plants for transgenic resistance (Yoneyama and Anazi, 1991). In a similar way, *Phaseolotoxin-insensitive OCTase* (Ornithine carbamyl transferase) gene has been isolated from *Pseudomonas syringae* pv. *phaseolicola* and introduced to beans to observe total resistance (Herrera Estrella and Simpson, 1995). This simple and useful strategy for molecular breeding of disease resistant plants by detoxification of pathogenic toxins has a great potential to be exploited against bacterial diseases in vegetable crops. The successful transgene (s) in vegetables against bacteria are given in Table 13.

Table 13: Examples of successfuls engineered vegetables against bacteria

Resistance against	Transgene product	Origin of transgene	Transformed plant
Erwinia carotovora	Lysozyme	T4 bacteriophage	Potato
Erwinia carotovora	Tachyplesin	Horseshoe crab	Potato
Pseudomonas syringae pv. *Phaseolicola*	Phaseolotoxin insensitive *OCTase*	*Pseudomonas syringae* pv. *phaseolicola*	Bean
Erwinia carotovora	Pestate lyase	*Erwinia carotovora*	Potato
Erwinia carotovora	Glucose oxidase	*Aspergillus niger*	Potato

Source: Singh *et al.* (2001).

Engineering for Insect-pest Resistance

Crop protection from insects plays a very important role in agricultural productivity. Progress in developing insect control in transgenic plants has been initially achieved through the expression of insect control protein genes of *Bacillus thuringiensis.* The insecticidal protein gene from *Bt* conferred for reaching resistance to classes of certain lepidopteron species when transgenic tomato, potato, maize and cotton plants have been produced. There are several *Bt* genes encoding proteins that are largely insecticidal. These proteins are not harmful to mammals but kill insect by interfering with the ion transport in the midgut. Different types of proteins are available that selectivity kill specific groups of insects raising the possibility of producing plants resistant to particular pests.

Plants also have endogenous defense mechanisms against insect attack. Several plants have been shown to synthesize systematically protease inhibitors in response to wounding. These proteins are believed to provide some defense against feeding insects. Works have shown that cowpea seeds containing a high concentration of chymotrypsin inhibitor are resistant to insect attack (Hilder *et al.,* 1990). Transgenic tobacco plants have been generated that express constitutively different amounts of cowpea seed chymotrypsin inhibitor. Feeding experiments have shown that there is a direct relationship between the amount of chymotrypsin inhibitor produced and the degree of protection afforded against a range of insects. Secondary metabolities produced by plants have also been implicated in the resistance to insect attack. However, biosynthesis of each of these metabolites involves a series of steps (sometimes even more than one biosynthesis pathway), each control by a separate gene. Furthermore, these genes are tissue specific in expression. These features make the production of transgenic plants difficult in this case. CpT_1 is a secondary metabolites and its transfer is easier because of it is single gene controlled. The production of transgenic plants by transfer of genes for entire multi-enzyme biosynthesis pathway (or for its augmentation) is though not yet achieved, but seems possible in future. For this purpose not only the transfer of genes is required, but the gene expression needs to be regulated, otherwise it leads to 'yield penalty' and also to toxic effect when consumed by human and livestock. Defined promoter sequences are now available, which will, allow in the transgenic plants, a control over the temporal and a partial expression of genes for biosynthetic pathway of secondary metabolites. Utilizing these facilities the transgene (s) expressing resistant to insects are given in Table 14.

Table 14: Engineered vegetables expressing insect resistant gene.

Transgenic Product	Organic of Transgene	Target Insect	Transformed Plants	Reference
Cry 1 AB	*Bacillus thuringiensis*	Lepidoptera	Tomato	Delannay *et al;* Noskin1989
Cry 1 AB	*Bacillus thuringiensis*	Lepidoptera	Tomato	Kumar *et al.*, 1998
Cry 1 H	*Bacillus thuringiensis*	Lepidoptera	Corn	Duck *et al*; 1997
Cry 3 A	*Bacillus thuringiensis*	Coleoptera	Brinjal	Jelenkovic *et al;* 1998
CpT_1(cowpea trypsin inhibitor)	Cowpea	Coleoptera Lepidoptera	Tomato	Gatehouse *et al;* 1992
Tomato proteinase inhibitor I	Tomato	Lepidoptera	Tomato	McGurl *et al.*, 1994
Tomato proteinase inhibitor II	Tomato	Lepidoptera	Tomato	McGurl *et al.*, 1994
α-amylase inhibitor of the common bean	Common bean	Coleoptera	Common bean	Schroeder *et al.*, 1995
Snowdrop lectin (GNA)	-	Lepidoptera	Tomato	Gatehouse *et al;* 1992
Bovine pancreatic trypsin inhibitor	-	Lepidoptera	Lettuce	Schuler *et al.*, 1998

Source: Singh *et al.* (2001).

Engineering for Parthenocarpic Plants

The plants able to develop fruits in the absence of fertilization are said to be parthenocarpic. In such fruits, seeds are absent which can increase fruit acceptance by consumers. Parthenocarpy has been achieved by genetic engineering in brinjal and tomato. For parthenocarpic development of fruit, the heterologous gene, def H9-iaaM, is used in transformation. Line transformed with this gene gave normal sized parthenocarpic fruits even from the emasculated flowers. It has been observed that in a protected cultivation trial, three transgenic plant hybrids yielded more than the corresponding non-transgenic control even when the latter used treated with plant growth regulators (Rotino *et al.,* 1998). The transgenic vegetables for parthenocarpic fruits are given in Table 15.

Table 15: Successful engineered vegetables for parthenocarpic fruits

Transgenic aganist	Transgene	Origin of transgene	Transformed plant	Reference
Tomato	*Rol B gene*	*A. rhizogens*	Tomato	Carmi *et al.*, 1997
Brinjal	*Iaah gene*	*Ps. Syringae* pv. *savas*	Brinjal	Rotino *et al.*, 1997

Source: Singh *et al.* (2001).

Engineering for Post-Harvest Traits

In tomato plants have been engineered for reduced synthesis of polygalacturonase, which is responsible for softening the cell wall during ripening of some fruits (Fig. 20). Polygalacturonase degraded the pectin fraction of the fruit cell wall thus plays a role in fruit softening. It has been shown that a single antisense gene could inhibit expression of tomato *Pgu* gene during ripening. Using the antisense technology, antisense RNA from polygalacturonase has been inserted. When normal PGU mRNA is synthesized the antisense RNA binds to it and prevents synthesis of *Pgu.* Antisense RNA against ACC synthase (or ACC oxidase. It has been observed that when it expressed in tomato plants and this inhibited by ethylene biosynthesis by 99.5% production of which appears to be involved in the initiation, modulation and coordination of expression of many of the genes required for the ripening process. Genetically engineered tomatoes have become a reality of commercial company calgene is set to market genetically engineered '*Flvr-savr*' tomatoes using antisense RNA technology. These tomatoes have extended shelf life and increased resistance to brushing. The successful transgene (s) in vegetable crops for post-harvest traits are given in Table 16.

Engineering for Male sterility and Fertility Restoration

During 1990-1992, an exciting example of producing transgenic plants with male sterility and fertility restoration genes has become available in *brassica napus*. This fasciltate hybrid seed without manual emasculation and controlled pollination. In 1990, C. Mariani and other from Belgium, successfully used a gene construct having other specific promoter (from TA29 gene of tobacco) and bacterial coding sequence for a ribonuclease (barnase gene

Table 16. Examples of successful engineered vegetables for post harvest traits

Purpose of Genetic manipulation	Transgene Product	Origin of Transgene	Reference	Transformed Plant
Improved storage/shelf -life	Antisense polygalacturonase	Tomato	Sheely *et al.*, 1980	Tomato
Ripening	Antisense ACC oxidase	Tomato	Hamilton *et al.*, 1990	Tomato
	Antisense ACC synthase	Tomato	Oeller *et al.*, 1991	Tomato
Fruit pigmentation	Phytoene synthase gene	Tomato	Bird *et al.*, 1991	Tomato

from *Bacillus amyloliquefaciens*) for production of transgenic plants in *B. napus*. The same group of workers (1992) used another gene construct later involving the same anther specific promoter *i.e.* TA29 and the barstar gene from *B. amyloliquefaciens*, for production of transgenic plants in *B. napus*. The product of *barstar* gene is a *riboinuclease-inhibitor*. It forms a complex with ribonuclease and neutralizes its cytotoxic properties. In *Bacillus, ribonuclease* is active extracellularly and bacterium itself is produced by a *ribonuclease inhibitor* protein coded by *barstar gene*. When transgenic male sterile plants (with *barnase* gene) are crossed with transgenic male fertile plants (with *barstar* gene), the F_1 plants expressed both genes so that male fertility is restored due to suppression of cytotoxic ribonuclease activity in the anther by the formation of cell specific Rnase/ Rnase – inhibitor complexes. This system known as *'Barnase-Bar Star'* systems has also been proved successful in tomato and corn.

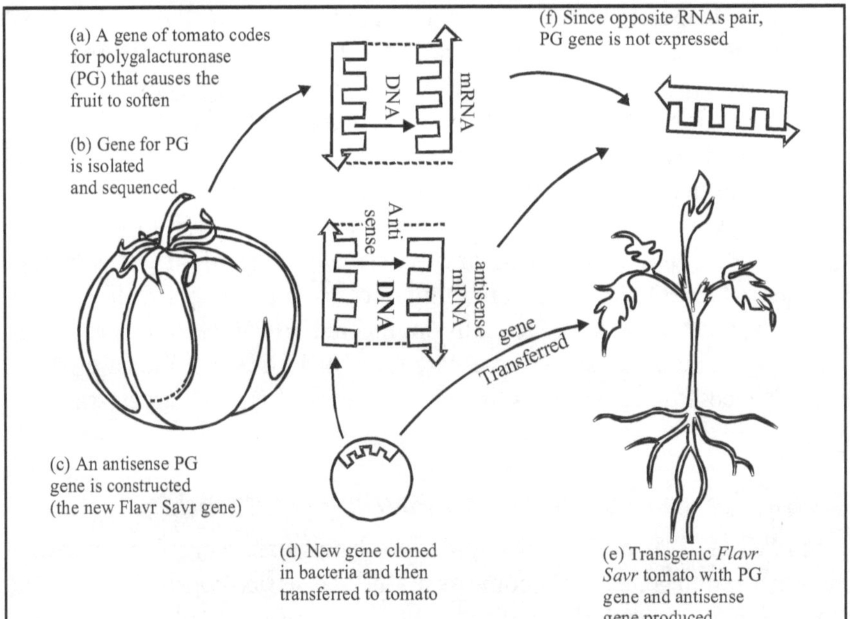

Figure 20: Production of transgenic *Flavr Savr* tomato using in antisense RNA gene.
Source: Gupta, 2001, p. 378.

Engineering for Herbicide Tolerance

Due to increasing concern about contamination of environment by herbicides, new herbicides are being develope that are safer and biodegradable. This has necessitated the development of resistance in crop plants against safer herbicides. Pre-and post-emergence herbicides are used to control weeds in vegetable crops. However, certain herbicides are not selective to the crop. Herbicides, normally affect process like photosynthesis or biosynthesis of amino acids (Table17). Two approaches have, therefore, been used for the development of resistant plants. **First** approach is modification of the target in which target has been modified for developing resistance against at least three herbicides (glyphsate, sulphonylureas and imidazolinones). The Transgenic petunia plants resistant to glyphosate (active ingradient for Roundup herbicide) have been developed by transfer of a gene for EPSPS (5- enol-pyruvyl- shikimat-3- phosphate synthase), that overproduces this enzyme. This over expression gene is isolated from plants selected for herbicide resistance. In some cases, a gene *aroa* is isolated from the bacteria, *Salmonella typhimurium* or *E. coli*, and was transferred to tomato and/ or tobacco. Another class of herbicides includes sulphonyl urea compounds (active ingredients of Glean and Quist herbicides) and imidazolinones that inhibit the enzyme acetolactate synthase (ALS), which is involved in biosynthesis of branched chain aminoacids like leucine, isoleucine and valine.

Transgenic tobacco plants expressing a mutant ALS gene from tobacco or Arabidopsis, have been produced that are tolerant to sulphonylurea herbicides. Similarly, transgenic tobacco plants have been produced by incorporation of genes for resistance against (i) *L-phosphinothricin* (PPT), which is the active ingredient of herbicide 'Basta' and inhibits glutamine synthase (GS) and (ii) atrazine, which inhibits photosynthesis. These genes are isolated from *Medicago sativa* and *Amaranthus hybridus,* respectively. The transgenic tobacco plants expressed low level of resistance against corresponding herbicides. A German company, Hoechst, isolated a gene from '*Streptomyces hygroscopic* which encodes an enzyme, capable of inactivating the herbicide ' Basta'. Transgenic plants with this gene have been produced and field tested, demonstrating effectiveness of this gene for protection against the herbicide 'Basta.'

Second approach is detoxification by degradation of herbicide in which a number of detoxifying enzymes have been identified in plants as well as in microbes. Some of these enzymes include: (i) glutathione-S transferase or GST, which detoxifies the herbicide atrazine; (ii) nitrilase (coded by gene *bxn* in *Klebsiella pneumoniae*, which detoxifies the herbicide bromoxynil, and (iii) phosphinothricin acetyl transferase or PAT (coded by *bar* gene in *Streptomyces* spp.), which detoxifies the herbicide PPT (*L- phosphinothricin*). Transgenic tomato plants using *bxn* gene from *Klebsiella* and *bar* gene from *Streptomyces* and transgenic plants in potato, oilseed rape (*Brassica napus*) and sugarbeet using *bar* gene from *Streptomyces* have been obtained and are found to be herbicide resistant. Tissue culture technique has been used to isolate herbicide tolerant mutants. A paraquat tolerant mutant has been identified in *Lycopersicon*. In this case, an interspecific hybrid strain L_2 is used. The callus cells of this genotype are kept on an agar medium containing a lethal concentration of paraquat. The tolerant clone grew under these conditions. Plants produced have about

30-folds increase in paraquat tolerance as compared to strain L_2 (Thomas and Pratt, 1982). Similarly, cultures have been used to select and isolate mutants tolerant to various chemicals *e.g.* metribuzin in *Lycopersicon* (Trebst and Wietiska, 1975; Ellis, 1978) and *D. carota* (Templeton, 1981) and 2,4-D resistant cells in *D. carota* (Gressed, 1979).

Table 17: Mechanism of action of different herbicides and basis of achieving resistance against them in transgenic plants

Active Principle of Herbicide	Inhibited Pathway	Target Product	Use	Basis of Resistance
Amino acid biosynthesis inhibitor				
Glyphosate (Roundup)	Aromatic amino acid biosynthesis	EPSPS	Broad spectrum	Over expression of *EPSPS* gene; bacterial *aroA* gene
Sulphonylurea and Imidazolinones	Branched chain amino acid	ALS	Selected crops	Mutant *ALS* gene
Phosphinothricin	Glutamine biosynthesis	GS	Broad spectrum	Gene amplification *bar* gene: detoxification
Photosynthesis inhibitors				
Atrazine (lasso)	Photosystem II	QB (32k Da protein)	Selected crops	Mutant *PsbA* gene: *GST* gene:; detoxification
Bromoxynill (buctril)	Photosynthesis	-	Selected crops	*bxn* gene: detoxification

Source: Gupta (2001).

Engineering for Salt Tolerance

Salt is one of the most important limiting factor for the production of vegetable crops. A wide range of variation in plants for tolerance to salt has been reported. Traditional breeding is expensive, time-consuming and sometime may not be successful. Therefore, the plant tissue culture technique may be used to obtain salt tolerant cell lines. Salt tolerant lines have been isolated in *Lycopersicon* by the tissue culture technique. *L. peruvianum* and *L. pennelli* have been found tolerant to sodium chloride. Similarly, applying selection pressure to suspension cultures has isolated salt tolerant lines of *Capsicum annum*. Such lines could grow in liquid media containing 1-2 w/v sodium chloride.

Engineering for Abiotic Stress

Crop plants are subjected to a variety of stress conditions like, drought, high salt concentration of soil, excessive moisture, cold *etc.* identification and isolation of the most important gene involved in protection against stress conditions could help in development of transgenic plants having resistance/tolerance to different stress conditions. Studies are also being conducted on metabolites like proline and betaines, that are implicated in stress tolerance. With this background, transgenic plants resistant to a variety of stresses will produce in future. Exposure of plants to water deficit caused by either drought, salinity or freezing is manifested as an array of changes in cellular process (Bray, 1993). The transgenic strategy for drought tolerance came from analysis of plants that grow under desert (xerophytes) or extreme saline (halophytes) conditions. These plants survive water stress by accumulating sugar alcohol like mannitol, sorbitol, myoinositol *etc.* that may act as osmolytes and help in osmotic adjustment. They also help in osmoprotection serving as scavengers of active oxygen generated during water stress. Transgenic tobacco plants expressing bacterial mtl D gene that codes an enzyme for mannitol synthesis showed tolerance to high salinity. Similarly, better drought tolerance in tobacco was achieved by the introduction of a gene that helps in the synthesis of fructan. Transgenic tobacco producing fructans in leaves and tubers have been produced (Rober *et al.*, 1996) but their poor agronomic performance in terms of tuber size and number so far has further investigations. It has been shown that one small heat shock protein (*hsp*) gene encoding HSP 17.7 plays an important role in the ability of carrot cells and plants to survive thermal stress (Malik *et al.*, 1999). A protein purified from carrot shares these functional features with antifreezing proteins of fish (Worrall *et al.*, 1998). Myer *et al.* (1999) isolated a gene encoding an antifreezing protein (AFP) from carrot. This protein is highly similar to polygalacturonase inhibitor protein (PGIP) family of apoplastic plant leucine rich repeat (LRR) protein. Low temperatures rapidly induce expression of *af* gene. Further, more expression of the *afp* gene in transgenic *Arabidopsis thaliana* plants lead to an accumulation of antifreeze activity. It is concluded that this type of plant antifreezing protein has been recently evolved from PGIPS. In tobacco cold resistance has been achieved by transferring a gene from *Arabidopsis thaliana.* In tomato and tobacco, anti-freezing genes has been transferred from fish winter flounder.

Some Achievements

China was the first country to commercialize transgenic in the early nineties with the introduction of virus-resistant tobacco, which was later followed by a virus-resistant tomato (James, 1998). Subsequently, total area under transgenic increased 4.5 fold from 2.8 million ha in 1996 to more than 12.8 million ha today, with seven crops in six countries. Commercially released transgenic vegetable varieties are given in Table 18. In India, genetically altered seed has yet not been commercialized as the transgenic varieties developed are undergoing field trials and evaluation. The private sector is emerging as a major player in the area of transgenic research and commercial field releases in view of the fact that most Indian companies have now tie ups with multinational companies involved transgenic research (Table 19). Pro Agro-PGS have obtained permission to carry out controlled field trials with male sterile cauliflower and transgenic tomato, brinjal, cauliflower and cabbage carrying *Bt* gene. Rallis India Ltd. has been given permission by DBT to carry out transformation of chilli and bell pepper with GNA gene, pepper vein banding virus (PVBV), coat protein gene and its modified sequences, and also to conduct greenhouse trials of these transformants. Transgenic brinjal carrying *Bt* gene developed by the NRC for Plant Biotechnology New

Table 18: Commercial release of transgenic vegetable varieties

Trait	Crops	Name	Company	Product status
Quality (vine-ripen flavour, shelf-life)	Tomato	Flavr- Savr	Calgene	Released 1994
Quality (vine-ripen flavour, shelf-life)	Tomato	Endless Summer	DNA Plant Technology	Blocked by patent claims
Quality (paste consistency)	Tomato	-	Zenca	Released in 1995
Virus resistant	Tomato, capsicum	-	China	Released in 1993-94
Virus resistant	Squash	Freedom II	Asgrow	Released in 1995

Source: Anand (2001).

Table 19: Vegetable transgenic developmental activities in India

Institute	Vegetable crop	Transgenic	Target
Public Sector			
CPRI, Shimla	Potato	*Bt* cry 1 Ab	Develop plant resistant to potato tuber moth
		Osmotin	Develop water stress tolerant plants
		Coat protein	Develop plants resistant to potato virus Y (PVY)
Delhi University, Delhi	Tomato	Ctx-B and Tcp antigens of *Vibrio cholerae*	Edible vaccine development
	Brinjal	Chitnase, glucanase and thaumatin encoding genes	Create disease resistant plants
IARI, New Delhi	Brinjal	Cry 1Ab	Generate lepidopteron resistant plants
	Tomato	Cry 1Ab	Develop lepidopteron resistant plants
	Cauliflower	Cry 1Ab	Develop diamond back moth resistant plants

	Cabbage	Cry 1Ab	Develop diamond back moth resistant plants
	Tomato	ACC synthase	Controlling fruit ripening
		Replicase gene	Develop tomato leaf curl virus resistant plants
IIHR, Bangalore	Muskmelon	Rabies glycoprotein	Double edible vaccines
	Tomato	Leaf curl virus	Generate leaf curl virus resistant plants
		Chitinase and glucanase	Develop plant resistant to fungal disease
JNU, New Delhi	Tomato	OXDC	Develop fungal diseases resistant plants
	Potato	Ama 1	Develop protein rich potatoes
Private Sector			
IAHS, Bangalore	Tomato	Alafa glucanase and tomato leaf curl virus genes	Generate plants resistant to viral and fungal diseases
Proagro PG-S (India) Ltd;Gurgaon	Tomato	Cry1 Ab	Lepidopteran resistant plants
	Brinjal	Cry1 Ab	Lepidopteran resistant plants
	Cauliflower	Cry1H/ Cry 9C	Lepidopteran resistant plants
		Bar, barnase, barstar	Superior hybrid culture.
	Cabbage	Cry1H/ Cry 9C	Lepidopteran resistant plants

Source: Kalia (2004).

Delhi, is also being evaluated. Besides, several programmes on developing transgenic vegetable are also going in Government-funded research laboratories in India, These include: (a) development of insect- resistant brinjal, tomato, cauliflower and cabbage through transfer of *Bt* gene, (b) development of transgenic tomato resistant to TLCV, CMV, TSWV and early blight, (c) development of transgenic watermelon resistant to watermelon bud necrosis virus, (d) genetic manipulation of ripening in tomato and (e) development of transgenic amaranths with low oxalate and nitrate level.

Use of Molecular Markers

Molecular markers may be used for following purposes:

GeneTagging

The most interesting application of molecular markers at present is the ability to facilitate the method of "Conventional" gene transfer. In plant breeding, gene tagging is a term commonly applied where a genetic marker is used as the selection criteria for a closely linked agronomic trait. Example of some markers linked to disease resistant genes are summarized in (Table 20). A classic example resulted from the discovery of tight linkage between the isozyme markers *Aps 1* and *Mi*, a gene encoding root-knot nematode resistance in tomato (Rick and Fobes, 1977). As a result, many seed companies now use isozyme analysis as the preferred method for the incorporation of this trait into elite tomato lines. More than 1,000 markers have been mapped onto the tomato molecular linkage maps and these markers cover a total of 1,276 map units (Tanksley *et al.*, 1992). Many of the markers of tomato map have also been mapped onto the potato and pepper molecular maps. All these Solanaceous species share the same chromosome number and by having molecular maps based on the same RFLP probes; it has been possible to determine chromosomal homologies with a fairly high degree of precision (Bonierbale *et al.*, 1988; Tanksley *et al.*, 1988 and Tanksley *et al.*, 1992).

Another area where molecular markers facilitate gene transfer is the analysis of linkage drag. Linkage drag frequently encountered when wild species are used as a trait donor source; while backcrossing the trait, considerable excess linked chromosomal DNA may inadvertently be incorporated. A study by Young and Tanskley (1989) demonstrates the power of RFLP in these areas. The tomato cultivar "Craigella" has been found to carry approximately 50 cM piece of DNA derived from the disease resistant donor parents following 11 backcross generations. In a single generation of selection, Young and Tanksley (1989) were able to reduce the size of this fragment to less than 7 cM.

QTL Mapping

Many quantitative traits of economic value under polygenic control are selected directly. Such a selection is often ineffective, since the effect of each gene is small and is also influenced by the environment. Therefore, one would welcome a procedure for indirect selection, which is not influenced by the environment. For this purpose, linkage between quantitative traits and major marker genes should be known. Such marker genes involving morphological traits have been identified and mapped in several crop plants. Isozyme marker loci linked to quantitative traits of economic value had also been earlier identified in tomato and maize. However in recent years, it became possible to identify and map quantitative traits loci (QTLs) using RFLPs markers particularly in some crop plants like tomato and maize. With QTL analysis, hypothesis about the effects of a locus on a quantitative trait are tested at discrete positions along a map of genetic markers. The presence of a QTL is inferred when a significant amount of phenotypic variance can be attributed to estimate QTL parameters. This type of procedure was first described as "Simplified Interval Mapping

(SIM)". Several variations of "SIM" have been described by Knapp *et al.*, (1990) and Haley and Knot (1992).

All the SIM procedures search for a single "Target QTL" at positions throughout a mapped genome. When multiple QTL segregate, the sampling error associated with inferences about a target the effects of other QTL may inflate QTL. Furthermore, linked QTL can cause biased estimates of QTL position. These problems have been recognized for sometime (Knapp *et al.*, 1990), and have stimulated efforts to develop methods of QTL analysis that can simultaneously account for multiple QTL. Zeng (1993) described a method called "Composite Interval Mapping" (CIM) which permits searching for one QTL at a time while simultaneously accounting for the effect of other segregating QTL. This is accomplished by filling parameters for a target QTL in one interval, while simultaneously fitting partial regression co-efficients for "background" markers, outside of that interval. In principle, the genetic variance caused by QTL other than the target is absorbed by regression coefficients of the background markers. Other authors (Jansen and Stain, 1994) have also been described similar methods based on this principle. Another facet of QTL analysis concerns the fact that QTL effects may depend on environment. Methods of analysis of QTL x R interactions are not well developed. Tinker and Mather (1995) proposed a procedure of simplified composite interval mapping (SCM) for the analysis of QTL replicated in multiple environments.

The QTL mapping experiments are notorious for their unwieldy size. Most QTL experiments have involved more than 200 individuals, and in some cases as many as 2,000 or more. It is advocated that even 200 individual may be too few for reliable QTL detection, and addresses issues related to choice of population size in QTL mapping experiments. To extract more information from smaller population, there are three approaches for QTL mapping. The **first approach** selective genotyping is an excellent approach or efficiently mapping QTLs which influence a single phenotype only. This method is suitable if evaluation of an individual's phenotype can be done more cheaply and easily than evaluation of its DNA marker karyotype, a condition which is often true for plants. Specifically selective genotyping involves the identification of a subset of individuals from a genetic mapping population, which represent the most extreme phenotype in the population. These individuals harbour more information than phenotypically average individuals, since they are more likely to contain a high proportion of the + or – alleles, repectively, at QTL affecting the target trait. By phenotyping a large population, and then selecting only the most extreme individuals for genotyping, one can obtain greater or equal information about QTLs than from exhaustive mapping of randomly chosen individuals. A limitation of selective genotyping is the fact that it is suitable for analysis of only one phenotype at a time. This limitation often proves serious in applied experiments such as plant that requires evaluation for many independent characteristics. The **second approach**, Comparative QTL Mapping is an evolutionary approach to QTL mapping. It is becoming an increasingly powerful means to expedite QTL analysis. The finding that diverse taxa within common taxonomic families often share similar gene order over large chromosomal segments has been a basis for comparative mapping, alignment of chromosomes of different taxa based on common reference loci. The structural similarity of chromosomes in different taxa is often accompanied by functional similarity in locations of genes influencing common phenotypes. Data bases of genes previously reported to affect a phenotype are becoming increasingly powerful tools for predicting the chromosomal locations likely to account for phenotyping variation in new experimental crosses.

The **third approach** is DNA pooling. This approach has tremendously simplified genetic mapping of simply inherited traits in plants involving the *adhoc* pooling of DNA from individuals sharing a particular characteristics. In principle, this is similar to selective genotype, but treats the two extremes in a phenotypic distribution as a single DNA sample. While this approach is an effective way to map genes which account for 100% of variance in a trait, both theoretical and empirical results suggests that it has limited applicability to QTL mapping (Wang and Paterson, 1994 and Darvasi and Soller, 1994). Among phenotypically extreme individuals for a polygenic trait, rare QTLs with usually large effects may be fixed, and, therefore, may be detected as a chromosome segment which is polymorphic between the pools. However, the majority of QTLs, with much smaller phenotypic effects, will remain heterogeneous in the pools and therefore will escape detection. Complicating factors such as dominance and Non-Mendelian segregation reduce the feasibility of bulk segregant approaches to QTL mapping (Michelmore *et al.,* 1991). While rare QTLs of very large phenotypic effect might be reliably detected by bulk segregant analysis, sites of many QTLS with smaller effects on a trait are best evaluated by comprehensive mapping.

The QTL mapping experiments are conducted towards three basic objectives for the improvements of parental lines of hybrids. Many QTL mapping experiments are conducted for locating genes, which account for genetic variation in agriculturally important phenotypes, as a starting point for use of marker assisted selection in parental line improvement. By identifying DNA markers, which are diagnostic of a particular phenotype, the breeder can make selections among seedling grown in non-target environments and accelerate progress towards classical objectives. Significant advantages are for the breeders in having DNA markers for phenotypes that are difficult to measure, that can only be measured after plants have already contributed to the gene pool of next generation, or that require unusual environments to evaluate. The QTL mapping also has served as a starting point for introgression of exotic chromatin, delineating target genes and using DNA markers to reduce the portion of non-target donor chromatin transferred (Tanksley and Nelson, 1996). A long-term objective of some QTL mapping experiments is the molecular cloning of genes underlying specific phenotypes. Although, no QTLs *per se* have been clonned yet, in principle cloning of QTLs can use logical extensions of positional approaches used to walk to genes of agricultural importance in parental lines of hybrids. By making genetic stocks which are "near isogenic" for the chromosomal regions containing QTLs, individual QTLs can be rendered virtually discrete, mapped to precise locations, placed on megabase DNA coating, associated with candidate genes, and tested for mutant complementation. Because of small effects of individual QTLs on phenotype, single-plant tests of heterozygous primary transformants will usually be inadequate, rather progeny testing may be necessary. QTLs information may also serve as a supplement or adjunct to other molecular cloning approaches, providing a means to test whether candidate genes isolated by strategies such as tissue-specific expression subtractive methods, insertional mutagenesis, or other methods, co-segregate with the target phenotype. In future, availability of detailed physical maps and transcript maps of several key taxa, together with comparative maps collecting the chromosomes of a wide range of taxa, may offer a much more powerful database for identification and evaluation of candidate genes associated with QTLs.

The QTL mapping has been used in several instances to ask basic questions about evolutionary processes. The QTL analysis tends to support the generalization that many genetic phenomena are more discrete and more rapidly- evolving than classical genetics and evolutionary models would have anticipated.

Table 20: Identification of markers linked to disease resistant genes in vegetable crops

Host	Pathogen	Gene	Type of markers	Reference
Tomato	*Meloidogyne incognita*	*Mi*	RAPD	Williamson *et al.*, 1994
	Cladosporium fulvum	*Cf9*	AFLP	Thomas *et al.*, 1995
	Meloidogyne incognita	*Mi3*	RAPD, RFLP	Yaghoobi *et al.*, 1995
	Leveillula taurica	*Lv*	RFLP, RAPD	Chuwongse *et al.*, 1994
	Phytopthora infestans	*Ph2*	RFLP	Moreau *et al.*, 1998
	Verticillium dahlie	*Ve*	RAPD	Kawchuck *et al.*, 1994
	Yellow leaf curl virus	*Ty1*	RFLP	Zamir *et al.*, 1994
	Tospovirus	*SW-5*	RFLP	Brommonschenkel and Tanksley, 1997
	Fusarium oxysporum f. sp *lycopersici*	*I-2*	RFLP	Scott *et al.*, 2004
Potato	*Phytopthora infestans*	*R1* and *R3*	RFLP	El-kharbotly *et al.*, 1994
Lettuce	*Bremia lactuceae*	*Dm 17* and *18*	RAPD, SCAR	Maisonneuve *et al.*, 1994
		Dm 8 and *10*	RAPD, SCAR	Witsenboer *et al.*, 1995
	Plasmopara lactucae - radicis	*Plr*	RAPD, RFLP	Kesseli *et al.*, 1994
	Turnip mosaic virus	*Tu*	RAPD, RFLP	Robbins *et al.*, 1994
Bean	Common bean mosaic virus	*I*	RAPD	Melatto *et al.*, 1996
Pea	Pea common mosaic virus	*Mo*	RFLP	Direlewanger *et al.*, 1994
	Erysiphe polygone	*Er*	RAPD	Direlewanger *et al.*, 1994
	Fusarium oxysporum	*Fw*	USAT	Dirlewanger *et al.*, 1994
Chinese cabbage	*Plasmodiophora brassicae*	-	RAPD	Kugimiki *et al.*, 1997
Watermelon	Fusarium wilt races 0, 1 and 2	-	RAPD	Rebing *et al.*, 2004

Source: Singh *et al.* (2001).

Gene Cloning

The advent of genome mapping at the DNA level has provided a method for localizing genes of economic importance to specific chromosomal positions. The ability to map any gene of economic importance to a defined chromosomal site opens the possibility of isolating genes via chromosome walking. This method is sometimes referred to as 'map-based cloning' or the 'top down approach'. Since it starts with an interesting phenotype observed at the whole plant level (*e.g.* diseases resistance or high yield) and proceeds to clone the gene via genetic mapping and chromosome walking (Table 21). Chromosome walking presents serious difficulties in crop species. They tend to have relatively large genome. For example, 'average crop species has a nuclear genome 10-100 times larger than yeast, *Drosophila* or *Arabidopsis*. Nevertheless, the problems with chromosome walking the large genome is rapidly being overcome by new methods for manipulating large pieces of DNA (*e.g.* pulsed field gel electrophoresis and yeast artificial chromosome cloning). In tomato, the availability of a high density molecular linkage map and a yeast artificial chromosome library potentially provide the foundation on which to initiate map-based gene cloning for genes underlying any trait that can be genetically mapped (Martin *et al;* 1992). A critical prerequisite for map-based gene cloning is that target gene must be localized definitively with respect to

Table 21: Gene mapping in vegetable crops for various polygenic traits

Crop	Trait	Molecular Marker	Reference
Tomato	I-2 *Fusarium oxysporum*	RFLP	Segel *et al.*, 1992
	Male sterile 14 gene and a hotspot for recombination	RFLP	Gorman, 1992
	Regulation of iron - metabolism	RFLP	Ling *et al.*, 1996
	Sequencing of cDNA clone	RFLP	Ganal *et al.*, 1998
	Integration of the classical and molecular linkage maps of chromosome	RAPD	Weide *et al.*, 1993
	Soluble solids, fruit weight, fruit pH, fruit size and shape, fruit ripening and yield	RFLP	Tanksley *et al.*, 1996
	Ripening inhibitor and non ripening loci	RFLP	Givanmoni *et al;* 1995
	Root knot nematode resistance loci	AFLP and RFLP	Kaloshian, *et al.*, 1998
	Salt tolerance	RAPD	Foolad and Chen, 1997
	Golden nematode resistance gene (*Hero*)	RFLP	Ganal *et al.*, 1995
	Lateral suppressor (*Is*) loci	RFLP	Schumacher *et al.*, 1995

	Tandemly repeated telomeric DNA sequences	RFLP	Ganal *et al.,* 1992
	Physical location of *cf- 2* gene	RFLP	Dixon *et al.,* 1995
	Integration of the classical and molecular linkage maps of tomato chromosome 6	RAPD	Weide *et al*., 1993
	Close linkage between the *cf -2/ cf -5* and *Mi* resistance loci	RAPD and RFLP	Dickison *et al.,* 1993
	Carotenoid loci	RAPD	Zhang and Stommel, 1998
Potato	High resolution map of the vicinity of the R1 locus of chromosome V of potato	RFLP and AFLP	Meksem, 1995
	Gene pool similarities of potato cyst nematode population	AFLP	Folkertsma, 1996
	Quantitative resistance to blight	AFLP	Meyer *et al.,* 1998
	High resolution map of chromosome VII harbouring the nematode resistance gene	RAPD and AFLP	Ballvora *et al*., 1995
	Mapping of simple sequence repeat loci	RFLP	Millbourne, 1998
Cabbage	Linkage group alignment	RFLP	Hu *et al.,* 1998
	Location of self-incompatibility locus	RFLP	Camargo *et al*; 1997
	Comparative genome mapping	RFLP	Lagercrantz Lydiate, 1996
	Linkage group aligmnment from four independent- RFLP maps	RFLP	Hu *et al*., 1998
Cucumber	Plant height, sex expression and lateral branches	RFLP	Serquen *et al*., 1998
Melon	Construction of molecular map	AFLP	Wang *et al.,* 1997
Hot pepper	*Xanthomonas campestris*	RFLP and RAPD	Kang *et al.,* 1998
Brinjal	Bacterial wilt resistance	RAPD	Nunome and Hirai, 1998
Onion	Downey mildew and onion leaf blight	RFLP and RAPD	Kik *et al.,* 1995

Source: Singh *et al*. (2001).

markers in that region of the map. With the small genetic distances being considered, analysis of very large progenies is often required for such fine mapping. New protocols for rapid DNA isolation as well as new strategies for high resolution genetic mapping should enhance our ability to meet this prerequisite for map based cloning (Churchill *et al.,* 1994). It is anticipated that, in future, molecular linkage map will play a key role in cloning genes for a variety of economically valuable and/ or biologically interesting characters including disease resistance, growth habit, fruit ripening behaviour and even yield.

Transposon Tagging

Transposon tagging might be more straight forward approach for the molecular isolation of useful genes. Application of endogenous plant transposable elements has shown to be obtaining plant mutants. The insertion of a transposable element in to genetic locus can give rise to mutation. Using a transposable element from maize, the *Ac*-element, Federoff *et al.* (1984), were able to clone a locus, which until then was only characterized genetically. Their result indicates that in principle any locus in maize, identifiable with transposable element, can be cloned and analysed at the molecular level. This approach, called transposon tagging, has been successfully applied for the isolation of various maize genes. In vegetable crops, researches on transposon tagging are going on.

Synthetic Seed Technology

Concept

Among the advanced techniques of plant biotechnology that have relied on the concept of somactic embryogenesis, is synthetic seed or artificial seed technology (Litz and Gray, 1995). Synthetic seed ('synseed') is referred, in a narrow sense, to encapsulated somatic embryos which functionally mimic seeds and can develop into seedlings under suitable conditions. In a broader sense, it also refers to encapsulated buds, bulbs or any form of meristems which can develop into plantlets. Initially, synthetic seed preparations were initiated to using clumps of somatic embryos (*e.g.* celery-Kitto and Janick, 1982) or single somatic embryos (*e.g.* alfalfa-Redenbaugh *et al.* 1984). Presently, several laboratories are engaged in using somatic embryos for production of synthetic seeds. In addition to using somatic embryos, research has also been focused on the use of vegetative propagules like axillary buds, adventitious buds, shoot-tips and cormlets. Bulbs and protocorms for production of synthetic seeds (Rao *et al.,* 1986). The main utility of these alternative means would be in those crops where somatic embryogeny is not well-established and in such cases, synthetic seed systems can become a very useful tool for propagation, delivery of tissue cultured plants, germplasm storage and exchange.Synthetic seeds offer many advantages. Direct delivery of encapsulated material saves many subcultures to obtain plantlets and hence reduces cost. The encapsulated material may also carry pesticides, fertilizers, nitrogen-fixing bacteria and even microscopic parasite-destroying worms. A very important application is that synthetic seeds can be used for cryopreservation of germplasm (Tissereau *et al*., 1994 and Rao and Riley, 1994).

Potentialities of Synthetic Seed Technology

- Direct delivery of tissue-cultured plants to the field, avoiding laborious and time-consuming steps of transplantation.
- Propagation of desirable genotypes with genetic uniformity.
- Reduction in costs of vegetatively propagated elite plants.
- Delivery of genetically unstable genotypes such as transgenic plants or genetically abnormal plants.
- Effective handling of fragile somatic embryos or other *in vitro* propagules, and
- Preservation of plant germplasm and convenience in germplasm exchange.

Steps for Production of Synthetic Seeds

Several plant species undergo somatic embryogenesis *in vitro* and somatic embryos proceed through similar developmental stages like that of zygotic embryos to form a complete plant. In plant systems like carrot, celery and alfalfa, somatic embryogenesis has been well-characterized and this has led to the development of successful synthetic seed technology using encapsulated somatic embryos. In addition, plants also have been produced from encapsulated embryos of *Brassica*, cotton, lettuce, rice, soybean, sandalwood, brinjal and horseradish. The synseed technology has earlier reviewed by Redenbaugh (1990 and 1993). For synthetic seed system to be successful, the quality of somatic embryo is very important for conservation and production of plants. For achieving uniform quality embryos, it is necessary to have a synchronously developing culture from which embryos at a specific developmental stage can be harvested. For large-scale application, it is also important to develop bioreactor cultivation. Redenbaugh *et al.* (1988) identified some critical steps (Fig. 21).

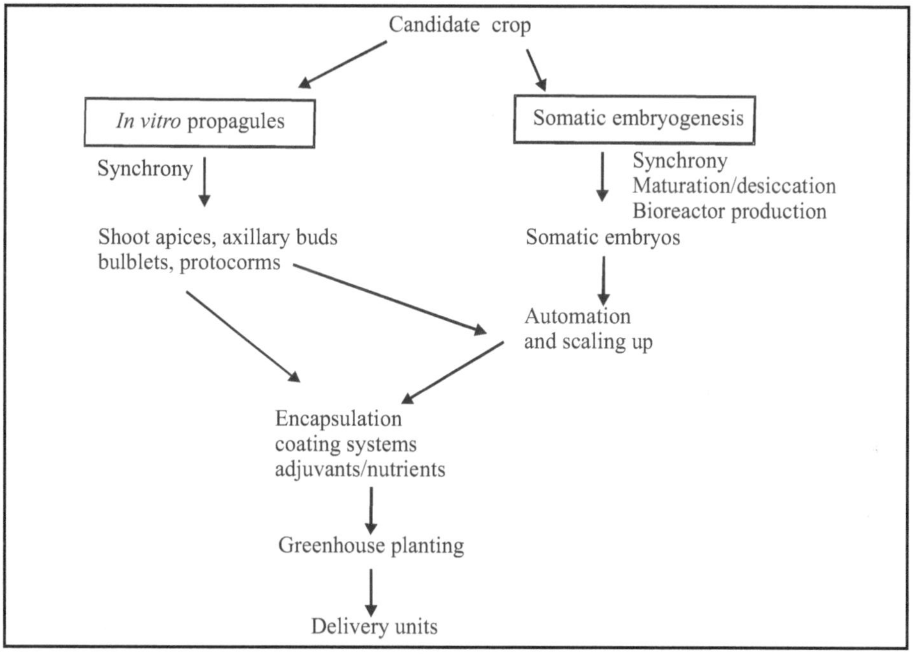

Figure 21: Critical stages in successful production of synthetic seeds.

Method of Synthetic Seed Production

Encapsulated Methods

Somatic embryos (or any other kind of *in vitro* propagules) are encapsulated to produce units that are analogous to seeds. The encapsulated matrix should give physical protection and play the role of artificial endosperm (like that of an endosperm or cotyledons in case of normal seeds) during early stages of growth of encapsulated material. In such a case, coat should be non-toxic to propagule, protect it from mechanical damage during handling and allow embryo development and plant conversion. Several gelling agents such as sodium alginate, sodium alginate with gelatin, potassium alginate, *guar* gum, ager, tragacanth gum, sodium pectate, carboxymethylcellulose, carrageenan with locust bean gutm and Gel-rite™ are useful for endocapsulation (Table 22). Sodium alginate, a food thickener from brown algae, has been used in most studies due to its moderate viscosity, low embryo toxicity and ease in forming capsules around embryos (Redenbaugh *et al.,* 1986). Maruyama and Kinoshita (1997) tested double layered alginate beads (by mixing explants and calcium alginate beads with 3% sodium alginate) for encapsulation. The ability of several polysaccharides and galactomannans to combine with each other and with starch to produce encapsulation beads has also been investigated (Sarvari *et al.*, 1997). These compounds are found to be better as compared to alginate for germinating beads.

Table 22. Gels used for encapsulation

Gel	Conc. % w/v	Complexing agent	Conc. In mM
Sodium alginate	0.5-5.0	Calcium salts	30-100
Sodium alginate with gelatin	2.0-5.0	Calcium salts	30-100
Carrageenan with locust bean gum	0.2-0.8 0.4-1.0	Potassium or ammonium chloride	500
Gelrite	0.25	Temperature lowered	-

In addition, polyethylene oxide homopolymers, acrylic co-polymer containing carboxyl groups, potassium propenoate-propenamide co-polymers, carboxyl methylized cold soluble swelling starch, synthetic trioctahedral smectite, synthetic sodium-magnessium lithium silicate, starch plus synthetic polymer of acrylamide and sodium acrylate (an anional flocculent) have also been tested as coating agents (Janick *et al.,* 1993). Even after coating with a suitable agent, it is often necessary to add nutrients and fungicides or pesticides to enhance the survival of encapsulants once they are planted *in vivo*. Ogunwolu and Idowu (1994) used powdered root bark of *Zanthoxylum zanthoxyloides* and seed extract of neem for the control of insects and considered that these could well be alternatives in place of insecticides for synthetic seeds.In general, two types of synthetic seeds, hydrated and desiccated, are produced. Hydrated synthetic seeds are somatic embryos encapsulated in hydrogel such as calcium alginate. Coating a mixture of somatic embryos in polyoxyethene glycol or polyethyleneoxide, on the contrary, produces desiccated synthetic seeds. In this method, dried mixture is allowed to dry for several hours on a Teflon surface in sterile hood. The dried mixture is then hydrated and placed on a medium for scoring embryo survival. Neither of the two types of synthetic seeds is completely satisfactory for embryo survival

(Redenbaugh *et al.*, 1991). Hydrated capsules are difficult to store because of requirement of the embryo for respiration and also they dry out unless kept in a humid environment or coated with a hydrophobic membrane. In case of desiccated capsules, desiccation itself may damage the embryo.

The propagules (somatic embryos/axillary buds/shoot tips) are carefully isolated from *in vitro* cultures, bolt dried on filter paper, and are then dipped in 100 ml gel of 3% sodium alginate (Sigma Co., USA) prepared a nutrient medium. The propagules can then be picked up by forceps or by pipette method and dropped into a solution of 2.5% calcium chloride. The drops (beads) are then left in this solution for 30-40 minutes with occasional gentle agitation. After the incubation period, the beads are recovered by decanting the calcium chloride solution and washing them 3-4 times in hormone-free nutrient medium. Such encapsulated propagules can then be placed on different nutrient media and substrates like filter paper, cotton, soilrite and soil for development and conversion to plants.

Applications

Applicaions of synthetic seed vary from crop to crop depending on the existing production systems, cost involved and necessities for improvement. A list of potential crops for synthetic seed development is presented in Table 23. In general, for seed propagated crops, outstanding hybrids can be multiplied through tissues culture and propagated crops; outstanding hybrids can be multiplied through tissues culture and propagated using synthetic seeds. In case of vegetatively propagated crops, particularly those with a high-production value, very efficient embryogenic systems (involving high quality somatic embryos and a high rate of plant conversion) coupled with synthetic seed technique can be very effective. Examples include, potato, sweet potato, fruit trees and woody plants, dioecious plants where either sex alone would be useful from an agricultural view point such as asparagus and kiwi fruit, and tetraploid plants which can be hardly bred such as potato and alfalfa (Kamada, 1985). Successful reports of synthetic seed research in vegetable crops are given in Table 24.

Table 23. Potential horticultural crops for artificial seeds

Category I (strong technological basis)		Category II (strong commercial basis)			
Alfalfa	Oilpalm	Asparagus	Impatiens	Cucumber	
Sandalwood	Banana	Begonia	Cyclamen	Soybean	
Caraway	Orchardgrass	Broccoli	Lettuce	Doglas fir	Spinach
Carrot	Orange	Cardamom	Loblolly pine	Garlic	
Celery	Panicum	Cauliflower	Petunia	Geranium	
Coffee	Pennisetum	Corn	Potato	Gerbera	Tomato
Brinjal	Walnut			Grape	Watermelon

(Modified after Redenbaugh *et al.* 1988).

Table 24: Reports on synthetic seeds in vegetable crops

Crop	Parts Used	Reference
Asparagus	Somatic embryos	Ghosh and Sen (1994)
Carrot	Somatic embryos	Kitto and Janick (1982)
Solanum spp.	Shoot tips	Fabre and Dereuddre (1990)
Solanum melongena	Somatic embryos	Rao and Singh (1991)
Tomato	Seeds	Garret *et al.* (1991)

Use of Polymerase Chain Reaction (PCR)

General Concept

Polymerase chain reaction technique is also known as people choice reaction which replaces the completely the gene cloning with vector in due course of time. This technique was discovered by Kary Mullis in 1985. In PCR technique, DNA replication, the reaction takes place in an 'eppendorf tube' where the primer strand is added from outside in the form of a deoxyoligo-nucleotide, and DNA polymerase enzyme is added to help in polymerization. Unlimited supply of amplified DNA is obtained by repeating supply of amplified regular denaturation of freshly synthesized double stranded DNA molecules by heating it to 90-98°C, the mixture of two primers, recognizing the two standards and bordering the sequence to be amplified is cooled to 40-60°C. This allows the primers (which are excess) to bind to their complementary strands through renaturation. The presence of Taq DNA polymerase enzyme and all the four essential nucleoside triphosphates allow synthesis of complementary strands in the usual manner. In a thermal cycler, this process is automatically repeated 20-30 times (as predetermined by a computer device), so that in a single afternoon, a billion copies of the sequence flanked by the left and right primers can be produced. In order to continue the synthesis, the temperature of the mixture is alternatively increased (for denaturation) and decreased (for renaturation) once every 1-3 minutes (as fixed by the computer device). Therefore, during temperature rise the enzymatic activity of DNA polymerase should not be destroyed, otherwise one may have to add a fresh aliquot of enzyme in each cycle of amplification. This became possible only by the discovery of thermostable enzyme Taq DNA polymerase isolated from *thermus aquaticus* growing in hot springs. This enzyme acts best at 72°C and the denaturation temperature of 90°C does not destroy its enzymatic activity. Later, other thermostable enzymes like Pflu DNA polymerase isolated from *Pyrococcus furosus*, were discovered and found more efficient. These enzymes allowed automation of the entire process and automatic PCR thermal cyclers are now available (each for a price of around Rs 2.00 lakhs), which can amplify DNA sequences at least speed unattended. The major advantages of PCR technique are:

- PCR may eventually replace cloning technique.
- This technology is much easier and requires much smaller quantity (nanogram or ng = $10^{-9}{}_{g}$ of DNA).

Comparision of PCR and Gene Cloning

Parameter	PCR	Gene Cloning
Final results	Selective amplification	Selective amplification of specific sequence
Manipulation	*In vitro*	*In vitro* and *in vivo*
Selectivity of specific segment from complex DNA	First step	Last step
Concentration of starting material	Nanogram (µg)	Microgram (µg)
Biological reagent required	DNA	Restriction enzyme, ligase, vector, bacterial cell/host
Automation	Yes	No
Labour intensive	No	Yes
Error probability	Less	More
Application	More	Less
Cost	Less	More
User's skill	Not required	Required
Time for a typical experiment	Four hours	2-4 days

Source: Gupta (2001).

Applications of PCR

- Using PCR, DNA polymorphism can be studied at loci with known sequence (using primers based on known sequence) or at random sites (using primer having unknown but random DNA sequence – these are called random primers).
- It is used for DNA polymorphism, however, involves the use of random primers giving Random Amplified Polymorphic DNA (RAPD).
- PCR technique can also be used for study of Random Amplified Polymorphic DNA.
- It is also useful for study of DNA polymorphism at VNTR (variable number tandem repeats, also known as minisatellites.
- It is useful in molecular mapping
- Sequence tagged microsatellite sites (STMS) or simple sequence repeat (SSR) loci examined and mapped by PCR technique.
- PCR technique has also been utilized for developing molecular markers closely linked to specific genes of economic importance.
- PCR also confirmed presence of transferred gene.

9

INTELLECTUAL PROPERTY RIGHTS - RELATED ISSUES IN HYBRID TECHNOLOGY

Definition and History

The dictionary meaning of property is 'estate whether in lands, goods or money'; such property is often referred to as tangible, material or physical property. The laws of the land protect the ownership of and the associated rights to physical property. In contrast, "intellectual property is an idea, a design, an invention, a manuscript, *etc.* which can ultimately give rise to a useful product/ application. The development of such property, as a rule, requires intellectual inputs, ingenuity and innovativeness; it also demands considerable monetary and other resources. Therefore, inventor of an intellectual property would like to ensure at least a fair reward for copied, imitated or reproduced; this minimizes the returns to the original inventor. Intellectual property right recognizes the right of an inventor to derive economic benefits from his invention (*i.e.* intellectual property); this right is called intellectual property right (IPR), (Singh, 2002). The governments, however, recognize the IPR, only so long as it is not to the detriment of the society. In India, innovation and novel techniques were retained within the families / small social groups that developed them, and there was no other system of protecting their rights to the knowledge so generated. The brief history of intellectual property right in India is:

- In 1856, the British Government in India introduced the Act of Protection of Inventions. This act was based on the British Patent Law of 1852.
- In 1872, Patents and Designs Protection Act was passed.
- In 1883, the Protection of Inventions Act was introduced. It was consolidated as Inventions and Designs Act in 1888.
- On 15 August 1947, the Indian patents and designs came under the management of Controller of Patents and Designs.
- The Indian Patent Act (1970) was introduced in the parliament in 1965, was modified in 1967 and passed in 1970.
- Protection of designs is covered by the Indian Patent and Design Act (1911) with amendments in 1978 and amended rules in 1985.
- Trademark protection is in force since 1 June 1948 under the 1940 Act. This act was amended as Indian Trade and Merchandise Marks Act (1958), which became effective on 25 November 1959. The copyright laws in India are as per international standards.

Intellectual property protection of plant varieties is some what different from that of other forms of technology because a new plant variety is not out of a single innovative step.

Even though, a new variety may not involve an inventive step, its economic usefulness to society is unquestionable. It hence requires the much needed encouragement. The development of new variety by providing incentives such as Plant Breeders' Right over exclusive production and marketing of seed of an improved variety. Many countries recognized plant varieties as an intellectual property and grant a protection to them through a patent (in USA) or a suitable form of plant breeders (PBR) (USA, Japan, Australia, European Union, Israel, *etc*.). In USA, three different systems are available for protection of IPR related to plants: (i) Plant Patents Act (1930) covers varieties of asexually propagated crops, *e.g.* ornamental and fruit trees. The application procedure requires a detailed description of the variety, and is usually handled by patent lawyers, (ii) The Plant Variety Protection Act of 1970 is US version of the plant breeders' rights system followed by the European Union and several other countries, (iii) The Utility Patents Act 1985 was originally meant to cover manmade industrial inventions and processes. 'No obviousness' is the main criterion of utility patents. Patents to plant varieties are now being granted under the provisions of this act (during 1997-98, 55 patents were granted to maize varieties and 40 for soybean varieties). Utility patents are considered to be the most powerful and most expanded in scope of their coverage; a single patent may cover several varieties, an entire species/ genus, genes/ proteins or technology and processes.

Benefits From IPR

- It encourages and safeguards intellectual and artistic creations.
- It enables the dissemination of new ideas and technologies quickly and widely. This is achieved by the requirement of disclosure for grant of patents *etc.*
- It encourages investment in research and development efforts. It provides consumers with the results of creations and inventions.
- It provides increased opportunities for the distribution of above effects across countries in a manner proportionate to the national levels of industrial and economic development.

Problems From IPR

- IPR has encouraged monopolies; many takeovers have been motivated by access to an IPR.
- The IPR situation in many cases (Section 37.6.7) is quite complicated. Monitoring and tackling the IPR aspects of inventions (a) enhances cost, (b) demands time, attention and effort, and (c) may act as a disincentive for research and development efforts.
- It may be treated as a threat to food security.
- It may adversely affect biological diversity and ecological balance.
- It may be detrimental to the livelihood of the poor in developing countries.

Management of IPR

IPR management involves following activities:

- Transfer of IPR appropriately and optimum value to obtain attractive returns for the expense involved in generating the intellectual property.
- Establishment of collaborative linkage to facilitate exploitation of the IPR.
- Monitoring infringements of the IPR and enforcing one's rights wherever and whenever necessary.
- Renewal of patents and designs periodically in every country where they have been granted.
- Quantitative evaluation of factors as research and development investments to the royalty ratio will have to be done periodically. This will enable the IPR holder to decide as to which IPRs are to be renewed in the different parts of the world.

Patent

A patent is the right granted by a government to an inventor to exclude others from imitating, manufacturing, using or selling the invention in question for commercial use during the specific period. Patents are granted for:

- An invention (including a product)
- Innovation / improvement in an invention
- The process of an invention/ product
- A concept.

Patent Requirements

The chief requirements for the grant of a patent are:

- *Novelty:* The invention must be new and should not be already known to the public.
- *Inventiveness:* The invention should not be obvious to a person skilled in the art, and should represent an innovation.
- *Industrial Application and Usefulness:* The subject matter of the patent must have an industrial application, either immediate or in future, and this application should be useful to the society/ nation.
- *Patentability:* The subject matter of patent must be patentable under the existing law and its current interpretation. This criterion, at present, varies from country to country and time within the same country.
- *Disclosure:* A patent may be viewed as a contract between the society and the inventor wherein the inventor discloses his invention in return for the protection granted to him by the society to control the commercial aspects of his invention to the extent that it is detrimental to the society. The disclosure of an invention gives an opportunity to other inventors to improve upon various features of invention so that it becomes more efficient and / or useful. This in turn, results in scientific and economic progress of society / nation.

Limits of Patents

- *Limitation of Time:* It is valid for a specific period of time from the date of award. In most countries, this period is 15-20 years. The Indian Patent Act (1970) grants protection for 7 or 14 years. There is a strong argument for an adequately longer protection period on the grounds that in several cases, *e.g.* pharmaceuticals, it may take (in various tests and scaling up) up to 10 years from the time a patent is awarded to the product reaches the market. The patent holder, therefore, should be allowed sufficient time to benefit from his invention.
- *Limitation of Space:* A patent is valid only in the country of its award. It is not valid in other countries. A group of nations may agree to honour the patent awarded by any member country, *e.g.* in European Economic Community. World Trade Organization (WTO) has a similar provision in that a patent awarded by WTO will be valid in all member countries.

Procedure of Patenting

An inventor files a properly prepared application (according to the prescribed proforma) with the patent office of the concerned country. Patent officials scrutinize the application. If found unsuitable for patenting, it is returned to the inventor. But if it is considered suitable for patenting, the invention along with adequate details of the desired patent are published for the information of all concerned. Anyone who wishes to challenge the award of patent can do so within a specified period of time, *e.g.* 18 months in Germany. In case a patent application is not challenged, the patent is awarded immediately after the expiry of this period. But if a patent is challenged, competent authority of the patent office hears the arguments and counter arguments of both applicant and the person challenging the application and a final decision is taken on the award of the patent. Thus, if a patent application is rejected due to contest following its publication, the main features of invention stand disclosed. The inventor cannot now resort to the option of trade secret (when the individual / organization owing an intellectual property does not disclose the property to anyone and keeps it as a closely guarded secret to promote his business interests, it is called trade secret.). Therefore, it is in the interest of an inventor to ensure, before filling a patent application, that the application is not likely to be rejected at least on account of a contest by a third party.

Plant Breeders' Rights (PBRs)

Concept and General Information

These are the rights granted by the government to a plant breeder's originator or owner of a variety to exclude others from producing or commercializing the propagating material of that variety for a period. The history of IPRs in plants date back to 1833 when they were applied in one form or the other at a primitive stage. However, after the Second World War, patent/ PVP PBR Acts were enacted to attract greater investment in plant breeding efforts. With the establishment of WTO on 1 January 1995, all the member countries are in process of developing their own system of protection for new varieties of plants where Union International Pour La Protection Des Obtentions Vegetables (UPOV) is quite active to have more and more countries joining their system of PVP. Union International Pour La Protection Des Obtentions Vegetables was formed in early 70s (Anon, 1993 a and b). It cooperates in the matter of Plant Breeders Rights with the WTO. The Union comprises 37 member countries. All party to its 1978 Act, with the exception of Belgium and Spain, who are party to its 1961 Act as amended by its additional Act of 1972. However, 15 countries have already signed its 1991 Act. These countries are Belgium, Canada, Denmark, France, Germany, Ireland, Israel, Netherlands, New Zealand, South Africa, Spain, Sweden, Switzerland, The United Kingdom and The USA. Member States of UPOV are required to modify their national law before being in position to accede to the 1991 Act. It requires five States to accede to it, of which at least three must be State, which are party to the 1878 or earlier Acts of the convention. Any non-member states given the option to accede to 1978 Act at any time until the 1991 Act comes into force. The UPOV 1978 Act is now closed and UPOV Act has come into force since several member countries have ratified the latter. The differences between UPOV Act 1978 and UPOV Act 1991 are given below:

Issues	1978 Act	1991 Act
Membership	A state only is the party	Inter-governmental organization competent for enacting and implementing legislation can also be the party
Discovery	Breeder is entitled to protection as 'discover' irrespective of the origin, artificial or natural of the initial variation	A mere discovery is not sufficient. The breeder must also have developed the variety
National treatment	Member state may limit the right on a new variety to nationals of states. A similar reciprocity may also be applied by a member state granting more extensive rights	Reciprocity rule does not apply. Operation of the principle of national treatment to one all without qualification
Scope	Authorization of breeder is required for : • The production for purpose of commercial marketing of the propagating material, the offering, for sale of the propagating material. The marketing of such materials, the repeated use of the new plant variety for the commercial production of another variety, say as parents in case of hybrids • The commercial use of ornamental plants or parts, therefore as propagating material in the production of ornamental plants or cut flowers • Does not require authorization for use of material for further research. • Farmers have privilege to use, reuse his farm as seed and can dispose of his produce	Authorization of breeder is essential for : • Production or reproduction conditioning for the purpose of propagation. Offering for sale selling or other marketing, exporting importing and stocking for any of these purposes. Furthermore, the 1991 Act specifies four subject matters to which the breed ers right extends • Varieties, which are not clearly distinguishable from the protected variety • Varieties, which are essentially derived from the protected variety • Varieties whose production requires repeated use of the protected variety
Minimum number of genera or species	• At least 5 to start with • At least 10 within 3 years	• All after 5 years • transitional period

Issues	1978 Act	1991 Act
to covered	• At least 18 within 6 years • At least 24 with in 8 years	• All after 10 years if only bounded by 1991 Act to start with while acceding 15 plant genera/species
Period Protection	• 18 years for grape vines and trees including root stock • 15 years for all other	• 25 years • 2.20 years
Special title/patent	Each State is free for any of the two forms of protection except those where it was a practice before 31 October 1979	No alternative forms of protection

Source: Rai (2001).

Requirements for PBR

To qualify for PBR protection a variety / hybrid has the following qualities:

Novelty: The criterion of novelty replaces that of 'non obvious' which is the requirement for protection of a new invention. It is used in the sense that the new variety should not have been commercially exploited for more than a year before granting protection.

Distinctiveness: The new variety must possess a trait that makes it distinguishable from other known varieties by one or more identifiable morphological, physiological, or other characteristics.

Uniformity: The new variety must be uniform in appearance under specified environment of its adoption.

Stability: The new variety must be stable in appearance and its claimed characteristics over successive generations under specified environment.

Conditions of Protection by PBR

- Production for commercial purposes, offering for sale and selling propagating material becomes the exclusive right of the holder of the PBR-title
- Production, sale *etc.* of material for consumption or industrial use does not fall under the protection
- The protection is limited to the propagating material produced for commercial purposes, which means that a grower is allowed to reserve a portion of his harvest for use as seed for his own next crop without the permission of the holders of the right. This is called farmers exemption, but a farmer who does this is not allowed to sell such material to anyone else
- Exchange of similar material of different cultivars among farmers is illegal
- The minimum period of protection prescribed by the UPOV convention is 15 years, with 18 years for grape and forest, and ornamental trees. Several member states have established a longer period of protection of 20-25 years, with 30 years in France for inbred lines of maize and for clovers and a few grasses

- Protection is granted for an unlimited period
- The use of propagating material from protected cultivars for scientific purposes is not dependent on permission of the holder of the right. Breeders are, however, unhappy with this
- The use of protected cultivars for creation of genetic variability for plant breeding purposes is not dependent on permission of the holder of the right.

Advantages from PBR

- The opportunity to breeders of obtaining profits from varieties / hybrids developed by them will act as an incentive in promoting plant breeding research
- It encourages private companies to invest in plant breeding activities
- It will enable access to varieties/ hybrids developed in other countries and protected by IPR laws
- Increased competition among various organizations engaged in plant breeding is likely to be beneficial to both the farmers and the nation.

Disadvantages from PBR

- Plant Breeders' Rights will encourage monopolies in genetic material for specific traits
- It suppresses free exchange of genetic material and may encourage unhealthy practices
- The holder of PBR-title may produce less seed than the demand in order to increase price for achieving more profit
- Farmers' privilege to re-sow the seed produced by him may become gradually diluted/ eliminated
- PBR may result in increased cost of seed, which will be burdensome to the poor farmers of India, and would limit the benefits from new varieties to small segment of rich farmers
- PBR would delay the spread of new varieties among Indian farmers since it prohibits exchange/sale of seeds of protected varieties among farmers.

Proposed PBR in India

India is trying to evolve a *sui generis* system of PBR. A *sui generis* system simply means 'a system of their own *i.e.* designated by them , India in this case. The essential features of UPOV 1978 Act are being considered to be adopted. Some important features of the proposed Indian *sui generis* system are as given by Rana (1995) are:

- Farmers' Right
- Researchers' use right
- Protection period 15 years for annuals and 18 years for perennials (fruit crops and vines)

- Compulsory deposit in national gene bank
- Compulsory certification
- Compulsory licensing
- Establishment of a National Authority for the protection of breeders', farmers' and researchers' use rights.

Farmers' Right

Agriculture began some 10,000 years ago. During this long period of time genetic resources have been selected, developed, used and conserved by farmer families and farming communities of, particularly, the gene- rich developing countries. Thus, the same materials have been and are being collected, conserved and used as raw materials to evolve the modern high-yielding varieties of various crops. Seed sales of these improved varieties earn huge profits for seed corporations. It has been argued that farmers should be allowed a share in this profit in recognition of their contribution by way of development of germplasms of various crops. This has been recognized by FAO Resolution no. 5/89 as farmers' rights, which arise from the past, present and future contributions of farmers in conserving, improving and making available plant genetic resources, particularly in center of origin/ diversity. It has been emphasized that farmers' rights should be obligatory and should not be relegated as privileges. The act contains the following things :

(1)
- A farmer who has bred or developed a new variety shall be entitled for registration and other protection in like manner as a breeder of a variety under this act
- The farmers' variety shall be entitled for registration if the application contains declaration as specified in clause (*h*) of sub-section (1) of section 18
- A farmer who is engaged in conservation of genetic resources of land races and wild relatives of economic plants and their improvement through selection and preservation shall be entitled in the prescribe manner for recognition and reward from the Gene Fund; provided that material so selected and preserved has been used as donors of genes in varieties registrable under this Act
- A farmer shall deemed to be entitled to save, use, sow, resow, exchange, share or sell his farm produce including seed of a variety protected under this Act in the same manner as he was entitled before coming into force of this Act; provided that the farmer shall not be entitled to sell branded seed of a variety protected under this Act. Explanation-For the purpose of *clause*-iv, 'branded seed' means any seed put in a package or any other container and labeled in a manner indicating that such seed is of a variety protected under this Act.

(2) Where any propagating material of a variety registered under this Act has been sold to a farmer or a group of farmers or any organization of farmers, the breeder of such variety shall disclose to the farmer or the group of farmers or the organization of farmers, as the case may be, the expected performance under given conditions, and if

such propagating material fails to provide such performance under such given conditions, the farmer or a group of farmers or any organization of farmers as the case may be , may claim compensation in prescribed manner before the authority and the authority shall after giving notice to the breeder of the variety and after providing him an opportunity to file opposition in the prescribed manner and after hearing the parties, it may direct the breeder of the variety to pay such compensation as it deems fit, to the farmer or the group of farmers or the organizations of farmers, as the case may be

- A breeder or the person making application for registration of any variety under chapter III shall disclose in the application the information regarding the use of genetic material conserved any tribal or rural families in the breeding or development of such variety
- If the breeder or such other person fails to disclose any information under sub-section (I), the Registrar may, after being satisfied that the breeder or such person has willfully and knowingly concealed such information, reject the application for registration.

(3) Any person, group of persons or any governmental or non-governmental organization may on behalf of any village or local community in India, file in any centre noticed, with the previous approval of the Central Government by the Authority in the Official Gazette, any claim attributable to the contribution of the people of that village or local community, as the case may be, in the evolution of any variety for the purpose of staking a claim on behalf of such village or local community.

(4) Where any claim is made under sub-section (1), the centre noticed under the sub-section may verify the claim made by such person or group of persons of such village or such governmental or non-governmental organization in such manner as it deemed fit and if it is satisfied that such village or local community has contributed significantly to the evolution of the variety which has been registered under this Act, it shall report its findings to the Authority.

(5) When the Authority on a report under sub-section (2) is satisfied after such enquiry as it may deem fit, that the variety with which the report is related has been registered under the provision of this Act, it may issue notice in the prescribed manner to the breeder of that variety and after providing opportunity to such breeder to file objection in the prescribed manner and of being heard, it may subject to any limit noticed by the Central Government, by order, grant such sum of compensation to be paid to a person or a group of persons or governmental or non-governmental organization which has been made claim under sub-section (1) to the Authority, as it may deem fit.

(6) Any compensation granted under sub-section (3) shall be deposited by the breeder of the variety in the Gene Fund.

(7) The compensation granted under sub-section (3) shall be deemed to be an arrear of land revenue and shall be recoverable by the Authority accordingly.

Protection of Plant Varieties and Farmers' Rights Authority and Registry

(1) The Central Government shall, by notification in the official Gazette, establish an Authority to be known as the protection of plant varieties and Farmers' Rights Authority for the purpose of this Act.

(2) The Authority shall be a corporate by the name aforesaid, having perpetual succession and a common seal with power to acquire, hold and dispose of properties, both movable and immovable, and to contract, and shall by the said name sue and be sued.

(3) The head office of the Authority shall be at such place as the Central Government may, by notification in the Official Gazette, specify and the Authority may, with the previous approval of the Central Government, establish branch offices at other places in India.

(4) The Authority shall consist of a Chairperson and 15 members.

(5) (a) The Chairperson , to be appointed by the Central Government, shall be a person of outstanding calibre and eminence with long practical experience to the satisfaction of that Government, especially in the field of plant varietal research or agricultural development.

(b) The member of Authority, to be appointed by the Central Government, shall be as follows, namely:

- The Agriculture Commissioner, Government of India, Department of Agriculture and Cooperation, New Delhi, ex-officio
- The Deputy Director-General in charge of Crop Sciences, Indian Council of Agricultural Research, New Delhi, ex-officio
- The Joint Secretary Incharge of Seeds, Government of India, Department of Agriculture and Cooperation, New Delhi, ex-officio
- The Horticulture Commissioner, Government of India, Department of Agriculture and Cooperation, New Delhi, ex-officio
- The Director, National Bureau of Plant Genetic Resources, New Delhi, ex-officio
- One member not below the rank of Joint Secretary to the Government of India, to represent the Department of Biotechnology Government of India, ex-officio
- One member not below the rank of Joint Secretary to the Government of India, to represent the Ministry of Environmental and Forests, Government of India, ex-officio
- One member not below the rank of Joint Secretary to the Government of India, to represent the Ministry of Law, Justice and Company Affairs, Government of India, ex-officio
- One representative from a National or State level farmers' organization to be nominated by the Central Government
- One representative from a tribal organization to be nominated by the Central Government

- One representative from the seed industry to be nominated by the Central Government
- One representative from an agricultural university to be nominated by the Central Government
- One representative from a National or State Level women's organization associated with agricultural activities to be nominated by the Central Government
- Two representative of State Governments on rotation basis to be nominated by the Central Government

(c) The Registrar-General shall be the ex-officio Member-Secretary of the Authority.

(6) The term of office of the Chairperson and the manner of filling the post shall be such as may be prescribed.

(7) The Chairperson shall appoint a standing Committee consisting of five members, one of whom shall be a member who is a representative from a farmers organization, to advice the Authority on all issues including farmers' rights.

(8) The Chairperson shall be entitled to such salary and allowances and shall be subject to such conditions of service in respect of leave, pension, provident fund and other matters as may be prescribed. The allowances for non-official members for attending the meetings of the Authority shall be such as may be prescribed.

(9) The Chairperson may resign his office by giving notice thereof in writing to the Central Government and on such resignation being accepted, he shall be deemed to have vacated his office.

(10) On the resignation of the Chairperson or on the vacation of the office of the Chairperson for any reason, the Central Government may appoint one of the members to officiate as Chairperson till a regular Chairperson is appointed in accordance with clause (a) of sub-section (5) 4.

(a) The Authority shall meet at such time and place and shall observe such rules of procedure in regard to the transaction of business at its meetings [including the quorum at its meeting and the transaction of business of its Standing Committee appointed under sub-section (7) of section 3] as may be prescribed.

(b) The Chairperson of the Authority shall preside at the meetings of the Authority.

(c) If for any reason the Chairperson is unable to attend any meeting of the Authority, any member of the Authority chosen by the members present shall preside at the meeting.

(d) All questions which come before any meeting of the Authority shall be decided by a majority of the votes of the members of the Authority present and voting and in the event of equality of votes, the Chairperson of the Authority or in his absence, the person presiding shall have and exercise a second or casting vote.

(e) Every member who is in any way, whether directly or personally, concerned or interested in a matter to be decided at the meeting shall disclose the nature of his concern or interest and after such disclosure, the member concerned or interested shall not attend that meeting.

(f) No act or proceeding of the Authority shall be invalid merely by reason of-

- Any vacancy in, or any defect in the constitution of, the Authority; or
- Any defect in the appointment of a person acting as the Chairperson or a member of the Authority; or
- Any irregularity in the procedure of the Authority not affecting the merits of the case.

(g) • The Authority may appoint such committees as may be necessary for the efficient discharge of its duties and performance of its functions under this Act.

• The persons appointed as members of the committee under sub-section (1) shall be entitled to receive such allowances or fees for attending the meetings of the committee as may be fixed by the Central Government.

(h) Subject to such control and restriction as may be prescribed, the Authority may appoint such officers and other employees as may be necessary for the efficient performance of its functions and the method of appointment, the salary and allowances and other conditions of service of such other officers and employees of the Authority shall be such as may be prescribed.

(i) It shall be the duty of the Authority to promote, by such measures as it thinks fit, the encouragement for the development of new varieties of plants and to protect the rights of farmers and breeders.

In particular, and without prejudice to generality of the foregoing provisions, the measures referred to in sub-section (1) may provide for-

The registration of extent new plant varieties subject to such terms and conditions and in the manner as may be prescribed

Developing characterization and documentation of varieties registered under this Act

Documentation, indexing and cataloguing of farmers' varieties

Compulsory cataloguing facilities for all varieties of plants

Ensuring that seeds of the varieties registered under this Act are available to farmers and providing for compulsory licensing of such varieties if the breeder of such varieties or any other person entitled to produce such variety under this Act does not arrange for production and sale of the seed in the manner as may be prescribed

Collecting statistics with regard to plant varieties, including the contribution of any person at any time in the evolution or development of any plant variety, in India or in any other country, for compilation and publication

Ensuring the maintance of the Register.

(11) All orders and decisions of the Authority shall be authenticated by the signature of the Chairperson or any other member authorized by the Authority in his behalf.

(12) The Authority may, by general or special order in writing, delegate to the Chairperson, any member or officer of the Authority subject to such conditions or limitations, if any, as may be specified in the order, such of its powers and functions (except the power to make regulations under section 94) under this Act as it may deem necessary.

(13) In all proceeding under this Act before the Authority or the Registrar

(a) The Authority or the Registrar, as the case may be, shall have all the powers of a civil court for the purposes of receiving evidence, administering oaths, enforcing the attendance of witnesses, compelling the discovery and production of documents and issuing commissions for the examination of witnesses

(b) The Authority or the Registrar may, subject to any rules made in this behalf under this Act, make such orders as to cost as it considers reasonable an any such order shall be executable as a decree of a civil court.

Registry

(14) (a) The Central Government shall establish for the purposes of this Act, a Registry which shall be known as the Plant Varieties Registry.

(b) The head office of the Plant Varieties Registry shall be located in the head office of the Authority, and for the purpose of facilitating the registration of plant varieties, there may be established, at such places, as the Authority may think fit, branch offices of the Registry

(c) The Authority shall appoint a Registrar-General of Plant Varieties who shall be entitled to such salary and allowances and shall be subject to such conditions of service in respect of leave, pension, provident fund and such other matters as may be prescribed

(d) The Authority may appoint such numbers of Registrar as it thinks necessary for registration of plant varieties under the superintendence and direction of the Registrar-General under this Act and may make regulations with respect to their duties and jurisdiction

(e) The term of office and the conditions of service of the Registrars shall be such as may be provided by regulations

(f) The Authority may, by notification in the Official Gazette, define the territorial limits within which a branch office of the registry may exercise its functions

(g) There shall be a seal of the Plant Varieties Registry

(15) (a) For the purpose of this Act, a register called the National Register of Plant Varieties shall be kept at the head office of the r egistry, wherein shall be entered the names of all the registered plant varieties with the names and addresses of their respective breeders, the right of such breeders in respect of the registered variety, the particulars of the denomination of each registered variety, its seed or other propagating material along with specification of salient features thereof and such other matters as may be prescribed.

(b) Subject to the superintendent and direction of the Central Government, the Register shall be kept under the control and management of the Authority.

(c) There shall be kept at each branch office of the Registry a copy of the Register and such other documents as the Central Government may, by notification in the Official Gazette direct.

Related Issues in Hybrid Technology

Registration of Germplasm / parental Lines of Hybrids

There is a system of plant germplasm registration operating at the National Bureau of Plant Genetic Resources, under which several parental lines of hybrids of many crops have been registered.

The IPR and Transgenic/ GM Parental Lines/ Hybrids

The Indian PPVFR Act, 2001 provides for the protection of plant varieties, including the genetically modified new plant varieties as well as the essentially derived varieties. Plant varieties are not coverable under patent protection in India. However, scope of process patent is enlarged now. Applicants will also be required to furnish a detailed passport data of parental line material, clearly indicating their geographical origin, if drawn from Indian biodiversity, together with declaration on their lawful acquisition. This would facilitate consideration and admittance of all applicable benefit sharing cases by the authority functioning under the PPVFR Act, 2001. This will also facilitate establishing the corresponding 'initial variety' in case of the genetically modified (GM) variety is registered/ protected as an Essentially Derived Variety (EDV). The right holder of a GM variety will be required to disclose if expected performance under given conditions and the farmers or group of farmers who purchase and plant the said GM planted material but do not get the stated performance will have holder in a prescribed manner. The authority, under the Plant Protection Variety Farmers' Right (PPVFR) Act, 2001 will receive complaints, may hear the defendant or both sides and where applicable require the right holder to give compensation. In case of a transgenic hybrid/parental lines, the owner may have to deal and compromise with Intellectual Property Right (IPR)-protections comprising several patents/ protection titles, such as for the genetic base of variety, gene construct, marker, promoter, transformation method *etc*; such person may also seek protection for relevant product and process patents transgenic plant variety. For commercialization of transgenic, IPR has become a more important and complex issue than proprietary components/protections. It may not be agreements with the owners of these components. For example, a transgenic insect-tolerant plant may involve a large number of patents/protections as given in Table1.

All these patents may not be available with one institution /company. An IPR holder of even one patent could block the commercialization of the product. Therefore, it requires partnership with the IPR holders to commercialize the transgenic. Also, forging alliances between the CGIAR Centres and the private sector has become a more compelling issue because of the IPR.

Table 1: IPR required commercializing GM hybrids

Subject	Components	Examples	IPR
Parental lines	Germplasm/variety to be used as parental lines	Protected or patented variety/registered germplasm	Patent/PVP/pri or art for benefit sharing
Selectable marker gene trait	Promoter	34-S	Patent
	Coding sequence	*Npt-II*	Patent
	Promoter	TR	Patent
	Coding sequence	*Cry-1 Ab*	Patent
	Ti-plasmid	*Pgv-2260*	Patent
		Viral leader	Patent
Transformation technology gene expression technology	Transcription initiation *etc.*	Crop specific technology	Patent
	Translation initiation, codon usage	At GC	Patent

Source: Tewari (2004).

Meeting High Cost through Consortium Approach

Owing to high cost of hybrid technology and particularly of GM hybrids, a consortium approach is being followed by group of interested agencies. For example, seven Indian manufacturing companies came together under a consortium Swarna Bharat Biotechnics Pvt Ltd (SBBPL) and signed an agreement to acquire the Indian *Bt* techonology from the National Botanical Research Institute (NBRI), Lucknow.

Characterization of Hybrids and Parents

Varieties are now being characterized by not only distinguishing / diagnostic morphological characteristics but also by using biochemical molecular markers as well as DNA fingerprinting which would help in staking the claims of ownership on the parental lines and hybrids in the new IPR regime.

New Hybrid Technologies and IPRs

There are some new technologies causing IPR. For example recent advances made in apomixes include the following:

- Discovery of sexual plants in apomictic species to produce apomictic hybrids.
- New information on the genetics of apomixes.
- Progress in transforming apomixes from wild to cultivated species.
- Progress made in molecular mapping of the genes controlling apomixes.

It is to be noted that several of the above stated discoveries and technologies have been patented.The most significant among these is the patent obtained on apomixes for producing true-breeding plant progenies by W.W. Hanna and his group.

Geographical Indications and IPR

Act 22 under Sec.3 of GATT states geographical indications as "identifying a good as originating in the territory of a member or a region or locality in that territory with a given quality, reputation or other characteristics of the good". Champagne, a sparkling wine of France is an example of geographic indication. A Geographical Appellation Act safeguards specific products like the Basmati grain from encroachment by foreign enterprises, in much the same way the term "champagne" is reserved for a product from a particular region in France. In India, Geographic indication is in place and product like Basmati rice are already being covered. Further, Indian geographic indications such as Darjeeling tea, Dashehari mango *etc.* is to be increasingly covered.

Hybrid Seed Production and IPR

A change in modes of hybrid seed production and supply is on the anvil owing to new emerging trends like;

- Transgenic or GM group varieties.
- IPRs, which would make seed increasingly a proprietary item.
- Globalization of hybrid development and supply systems.
- Advent of new technologies.
- Consumer demand for making seed as ready to use packaging inputs.
- Increasing privatization.
- Increasing corporate and commercial management of even the public sector institutions.

The Protection of Plant Varieties would make the Parental Lines / hybrids and their Production a Proprietary Endeavour

This is expected to bring about change in seed production modes. The seed especially in case of high volume low value crops has yet been largely in the public domain. The scenario, however, is changing fast. The PPVFR Act, other IPR regulatory systems such as geographic indication, biodiversity, biotechnology/ transgenic *etc.* are going to change the present scenario. A system of contractual production, sharing of rights, licensing and sub-licensing has already commenced resulting in such linkages, which did not exist hitherto. The system has become more dynamic than ever and desired changes are being swiftly effected accordingly.

Information Technology and Hybrid Technology

With the increasing use of information technology, new avenues of hybrid promotion particularly hybrid seed promotion have been opened up. There are websites that, through the internet, give update status of prevailing availability, domestic and global prices, demands including global demands, acceptable conditions of trade and related aspects and facilitate E-commerce. "Seed Quest ™ Online" is an example of the kind. It would be build up a further robust strengthening hybrid seed sector and also to have an increasing share in global trade. The present global trends to limit the access to genetic resources and scientific knowledge in contrast to former free availability. The challenge, more so for public sector, is to produce public goods in an environment that is increasingly characterized by proprietary technology to Indian farmers in a scale-neutral mode.

Biosafety Regulations for Transgenic Hybrids

The first biosafety regulation was prepared by the National Institute of Health in the USA in 1976. In India, testing of transgenic and biosafety issues in regard to transgenic are being dealt in the Department of Biotechnology in collaboration with ICAR (DARE, MoA). For research activities, the following three categories have been formed.

Category I

It involves known genes like *Bt* where safety is established through known experience / literature. Experiments involving such genes are exempt but the respective Institutional Biosafety Committee should be informed about it.

Category II

It comprises of genes those for herbicide resistance, which require prior permission for research.

Category III

These involve number of plasmic toxin genes. These require review and approval of competent authority before commencement.

Institutional Committees

Recombinant DNA Advisory Committee (RDAC)

The committee takes note of developments at international levels in the field of biotechnology towards the safety regulation for India on recombinant research, use and applications.

Institutional Biosafety Committees (IBSCs)

These committees are to be constituted in all centers engaged in genetic engineering research and production activities.

Review Committee on Genetic Manipulation (RCGM)

The RCGM function to establish procedural guidelines for testing and regulatory process and related activities in relation to environmental safety. It also serves to recommend the type of containment facilities and conditions for experimental trials and other experiments. It covers activities related to these aspects.

Genetic Engineering Approval Committee (GEAC)

It is statutory body for reviewing and approval of activities involving large-scale use of genetically engineered organisms and their products in search and development, industrial production, environmental release and field applications. The ICAR, New Delhi, in collaboration with DBT, has finalized the testing procedures including field experiments for transgenic. The NBPGR, New Delhi, in association with DBT serves as a nodal agency for import, quarantine and early handling of transgenics.

Impact of Patents and PVP on Hybrid Seed Production

- The investment of private sector in research and developmental efforts on hybrid technology is expected to increase with the availability of patent and PVP provisions. Apart from hybrids, the transgenic will be much vogue.
- The gene protection technology or terminator technology, which will be feasibly employed in the coming four or five years, will extend the private sector activities in such self-pollinated crops also where farm-saved seed is coming in the way of commercial grains of the MNCs. However, in India, where farmers would have complete freedom and flexibility to use, re-use and sell the seeds the way they choose and where terminator technology may not be allowed, private sectors thrust on self-pollinated crops would be less as compared to hybrid research and development. Further, the private sector would be more interested in high-value, low-volume seeds where investments expected to be highly rewarding.
- Big MNCs are acquiring small companies all over the world. Depending on the likely access to external markets, some of the Indian companies would find it worthwhile to have global liaison / network in tropical and sub-tropical regions in the first instance due to better adaptability of their likely varieties/ hybrids and would enter in competition in other parts of the world in the second phase.
- Seed companies are learning how to deal with IPRs but at a high cost. Increased expenses include those for more legal counseling, for extensive record keeping on the part of researches and the cost of delays in implementing improvements until legal safeguards are in place. Public sector may increasingly indulge in basic and strategic research such as development of sterility sources and parental lines *etc*. As a result, commercial firms working with plants and plant products would need to deal with these public institutions. New terms, linkages and operational modalities will emerge. There will be more formal modes mostly through specific Material Transfer Agreements (MTAs).
- As per an analysis made by Duvic (1996), plant product will need to compete with non - plant products already on the market in the cost of production and product quality. In theory, producing industrial products in plants should cost less because the plants can use sunlight and air, and relatively low priced land and water as raw materials. But the small size and dispersed and variable nature of many potential markets for plant- produced industrial products will affect the profitability of producing these products in plants. The fractured nature of the market for plant-produced industrial products contrasts sharply with the traditional markets for farm seeds.
- Under UPOV 1991, provisions where essential derivation clause is brought above would further help the investment in horticulture specially in vegetable sector as the

profit of a variety / hybrid evolved through biotechnology by changing minor elements controlled by one or limited number of genes such as colour, aroma, fragrance, texture, keeping quality, resistance to biotic and abiotic stresses *etc*; would be receiving further greater attention, as arrangements will have to be had with the original breeder developing the varieties/ hybrids in which minor changes have been brought about. In fact, biotechnology techniques could go a long way in this direction.

- With PVP in operation, private sector involvement would go up and continue with greater exclusive, marketing rights of seed and seedling of vegetables. Even the farmers saved seeds would not be much detrimental, as farmers are not in a habit of producing and keeping the seed for further propagations'.

Terminator and Verminator Technology

Terminator Techonology

Concept and History: The terminator technology involves creation of transgenic plant that contains a gene whose expression can be controlled by external stimuli, introducing in plant genome, a series of functionally interrelated DNA sequences containing the following basic units, can produce such a system.

- A plant active promoter that is active at a particular stage in plant development referred to as 'transiently active promoter'.
- A gene whose expression results in altered phenotypes, which is linked to transiently active promoter by a blocking sequence separating the two.
- Unique specific excision sequences flanking the blocking sequences.
- A site-specific recombinase gene, which recognizes the flanking sequences of blocking DNA.
- An alternative gene encoding a repressor specific to repressible promoter, the action of repressor being responsive to an applied or exogenous stimulus.

In March 1998, a US patent (N) 5723765 was issued jointly to United States Department of Agriculture (USDA) and Mississippi-based Delta & Pineland Company, a major cotton supplier in USA. The patent issued to its inventors Melvin Oliver and Co-worker. On March 3, 1998 titled " Control of Plant Expression " was initially hailed by its Inventors as a major breakthrough in agricultural biotechnology. This Patent, which is broad in scope and is applicable to all plants and seeds of all species including transgenic, has also been published by the World Intellectual Property Office (WO: 96044393). This invention describes a technical method of enabling controlled non-availability of second-generation seeds and provides a technological means to ensure that the intellectual property contained those seeds protected, thus affecting a form of genetic copy protection. Such kind of genetic system has worked well with tobacco and cotton. After that, the Rural Advancement Foundation International (RAFI), a Canada based NGO has dubbed the technology as 'terminator technology', since, they believe that such a technology is unethical and terminates independence and threaten food security of over a billion resource poor farmers in developing countries. Seeds, which obtained by this techonology, are known as "devilish seeds or suicidal seeds". The basic purposes for inventing of this techonology are:

- This technology can be effectively used to prevent farmers from using saved seeds for sowing next crop. This will be done particularly in case of transgenic crops having desirable transgene(s) for example Bt for insect resistance or Roundup for the herbicide resistance, *etc.* From this angle, the technology is quite relevant to pure lines and open-pollinated varieties, but not so relevant to hybrid varieties, because in case of hybrids, farmers otherwise also change seed with each year. Hence, this technology will enforced farmers to buy seed fresh each year.
- This technology is also useful to protect the transgenes from being pirated by researchers elsewhere. That is why terminator technology is also known as the Gene Protection Techonology, The identification, isolation, modification and use of transgenes in developing transgenic crops are a time taking and highly expensive process. Once this transgenic plant developed and reach to farmers, the transgene(s) present in these plants can easily be transferred through hybridization to a normal crop variety. Thus, the monopoly of the inventors on transgene(s) is lost. With use of terminator technology the seed obtained after hybridization will also not germinate and thus the transgene(s) will be protected from being pirated.
- The terminator technology is basically an inter-play of genes, promoters and gene control mechanism. It is essentially developed for value capture of pure line in self - pollinated crops, where a transgene conferring an agronomically valuable trait may be incorporated. Pureline have the capacity to breed true to type generation after generation and can thus be saved and used by farmers for re-sowing. In case of hybrid varieties, there is a built-in protection, which compels farmers to purchase seeds every year. This is because heterotic vigour contributing to increased yield is exhibited in F_1 only.

How Terminator Technology Works in Hybrid Varieties

The way in which terminator technology will be used in hybrid variety is differed from pure line varieties. In a single-cross hybrid one of parent (preferably other parent (male) with gene B, there is a no need to use gene C in this case). Hybrid seed produced by crossing two parents will be treated with *tetracycline*. Soon after germination of hybrid seed, gene becomes active producing recombinase, which will remove blocking sequences (BS) between genes A and its promoter late embryogenensis abundant (LEA), thus activating A to produce ribosomal inactivating / inhibitor proteins (RIP) at late embryogenensis stage render embryo dead. Gene and promoters package consists of three genes namely DNA sequence I, DNA sequence II and DNA sequence III, which will be referred to as gene A, B and C, respectively. Gene A produces ribosomal inactivating/ inhibitor proteins (RIPs), which kill the embryo, thereby making seed incapable to germinate. Gene A is hooked with its promoter PA, which is transiently active promoter called late embryogenesis abundant (LEA) because, it is active only at the embryogenesis stage of seed development. The gene A is separated from its promoter PA by blocking sequence (BS), which is flanked on its both sides by an excision sequence (ES). As long as BS is present between gene A and PA, gene A is not expressed. Gene B is responsible for producing an enzyme called recombinase, which cleaves DNA at

ES, removing thereby the BS and ligating PA with gene A and making it functional. The promoter gene B is a repressible promoter, which is repressed by the repressor produced by gene C. The master control of the entire process is through an antibiotic tetracycline, which acts as inducers to PB by interacting with repressor molecule bound to PB.

Advantages

- This technology will induce private sector to make investment in research and development of pure line varieties and open-pollinated varieties, which they are not doing so far, because in these varieties farmers did not change seed, fresh each year. With the use of terminator technology, farmers will have to change seed fresh each year. Therefore, seed companies will get more profit. This will also result into stiff competition between public and private sector institutions to show a better performance and ultimately farmers will be benefited.
- With terminator technology, farmers will been forced to buy seed each year, therefore seed replacement rate will go up and this will increase the production at national level. Hence, our country will fulfill the demand of quality seed of farmers at the national level.
- Since, pre-harvest sprouting of grains due to rains at maturity is a serious problem in many crops. Using terminator technology can prevent this sprouting.
- Escape of transgene(s) for example transgene conferring resistance to herbicide, to wild weedy relatives of the crop through cross-pollination may lead to development of resistance in weeds against the herbicides. This gene flow can be effectively by hooking all the transgene(s) with terminator gene. Similarly, in future plants are likely to prevent the escape of pollen from such transgenic plants to normal crop varieties. Terminator technology can be effectively used for this purpose.

Disadvantages

- This technology will give serious setback to hand core molecular biology research particularly in the developed world.
- If this technology is used, on one hand farmers will be prevented from the age-old privilege of using saved seeds and on other hand it will become virtually impossible for seed producing agencies to meet the seed demand of the farmers.
- This technology will give a serious setback to transgenic research in India. Hence, our hope on transgenic crops may shatter. If we decide not to use terminator and transgene(s) from MNCs and other countries we will have to make our research and development programme more sound than what we have today, so that, we can develop useful transgene(s) ourselves and use them.
- In view of the increasing pressure to generate resources, public sector institution may decide to use terminator technology in the crop varieties developed by them, so that, they can sustain themselves using the revenue generated by sale of these varieties. This will ultimately lead to suffering of poor farmers.

- If this techonology is used, we will be deprived of enjoying the benefit of the sprouted seeds, as the seed will not germinate. Even if the seeds are obtained from the companies and allowed to sprout, they will not fit for consumption as they are already treated with an antibiotic *tetracycline*.
- If the terminator technology along with a useful transgene is used for developing new crop varieties and these varieties do extremely well on the field, farmers would like to use these in spite of high cost. This way, the new package will become extremely popular among the farmers, and the companies will also be happy due to increased turnover. However, this will lead to our dependency on MNC's for seed. Imagine, if at the peak sowing time the MNC's decide to withdraw the seed supply or they are forced to do so, we will not get seeds to sow our field at that time, eventhough our godowns are full of seeds, which will not germinate due to terminator seeds. This will lead to complete failure of crop and ultimately to mass starvation, enormous sterility and loss of human life. This is how seeds may be used as new weapons.
- In any case the seeds formed by pollination with terminator pollen will not germinate because of terminator effect. Thus, the terminator gene is responsible for its own death and there is no way by which this gene will perpetuate in environment. If one seed of a thousand seeds of a wild by which this gene will perpetuate in environment. If one seed of a thousand seeds of a wild or weedy species does not germinate, the species will not face the danger of extinction. Because, otherwise, these species have mechanism of over production in order to survive. In case of normal crop varieties, the loss in seed germination resulting from pollination with terminator pollen can easily be compensated by increasing the seed rate proportionate to out crossing, if saved seeds are used for raising next crop. Therefore, the fear expressed earlier about enormous sterility and loss of the entire biodiversity needs to be examined.
- The product of so-called terminator gene is a protein called "ribosomal inactivating protein (RIP). As indicates RIP inactivate ribosome's, the site of protein synthesis. Hence, expressing RIPs in seed embryo may have deleterious effect on seed protein particularly in Non-endospermic seed and seed protein quantity/quality may be reduced.

Verminator and Triator

On similar concept of terminator a UK based biotech firm 'Giant Zeneca' has used rodent fat gene to block the normal plant growth. Unless the blocking process is de-activated by chemicals, the plant carrying novel transgene(s) hooked up with rodent fast gene will not grow normal. As a result farmer will have to use secret chemical each year to de-activate the rodent fat gene technology has been dubbed as "Verminator A by RAFI". On same footing "Traitor technology" offers the opportunity to use a number of commercial characteristics into a plant variety, which the company can choose to either activate or de-activate external chemical sprays or soaking of platform seed. The Gene Giants want to tie their traitor seed to their property chemical so that one is useless without other.

10

ECONOMICS, IMPORT AND EXPORT

Introduction

Hybrid is the first generation cross-produced through crossing between two genetically different plants, lines or varieties. The seeds produced in this way are called hybrid vigour, which serves as a key for increasing the yield of many self, and cross-fertilized crops. With limited possibilities of horizontal expansion in area and to meet the ever-increasing demand of quality vegetables by growing populations, the exploitation of hybrid vigour is best option for increasing both productivity and quality. Vegetable hybrids under optimum crop production and protection management produce economically more yield than improved, open-pollinated varieties. The uniform size, early maturity, improved nutritional quality, adaptability to adverse environments and resistance to various biotic and abiotic stresses, are additional advantages of hybrids resulting in better monetary returns to growers. Development of hybrids in vegetables brought about a near revolution but still India has not been able to harvest full potential of hybrid technology as compared to USA, Japan, Canada, Europe and Israel. About 10-12% of the total vegetable area is under hybrids in India as against 75-90% in Japan or USA. In India, only tomato and cabbage are leading vegetable crops, occupying 30-35% area under hybrids while rest of the vegetables cover less than 15%. The hybrid seed production and their cultivation, therefore, should be popularized on warfooting to achieve the targets. Many hybrids have been developed both by public and private organizations as given in Chapters 6 of Table 6, which are competing among themselves to produce best hybrids to compete in world market and selling seeds on very high rates ranging from Rs 5,000 to 50,000/ kg of seed. Though, they are making very huge profits but their contribution in promotion of hybrid technology is exemplary and cannot be overlooked.

Recently, a major breakthrough has been realized in cabbage, tomato and brinjal production using hybrids. The entire cabbage and tomato production in Maharashtra, Karnataka, Tamil Nadu and West Bengal are under F_1 hybrids and extending even in Punjab, Gujarat, Delhi, Haryana, Jammu & Kashmir, Uttar Pradesh and Rajasthan. Growing of hybrids not only fetches higher income to growers but also have better price in the market for their uniform quality produce as compared to commonly-grown varieties. The average yields of hybrid tomato in these states are 75-80 tonnes/ha as against 20-25 tonnes/ha of the other varieties. Generally, vegetables are sold at a premium price, about Rs 1 or 2 more a kg than others. For cultivation of hybrids, advance production technology and adequate inputs are necessary for getting good yield. Without this, it is not possible to exploit the full potential of hybrid vigour and obtain maximum return to cover the extra expenses in production. Unfortunately in our country, the progress in development and popularization of hybrids are very slow and mainly due to high cost of hybrid seed and their non-availability. An average vegetable grower is not in a position to afford such a high price for purchasing hybrid seeds. The high cost of hybrid seeds which is mainly because of expensive manual operations involved in its production. However, cost of seeds is less than 10% of the total expenditure incurred on cultivation of one hectare hybrid crop. The increased price of hybrid seed is compensated by higher yield, which is generally 3-4 times more than open-pollinated varieties. Therefore, there should be a strategy for development of large-scale production of hybrid seed at low cost. For this, hybrid production based on male sterility, self-incompatibility, gynoecious or monoecious lines is important and is being utilized for economical hybrid seed production. Besides, use of growth substances like MH, ethrel, gibberellic acid or silver nitrate may also be used after proper standardization. Tapetum based stable transgenic male sterile system for production of hybrid seeds and mass propagation of parental lines especially self-incompatible, male sterile and gynoecious lines *in vitro* are some of the important biotechnological areas, which can play a vital role for making the hybrid technology successful. To produce larger quantity of hybrid seeds at a reasonable cost different strategies involving natural and man made mechanisms, which ease the hybrid production, should be considered and exploited.

Mass Production of Hybrid Seed in India

Hybrid seed production is a labour-intensive activity. Involvement of farm men, women and rural youth not only helps in mass production of hybrid seed but labourers are comparatively cheap which reduce the total cost involved in its production beside providing employment to rural poors. Many international companies from USA, Holland and Japan are employing cheap labourers and producing hybrid seed in India, marketing it in different parts of the world round the year and making India a partner of this activity. There are following stratigies for mass production of low-cost hybrid seeds in India.

- To make hybrid seed production attractive and cost-effective, farm families and rural youths producing hybrid seeds with the help of their own members and supplemented labour should be given practical training on all aspects of hybrid production such as crop raising, crossing, harvesting and processing of hybrid seeds which would improve their efficiencies in handling of hybrid activities. Many ICAR institutes, State Agriculture University and even private seed companies are organizing such trainings for popularizing and increasing seed production at a reasonable cost.
- Another factor which is important for reducing the cost of seed is through production of seed under natural condition and in the areas where congenial climatic conditions, cheap labour, fertile soil and assured irrigation are available for production of quality seed. For this, areas in different parts of the country have to be identified so that high costs incurred on production of seed under protected artificial conditions like glass or greenhouses are overcome and quality seed is produced on affordable cost and made available to farmers' round-the-year.
- In India, hybrid seed production by private companies has been concentrated mostly in and around Bangalore, Kolar, Haveri and Dharwad in Karnataka; Pune, Nasik,

Aurangabad and Jalna in Maharashtra and in a few districts of Andhra Pradesh and Punjab. Due to mild climate in these areas, seeds of tomato, brinjal, capsicum, okra, melon, cucumber, onion and gourds are produced in open, while hybrid seeds of cabbage, carrot, radish late cauliflower and capsicum are produced in hills of Jammu & Kashmir, Himanchal Pradesh and Uttaranchal but the quantity of seed produced in these hills is very less and mostly for temperate crops, we depend on imported seeds. In most of these seed-production zones, seed companies, including multinationals do not directly involve in seed production. They arrange contracts with local farmers through middlemen called 'Seed organizers' who mediate between companies and farmers for production of seeds. Each seed organizer is given some area and target of production. It is the responsibility of the seed organizers to identify the farmers who are willing to undertake seed production. They supply seed of parental lines to farmer, determine method of cultivation and set price of seeds at harvesting before cultivation begins. They collect seeds from these farmers and hand over to companies. In agreement, it is specified that farmers must sell their seed only to that company at the agreed price. Companies also give advance about 30-50% of production cost as initial capital to farmers. If quality of seed is not up-to-the mark, then companies do not buy seed from farmers and contracts are cancelled. The total responsibility of producing genuine seed lies with farmers. These limits put on seed producers; the seed companies impel farmers to have complete control in production of quality hybrid seed. Similarly, collective efforts can also be made by Government agencies like NSC, State Seed Corporations, SAUs and ICAR institutes.

- The hybrids seeds can be produced in larger quantity, wherever conditions ideal for hybrid seed production are available. In India, other than already existing zones, hills, which constitute Kashmir in Jammu & Kashmir and Katrain in Himanchal Pradesh. Once they are known for seed production, are neglected particularly with the introduction of liberalized seed policy. These places still have tremendous potential not only for production of hybrid seed of temperate vegetables but also for production of hybrid seed of other crops like capsicum, tomato, brinjal, chilli, cucumber, okra *etc*; because ideal weather and availability of cheap labour, fertile land, and assured irrigation accompanied by greater seed production skills and expertise. The hybrid production in hills not only results in high F_1 seed yield and quality but cost of hybrid seed produced in these areas would also be very low as compared to seeds produced in other parts of the country and imported hybrids. The hills could have monopoly on production of hybrids and can become Holland of Europe from where large quantity of seed can be made available for sale to other parts of the country at very reasonable rates. Another approach would be to produce seed directly through progressive farmers. If hybrid seed production is taken up on large scale by farmers themselves through seed village concept under the auspices of Government agencies and SAUs, this would not only eliminate middleman and profit-making companies but would produce large quantity of quality hybrid seed at much cheaper cost. Production of quality seed through seed villages is already successful in cereals and can be applied in vegetable hybrid seed production with a caution of maintaining utmost genetic purity for producing genuine seed of highest quality.

- Either public or private sector alone can not suffice the need of seed for millions of hectares. Both agencies should come forward with a hybrid programme involving gross root farmers for producing hybrid seeds at affordable cost, so that, hybrid technology could popularized.

Factors Affecting Production and Popularization of Hybrid Seeds

In India, neither the actual seed grower nor the vegetable grower harvesting the full benefits. It has been generally seen that middlemen and companies have shared the maximum benefits. We have to develop such systems where prices of hybrids seed are reasonably low so that vegetable grower could not face difficulties in purchasing the seed of high-yielding hybrids and simultaneously the seed producer could get a maximum share and profit for the hybrid seed production. Once, seed producer and vegetable growers both are satisfied, the system of seed production will be economically-viable and lost till the balance is in equilibrium. Successful hybrid seed production programmes depend upon a large number of factors such as:

- Cost of seed production
- Demand of a particular crop
- Productivity per unit area
- Share of profit on sale by seed producer
- Economic status, knowledge and expertise of the farmer
- Sale price and quality
- Additional productivity of vegetables to grower through hybrid seed
- Additional returns by use of hybrid seed.

Cost of Hybrid Seed Production

Production of hybrid seed in vegetables is a complex and expensive venture. It requires about 3-4 times more capital than ordinary commercial vegetable cultivation. The cost arises from major three components:

- Raising of healthy crop for hybrid seed production
- Cost of emasculation and pollination
- Post-harvest operations.

Raising of Healthy Crop for Hybrid Seed Production

Cost of raising crop varies on type of vegetable, season, region, availability of inputs and facilities like irrigation, soil, labour, *etc*. Various cost included in this component are:

- Cost of parental lines seed
- Cost of land rent, land preparation and labour
- Cost of seed sowing, nursery raising and planting

- Cost on fertilizing, weeding and pesticide treatment
- Cost of cultural practices and other special operations of seed production such as rouging, staking, stackling preparations over wintering, *etc*
- Cost of hiring of farm implements and machinery and other material cost involved in crop raising.

The cost of raising of crop for seed production under Kashmir conditions, ranges from Rs 0.5 lakh in okra to Rs 1.0 lakh in tomato. In chilli, capsicum and brinjal is around Rs 0.8 lakh, while in cucumber and bottle gourd, it is about Rs 0.9 lakh including stacking, (Ahmed and Narayan, 2005). This cost of raising crop is much less than the cost worked out for emasculation and pollination performed in hybrid seed production.

Cost of Emasculation and Pollination

The production of hybrid seed involves two basic procedures, emasculation and pollination. These procedures are carried out either manually or by certain mechanisms depend upon vegetable crops. The hand emasculation and pollination of individual flowers is most expensive but it is economical in tomato, brinjal, chilli, capsicum and cucurbits where single fruit gives large number of seeds and where small quantity of seed is required for planting of one hectare land. Such a procedure is not worthwhile in cabbage, cauliflower, knoll-khol, Brussels' sprouts, sprouting broccoli, radish and turnip where single pollination results in a limited number of seeds and in crops like onion and carrot where floral parts are so arranged that selection of florals in umbels and their emasculation is difficult. To ease the procedure and to produce hybrid seed at an economically-acceptable level, a number of morphological and physiological mechanisms operating in plants such as sex expression, male sterility, self-incompatibility, heterostyly *etc.* and certain pollinator have been exploited commercially to overcome expensive emasculation and hand-pollination. Removal of male plants in dioecious crops like spinach and asparagus or pinching of male flower in monoecious cucurbits before flowering facilitates large-scale emasculation and subsequent natural pollination results in production of hybrid seed in female plants or flowers, while in cabbage, cauliflower, kale, Chinese cabbage, turnip, radish *etc.* where self-incompatibility mechanism is operating both the male and female parental lines are producing hybrid seed by pollinating each other. In tomato, brinjal, capsicum, onion, carrot and melon, male sterility arises either due to genes or cytoplasm or induced through gameticides used as self-emasculating mechanism. The hybrid seed is produced by planting alternatively male sterile female parent and with pollinating with fertile male parent either by hand as in tomato, brinjal, chilli, capsicum or by natural pollination through pollinators as in onion, carrot or melons. Even heterostyly condition has been used in production of hybrids by pollinating long-styled flowers of some varieties of tomato and brinjal.

Cost of Post-harvest Operations

Hybrid seed production is a highly capital-intensive activity, requiring utmost care from harvesting to marketing. Fruits after attaining physiological maturity are harvested carefully from female parent and seeds extracted using standard procedures as applicable in different crops. After seed extraction, seeds are dried either in sun or through mechanical driers up to a moisture level of 7-8%, depending on crops. After seed drying, seeds are properly treated with fungicides in vaccume-packed preferably in small units. Packaging in small units makes identification, transportation, handling, marketing and distribution easier. A wide range of materials and types of container are used for seed packing. The choice of material and size of seed packing depends on seed, value of seed /kg/unit, destination, type of market, mode of distribution, protection required from biotic and abiotic hazards. Moisture-proof material is generally used for this purpose. Labelling is must on seed package containing all particulars as per seed certification standards like name of the crop hybrid, percent germination, date of packing, date of expiry, weight of seeds *etc.* Packed and labelled seeds are then stored in different structures like ordinary well-ventilated stores, air-cooled storages, dehumidifiers or in vapour-proof containers on short or long-term basis depending upon the requirement and after that seeds are marketed through authorized whole-sale and retail outlets on a pre-fixed prices. All post-harvest operations of hybrid seed production are labour intensive and based on modern techniques, therefore, cost of post-harvest operations are generally more than ordinary seeds resulting in additional costs.

Methods of Hybrid Seed Production and their Production Cost Analysis

On the basis of sex expression, flower structure, flower position, pollination mechanisms, male sterility and self-incompatibility and ease with which seed can be produced, following methods of hybrids seed production are described along with economics of production of hybrid seed.

Hybrid Seed Production Using Hand Emasculation and Hand Pollination

In bisexual crops like tomato, brinjal, chilli, capsicum and okra and in andromonocious varieties of muskmelon male part in bisexual flower is emasculated and pollinated manually by hand. This method was first practised in Japan in 1935 for production of hybrid seed in brinjal and subsequently in tomato (Obra *et al* 1945; Nishi, 1955 and Hojby 1958). It involves different steps like emasculation, bagging or caging, removal of bags/cages, pollination and subsequent bagging which takes about 50-60 seconds by a trained worker (Nishi, 1967) In a day, it is possible to handle only about 300-350 flowers, therefore, hybrid seed produced by this method is very labourious and expensive. Obra *et al.* (1945) produced 28 g of hybrid tomato seed from 3-4 hours of pollination, which was sufficient for planting in half an acre of land. Hojby (1958) obtained 24,000 seeds from 8.1 hours while Hafen and Stevenson (1956) produced 4,000 hybrid seeds from 51 minutes through emasculation and pollination. It reduced to 12 minutes time when male sterile plants were used as female parents. Hojby (1960) again worked out economics of hybrid seed production in tomato variety Pearson through natural pollination, hand pollination and hand emasculation and hand pollination with pollen of long-fruited San Marzano variety. The maximum hybrid seed quantity (108.0 lbs) was obtained with hand-emasculation and hand pollination than either hand pollination or natural pollination when male sterile seed parent was used. Both hybrid seed as well as quantity of hybrid seed, however, reduced in fertile seed parent than male sterile. Ahmed and Narayan (2005) worked out labour requirement and economics of hybrid seed production in chilli, capsicum, tomato, brinjal, okra cucumber and bottle gourd following hand emasculation and hand pollination. The number of flowers emasculated and pollinated/hour/day was 360 for chilli, 320 for capsicum, 400 for tomato, 340 for brinjal, 280 for okra and 240 each for cucumber and bottle gourd. The approximate number of plants which one labour was handle about 14 for Chilli 53 for capsicum, 20 for tomato, 34 for brinjal, 28 for okra, 48 each for cucumber and bottle gourd (Table 1).

Table 1: Emasculation and pollination per day/labour (4 hours, emasculation + 3 hours pollination + 1 hours preparation)

Crop	Average number of flowers emasculated/bagged and tagged in 4 hours.	Number of flowers pollinated in 3 hours	Number of flowers emasculated and pollinated / labour / day	Approximate number of plants handled / day/labour
Chilli	360	360	360	15.0
Capsicum	320	320	320	50.00
a) Tomato (fertile)	400	400	400	20
b) Tomato (male sterile)	600	600	600	30
Brinjal	340	340	340	34
Okra	280	280	280	28
Cucumber	240	240	240	48
Bottle gourd	240	240	240	48

Source: Ahmed and Narayan (2005).

On the basis of number of plants a labour can handle, total number of labours required per hectare for doing emasculation and pollination have been worked out as 1580 for chilli, 474 for capsicum, 741 for tomato, 436 for brinjal, 1058 for okra, 124 for cucumber and 93 for bottle gourd (Table 2). The total cost of hybrid seed production per hectare which includes labour cost of emasculation and pollination and the cost of raising of crop in one hectare of land ranged from 1.0389 lakhs for bottle gourd to 3.17 for chilli (Table 2). For capsicum, it is Rs 1.711 lakhs, and for tomato, brinjal, okra and cucumber, it is 2.115, 1.454, 2.087 and 1.086 lakhs, respectively. Involvement of more labour hours for emasculation and pollination per hectare resulted in high production cost as compared to cucurbits or brinjal whose labour requirement is almost 5-10 times lesser. Nearly, 74-76% of the total cost in chilli and okra and 42-52% in capsicum, brinjal and tomato go towards the cost of human labour who do the work of emasculation and pollination. For cucumber and bottle gourd, the cost of emasculation and pollination is only 13-17% of total cost of production. It is because, the labour requirement for pinching of male flowers and pollination of large female flowers is comparatively less, easy to handle and economical. The hybrid seed production using hand emasculation and hand pollination in other than cucurbits is therefore very expensive requiring 2-3 fold more capital than ordinary commercial seed production.

Hybrid Seed Production Using Hand Pollination Without Emasculation

This method is followed in crops where two sexes, male and female flowers, are separate as in cucurbits or crops where male part of bisexual flowers are non-functional, *i.e,* male sterility, which is exploited commercially in capsicum, tomato, brinjal, squash, muskmelon, cucumber, pumpkin *etc*. For production of hybrid seed in monoecious cucurbits, female flowers of seed parent and pollen flowers of male parents are bagged separately before opening and hand pollination, is done subsequently after opening of female flowers of seed parent. Without pinching of male flowers from seed parent in this method saves lot of labour resulting in low-cost seed production . Nishi (1955) worked out number of workers for pollination in cucumber and reported handling of about 350-400 flowers/day as against 240 flowers in cucumber and bottle gourd when both hand pollination and pinching of male flowers are done. It means 30-40% additional labour hours are required when hand emasculation, *i.e*, pinching of male flowers of seed parent is carried out. But this labourious emasculation can be overcome if some self-emasculating genes like male sterile genes are transferred to commercial seed parents using conventional methods or biotechnological techniques. Ahmed and Narayan (2005) compared the efficiency of hybrid seed production in tomato using hand emasculation and hand-pollination and no emasculation using functional male sterility and hand pollination. With the use of male sterility, it is possible to handle 600 flowers by one labour as against 400 when both hand emasculation and hand pollination is used. Male sterility therefore not only increase labour efficiency but also reduces cost of production of hybrid seeds. (Tables 2 and 3).

Table 2: Labour requirement, labour cost and cost of cultivation/ha

Crop	Approximate number of plants/ha at 4:1 (female male) ratio	Number of labourers /ha requiredfor emasculation and pollution	Labour cost (Rs) 150/labourers for emasculation and pollution (lakh)	Cost of cultivation /ha (lakhs)	Total cost of production /ha (lakh)	% total cost for emasculation and pollination
Chilli	23700 (Female) 5925 (Male)	1580	2.3700	0.80	3.170	74.76
Capsicum	23700 (Female) 5925 (Male)	474	0.711	1.00	1.711	41.55
Tomato	14815 (Female) 3704 (Male)	741 494 (MS)	1.1115 0.741	1.00 1.00	2.115 1.741	52.55 42.56
Brinjal	14815 (Female) 3704 (Male)	436	0.6540	0.80	1.454	44.97
Okra	29628 (Female) 7407 (Male)	1058	1.5872	0.50	2.087	76.05
Cucumber	5924 (Female) 1481 (Male)	124	0.1860	0.90	1.086	17.12
Bottle Gourd	4444 (Female) 1111 (Male)	93	0.1389	0.90	1.0389	13.36

Source: Ahmed and Narayan (2005).

Table 3: Cost benefit analysis of hybrid seed production of different vegetables under kashmir condition

Crop	Average number of seeds/ fruit	Seed weight/ fruit (g)	Seed number (g)	Number of fruits / plant	Seed weight / plant (g)	Seed yield/ha (kg)	Total cost of production of hybrid seed/ha	Average cost of production 1kg hybrid seed (Rs.) excluding cost on processing and marketing	Average sale price of hybrid seed/kg in India (Rs)	Cost benefit ratio
Chilli	87	0.59	147	25	14.75	349.57	3.170	906.82	10000	1:11.02
Sweet Pepper	281	2.72	103	6	16.32	386.78	1.711	442.37	15000	1:33.90
Tomato	195	0.63	309	20	12.60	186.67	2.115	1133.01	12000	1:10.59
(female)	203	0.65	298	20	13.00	192.59	1.741	903.99		1:13.27
Tomato										
(male Sterile)										
Brinjal	782	3.89	201	10	38.90	576.30	1.454	252.29	3500	1:13.87
Okra	50	3.50	14	10	35.00	1036.98	2.087	201.25	500	1:2.48
Cucumber	355	7.80	46	5	39.00	231.04	1.086	470.04	2000	1:4.25
Bottle gourd	426	52.00	8	5	260.00	1155.44	1.039	89.92	1800	1:20.01

Source: Ahmed and Narayan (2005).

Hybrid Seed Production Through Emasculation and Wind Pollination

This method is generally used in dioecious crops like spinach and asparagus and in monoecious sweet corn. The abundant quantity of light weight pollen produced in male parent is carried to female line by wind and the seed collected from female parent is the hybrid seed. All these procedures in this method are carried in isolation. As compared to hand emasculation and hand pollination, the hybrid seed production through this method is easy and economical as no labour is required for pollination. However, identification and removal of 50% male plants of seed parent in dioecious crops and detasslling of male part of seed parent of monoecious sweet corn before flowering are only major labour-involving procedure. No laborious bagging and tagging are also involved as in hand-emasculation and pollination. The efficiency can be further improved if some marker trait is associated either with male or female plant line of dioecious crop. It is identifiable before flowering as it facilitates easy identification and removal of males from female parent in a hybrid seed production plot. The efficiency also depends upon the pollen-producing capacity of male parent, the prevailing weather conditions especially dry sunny day, direction and speed of wind. Planting of male first followed by seed parent in the direction of wind is very important for obtaining higher and better seed setting. Due to very less labour requirement, the cost of production of hybrid seed through this method is very cheap as compared to other methods. Data presented in Table 4 showed that cost of production of one lb hybrid seed is just \$11-18 in sweet corn and \$12-13 in spinach as against \$1917-2109 in tomato and \$1246-1534 in pepper through hand-emasculation and hand-pollination. This method is although very easy and economical but limited only to vegetables crops, which are wind-pollinated.

Hybrid Seed Production Through Hand Emasculation and Use of Insect-pollination

This method is used mostly in monoecious cucurbits like cucumber, melons, squashes and gourds where male and female parents are planted alternatively in commercial hybrid seed production plots in a 1:2 ratio and male flower of female parent are removed before opening regularly and allowed insect pollination in isolation of which pollen of male parent are transferred to female parent through insect pollinators. The seed harvested from female is collected as hybrid seed. It is one of the cheapest methods, which does not involve laborious hand emasculation and hand-pollination. The identification and pinching of male flowers of female parent can be done very easily at bud stage and before opening by very few labourers . Nishi (1955) reported that only 30 workers are required in one hectare of cucumber crop for carrying out emasculation work as against 124 and 93 labourers required in cucumber and bottle gourd, respectively when both emasculation and pollination is carried out (Ahmed and Narayan, 2005). This method saves about 50-75% of labour cost involved in emasculation and pollination. The cost of hybrid seed thus produced is very less $ 127/lb. The cost would further reduce if hand-pinching is replaced by elimination of males through application of ethrel or by using gynoecious lines. By using gynoecious lines and ethrel elimination one can produce one pound hybrid cucumber seed in just $ 20 as against $ 97-127 when hand-emasculation is used (Table 4). Nowdays, gynoecious lines and ethrel in cucumber, ethrel in pumpkin and squash being commercially used for economical production of hybrid seed in world wide.

Table 4: Cost comparison analysis of hybrid seeds (produced through different methods/ techniques) vs non-hybrid seeds (open-pollinated or pure line cultivars)

Crop	Method	Non-hybrid cost /lb ($)	F_1 Hybrid cost /lb ($)
Sweet corn	Detasseling wind pollination	10	11-18
Summer squash	Elimination of males by ethrel, insect pollination	19	119-255
Onion	CMS, insect pollination	100-208	167-638
Spinach	Hand rouging of male plants	11	12-13
Muskmelon	Hand emasculation, insect pollination	-	255-639
Cabbage	Self incompatible, insect pollination	79-102	351-682
Tomato	Hand emasculation and hand pollination	534-829	1,917-2,109
Pepper	Hand emasculation and hand pollination	134-421	1,246-1,534
Brinjal	Hand emasculation and hand pollination	-	509-1085
Carrot	CMS, insect pollination	63-104	175-534
Cucumber		-	-
Slicer	Hand emasculation or ethrel treatment	43	97-127
Pickler	Ethrel treatment or hand- emasculation	60	-
Slicer	Gynoecious flowers	-	20
Pickler	Gynoecious flowers	-	206-271
Broccoli	Self-incompatible, insect pollination	159	359-508
Pumpkin	Ethrel elimination of males	47-143	95-319
Watermelon	Diploid	57-58	278-444
	Triploid	-	700-780

Source: (http://hort.science.orst.edu/233/sample.htm).

Hybrid Seed Production through Free Insect or Wind Pollination Without Emasculation

It is one of the most important method particularly in crops where emasculation is tedious as in onion and carrot due to umbellate inflorescence and in crops where each pollination results in only a small number of seeds as in cole crops and radish. The presence of self-incompatibility in cruciferous vegetables and male sterility in onion, carrot and beet have been exploited to overcome laborious emasculation. While insect-pollinator and wind have been used for carrying out pollination. Since both emasculation and pollination have been done utilizing plants inherent mechanisms like self-incompatibility male sterility and natural pollinating agencies like wind and insect, hybrid seed production may be more economical in comparison to methods where manual emasculation and pollinations are carried out. The cost of production of hybrid seed of onion and carrot using male sterility and free insect pollination per lb ranged from $167-638 and $175-534, respectively as compared to cost by hand-emasculated and hand-pollinated seeds of tomato or pepper whose production cost for one lb seed ranged from $ 1246, to 2109 (Table 4).

Cost, Benefit Analysis of Vegetable Hybrid Seed Production

The hybrid seed yield per hectare varies from one crop to another crop and from one variety to another and also from one area to another, depending upon the quality of soil, topography, weather conditions and labour availability. Hojby (1960) recorded hybrid seed yield ranging from 95.1 in fertile seeded parent to 107.5 lb/acre in male sterile tomato under South African conditions. Similarly, Ahmed and Narayan (2005) under temperate Kashmir condition recorded 186.67 kg/ha hybrid seed yield in fertile seeded parent to 192.59 kg/ha in functional male sterile tomato, indicating that Kashmir climate is more suitable for producing hybrid seed of tomato. The availability of cheap labour, fertile soil and congenial weather, the cost of production of per kg hybrid seed under Kashmir condition is just Rs 906.82 for chilli, Rs 442.37 for sweet pepper, Rs 904-1133 for tomato,Rs 252 for brinjal,Rs 201 for okra, Rs 470 for cucumber and only Rs 90 for bottle gourd. On the basis of average sale price the cost: benefit ratio of hybrid seed production for important crops as reported by Ahmed and Narayan (2005) under Kashmir conditions, ranged from 1:2.48 for okra to 1:33.90 for sweet pepper. For chilli it is 1:11.02, tomato 1:10.59 to 1:13.27, brinjal 1:13.87, cucumber 1:4.25 and bottle gourd 1:20.01 (Table 3). This indicated that except okra, hybrid seed production of rest of the vegetables highly remunerative with sweet pepper being highly profitable followed by bottle gourd, brinjal, tomato, chilli and cucumber. For every one rupee spent on cost of production we can get net returns of Rs 33.90, 20.01, 13.87, 13.27, 11.02 and 4.25, respectively in these crops. It is the assurance of the profits that attracts a number of private Indian and multi-national companies to venture in this highly profitable seed business.

There is a buoyant market for hybrid seeds. Unlike seeds of open-pollinated varieties, hybrid seeds are used for only one crop and therefore purchased every year. The demand for hybrid seed is increasing both in India and also in international markets. To meet the growing demand, new hybrid seed production zones should be identified. Temperate hills in this regard can compliment in meeting the requirement of hybrid seed. Ideal cool summer, fertile soil, adequate irrigation, cheap labour and seed production expertise available in temperate hills could be exploited for production of high quality hybrid seed at much lower cost than other traditional seed growing areas of southern India. Hybrid vegetable seed production in temperate hills, besides contributing to national seed pool can help in receiving sick seed industry in this area and also helps in generating employment to rural men and women.

Seed Import and Export

The biggest seed trader (import + export) is Japan with \$ 80 million, followed by Korea with \$ 23 million, and Thailand \$ 20 million. China, and India seed sector are in full growth, and their export capabilities will presumably increase. All other countries are importing vegetable seeds only. It has been observed that even countries with a very strong vegetable seed industry, have high volumes of import, of which the value is compensated by the value of export. The high imports include seed for domestic market which is produced abroad, because, either for economic, or climatic reasons, the production is not possible locally. This practice of custom seed production abroad has become very common for developed seed companies. A lot of developing countries are concerned about their dependency on foreign seeds, and the loss in their trade balance. Some had even taken measures to restrict import of seed. The fact is that some countries with tropical climate will always need to import seed of certain crops such as cabbage and cauliflower for which seed can not be produced locally. Moreover, India, which has restricted import of seed for a long time, and revised the decision in 1988, realized the positive change. For this reason, vegetable seed import should not be restricted nor banned anywhere, so that the best material make available to farmers. These policies should be accompanied by proper quarantine measures, to avoid the spread of unwanted diseases. India imported 170 tonnes of vegetable seeds in the year 1993, of which the main items are cabbage, capsicum radish, tomato, cucumber, squash, Chilli, and cauliflower. Most popular cabbage F_1 varieties is still imported from Japan. With the development of seed sector in the past decade, India has been exporting officially or unofficially a lot of vegetable seeds to neighbouring countries.

SECTION II

TECHNIQUES OF HYBRID SEED PRODUCTION

(PAGES 259-410)

11

SOLANACEOUS CROPS

Tomato

Genetic Resources

Tomato, a warm season crop, sensitive to frost, is usually cultivated in sub-tropical and mild cold climates. The National Bureau of Plant Genetic Resources (NBPGR), New Delhi, is the main centre for maintenance of tomato germplasms in India. The Bureau has systematically assembled more than 2,900 germplasms. Some of the promising germplasms identified for different traits on the basis of their morphological variations (Fig.1; see Plate-1) are given in Table1.

Table 1: Promising germplasms of tomato

Attribute	Promising Accessions
High yield	EC-106285, 118292, 4551, 163683, 173856, 170662, 110578, 1127 129081, 129353, 168084, 50357, 122063, 1301, 251636, 6504, 315481, 315486, 339059, 337827, 339074, 357829, 315483, TIO, P-388, 1, 501, NIC-1122231, EC-269288, IHR-2034, IHR-2030, IHR-2028, KM-1, KM-2, IC-6, Sel-148-6, EC-378645, 368838, 378639, 377983, 378833, 368830, 368855, IHR-1885, IHR-190, EC-257749, EC-1127, 130053, 163598, 329560, RT-6, CLN-2112-86-2-10-1-7-0, CLN-1558B, L0-2455, CLN-1463-A, EC-326142-A, BL-985, ECC-339061-1-1, CH-193, CLN-2413R, DR-2, HAT-142, EC-357841, 238508, 78836, 398702, 398715, 488370, 490119, 318193, 251613, 251566, 141887, 538408, 538411, 538416, Ovegen-II, Santian, Siletz, Oregan Pride, CLN-1466B, CLN 1460-A, S-1-96-66, Equinox, EC-400504, Arka Meghali, CH-24, EC-385654, EC-388943, EC-174045, NO-187 T-777.
Cold tolerant	Poter Narcariang, Early North, Puck, Fireball, Red Cloud, Cornell Breeding Lines 54-149, 54-179, P1-204050, 280597, Cold Set, Avalanche, Illalihin, Severianin , *L. pimpinellifolium*, Outdoor Girl, Immuna, Pusa Sheetal, Tempo, Oregon and AC-142, 248.
Drought tolerance	*L. pennelli,* IIHR 14-1, 146-2, 383, 553, 555, EC-130042, 35992 and Sel-28
Salt tolerance	*L. cheesmani* and Sabour Suphala, EC-251706, 151628
Heat tolerant	EC-41824, 7764, 164855, 169308, 164653, 122063, 37311, 31515, 126955, 130042 35220, 163709, 164322, 154660, 164677, Sparton Red-8, OK-7-2, 7-3, Saladette, TAMU Saladette, P126934, BL-6807, S-6916, C1A-5161, VF-36, Punjab Tropic, Marzano, P-4,

	Avalanche, Chiko-III, Ment, C-28, Red Rock, Narcaring Breed ing line-165, Red Cherry, KS-1, 2, Walter, *L. cheesmani,* EC-162935, AC-325, 326, PS-1, EC-1127, 130042 and EC-168064.
Transportation	EC-161252, 7352, 100845 and 110268.
Large fruit	EC-163708, 41340, 267726, 267728, 231751, 168070, 267725, 367857, 361674, 362945 and 367856.
Early-maturing	EC-357833, 357827, 362956, 362940, 362944, 362947, 362948, 362950, 99935 and 129602.
Late-maturing	EC-357829, 357830, 362941, 362942, 362951, 241157, 163594, 212469 and 241134.
High ascorbic acid	*L. pimpinellifolium,* Double Rich and Hisar Arun.
High TSS	A-276, A-408 and Red Cheery.
High Alcohol insoulable solids (AIS)	VF-109 and 9039 M.
Red colour (lycopene and β-carotene)	High pigment and Crimson.
Provitamin A (β-carotene)	Caro Red
High volatile compounds (flavour characteristics)	Campbell-146, Rick High Sugar, Cherry and LA-1563.
Anthracnose resistant	PI-272636.
Early blight resistant	*Lycopersicon hirsutum f. glabratum,* 68B134 and Southland (Collar rot).
Fusarium wilt	Rustrel (race 2), No.126915-1-8 (race-1 and 2), Pan American, *L. pimpinellifolium, L. hirsutum,* PI-13448, *L. peruvianum*- EC-148898, Roma, Columbia, HS-110 and EC-66007.
Late blight resistant	*Lycopersicon pimpinellifolium, L. esculentum* var. *cerasiforme,* Otawa-30, 31, Fla-P14, WV-36, West Virginia- 63 and TRB-1, EC-326139, 315481, 162508, 2512780 and 165700
Fusartum wilt and fruit rot resistant	EC-160195, 122527, 21626, 103598, 116874, 117008, 148898, 115947, 117671, 117163, 128769, 174023, 174041, SGP8M2, Columbia, Pant Bahar, Marglobe, Pan America, Mangla, HS-101, *L. pimpinellifolium, L. hirsutum,* PI-13448, *L. peruvianum* and Roma.
Bacterial wilt resistant	*Lycopersicon pimpinellifolium,* PI-127805A, 126408, Venus, Saturn, BT-1, LE-79 (Shakti), VC-48-1, BWR-1, BWR-5, Arka Abha, Arka Alok, Pant Bahar, BT-10, BT-1, Sonali and Rodade.
Bacterial cancker	Bulgarian -12 and Utah-737.
Verticillium wilt	VR Moscow
Tomato mosaic	Ohi M-29
TMV (Strain – 1, 2, 3 and 4) resistant	*Lycopersicon chilense* and Ohio MR-9.
Leaf curl virus *resistant*	*Lycopersicon pimpinellifolium*, A–1921 and *L. hirsutum. f. glabratum,* B-6013, Hisar Anmol, H-36, H-86, H-88, CLN-2123A, CLN 2116-B, CLN-212-E, CLN-5915-206, AC-400 and AC-441.

Yellow leaf curl *virus resistant*	*Lycopersicon pimpinellifolium, L. hirsutum f. glabratum, L. peruvianum, L. chessmanii* and *L. chessamanii* sub sp. minor LA-1401
White fly	IVT 74453 (*L. hirsutum*)
Fruit borer resistant	*Lycopersicon hirsutum, L. hirsutum f. glabratum* EC-110264, 101552, 124406, 129599, 129597 129698, 129791, 121451, 160191, 161254, 119121, PI-126449 (*L.hirsutum*), L-633, EC-129354 and 124406
Leaf miner resistant	*Lycopersicon hirsutum, L. hirsutum, f. glabratum* and Pearson VF-6
Meloidogyne incognita, *M. javanica* and *M. arneria* resistant	Ronita, Rossol, Sanatacruz, Venus, Satumo, Atkinson, Manalucie, *L.peruvianum,* Pusa-120, PNR-7, NRT-3, NRT-8 NRT-12, TH-2312, and TH-802

Production Objectives

Tomato is the most important crop in respect to development of hybrids and production of hybrid seeds in the world. Hybrid vigour in tomato was realized first time in 1908 and first commercial hybrid developed in 1940 in Japan. In India, Indo American Hybrid Seed Company took the lead in releasing the first hybrid in 1973. Presently, more than 75 thousands ha of area is covered under *hybrid varieties* and every year area is increasing considerable. Both the private and public sectors are actively associated for development of hybrids with following objectives:

- To develop early and high-yielding hybrids.
- To develop hybrids resistant to devastating diseases and insect-pests of the areas.
- To develop hybrids resistant to root-knot nematodes.
- To develop hybrids for cold and high temperature conditions.
- To develop hybrids for prolonged storage and transportation.
- To develop hybrids for processing.
- To develop hybrids for multiple diseases and insect pests resistant.

Botany and Floral Biology

Tomato produces a strong tap root but very often it is damaged at the time of transplanting, consequently, the lateral adventitious roots develop profusely. The adventitious roots develop at the rate of about 33 mm/ day, penetrating 1m depth and attaining a length of 1.5 m (Ward, 1964). However, root development largely depends upon the cultivar, the growing media and the moisture in it. The stem is soft, brittle, hairy and when young hard, woody, and conspicuously branched mature. Its organization is in the sequence of two or three leaved sympodia. It is erect to semi-erect. Epidermal hairs with oil glands are present at the early plant growth stage. The leaf is alternate, petiolate and 2.15 inch wide. The leaflets are unequeal odd-pinnate. The apex is narrowed or acuminate, acute and irregularly serrate. The number of leaflets varies from four to several and the length of leaf from 6 to 12 inch on leaves. Hairs or trichomes are formed by outer growth of epidermal cells.

The hairs may be unicellular or may become multicellular by subsequent cell division. Inflorescence is extra axillary cymes with dicotomous or polycotomous branching. It is lateral and arises from the stem opposite to leaf. It may be terminal or may continue with vegetative shoot (Muller, 1940). The number of flowers per cluster varies from three to several.

The early flowers that open in cluster are large-sized as in typical cymes. The abscission of flowers may take place due to high or low temperature or some other internal and external factors. The flower drop is due to development of abscission layer at the mid point of pedicel. The plant produces bright yellow flowers. The flowers are pentamerous, bisexual, regular, complete, ebracteate, and hypogynous. The lateral section of flower in (Fig. 2) shows the details of sepals, petals, stamens, style ovary and ovules. The pistil has several carpels. The anther connate appearing in the throat of corolla and they dehisce through the poricidal dehiscence. Usually, style is shorter than anther cane. Therefore, there is a high degree of self- pollination. The fruit is soft berry, glandular hairs and glands are usually present on small fruits, which degenerate in advanced stage. The fruits of some varieties or species do not have glandular hairs and glands. The seed is oval flattened buff to pale brown, 3-5 mm in length. It is derived from the anatrosperm are enclosed in the seed.

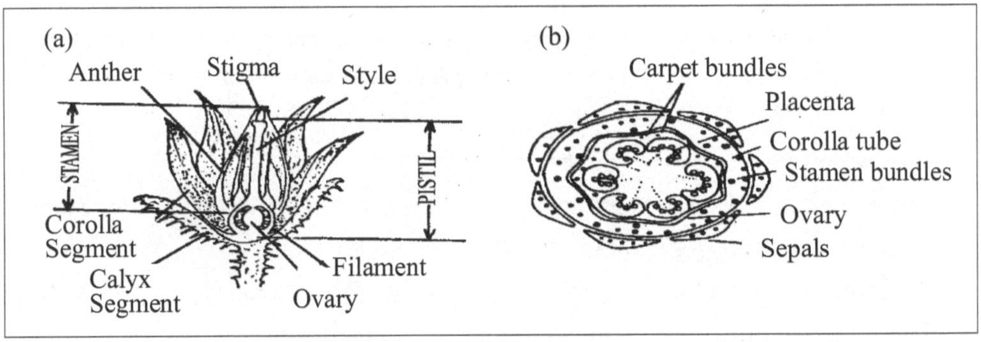

Figure 2: Tomato Flower: (a), Vertical section, (b), Transverse section. *(Source*: Kalloo, 1986).

The anthesis of flower starts in morning at about 6 AM with maximum flower opening in the late morning. The maximum amount of anther dehiscence takes place from 8 to 11AM, depending upon the season. The stigma becomes receptive 16 hours before anthesis and remains thus for five days after anthesis. Pollen grains germination takes place approximately two hours after pollen when applied on the stigma. Ovules ripen over two days. Pollen can be stored upto three days under ordinary conditions. At 0°C, it can store for 2-3 months. While, it remains viable for six months at 5°C in a desiccator. Pollen from one day-stored flower provides best seed setting. The optimum temperature for pollen germination is 18-25°C. Tomato pollen germinates best in a liquid solution of 20% sucrose and 10 ppm boron.

Method of Hybrid Seed Production

There are two methods used for hybrid seed production in tomato. These are:

- Hand emasculation and pollination
- Use of male sterile lines

Hand Emasculation and Pollination

For hybridization, the selection of appropriate size and kind of bud for emasculation is important for success of crossing to get maximum hybrid seed yield. Only first formed flower bud from each flower cluster (inflorescence) should be selected for emasculation, which are expected to open (anthesis) the next day. The remaining buds in cluster should be removed with the help of forceps or scissors. Mature buds are emasculated 12-16 hours. before their opening in the afternoon or evening and pollinated in next morning. Those

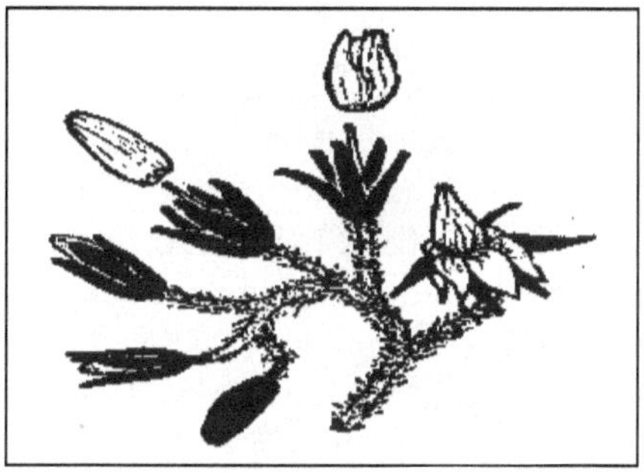

Figure 3: Tomato flower for emasculation.

flowers whose sepals are opened at 30-45°C with axis of flowers are taken for emasculation (Fig. 3). For emasculation, the stamens are removed with the help of a forcep. The corolla and androecium, which form the cap, are removed with the help of two forceps. Sometimes, emasculation and pollination are completed on the same day. But in this process, there is a chance of getting selfed seeds. Repeated pollination has been found better for fruit setting. Parachlorophenoxy acetic acid (PCPA) at 0.05% can be applied on the pollinated flowers as it increases hybridization. In a small scale of hybrid seed production, pollen grains are collected with a needle or a forcep. In a large scale programme, pollen grains are collected more efficiently and economically with some transfer tool. Sometimes, pollen grains are used directly from flowers, but this method is not efficient. A pollen collecting apparatus (vibrating pollen-collection) is also used which collects pollen from 1,000 flowers /hour. In morning, stamens are hygroscopic, therefore, under wet conditions pollen grains do not come out. The flower should be picked in the afternoon for pollen collection or they should be dried when collected in morning.

Pollen can be stored for certain duration at low temperature. At room temperature and 70-80% relative humidity pollen can be stored for 5-6 days and flowers for about 2 days. For pollination, a needle, a brush, or a force are generally used but small-blackened dental spoons and J-shaped glass tubes are more useful for direct transfer of pollen to stigma. Pollination instruments should be sterilized in ethanol before and after use. Pollen of one male flower is sufficient for pollination of 12-18 flowers. Efficient pollination is affected by temperature, humidity and wind velocity. High wind velocity reduces its efficiency. The receptivity of stigma and nutritional status of plant are some other factors. The stigma should be receptive and the plant should have normal C/N ratio. Soon after pollination, a tag, wire or any other convenient marker should mark the pollinated flower. The ratio of male and female plants is 4-5:1. Normally, fertilization takes place 24-50 hours after pollination. The maximum fertilization and fruit setting take place when pollination takes place at the time of anthesis. The time taken from pollination to fertilization depends mostly upon temperature, light condition and genetic factors. 'Pinching off', a part of two sepals of calyx so as to identify crossed fruit marks the pollinated flowers. Another way for identification of pollinated flowers is to tag with price luggage level with the information of name of parents and date of pollination. After completion of hybridization programme, all the non-crossed flowers of female plants are removed, which facilitate vigorous development of crossed fruits and seeds.

About 50-60 small or 30-40 medium sized fruits can be crossed on a plant. The first signs of ovary enlargement can be detected after 4 days of pollination. Fruits of tomato ripe usually at 50-60 days after pollination. The maturity index is red ripe fruits. Before harvesting of crossed fruits, all the open-pollinated (non-hybrid) fruits are removed to eliminate chance of contamination in hybrid fruits. Hence, it should be secured that only clipped sepal fruits (crossed) are harvested. The fruits should not be collected in metal container because it may react with acid present in tomato juice and affect seed viability and colour. It is also possible to harvest fruits and seeds of male parental lines from hybrid seed production field, for their utilization in next season. Similarly, some plants of female lines can also be utilized to multiply seeds in hybrid seed production. There are several methods of extraction of tomato seeds. Fermentation and acid methods are most commonly practised. In fermentation method, red ripe fruits are collected and crushed in wooden or plastic container (metallic container should not be used). Fermentation (release of CO_2 in the form of forth) starts after 1-2 days in summer and 6 days in winter season and seeds are settled in the bottom, while pulp floats on the top of container. In any case fermentation should not be allowed for more than 3 days, as more fermentation may spoil seed quality. Immediately after fermentation, pulps are removed and seeds are washed thoroughly in clean water.

Seed should not be dried under shade up to 8-10% moisture levels. In acid treatment method, red ripe fruits are cut into pieces and slimy masses are separated. The slimy masses are subjected to hydrochloric acid (HCl) treatment (@ 6.5 ml / kg slime). Due to HCl action, seeds rapidly start separation from the silmy mass within 15-30 minutes. It is very important to remove acidified liquid immediately (30 minutes) after the separation of seeds to avoid possible adverse affects of acid on quality of seeds. After extraction of seeds, they are kept in partial shade. After two or three days, seeds are dried depending upon weather condition. Dry blower can also be used for drying the seeds. It is observed that slow drying is beneficial for obtaining good quality seeds. The seeds having 8% moisture are ready for packing. The seeds can be packed in perforated polythene bags or plastic coated paper bags.

Use of Male Sterile Line

Genetic Male Sterility

Male sterility is an inherent in ability of the plant to produce viable pollen. This abnormality was earlier considered as an undesirable phenomenon and breeders preferred to eliminate it by selection in segregating generations. However, as the importance of heterosis breeding increased male sterility proved an asset. Large scale hybrid seed production sometimes remains handicapped because of high labour cost, unavailability of trained labour at crucial times, *i.e.* at blooming, bad weather conditions like continuous rains, *etc*. The discovery of certain male sterile mutants which help to eliminate more laborious operations of emasculation combined with various marker genes, further facilitate identification of undesirable types even at seedling stage, has widened the every basis of hybrid seed production. This technique has also reduced the total cost of hybrid seed production. Crane (1915) was first reported male sterility in tomato, and genic male sterile lines are more economical and efficient. This can be achieved through: (a) pollen sterility in which pollen grains are unfunctional, (b) functional male sterility in which pollen grains are functional but flowers are cleistogamous with indehiscence anthers, (c) antherless flowers in which stamens show only vestigial development and (d) positional sterility (long stigma). All of these modes of male sterility and their inheritance are given in Table 2.

Table 2: Male sterility and their inheritance in tomato.

Type of Sterility	Description	Inheritance
Pollen	Pollen abortive	Monogenic recessive
Positional	Stigma exerted	Monogenic recessive
Functional	Anthers do not dehisce	Monogenic recessive
Staminal	Stamens absent	Monogenic recessive

The pollen sterility type of male sterility is maintained by crossing hetrozygotes with male sterile line resulting in two types of progeny-half heterozygous male fertile and half-male sterile (Fig. 4). Some male sterility genes show the tendency of association with other genes such as *ms*^{32} with au and *ms*^{35} with S, and *aw* and *d* genes, but association are not of any practical significance for hybrid seed production. Although, several genic male sterile lines have been reported, their use include problems of purity of *ms* lines. In this case, recessive marker seed parents like potato leaf type, anthocyaninless (aa) and brown seeded can be used to detect the purity of hybrid seed.

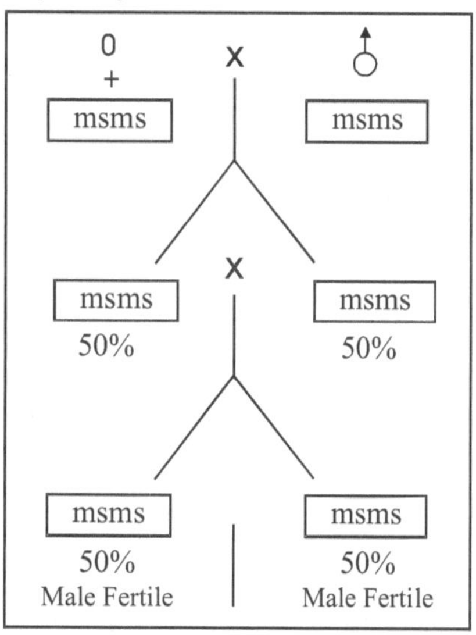

Figure 4: Maintenance of genic male sterile line in tomato (*Lycopersicon esculentum* Mill.).

Figure 4 illustrated the following points

- Identification and/or incorporation of male sterile.
- Transfer of male sterile mechanism into otherwise desirable female parent.
- Maintenance of male sterile lines.
- Maintenance is worth the consideration and possibly is the reason behind that is why this type of male sterility has not been exploited commercially in heterosis breeding to a greater extent. The scheme of maintenance of this kind of male sterility involves.

(a) Identification and isolation of heterozygotic fertile plants from mixed population.

(b) Crossing male sterile line with heterozygote pollen fertile parent of selfing the heterozygotes.

(c) Identification of male sterile plants from the mixed population of fertile and sterile for use as a female parent in heterosis breeding.

Here again, the last step poses quite a number of problems. The progenies consist of fertile and sterile plants, the ratio of 1:1 and 3:1 when backcross and selfing techniques are used respectively. The former being more efficient as far as continuity of sterility is concerned. Roguing out of fertile plants can be taken up after the first bloom. Therefore, the hybrid seed production must wait until anthesis and after those plants are allowed blooming in the female parent for hybrid seed production. The efficiency of this method can considerably be increased if the plants are identified early at the seedling stage and only sterile ones are transplanted into seed production block. This can be accompanied only using marker gene such as potato type leaf, anthocyaninless and brown seed. While using male sterility, certain measures have to be applied carefully for the production of hybrid seeds.

Generally, two lines are involved in crossing which should be entirely pure in order to produce homogeneous hybrids. Care should be taken in maintaining the male sterile line and one should be sure of one hundred per cent sterility in male sterile lines used. While planting very close spacing, which may be almost, double is required to maintain proper plant stand after removing the fertile heterozygotes. Soon after initiation of flowering, fertile plants are removed. Lines before using for crossing are thoroughly tested for their specific combining ability and expression of degree of heterosis. A comparative study using male sterile and male fertile line has been done at IARI, New Delhi, and the following observations were recorded :

- Time saved by male sterile line was 63 %
- Average fruit setting by using male sterile line was 79.35 %
- Average fruit setting using male fertile line was 65.40 %
- Time saved by using male sterile lines in recovery of seed was 60 %
- Hybrid seed cost by using male sterile line per kg was Rs. 466
- Hybrid seed cost by using male fertile line per kg was Rs. 3691

Although, cost of hybrid seed production is much lesser by using male sterile lines but the percentage increase in term of productivity has been found much higher by using male sterile lines as compared to male fertile lines. It does not compensate with the advantage contained in cheap F_1 hybrid seed production. Using genetic male sterility in the development of tomato hybrids has the following drawbacks:

(a) Maintenance of male sterile stock is cumbersome. Since, the progeny consists of both sterile and fertile plants, roguing of the later increases cost of hybrid seed production

(b) Since, tomato is highly self-pollinated, free outcrossing is prohibitive, thus, leading to poor seed and/or fruit setting. In tomato, male sterile line produced only 0.3-1.5 kg hybrid seed/ acre as a result of natural outcrossing (Yordanov, 1983). This method could gain popularity in practice only if suitable insect pollinators or other means are found to raise the percentage of cross-pollination to the extent of at least 50 %. In this context, Frankel (1977) offered five suggestions to alleviate problem of poor outcrossing

- Manipulation of male: female ratio
- Manipulation of spacing and position of male and female parents
- Choosing adequate plants of the female parents
- Choosing pollen parents that attractive to insects and is good pollen shedder
- Choosing an appropriate seasons and location to work in.

Functional Male Sterility

In tomato, it has been reported by Larson and Pour (1948), Lesley and Lesley (1958), Mittal and Thomas (1961), Mittal *et al.* (1962), Singh *et al.* (1966) and Brezhnev (1976). Such type of male sterility is governed by single recessive factor. This type is maintained by selfing because of fertile pollen. For the production of hybrid seed, one has to make sure that selfing is absolutely controlled, otherwise, selfed seeds are produced. If the stigma position is slightly exerted, hybrid seed production becomes more efficient and the chances of selfing are also avoided. Antherless plants are controlled by a single recessive factor and their expressions are influenced by temperature. Dorosiev (1970) has also been suggested the possibility of using of stamenless type for production of hybrids.

Longistyly- (ex) Male Sterility

It is a heritable character, constitutes a barrier against self-pollination and imposes a high degree of allogamy (Butnaru, 1979). Longistyly forms with ps and with ps_2 have been found effective for tomato hybrid production (Tikoo and Anand, 1980). Exerted condition can be induced through GA application. This mechanism is not stable enough and is influenced considerably by changing ecological conditions. So, care must be taken that the machanism is strong enough to ensure complete cross-pollination. No difficulties are evident in maintaining this type of recessive mutant. Earlier, Gueorguiev and Atanassova (1981) concluded that out of longistyly and stamenless forms, the former possing ps_2 gene has been found the most useful and dependable for hybrid seed production. Line possessing ps_2 gene includes start BR and Start-24, both having low rate of selfing.

Cost of Hybrid Seed Production

In tomato single-cross normally produces 40-50 seeds depending upon parents used. Selection of site for F_1 hybrid seed production is also an important consideration as it affects economics of hybrid seed production. Moderate climate with good soil conditions as prevailing in Dharwar area of Karnataka which is an ideal location for mass seed production. The labours are also easily available at moderate rates. There is a misplaced belief that hybrids are too expensive for the Indian farmers who have inadequate resources. Improvement of agro-techniques especially in nursery management and raising of tomato seedlings in particular, has brought down the seed requirement of hybrid seeds per hectare from 500 to 165 g. Further, availability of quality seed with high germination percentage will reduce the wastage of seeds. Use of protrays and wooden flats with better nursery management, have vastly improved the technology of hybrid varieties. A leading processing company in Punjab has successfully adopted mechanical sewing of defused seeds of hybrid tomato through drill and raising of seedlings in low tunnels covered with nylon net. The cost of hybrid seeds works out hardly 10 % of the total expenditure of crop production in tomato. The income generated by 1.0 kg of tomato seed is given below:

Open-pollinated variety	Hybrid
Area covered 2 ha @ 500 g/ha	6 ha @ 165 g/ha
Yield 50 tonnes @ 25 tonnes/ha	450 tonnes @ 75 tonnes/ha
Value Rs. 50, 000/ha @ Rs. 1.0/ kg	Rs. 9, 00,000 @ Rs. 2.0/kg
Production cost Rs. 10, 000/ha	Rs. 32, 000/ha
Profit to grower Rs. 30, 000 (15,000/ha)	Rs. 7, 00,000 (or 1,60,000/ha)

Source: Swarup *et al.* (1992).

National Released Hybrids

Arka Abhijit

Developed and released from IIHR, Bangalore. Its plants are semi-determinate having 75 cm plant height and takes 55-60 days to first fruit maturity. Fruits are round having 70 g average fruit weight with green shoulder. TSS is 4.3 % and 17 days keeping quality at room temperature (from breaker to softening). This hybrid yielded 650 q/ha in 140 days of crop duration. It is resistant to bacterial wilt and suitable for fresh marketing.

Arka Shreshtha

Developed and released from IIHR, Bangalore. Its plants are semi-determinate, 79 cm tall and takes 55-60 days for first fruit maturity. Fruits are round having light green shoulder and 75 g average fruit weight. The TSS is 4.7 %. Fruits can be stored for 17 days (from breaker to softening stage) at room temperature. It gives 830 q/ha fruits in 140 days of crop duration. It is resistant to bacterial wilt and suitable for both fresh marketing and processing purposes.

Arka Vardan

Developed and released from IIHR, Bangalore. Its plants are indeterminate. Fruits are flat round, firm deep red with green shoulder and average weight of 140 g. It is resistant to nematodes and recommended for Jammu & Kashmir, Himachal Pradesh, Uttranchal, Punjab, Bihar, Uttar Pradesh, Karnataka, Tamil Nadu and Kerala.

Arka Vishal

It has been developed at IIHR, Bangalore. Its plants are indeterminate. Fruits are round, firm, deep red with green shoulder and average weight of 140 g. It is tolerant to fruit cracking and matures in 160 days with an average yield of 750 q/ha. It is recommended for Punjab, Bihar and Uttar Pradesh.

ARTH-3

It has been developed at Ankur Seeds Pvt. Ltd., Nagpur. Plants are determinate with medium dark green leaves. Fruits are squarish round and deep red. Flesh is red with 4.5 to 5.0 % TSS. It is moderately tolerant to drought. Harvesting starts after 80-85 days after sowing and continues up to 160-180 days, which gives an average yield of 880-950 q/ha. It is recommended for Assam, West Bengal, Maharashtra, western parts of Madhya Pradesh, Karnataka, Tamil Nadu and Kerala.

ARTH-4

Ankur Seeds Pvt. Ltd., Nagpur, has developed it. Its plants are semi-spreading and indeterminate with dark green leaves. Fruits are deep red and flat round. Flesh is red with 4-4.5 % TSS. It is moderately tolerant to drought. Harvesting starts after 80-85 days after sowing and continues up to 170-190 days, which gives an average yield of 1000-1100 q/ha. It is recommended for Punjab, Bihar, Uttar Pradesh, Karnataka, Tamil Nadu and Kerala.

COTH-1

It is a cross between IIHR-709 and LE-812, developed at TNAU, Coimbatore. Plants are determinate and 75 cm tall. Fruits are round to slightly oblong, medium-sized (average weight 60 g) and bright red at maturity. First picking starts after 88-94 days, average yield being 850-950 q/ha in 140 days of crop duration.

Madhuri (BSS-39)

It has been developed at Beejo Sheetal Seed Pvt. Ltd., Jalna. Plants are determinate and early in maturity. Fruits are dark red, flat round with thick skin, average weight of 100 g, crack resistant and can with stand long transportation. It gives an average yield of 500-700 q/ha.

Manisha (BSS-99)

It has been developed from Beejo Sheetal Seed Pvt. Ltd. It has good self-life and transport quality. It gives an average yield of 800-1,000 q/ha.

Meenakshi (BSS-20)

Developed at Beejo Sheetal Seeds Pvt. Ltd., Jalna. Its plants are indeterminate. Fruits are dark red, round global with average weight of 80 g. It is resistant to fusarium wilt. It gives an average yield of 600-1,000 q/ha.

Megha (BSS-98)

It has also been developed from Beejo Sheetal Seed Pvt. Ltd., Jalna. Its plants are determinate. Fruits are red with thick skin, good shelf life and transport quality. It gives an average yield of 500-700 q/ha.

Pusa Hybrid-1

The plants are compact, determinate with good foliage cover having prolific-bearing. Fruits are round, smooth and attractive. It is ready for market from late May to mid-July. Pusa Hybrid-1 has an additional advantage of providing tomatoes form June to mid-July, the lean period for tomato in Northern-India. It can set fruit up to a temperature of 28 ^{0}C during night. The average fruit weight is 63 g and yielded 320 q/ha.

Pusa Hybrid-2

It has developed by combining desirable commercial characters. The IIVR, Varanasi, has recommended it for cultivation in humid Western Himalayan regions, sub-humid Sutluj-Ganga alluvial plains and Western plains and semi-arid plateau. The plants are compact, semi-determinate with good foliage cover having prolific-bearing. Fruits are round to flattish round, firm, smooth, attractive and develop a uniform red colour on maturity. It is highly tolerant to root knot nematode. The average yield is 592 q/ha.

Pusa Hybrid-4

Plants are determinate, compact with dark green foliage having prolific bearing. Fruits are attractive, round, smooth, medium-sized (70-80 g), 3-loculed, thick pericarp and uniform in ripening. It is highly tolerant to root-knot nematodes under field conditions. It takes 130 days from transplanting to fruit ripening in spring season and 120 days in autumn.

RHRTH-57

It is a cross of 32-2 and 85-1 and developed at MPKV, Rahuri. Fruits are oval, red, medium size and medium stout with 4.1 % TSS and uniform ripening. It gives an average yield of 730 q/ha.

RHRTH-92

It is a cross of Dhanashree and RHRT-1. It has been developed at MPKV, Rahuri. Fruits are round, orange-red, medium-sized, smooth surface, very stout with uniform ripening and 3-9 % TSS. It gives an average yield of 670 q/ha.

TH-802

It is a cross between Healani and Castle Rock. This is resistant to root-knot nematode hybrid and developed at PAU, Ludhiana. Plants are determinate, spreading type with broad leaflet and dark green foliage. Fruits are round, medium size (68 g of average weight) with thick pericarp (6.4 mm) and 4.75 % TSS. Gives an average yield of 631.8 q/ha in 140 days crop duration.

TH-2312

It is a cross of VFN-8 and Punjab Chhuhara. It has been developed at PAU, Ludhiana. Plants are dwarf and spreading with dense foliage. Fruits are oval-round, medium-sized (67.5 g average weight), medium firm with thick pericarp (5.4 mm) and 4.75 % TSS. It gives an average yield of 631.8 q/ha (6.46 % increase over Rupali) in 150 days. This is resistant to root-knot nematode.

VRTH-1 (Kashi Hybrid-1)

An early maturing hybrid having determinate growth habit. It has been developed at IIVR, Varanasi. The average yield is 1100 q/ha. This hybrid shows 87 % heterosis over standard check (ARTH-3). It posses field tolerance to TLCV and early blight.

VRTH-2 (Kashi Hybrid-2)

It is a short duration hybrid having maximum fruit production of 900 q/ha. Developed at IIVR, Varanasi, this hybrid has 54 % heterosis over standard check (ARTH-3).

Seed Certification Standards

I. Field Inspection

A minimum of four inspections should be made as follows:

- The first inspection is done at flowering to determine isolation volunteer plants, outcrosses and other relevant factors.
- The second and third inspections are done during flowering and fruiting stage to check isolation, off types and also fertile segregants in seed parent, if male fertile line is used, and other relevant factors.
- The fourth inspection is done at mature fruit stage and prior to harvesting to determine the true characteristics of fruits and other relevant factors.

II. General Requirements

1. Isolation Distance

Contaminant	Minimum Distance (m)	
	Foundation	Certified
1	2	3
Fields of other varieties including commercial hybrids of the same variety	200	100
Fields of the same hybrid (code designation) not conforming to varietal purity requirements for certification	200	100
Between blocks of the parental lines in case seed parent and pollinator are planted in separate blocks	-	5

III. Specific Requirements

Factor	Maximum Permitted (%)*	
Off types in seed parent	0.010	0.050
Off types in pollinator	0.010	0.050
** Fertile segregants (pollen shedding cymes) in seed parent	0.050	0.10
***Plants affected by seed borne diseases	0.10	0.50

*Standards for off types and fertile segregants (pollen shedding cymes) in seed parent should met at and after flowering and for seed-borne diseases at final inspection.

** It shall be applicable in case male sterile line is used for producing hybrid seed.

** *Seed-borne diseases should be: Early blight (*Alternaria solani* Sorauer), leaf spot (*Stemphylium solani* Weber), tobacco mosaic virus (TMV).

IV. Seed Standards

Factor	Standards for each class	
	Foundation	Certified
1	2	3
Pure seed (minimum)	98.0 %	98.0 %
Inert matter (maximum)	2.0 %	2.0 %
Other crop seeds (maximum)	5/kg	10/kg
Weed seeds (maximum)	None	None
Germination (minimum)	70 %	70 %
Moisture (maximum)	8.0 %	8.0 %
For vapour-proof containers (maximum)	6.0 %	6.0 %

All certified seed lots, which have been produced by adopting emasculation, and hand-pollination are subjected to grow out test and shall confirm to following minimum genetic purity requirements:

Class	Genetic purity (%) (minimum)*
Certified	90.0 %

*During grow out test, the off type plants (other than selfed plants) such as segregants, outcrosses and plants of other varieties should not exceed more than 1.50 % out of the 10.0 % plants earmarked for selfed plants. The minimum population size of 400 plants is maintained in two replicates of 200 each or four of 100 each throughout the test and each plant shall be examined individually. The rejected number shall be as follows:

Class	Genetic purity (%) (minimum)	Reject number
Certified	90.0 %	44

Brinjal

Genetic Resources

Brinjal or eggplant (*Solanum melongena* L.), a crop of Indian origin, has also been developed some secondary variability in China. In brinjal, more than 2,531 germplasms of *Solanum melongena* L. and 327 wild *Solanum* species are available at NBPGR, New Delhi. On the basis of morphological variations (Fig. 5, see Plate 1), following promising accessions have been identified for different traits (Table 3).

Table 3. Promising accessions/lines/varieties of brinjal for different traits.

Trait	Promising Accession (S)/ Lines / Varieties
Good cluster bearing	IC-201131, 201132, 201136, 201139 and EC-381023.
Cluster bearing	IC-112960, 99747, 111428, EC-304951, NIC-3654, 5874, 5877 and 5904
Very long fruits (>30 cm)	IC-201132, 201231, NIC-10459 and 6067.
Long fruits (20-30 cm)	IC-90087, 90103, 90125, 90148, 99747, 90982 and 99702.
Oval purple fruits	IC-112343, 127018, 99675, 99715, NIC-4283, 6878, 6885, 6886, 6960, 9430, EC-109083 and 305044.
Purple long fruits	IC-74234, 89905, 90108, 90508, 90862, 90764, 112312, 90975, 126869, 144038, 144084, 144166, 111025, 111083, EC-304975, 329327, 396466, 396505 and 400211.
Purple round fruits and NIC-5767	IC-89832, 89841, 126955, 137674, 143995, 144045.
Deep purple oval fruits	IC-90956, NIC-3603, 4064 and EC-305049
High Yield	BBSR-34, BBSR-51, BBSR-104, BBSR-109, BBSR-118, CH-575, CH-378, CH-450, CH-665, ACC-373, EC-379246, IC-90806, 90031, HAB-884, HAB-894, 677-10-8, BB-46, 586-4, 480-1, 624-5, BBSR-135, BBSR-127, EC-384565, EC-379246, 480-1, 498-8, 509-4, 540-8, 550-4-2, CGB-10, CGB-18, CCGB-24, CGB-28, IC-332506, IC-383190, VKG-2/228, IC-144060, BBSR-36, BBSR-54, BBSR-8, CH-573, CH-573, CH-665, IC-127241, 126727, 99674, 89974, 249326, EC-438678, 305039, 136240, IC-111087, 112342, 136306, G-12, G-13, P-4, G-8, G-10, G-11, G-13, AB-21C-74207, 74205, 137769, EC-282490, IC-113002, PB-17, PB-26, IHR-115, IHR-149, Rajendra Baingan-II, Rajendra Annapurna, EC-305043, NIC-5765, B-1 and B-40.
Bacterial wilt resistant	*Solanum nigrum, S. sisymbriifolium, S. torvum, S. melongena var. insanum, S. xanthocarpum, S. toxicarium, S. nigrum, S. melongena* (Sm-6-6, Sm-6-7, BWR-7), Comuy, Long green, Aroman, Sinampire, Taiwan Naga, L-7 and 19, Muktakeshi, Purple Long, Udippi, Pusa Purple Cluster, Improved Muktakeshi, Pusa Purple Round Annamali, Arka Nidhi, Arka Keshav, Arka Neelkanth, Utkal Tarini, JC-1 and JC-2.
Fusarium wilt resistant	*S. incanum.*
Phomopsis blight resistant	Pusa Bhairav, Pant Samarat, Florida Market, Florida Beauty, Pusa Purple Cluster, Line 119-12-2-264-1-2, PB-30, BC-1, JC-2, KS-352, IC-90922, 90047, 90970, EC-384565 and wild species like *S. xanthocarpum, S. gilo, S. khasianum, S. torvum S. sisymbriifolium and S. testiculatum.*
Little leaf resistant	*S. integrifolium* (H-6, 7,9 and10), Pusa Purple Cluster, H-8, S-212-1 and S-252-1.
Jassid resistant	Mangari Gota.

Shoot and fruit-borer resistant	*S. incanum, S. gilo, S. indicum, S. khasianum, S. sisymbriifolium, S. anomalum,* Aushey, Black Beauty, H-165, H-407, H-408, Annamalai, Pant Samarat, Bhagyamati, Aushay, Pusa Purple Cluster and AM-62.
Meloidogyne spp. resistant	*S. sisymbriifolium.*
M. incognita resistant	Black Beauty and Vijaya.
M. javanica resistant	Bhanta and Black Beauty.
Frost tolerance	*S. incanum.*
Drought tolerance	*S. microcarpon.*
High temperature	R-34.
Multi-resistant (root-knot nematode + phomopsis fruit rot + verticillium wilt)	Florida Market.

Production Objectives

Though, a numbers of hybrids have been developed both by the public and private sectors, therefore but they fail to fulfil our requirements therefore, following objectives should be considered while developing brinjal hybrids.

- To develop hybrids having straight plant growth habit.
- To develop hybrids having less seed and more pulp.
- To develop hybrids having good attractive colours.
- To develop hybrids having less fruit thickness.
- To develop hybrids resistant to phomopsis blight, bacterial and fusarium wilts and little leaf.
- To develop hybrids resistant to root-knot nematodes.
- To develop hybrids resistant to shoot and fruit-borer, and jassid.
- To develop hybrids resistant to frost.
- To develop hybrids which can give flowering and fruiting in very low and high temperatures.
- To develop multiple disease and insect-pests resistant hybrids.

Botany and Floral Biology

Plants are polymorphous, erect, aculeate or unarmed herb, woody at the base and 0.5-1.5 m in height. The crown is broad. Branches are light green or tinged with violet, rather thick, densely clothed with grey and stellate tomentum. Leaves are alternate, solitary or in pairs, on shorter or longer stalks, simple, ovate or oblong-ovate and elleptic. Base is often unequalsided and oblique, rounded, truncate or cordate, apex triangular, acuminate, obtuse or acute, irregularly undulate, angular or more or less deeply lobed, above dark green, on both sides dull, densely clothed with soft, grey, stellate tomentum, flaccid, 7-25 cm long, 5-15 cm wide and reticulate-veined. Veins are transparent. Petiole is robust, subterete, yellowish-green, densely white stellate tomentose and 1.5-10 cm long.

Flowers are large, stalked, lateral or leaf opposed. Pedicel is light green or strongly tinged with violet, terete, aculeate or not, densely stellate tomentose, patent or curved and 1.5-2 cm long. Calyx is tubular-campanulate, 5-lobed and less than half way down, on the outside greyish green, often strongly tinged with purple, on both sides densely stellate-tomentose and 1.5-2 cm long. Tube is ovoid-oblong, obtusely pentagonous, + 1 cm long and 0.5-0.7 cm diameter; segments narrowly oblong, acuminate, obtuse, obliquely erect and 0.5-1.5 cm long. Corolla is gamopetalous, deeply 5-lobed, stellately spreading, on the outside light violet, densely white stellate tomentose, within dark violet, glabrous, rugose and 2.5-4 cm diameter. Lobes are short, ovate-oblong and acute. Stamens are five, erect, ovoid-oblong, vitelline, opening at the top by two pores and 0.5 cm long. Berries are most variable as to shape, colour and dimensions, globose or oblong, white, yellow or more or less intensely violet, glabrous, 2.5-15 cm long and 2.5-4 cm diameter. Fruiting calyx is campanulate, much enlarged, bursting on one or two sides, pale green or tinged with purple, thinly stellate tomentose, unarmed or beset with scattered prickles, 3-6 cm long, 3-5 cm diameter, fruiting pedicel much elongated and thickened towards the apex, unarmed or with scattered small prickles, pale green, nodding and 3-7 cm long. Seed is numerous.

Though brinjal is self-pollinated crop, there is high degree of cross-pollination due to heteromorphic flower structure (called as heterostyly). The extent of cross-pollination has been reported as high as 29 % and hence, it is classified as often cross-pollinated or facultative cross- pollinator. There are four types of flowers in brinjal (Fig. 6) like (a) long-styled (b) medium-styled (c) short-styled and (d) pseudo-short styled. Fruit setting occurs only in long-and medium-styled flowers. Flowers generally emerge 40-50 days after transplanting. Anthesis occurs at about 6-8 am in August-September and usually between 9.30 and 11.15 am during winter (December-January). Peak time for anthesis is 8.30-10.30 am. The pollen dehiscence starts from 9.30 to 10.00 am. It is observed that daylight, temperature and humidity mainly influence anthesis and dehiscence. The receptivity of stigma can be observed from the pump and sticky appearance, which gradually turn brown with the loss of receptivity. Stigma receptivity is highest during anthesis, *i.e.* flower opening. The period of effective receptivity ranges from a day prior to flower opening until about four days after opening. Pollen is most fertile immediately after the anther dehiscence. Opening of anthers is mostly by a pore or slit at or near the apex. Pollen grains usually remain viable for a day during summer and 2-3 days in winter under field conditions. According to Deshpande *et al.* (1978) maximum anthesis occurs at 6 am in Arka Shirish and at 8 am in Arka Sheel and Arka Kusumaker.

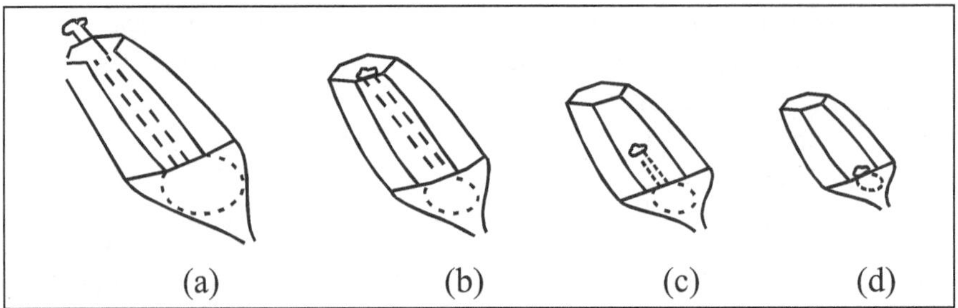

Figure 6: Types of flowers in brinjal: (a) long-styled (b) medium-styled (c) short-styled and (d) pseudo- short-styled.

Methods of Hybrid Seed Production

Hand Emasculation and Pollination

For production of F_1 hybrid seed on small scale, female and male parents of a cross are grown side-by-side but away from other genotypes. The matured long-styled flower buds are emasculated on previous day of their opening in the afternoon and covered with butter paper bag. Similarly, male parent long-styled mature buds are also covered with butter paper for collection of anthers. Hand pollination of emasculated flowers is done on the next day between 9 am and 1 pm in winter and 7 pm and 12 am in summer. Crossed flowers are covered again after pollination and tagged with 'x' sign and dated. Fertilized flowers ovaries get prominence in about 10 days of pollination. This procedure of pollination is practised when hybrid seed production required on a small scale. The crossed flowers grow into mature fruits in 30-55 days depending on the size of fruit and season of crop. Seeds can be extracted and properly dried soon after harvesting fruits or after a few days of storage under shade. In commercial hybrid seed production in brinjal, a professionally trained manpower is essential because of emasculation and pollination of flowers is done by hand. Male and female parents of a cross are grown 200 m in isolation from each other and also 500 m away from other genotypes. The emasculation of long-styled matured flower buds of female parent is done in the afternoon on a large scale which are pollinated an opening the next day till 14.00 hours. For pollination, the emasculated flowers, anthers from male parent mature flower buds are collected in good amount in large petridishes. These anthers are kept under light during night to dehisce the pollen grains. The pollen grains are sieved through muslin cloth and are collected into a container and used for hand pollination.

In about 8-10 days, fertilized flowers (ovaries) develop into prominence. After completion of hybridization programme, all the non-crossed flowers of female plants are removed, which facilitate vigorous development of crossed fruits and seeds. Fruits of brinjal take approximately 70-80 days after pollination to become suitable for harvesting and extraction of good quality seeds. Fruits are harvested at natural maturity for seeds. Usually during the natural maturity, brinjal fruit skin colour attains light yellow to deep yellow/ orange colour. Before harvesting of matured crossed fruits, all open-pollinated (non-hybrid) fruits are removed to eliminate chance of contamination in hybrid seed lot. Hence, it should be secured that only clipped sepal or tagged fruits (crossed) are harvested. For extraction of seeds, fruits are cut, crushed and macerated along with water. The seeds are separated from the pulp during washing of crushed fruits. Further the seeds are dried in partial shade up to 8 % moisture level. After cleaning, the seeds are packed in suitable moisture-proof packaging material and stored optimum moisture and temperature condition for better seed viability.

Parental lines of hybrids are maintained in isolation. Care is taken in growing a substantial population of each parent. Plants which are not matching to the parental characters particularly in early flowering and bearing, more number of fruits/plant, size of fruits, quality parameters and resistant factors to common insects and diseases are strictly roughed out. Plant type and other morphological characters are also given due consideration in maintaining the purity of parent. Strict roughing of undesirable plants is observed to keep the purity of parental lines. Seeds of parental lines are multiplied in large amount and properly stored for use in coming year. The technique of hybrid seed production in brinjal is not difficult. It also does not involve much of production cost. Normally, 150-200 g of hybrid seed is enough to cover one hectare of land. The labour cost of production of 200 g of seed is nominal. On an average, each crossed fruit yield 3 g seed. This implies that 70 crossed fruits shall yield 210 g hybrid seed. The labour cost of producing 70 healthy crossed fruits cannot be more than a few hours/person/day for a few days. Basavaraja (1986) reported that production of one kg hybrid seed takes 17.36 hours.

Use of Functional Sterility

The functional male sterile line (UGA MS-1) is available and is being utilized to develop experimental crosses. Commercial exploitation of this line is an attractive proposition, because 100 % male sterile plants are obtained after forced selfing of these plants. Hence, it is anticipated that in the future utilized for commercial seed production.

National Released Hybrids

ARBH-1 (SHYAMAL)

It has been developed by Ankur Seeds Pvt. Ltd., Nagpur. Plants are semi-spreading to spreading with vigorous growth. Fruits are purple and oblong. Flesh is milky-white. Shelf life is 4-5 days, hence it has good transport quality. The hybrid is tolerant to sucking-borer insects and diseases. First picking starts 100-110 days after transplanting and continues up to 240-270 days with an average yield of 700-750 q/ha.

ARKA NAVNEET

It has been developed at IIHR, Bangalore. Plants are indeterminate. Fruits are deep purple, black glossy, green calyx, weight of 350-400 g, white flesh, with less seeds, having good cooking and keeping quality. Matures in 120 days with an average yield of 700 q/ha.

BH-2

A cross of Punjab Neelam and Punjab Barsati. It has been developed PAU, Ludhiana. Plants are medium-tall (79.0 cm), erect and thornless with green leaves. Fruits are oblong, shining deep purple, large (237.2 g average weight) with 6.55 mm pericarp thickness. It gives an average yield of 442 q/ha.

COBH-1

A cross of ED-45 and CO2, developed at TNAU, Coimbatore. Plants are tall (90.8 cm) with average of 4.4 branches, each plant bears 55.4 fruits. Fruit size 8.94 cm x 12.43 cm and average weight of 64 g. It gives an average yield of 3.5 kg/plant.

MHB-10 (KALPATARU)

It has been developed at Maharashtra Hybrid Seeds Co. Ltd., Jalna. Plants are erect, bushy, compact and spiny. Fruits are round to slight oval, shining, greenish purple with white streaks (bicolour) and green spiny calyx. The average fruit weight is 60-70 g. First picking starts 60-65 days after planting and continues up to 125-130 days. Its average yield of is 400-500 q/ha.

MHB-39 (RAVAIYYA)

It has also been developed at Maharashtra Hybrid seeds Co. Ltd., Jalna. Plants are erect, compact and non-spiny. Fruits are oval, shining, reddish purple and non-spiny. Calyx bears in bunch of 4-6 fruits. It is an early-bearing hybrid and takes about 60-65 days for first picking, which continues up to 125-130 days. It gives an average yield of 350 q/ha.

NDBH-1

It has been developed at NDUA & T, Faizabad. Plants are semi-erect, non-spiny, hardy green stem with purple green upper portion. Leaves are purple green. Fruits are round purple, attractive, shining, soft, fleshy, less seeded with good flavour. The average fruit weight is 150-200 g. Fruits are harvested 70-75 days after transplanting.

NDBH-6

It has been developed at NDUA & T, Faizabad. Plants are semi-erect, non-spiny, green stem with light purple upper portion. Leaves are green with rough surface and purple veins. Fruits are long oval, attractive purple, soft, fleshy, weighing 150-200 g / fruit. Fruits bear in bunch, moderately resistant to shoot and fruit-borer, phomopsis blight and little leaf. Its average yield is 600 q/ha.

PUSA HYBRID-5

Plants are vigorous, non-spiny with semi-erect branches. Leaves are green with bright pigmentation towards growing tips and young leaves. Fruits are long, glossy, attractive, dark purple with partially pigmented peduncle. The average fruit weight is about 100 g. It takes 80-85 days from sowing to first picking and further pickings continue from October to December in central and northern plains, whereas in Kerala, Tamil Nadu, Karnataka and Andhra Pradesh from November to May. It gives an average yield of 525.6 q/ha at New Delhi.

PUSA HYBRID-6

It is also an early-bearing hybrid. The plants are vigorous, non-spiny with partial pigmentation on younger leaves and growing tips. Fruits are round, glossy, attractive, purple having partially pigmented peduncle. The average fruit weight is about 100 g. It takes 80-85 days from sowing to first picking and further pickings continue from October to December. It gives an average yield of 460 q/ha at New Delhi.

PUSA HYBRID-9

The plants are non-spiny with strong upright branches, all over green showing light pigmentation on younger leaves. Fruits are big-sized, oval round, glossy, highly attractive, dark purple with partially pigmented stalk and calyx. Each fruit weigh is about 300 g. Fruits are ready for first picking 85-90 days after showing. Its yield potential is 637.20 q/ha. The IIVR, Varanasi, has recommended it for cultivation in Junagardh and Jalna in Maharashtra, Anand in Gujarat and central plains of India in 1997.

RHRBH-1

It is a cross of RHRB-1 and JB-16, developed at MPKV, Rahuri. Fruits are non-spiny and green calyx, purple having white stripes, 8.5 cm length and 6.4 cm diameter. It gives an average yield of 576.0 q/ha.

RHRBH-2

It is a cross of RHRB-1 and RB-25, developed at MPKV, Rahuri. Plants are short and spreading type. Fruits are purple with white stripes, 5.5 cm length, and 4.5 cm diameter with spiny calyx. It gives an average yield of 407 q/ha.

RHRBH-3

It is a cross of RHRB-1 and Vaishali, developed at MPKV, Rahuri. Plants are tall and spreading type. Fruits are purple with white stripes, 7.5 cm length, and 5.5 cm diameter with spiny calyx. It gives an average yield of 440 q/ha.

VRBHR-1

It is a cross of Pant Rituraj and BR-SPS-14, developed at IIVR, Varanasi. Plants are of 70.8 cm tall. Fruits are round, light purple with 10.28 cm diameter, 11.6 cm length, and average weight of 569 g with green calyx. The average yield is 789.0 q/ha.

Seed Certification Standards

Following guidelines should be followed for seed certification of brinjal hybrids.

- A parental line to be eligible for certification should be from a source such that its identity may be assured and approved by the certification agency.
- Hybrid seed to be eligible for certification should be the progeny of two approved parental lines.
- A parental line should be a relatively true breeding strain.

- The certified class seed should be the hybrid seed to be planted for any use except seed production.
- Land to be used for seed production should be free of volunteer plants.
- The field inspection seed should be: (a) before flowering in order to determine isolation, volunteer plants and other relevant factors, (b) during flowering and fruiting stage to check isolation, off types and other relevant factors and (c) mature fruit stage and prior to harvesting to determine the true characteristics of the fruits.

The isolation distance should be as given below:

Contaminant	Minimum distance (m)	
	Foundation	Certified
Fields of other hybrids including commercial hybrid of the same variety	200	200
Field of the same hybrid (code designation) not confirming to varietal purity requirements for certification	200	200
Between blocks of the parental lines in case seed parent and pollinator are planted in separate blocks.	-	5

The specific requirements for seed certification are as given below:

Factor	Maximum permitted (%)*	
Off types in seed parent	0.010	0.050
Off types in pollinator	-	0.050
**Plants affected by seed borne diseases	0.10	0.50

*Standards for off types shall be met at & after flowering and for seed-borne diseases at final inspection.
** Seed-borne diseases shall be: phomopsis blight (*Phomopsis vexans* Sacc. and Syd. Harter).

The seed standards should be as given below:

Factor	Standards for each class	
	Foundation	Certified
Pure seed (minimum)	98.0 %	98.0%
Inert matter (maximum)	2.0 %	2.0 %
Other crop seeds (maximum)	None	None
Weed seeds (maximum)	None	None
Germination (minimum)	70 %	70 %
Moisture (maximum)	8.0 %	8.0 %
For vapour-proof containers (maximum)	6.0 %	6.0 %

Note: All certified seed lots should be subjected to grow-out test and shall confirm to the following minimum genetic purity requirements.

Class	Genetic purity (%) (minimum)*
Certified	90.0

*During grow out test, off type plants (other selfed plants) such as segregants, outcross and plants of other hybrids should not exceed more than 1.50 % out of the 10.0 % plants earmarked for selfed plants.

The minimum population size of 400 plants shall be maintained in row, replicates of 200 each or four of 100 each throughout the test and each plant shall be examined individually. The reject number should be as follows:

Factor	Genetic purity (per cent) Minimum	Reject number
Certified	90.0 (10 in 100)	44

Chilli

Genetic Resources

Chilli (*Capsicum frutescens* L) belongs to family Solanaceae. It is usually grown in warm to hot and humid climate throughout the country. The large-fruited bell pepper of today originated from tiny pungent and pointed-fruited wild species *C. annum L.* var. *avicular.* There are other species, *viz. C. chinense, C. baccatum* and *C. pubescens* which are commonly grown in Southern and Central America except *C. frutescens*, is also grown in some parts of USA. In India, it is grown in southern states and northeast region. Chilli species and their characteristics are given in Table 4.

Table 4: Important chilli and their characteristics.

Botanical Name	Common Name	Comments
C. annuum L.	Sweet pepper, chilli hot pepper	Extensively cultivated in South Asia, constitutes principal source of commercial dry chilli.
C. annuum L. var *avicular*	Bird pepper	Wild type, said to be progenitor of bell pepper.
C. annuum var. *grossum Sendt.*	Sweet pepper	Fruit contains less capsaicin, green colour used in salad.
C. annuum var. *longum sendt.*	Pepper	Used to produce condiment.
C. baccatum H.B.K. Syn. C. angulosum	Pepper	Commonly grown in Central America.
C. bacatum L. var *bacatum Syn. L. microcarpum Cav.*		Found in Peru, Bolivia, Paraguay, North Argentina and South Brazil.

Contd...

Table 4 Contd...

C. cardenasii Heisir & Smith		Wild type, self-incompatible, related to *C. pubescens.*
C. chinese Jacq	Pepper	Closely related to *C. frutescens.*
C. eximum Hunziker	Pepper	Wild type pseudo self-incompatible, related to *C. pubescens.*
C. frutescens L.	Tobasco pepper, bird chilli	Widely cultivated in dry regions of India, highly pungent fruits used for sauce preparation.
C. minimum Roxb. Syn. C. fastigiatum Bhumme	Bird-eye-chilli	Cultivated throughout India but at very limited scale, closely resembles *C. annum.*
C. pubescens Ruiz. and Paron	Pepper	Introduced in India for breeding purposes.

Due to the long history of cultivation, outcrossing nature and popularity of the crop, a large genetic variability including local landraces has evolved (Fig. 7, see Plate-2). Therefore, India is considered as a secondary centre of diversity for chilli. In hot chilli, great variability for several attributes occur throughout India, particularly in peninsular region, north-eastern region, in foothills of the Himalaya and Gangetic plains. In India, chilli improvement projects have excluded at almost all the SAUs and several ICAR Institutes For successful breeding programme, all the centers have collected and maintained a large number of genetic resources. The NBPGR, New Delhi, and Satellite Center of Amarvatti have main responsibility to collect and evaluate chilli germplasms. Locally collected lines are also being maintained and evaluated at Regional Station, Meghalaya. A list of promising accessions of chilli for different attributes are given in Table 5.

Table 5: Promising accession (s)/varieties/species of chilli.

Trait	Promising Accessions
High-yielding	EC-362918, 362922, 362930, 362998 and 339047, NIC-22076, 22058 and 19150, PIC-147681, 119193, 119257, 119676 and 19639, N-34, NP-46 A, Pusa Red, Kalyanpur Red, Kalyanpur Selection-1, Kalyanpur Yellow, Pant C-1, Pant C-2, G-2, G-3, G-4, G-5 Pusa Jawala, Arka Lohit, LCA-235, LCA-218, JCA-340, CA-138, NO-31, NO-27, NO-32, Suryamukhi, Beldanga Local, Dhanilanka, PDC-40, IC-119481, IC-119500, EC-345655, K-119457, K-119497, K-119412, BD-213, EC-345625, JCA-17, EC-339565, IC-204000, IC-204212, NIC-9577, KCS-2013, BC-25, LCA-353, Ajeet-6, KA-2, SH-KC-12, SH-KC-9, LK-100, PBC-3, C-5, C-6, C-7, CA-138, MC-111, INGR-04052, 04053, 04954, BSS-128, MSL/1/14, Baljuri, Dinhata Coll-1, Dinhata Coll-2, CCA-4261 (Cms line), CCA-4727 (Cms line), MS-12 (Gms line) PCB-77, LCA-15, MM-7, EC-126493, CA-83, CA-80, Jati Long and Round Chilli.
Early-flowering	IC-119490, 119497, 19267, 147637, 119264, EC-399566, 399568 and 399578.
Early-maturing	EC-399568.

Contd...

Table 5 Contd...

Yellow chilli	NIC-23829
Round cherry type fruits	NIC-23901
Paprika type	NIC-19953
High pungent type	EC-347263
Resistant to multiple diseases	EC-32333, *C. baccatum* and *C. frutescens*.
Fruit rot	*C. chinense,* Accession 1555, 1554 and 906, Chinese Giant and Spartan.
Bacterial leaf spot	P-163192, 260435, 163189, 163192, 2713222 and 322719.
Phytopthora root rot	PI-201234.
Root-knot	Santaka and 405B.
Root rot	Line 491, 493, Chinese Giant, Yolo, Spartan Lorai, S-27, S-41-1, *Capsicum chinenese* spp. and Waxy Globe.
Tobacco leaf curl virus	Pusa Sadabahar, Pusa Jwala, Punjab Lal, *Capsicum chinense* and P-159236.
Leaf curl virus	Pusa Jwala, Pant C-1 and PBC-535.
TMV	Sonnette, Keystone Resistant Giant, YoloWonder, Florida VR-2, VR-2-34 and XV R-3-25.
Fruit rot and bacterial wilt	Jati Long and Round Chilli.
CMV	LIC-13, LCA-315, G-4, LCA-305 and LCA-235.
Bacterial wilt resistant	Kali Cluster, White Kandri and Pant C-1.
Thrips	Caleapin Red, Chamatkar, NP-46A, BG-4, EC-113937, LCA-348, and EC-15916.
Mite	LEC-1, Kalyanpur Red, Punjab Lal, LIC-13, LCA-253, LIC-36 and EC-12896.
Aphid	Kalyanpur Red and X-1068.
Tolerant to excess moisture	LIC-7, LIC-9, EC-12275 and EC-120942

Production Objectives

Following objectives should be considered while developing the chilli hybrids.

- High genetic variability present in chilli has not been fully utilized in our country. This genetic variability can be successfully utilized in hybrid breeding programme.
- Since chilli is susceptible to various diseases and insects, emphasis should be given for development of multiple disease and insects resistant hybrids.
- Chilli is one of the most important crop for export point of view. Therefore, efforts should be on development of export purposes hybrids.

- Hand emasculation and pollination of hybrid seed production should be standardized, so that, hybrid seed can be produced cheaply and can reach average chilli grower of our country.
- Attempt should be made to develop male sterile lines linked with marker, so that male sterility can be commercially exploited for hybrid seed production.

Botany and Floral Biology

Plants are perennial, widely branched, erect and woody at the base. Plant height reaches up to 1.5 m. Leaves are solitary or 2-3 close together, very unequal having shorter or larger stalks, broadly ovate or oblong-lanceolate, base broad and abruptly contracted into the petiole. The apex is narrowed or acuminate, acute or obtuse, coarsely crenate or entire, shortly elliate, light green, feebly shinning, sparingly hairy, especially beneath, 1.5-10 cm long, 0.5-5.0 cm wide and having 0.5-3.5 cm long petiole. Flowers are small, solitary in leaf-axils or 2-3 or more together at the tip of branches and long pedicelled. Pedicels are erect, noding at the tip, light green shortly pubescent, terete and 1.5-2.5 cm long. Calyx is minute campanulate, minutely abtusely 5-toothed or subtruncate, yellowish-green, glabrous and not or hardy enlarged under the fruit. Corolla is gamopetalous, patent, flat, yellowish or greyish white deeply 5-partite and 0.5-1.0 cm in diameter. Segments are ovate-oblong, acute, 0.3-0.35 cm long and 0.2-0.25 cm wide. Stamens are five and erect. Anthers are bronze-green. Berries are small, narrow, variable as to length, usually erect, rarely pendulous, long stalked, ovate-oblong, narrowed upwards obtuse, glabrous, when ripe shinning blood red, orange coloured or yellowish white many-seeded, 0.75-2.5 cm long and 0.3-1 cm in diameter.

The extent of natural crossing in chilli is reported upto 80%. Flowers remain open for 2-3 days. The flowering begins 1-2 months after planting and it takes again one month for fruiting. The flower opening takes place during 5-6 am in the morning. Anther dehiscence takes place from 9 to 11 am. However, flower opening and anther dehiscence to a large extent depend upon the weather conditions. During cold as well as cloudy days, the opening of flower is delayed. Maximum fruit setting takes place when pollination is done at the time of opening of flowers. Since anthesis is completed by 8.00 am and anther dehiscence starts from 9 am, emasculation and pollination can be done simultaneously.

Methods of Hybrid Seed Production

Following methods are employed for production of chilli hybrids

Hand Emasculation and Pollination

Flowers are emasculated at bud stage. Emasculation may be done either early in the morning or in the previous evening before opening of flowers and calyx still covering the anthers and stigma. Pollen grains are transferred to stigma either from mature undehisced anthers by scooping it out through lateral sutures with needle or by touching a freshly dehisced anther to stigma with the forceps. With the help of a pair of forceps, the petals are easily parted and anthers are removed and bagged. If there is no attention of doing any emasculation on following one or two days all the operated buds, which are likely to open on these days must be removed and whole plant should be covered with a muslin cloth bag. Late in the morning or early in the afternoon of the following day, fresh flowers are plucked and intended male parents, which have been previously bagged, and pollen dusted on stigmas of the emasculated flowers. The pollen grains are dry and powdery which easily dislaged from the anthers with a gentle shake. To facilitate pollination, petals of pollen flower may be cut off. After pollination, the bag should be replaced and levels denoting details of parents and date of pollination should be tie outside of the bag. Crossing is found successful when the plants are in full bloom and at high natural setting. Crossing is avoided as far as possible during cloudy weather as setting is very low in such conditions. The ratio of seed parent to pollen should be 5:1. Pollinated flowers are periodically checked and the covering cloth bag removed in 4-6 days. After completion of hybridization programme, all the no-crossed (untagged) developing or developed fruits on female plants are removed, which facilitate vigorous development of crossed fruits and seeds. In case of Cms/ Gms based hybrid seed production, this practice is not required, as all the fruits developed on male sterile plants will be crossed fruits. Fruits of chilli usually mature 40- 60 days after pollination. The mature index is red ripe fruit. Before harvesting of crossed fruits, open-pollinated (non-hybrid) fruits are removed in order to eliminate chance of contamination in hybrid fruits. Hence, it should be secured that tagged fruits are harvested. For extraction of seeds, the harvested ripe fruits are dried and seeds are separated by maceration (commercial scale) or by longitudinal bifurcating (experimental scale) of fruits. After extraction, seeds should be dried up to 8 % moisture level. Before packaging in appropriate moisture proof packing material, seeds are cleaned on density gradient seed cleaner.

Use of Genetic Male Sterile Lines

Hand emasculation and hand pollination are most expensive methods of hybrid seed production in chilli because of high labour cost and very low fruit setting. Therefore, genetic male sterility mechanism is more economical and can be exploited for hybrid seed production. Genetic male sterility is maintained under heterozygous condition by backcrossing with recessive parent every year. For commercial seed production, heterozygous seed stock is grown which segregate into 50% heterozygous male fertile and 50 % homozygous male sterile as described in (Fig. 4, see Page-265). The farmer is removed from the population before flowering, where as the latter is kept for hybrid seed production. In this case, there is problem of identification of male sterile plant before flowering and for this morphological marker can be used.

Male sterile plants can be identified easily if male sterility is linked with recessive marker gene. Unfortunately, male sterility linked with marker has not been reported so far in chilli. Male sterile lines which are sensitive to temperature, should be selected for hybrid seed production. The induced male sterile gene in France (mc-509; Pochard, 1970; renamed ms-10 by Daskalov and Poulos, 1994) is allelic to ms allele, isolated spontaneously in Korea (Shifriss, 1997). The ms-2 line identified by Shifriss and Rylski (1972), is non-allelic to ms-1 isolated by Shifriss and Frankel in 1969. The ms-509 line (bell pepper type) of Pochard is introduced in India at PAU, Ludhiana and ms-10 is introgressed in three chilli, genotypes, *viz.* ms-12, ms-13 and ms-41 (Singh and Kaur, 1986). The ms-12 (ms-10) line has been utilized to produce hybrid seeds of CH-1 and CH-3 hybrids of hot pepper in Punjab (Hundal, personal communication). In Hungary, male steriles possessing ms-3 gene are being utilized to produce hybrid seeds (Kumar *et al.*, 2000). In the recent past development of stable cms lines of hot pepper has led to its increased utilization in hybrid development. In South Korea (Shifriss, 1997), China (Boaxi *et al.,* 2000) and India (Kumar *et al,* 2000), cms lines are being utilized to develop hybrids of hot pepper.

Use of Cytoplasmic-nuclear Male Sterile Lines

In chilli, stable cytoplasmic-nuclear male sterile (cms) lines are available, which can be utilized as female parent in hybrid seed production. Since, male part (pollen) is sterile, use of such line precludes huge cost on manual emasculation. Further, labour cost on the protection and tagging of crossed fruits is also not required. Thus, utilizing cms lines, the production cost of hybrid seed can be drastically reduced. The cms line (also called A line) is maintained by making crosses on it using pollen from maintainer line (also called B line). Through such crossing, 100 % male sterile seeds are obtained, unlike gms lines. The seedling prepared from such crosses can be transplanted as female rows (4) along with the selected restorer line (called R line) as male row (1) in hybrid seed production field. Since considerable amount of natural cross-pollination occurs in chilli, no hand-pollination is required. For development of cms line, the cms line (A) is developed by backcrossing of a selected maintainer line (B line) onto an already available A line for 6-7 generations. This generates a pair of A and B lines in the new genetic background. Restorer allele (Rf) is either introgressed into identified male parent (restorer breeding, or male parent is directly used for hybrid seed production on cms, if restorer (Rf) gene in homozygous state is already available in male parent (most of the chilli and sweet pepper genotypes possess Rf and rf genes, respectively). Similarly, if maintainer (rf) gene is not available in the line, which is to be developed as A line, then maintainer breeding would be required to transfer recessive gene (rf) in the desirable line.

National Released Hybrids

ACH-15

It is a cross of S-49 and CDPS-120, developed at Gujarat Agriculture University, Anand. Plants are 72 cm tall with average of four branches. Each plant bears 100 fruits. Fruits are of 10.4 cm x 3.4 cm size with 4.85 g average weight. First picking starts 107 days after sowing. It gives an average yield of 100 q/ha.

ACH-30

It is a cross of S 5-5 and Punjab Guchcedar, developed at Gujarat Agriculture University, Anand. Plants are 56.04 cm tall with an average of 4.2 branches. Each plant bears 78 fruits. Fruits are of 10.2 cm x 3.5 cm size with an average weight of 4.8 g. First picking starts 110 days after sowing. It gives an average yield of 85 q/ha.

ARCH-236

It has been developed at Ankur Seeds Pvt. Ltd., Nagpur. Plant is semi-spreading and profused branching. Fruits are elongated, undulated, pointed and green, turns red at maturity.

Hybrid has good shelf-life and good for fresh market. It has high capsicin content. The crop duration of this hybrid is 10-140 days. This hybrid is recommended for cultivation in Punjab, Bihar and Uttar Pradesh plains. The female parent of this hybrid is characterized by semi-spreading growth habit with green leaves. Fruits are green and elongated. It takes 75-80 days to flowering. While, male parent is spreading in growth habit with purple anther. Fruits are medium, green and long.

BSS-141

It has been developed at Beejo Sheetal Seeds Pvt. Ltd., Jalna. Fruits are dark green with high pungent and turn bright red at maturity. It gives very high yield. The female parent (JUBI) is characterized by medium-tall in growth habit. Fruits are dark green, smooth, mild pungent and dark red after maturity. While, male parent (CABE) is of medium-tall and spreading in growth habit. Fruits are wrinkled, green, medium pungent and turn red at maturity.

BSS-138

It has been developed at Beejo Sheetal Seeds Pvt. Ltd., Jalna. Plants are spreading type and 70-80 cm tall. Fruits are dark red, attractive, thick and medium pungent. The female parent (CIPAN) is characterized by medium-tall in growth habit. Fruits are wrinkled medium pungent and dark red at maturity. However, male parent (JUBI) is medium-tall in growth habit. Fruits are dark red smooth, mild pungent and smooth dark red after maturity.

CH-1

This male sterile based hybrid has been developed at PAU, Ludhiana. Plants are 100 cm tall, spreading type with light green foliage and each plant bears 200 fruits. Fruits are light green, 6.6 cm long and characterized by 23.05% dry matter, 0.82% capsicin, 136.5 ASTA unit colouring matter and 98.37 mg/100 g vitamin C. Plants are tolerant to diseases and pests. It gives an average yield of 95 q/ha red ripe fruits.

CH-3

This male sterile based hybrid has also been developed at PAU, Ludhiana. Plants are medium-dwarf (69.2 cm) having pendent fruiting and bears 220 fruits/plant. Fruits are dark green, 8.2 cm long and characterized by 22.55 % dry matter, capsicin 0.5 %, 145.1 ASTA unit colouring matter and 109.95 mg/100 g vitamin C. This hybrid has advantage over CH-1 in terms of earliness, better fruit colour, mild pungency and longer fruit suitable for processing. It gives an average yield of 95 q/ha (red ripe).

Seed Certification Standards

I. Field Inspection

A minimum of three inspections should be done, the first before flowering, the second during flowering and fruiting stage, and the third at fruit maturity stage and prior to harvesting.

II. Field Standards

General Requirements

Isolation: Seed fields should be isolated from the contaminants are given below:

Contaminant	Minimum distance (m)	
	Foundation	Certified
1	2	3
Fields of other hybrids, fields of the same hybrids not confirming to purity requirements for certification	400	200
Fields of chilli from capsicum and *vice versa*	400	200

III. Specific Requirements

Factor	Maximum permitted (%)	
	Foundation	Certified
Off types	0.10	0.20
**Objectionable weed plants	0.10	0.50

* Maximum permitted at and after flowering in the case of off types and the final inspection in case of seed-borne diseases.

** Seed-borne diseases shall be: Leaf blight (*Alternari solani* Sorauer.); Anthracnose (Ripe rot, Die back): (*Colletotrichum capsici* (Syd.) Butler & Bisby).

IV. Seed Standard

Factor	Standards for each class	
	Foundation	Certified
Pure seed (minimum)	98.0 %	98.0 %
Inert matter (maximum)	2.0 %	2.0 %
Other crop seeds (maximum)	5/kg	None
Weed seeds (maximum)	5/kg	None
Germination (minimum)	60 %	60 %
Moisture (maximum)	8.0 %	8.0 %
For vapour-proof containers (maximum)	6.0 %	6.0 %

Bell Pepper

Genetic Resources

Bell pepper belongs to genus *Capsicum* of the nightshade family Solanaceae comprising 20-30 species. The word *Capsicum* originated from the Greek world Kapso means to bite. In other version, it originated from Latin *Capsa* meaning box, which refers to the shape of fruit. *Capsicum* terminology is confusing. Various terms like pepper, chili, chile, chilli, aji, paprika and *capsicum* are used interchangeably for plants in the genus *Capsicum*. Both hot and sweet peppers have been originated from *C. annuum*. The *C. annuum* is distinguished by white flowers (rarely purple), anthers with anthocyanin, calyx with distinct points or teeth and longitudinal ridges. Flowers one /node (occasionally 2 at first flowering node), erect in wild form and pendant in many types. Seeds are tan, waxy, pale greenish yellow flowers, anthers with anthocyanin, calyx smooth without longitudinal ridges, characterize the *C. frutescens*. Flowers 2-5/node (usually 2), held stiffly erect with unopened flowers at an angle to peduncle. Cultivated types have less variability than *C. annuum*, but may have slender fruit up to 15 cm in length. Non-pungent forms are not known. *C. frutescens* requires a tropical climate, hence restricted in habitat.

The bell peper is highly polymorphic genus with much morphological genetic diversity both intraspecifically and interspecifically. Intra-species and inter-species improvement is possible to evolve better types. Since *Capsicum annuum* lacks the variability, hence studies on variability have been very few. Earlier, Arya and Saini (1977 a and b) found greater genetic variability for rind thickness. Indira *et al.* (1996) found genetic variability in fruit size (Fig. 8, see Plate-2). The promosing lines of bell pepper are :

• For high yield	EC-393214-2, EC-203583, SH-SP-1007, SH-SP-1004, Capsicum-2, Midway and EC-143570.
• For Paparika	SKAU-P-201, Arka Abhir, SH-P-1005 and SH-P-1006.

Production Objectives

Following objectives should be taken into consideration while developing bell pepper hybrids:

- To develop high-yielding hybrids.
- To develop hybrids capable of forming fruits in summer season.
- Attempt should be made to develop multiple disease resistance, i.e. hybrid resistant to fruit rot, anthracnose and viruses.
- Indian farmers do not like to have uniform maturity of sweet pepper due to market fluctuation. So, emphasis should be given to develop hybrid with indeterminate growth habit, so that, sequential or multiple harvesting can be possible.

- Seed production techniques for F_1 hybrids have to be standardized. Possibility of using genetic or cytoplasmic-genetic male sterility with marker genes should be explored.
- There is a strong need to build up adaptability/stability.
- Being a cool season crop, tropicalization is necessary, so that, its availability can be ensured easily in non-traditional areas.

Botany and Floral Biology

It is dicotyledonous plant behaving as a perennial in tropics and annual in temperate regions. Leaves are solitary or 2-3 close together, very variable as to size, on shorter or longer petioles, ovate-oblong or elliptic-lanceolate. Base is acute, decurrent along the petiole. Apex is narrowed, acute, coarsely crenate or entire, thin, above dark green, beneath yellowish green, on both sides dull and feebly shining, barring the nerves almost glabrous, 1.5-12 cm long and 0.7-5 cm wide. Petiole is 0.5-2.5 cm long. Flowers are solitary in leaf-axils, stalks long, patent, nodding or pendulous at the apex or rarely erect, yellowish green, shortly pubescent, longitudinally furrowed and 1-1.5 cm long. Calyx is small, campanulate, shortly 5-dentate, green glabrous, 0.2-0.25 cm long and much enlarged under the fruit. Corolla is gamopetalous, campanulate, white or yellowish white, rarely violet, deeply, 5-partite and 1-1.5 cm in diameter. Segments are spreading, oblong, acute, 0.5-0.7 cm long and 0.3-0.4 cm wide. Stamens are five, erect, anthers lilac or violet, afterwards turning bronze-green (Fig. 9). Fruit shape may be pointed, blunt or shrunken at the blossom end.

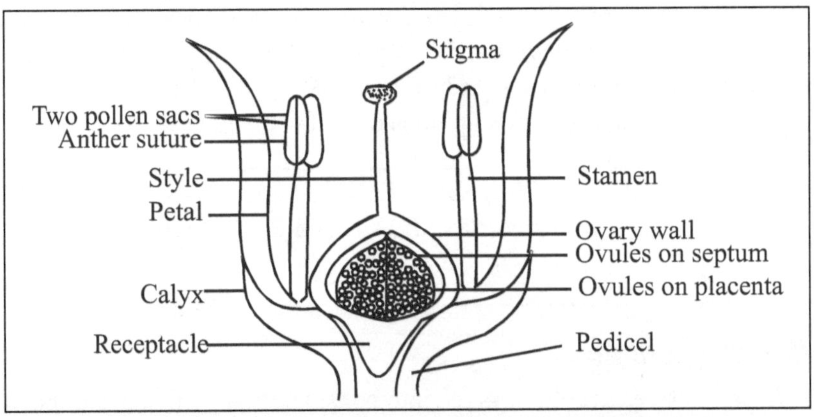

Figure 9: Bell pepper flower showing various parts.

Ripe fruits are shining blood red, violet, orange coloured or even green and many seeded. In bell pepper, anthesis starts at 7.15 am and continues up to 11.15 am with the peak at 7.15 (Vijay *et al.,* 1979). Anther dehiscence occurs after 30 minutes of anthesis. The stigma becomes receptive from the day of anthesis and remains receptive up to two days after anthesis. Fruit setting (80 %) has been found maximum on the day of anthesis and pollen fertility around 96 %. However, there is a considerable variability in time of anthesis and dehiscence. Fruit setting in flower buds pollinated a day prior to anthesis has been observed good. The extent of natural crossing in peppers has been reported to vary from 7.6

to 36.8 % in the field with a mean of 16.5 %. The extent of natural cross-pollination is estimated by planting a variety having recessive characters with another are having counterpart dominant traits. The seed setting on recessive parent are grown and crossed ones are identified by the dominant phenotype. In an open field, the minimum safe isolation distance for bell pepper is estimated at 600 fit.

Methods of Hybrid Seed Production

Hand Emasculation and Hand Pollination

Though bell pepper is self-pollinated but some cross-pollination does occurs between and within the cultivars of two species. It is observed that in bellpepper, natural cross-pollination occurs up to 68 %. The flower buds (whitish) one day prior to anthesis on the female line are selected for emasculation and it is done in the evening hours. The buds are bagged with perforated butter paper bag. The pollination on emasculated buds, with the pollen, collected from the male line (fully grown white buds, *i.e.* one day before anthesis), is done on the next morning of emasculation. Emasculated buds are tagged with the description of line A x line B, date of emasculation and date of pollination. After four days of pollination, the bags may be turn off from the closed end. The number of crossed fruits/ plant should be kept 5-8. After completion of hybridization programme, all the non-crossed (untagged) developing fruits or developed of female plants are removed, which will facilitate vigorous development of crossed fruits and seeds. Follow strictly fungicides and insecticides spray schedule to avoid fruit rot, virus and insect damages. Fruits of sweet pepper usually mature 40-60 days after pollination. The mature index is red ripe fruit. Before harvesting of crossed fruits, open-pollinated (non-hybrid) fruits are removed to eliminate the chance of contamination in hybrid fruits. Hence, it should be secured that only tagged fruits are harvested.

Hand Pollination without Emasculation

For this, genetic male sterility has been reported for hybrid seed production. In this case male sterile line (A) is maintained by crossing to a maintainer (B) line, heterozygous for recessive male sterility allele. Thus, half of the population will also be heterozygous and fertile, which must be removed. Genetic male sterility (GMS) is less useful for hybrid seed production than cytoplasmic male sterility (CMS) interacting with nuclear gene CMS was first discovered by Paterson (1958) in *Capsicum annum* PI-164835 from India. Cytoplasmic male sterlity system is advantageous over GMS system for mass hybridization. Because all of the female parent plants are male sterile as compared with ms gene method. The parent lines are required to produce F_1 hybrid seed by the CMS system (the male sterile or A line S $m^{+}s/ms^{+}$). The male sterile line is maintained by hybridization, in isolation with a maintainer line of the same cultivar. In the field, bees, supplemented by hand pollination during peak flowering, would do crossing. Crossing the S ms/ms line with a good combiner line or cultivar of similar type to generate heterosis would produce the commercial hybrid. Unfortunately, Peterson's S $m^{+}s/m^{+}s$ lines are unstable in fluctuating environments, producing pollen at lower temperature, and hence, could not be relied upon to produce F_1 hybrids.

Chemical Hybridizing Agents (CHAs)

This term is used to describe chemicals, which cause pollen abortion and render the treated plant male sterile while not affecting ovule fertility. Kohli *et al.* (1981) used on androcide GA_3 at 10,000 mg/litre in bell pepper thrice at the onset of flowering and once again after 10 days causing complete male sterility, which lasted throughout the season.

Seeds are extracted manually by bifurcating the fruits longitudinally. Crushing or squeezing the fruits when the seeds are extracted from fresh fully ripe fruits can extract seed, its moisture content is very high. It is therefore essential to clean and dry the seeds as quickly as possible. Seed must be thoroughly dried to moisture content below 8 % for proper storage. In well-ventilated place seeds can be stored up to two years. After drying seeds are packed in sealed container. Properly leveling of seed is very essential to maintain the identity of the seed lot.

National Released Hybrids

BHARAT

Indo-American Hybrid Seeds Company, Bangalore, released this hybrid during 1973. They are dark green, blocky, 4-lobed with thick flesh. They are large in size (150 g). Plants are vigorous. The yield potential is 200-250 q/ha in 120-140 days of crop duration.

KTCPH-3

It is a medium-maturity group hybrid, developed at IARI, Regional Station, Katrain. Fruits are blocky, smooth, green, 6 cm x 5.8 cm size with 80 g of an average fruit weight. Flesh is very thick, sweet and fines flavoured. It gives an average fruit yield of 276 q/ha.

KTCPH-5

It is a cross between Yolo Wonder and EC-143570. This hybrid is a medium-maturity type, developed at IARI Regional Station, Katrain. Plants are of medium maturity group and upright habit. Fruits are blocky, smooth, dark green, square shaped (6.0 cm x 6.0 cm), 100 g average weight with hard skin and thick flesh. It gives an average yield of 258 q/ha.

NS-436

Namdhari Seeds, Bangalore, developed it. Plants are characterized by medium plant height with good branching habit and very good foliage cover. Fruits are good in size, very firm dark green attractive fruits with four lobes and blocky in shape. It is a high yielding hybrid 180-200 g fruit size. Fruits are ready in 60 days after transplanting.

Seed Certification Standards

I. Field Inspection

A minimum of three inspections should be done, the first before flowering, the second during flowering and fruiting stage, and the third at the mature fruit stage and prior to harvesting.

II. Field Standards

A. General Requirements

Isolation: Seed fields should be isolated from the contaminants shown in column 1 of the Table below by the distances specified in columns 2 and 3 of the Table:

Contaminant	Minimum distance (m)	
	Foundation	Certified
Fields of other varieties	400	200
Fields of the same variety not conforming to varietal purity requirements for certification.	400	200
Fields of chilli from bell pepper and *vice versa*	400	200

B.Specific Requirements

Factor	Maximum permitted (%)	
	Foundation	Certified
Off types	0.10	0.20
** Objectionable weed plants	0.10	0.50

* Maximum permitted at and after flowering in the case of off types and the final inspection in seed-borne diseases.

** Seed-borne diseases shall be: Leaf blight (*Alternari solani* Sorauer.); anthracnose (ripe rot, die back): (*Colletotrichum capsici* (Syd.) Butler & Bisby).

III. Seed Standards

Factors	Standards for each class	
	Foundation	Certified
Pure seed (minimum)	98.0 %	98.0 %
Inert matter (maximum)	2.0 %	2.0 %
Other crop seeds (maximum)	5/ kg	None
Weed seeds (maximum)	5/kg	None
Germination (minimum)	60 %	60 %
Moisture (maximum)	8.0 %	6.0 %
For vapour-proof containers (maximum)	6.0 %	6.0 %

12

MALAVACEOUS CROP

Okra

Genetic Resources

Okra [*Abelmoschus esculentus* (L) Monch.] is a crop of the tropical and subtropical lowland regions of Africa, Asia, America and warmer temperate region of the Mediterranean. It is believed to be originated from tropical Asia or may have been existing in Africa and India as polyphylatic species. *A. tuberculatus*, one of the ancestors of okra occuring in India, has nine wild taxa, of which *A. esculentus* is popularly grown for pod production. The main region of cultivation and diversity of okra in India, Sri Lanka, Pakistan, Nepal and Bangladesh, are represented by various local cultivars of *A. esculentus* and *A. moschatus* (Table1).

A. tetraphyllus, is being grown in cultivated fields in the western region of India as many as 18 old-land races have been developed in this species. Two wild species, *A. crinitus* and *A. angulosus,* are exclusively of Asian origin. *A. moschatus* is present as wild form in Kerala, Maharashtra, Tamil Nadu, Karnataka and Uttar Pradesh. However, it is cultivated for its aromatic seeds in tarai belt of Uttranchal hills. Another wild species, *A. tetraphyllus,* is found in sugarcane field as well as barren lands of Uttar Pradesh, Bihar, Madhya Pradesh and Chhattisgarh. The rest species are fully wild taxa.

Table 1: Some characteristics of *abelmoschus* species and their distribution in South Asia

***Abelmoschus* Species**	**Somatic Chromosome Number**	**Wild (w)/ Cultivated (c)**	**Distribution**
A. angulosus	56	W	India and Sri Lanka
A. tuberculatus	58	W	India
A .moschatus	72	WC	India, Nepal and Sri Lanka
A. ficulneus	72	W	India and Bangladesh
A. tetraphyllus	138	W	India, Nepal and Sri Lanka
A. tetraphyllus var. pungens	138	W	India, Nepal and Sri Lanka
A. crinitus	9	W	India and Sri Lanka
A. caillei	196	WC	India
A. manihot	66	WC	India and Sri Lanka
A. esculentus	130	C	Worldwide

Source: Ghosh and Kalloo (2000).

In India, variability in okra (Fig. 1, see Plate-3) genetic resources includes wild *Abelmoschus* species, have been preserved at National Bureau of Plant Genetic Resources (NBPGR), New Delhi over the last five decades both by undertaking exploration within country and through introduction of promising material from abroad. Germplasm collections have been made both from indigenous as well as from exotic sources and are being utilized in different parts of the country. The NBPGR, New Delhi, maintains major collections. International Bureau of Plant Genetic Resources (IBPGR) has designated NBPGR with global responsibility of base collection of okra. Okra germplasms endemic to India and Bangladesh has also been given emphasis for collection and evaluation. Within India, okra germplasm collection programme is being carried out mainly through NBPGR, New Delhi, and Akola in Maharashtra. More than eight specific explorations have already been organized, exploring and collection from Punjab, Haryana, Rajasthan, Gujarat, central and western parts of Madhya Pradesh and Maharashtra, southern peninsular tract in Karnataka, Tamil Nadu and Kerala, in eastwards Andhra Pradesh, Orissa, Assam plains and adjoining hilly tracts of North-Eastern Region. The SAUs and ICAR institutes have also made contributions to NBPGR, which holds more than 2,386 accessions of okra, represented by nine species, with maximum variability in *A. esculentus*. Besides NBPGR, there are also some centres holding some accessions where okra breeding programmes are being continued. Such major centres are IIVR, Varanasi, IIHR, Bangalore, PAU, Ludhiana, GAU, Junagarh, KAU, Vellanikkara, Trissur, OUAT, Bhubaneswar and MAU, Parbhani. In India, NBPGR, New Delhi, has documented the information in publication of a catalogue on okra germplasms and identified promising lines for various attributes are given in Table 2.

Table 2. Identified promising lines of okra for various attributes

Trait	Accessions
Early flowering	IC-128062, 128065, 128072, 128088, 128096, 11479, 117218, EC-325356, 329364 and 329370.
Tall plant type	IC-128065, 128119, 128147 and 117005.
Long-fruited	IC-128037, 128045, 117018, 117095, 117212, 128061, 128069, 128121, EC-329375 and 329378.
More number of fruits/plant	IC-128064, 128067, 128069, 128071, 128119, EC-329372, 329375, IC-12876, 85588, 45893, 46995, 522988, 62738, 8536, 90049, EC-305616, 90195 and 90244; EC-305656, 305675, 30508, 305656, 305675 and 30508, TCR-899, EC-305749-C-4, EC-309722-A, EC-306722-A-4, NO-136, SC-25, NO-120, NO-122, NO-315, B-2, EC-329361, IC-11174, K-11173, EC-329422, KS-404, BBSR-12 and VB-9101.
Long duration fruiting	IC-264697, IC-2647, EC-305749, EC-30579-C-1, EC-306741-A-2 and IC-203863-24.
Resistant/ tolerent to damping off, *Rhizoctonia solani*	Red Ghana, Sel 7-1, BH- 27, IC-12096 and IC-17252.

Resistant / tolerant to powdery mildew	Nigeria, EC-32598 and IC-8248
Resistant / tolerant to cercospora blight	Sel-7-1, Round Selection, EC-32598, IC-10238, IC-1542, IC-8248 and IC-8248 EC-32598, EC-41292, IC-10238, IC-8248. *A. crinitus, A. moschatus and A. angulosus.*
Resistant / tolerant to fusarium wilt	IS-9273, IS-9857, IS-6653, Pusa Sawani Pusa Makhmali.
Resistant / tolerant to enation leaf curl virus	*A. ficulneus* and *A. manihot.*
Resistant/ tolerant to microphomina	IC-90186, U-4307, U-4365 and U-4579.
Resistant to powdery mildew	*A. tetraphyllus* and *A. angulosus.*
Resistant to yellow vein mosaic virus	*Abelmoschus manihot* ssp. *manihot, A. tetraphyllus,* Parbhani Kranti and Punjab Padmini and IC-1542. NIC-9303A, NIC-6308, NIC-3322, NIC-9408, NIC-3325, EC-329375, K-4409, ACC-49, ACC-52, SC-125, ACC-18, 37, 46, 56, 58, 60, 62, *A. crinitus* and *A. angulosus.*
Resistant to jassid	IC-7194, Sesswal Local, A. *moschatus,* IIHR 21, AE-30, IC- 7194, IC-8899, Crimson Smooth and *A.manihot* ssp *manihot.*
Resistant to fruit-borer	*A. tuberculatus,* Red-I, Red-II, Red Wonder-I, Red Wonder-II, *A. manihot,* Pusa Sawani and Long Green.
Resistant to nematode	Long Green Smooth (Birat, 1964).
Tolerant to mite	*Abelmoschus angulosus.*

Production Objectives

The following traits should be given major emphasis while developing hybrids in okra:

- There is a strong need to build-up adaptability/stability.
- To develop high-yielding hybrids capable of giving more marketable yield of dark green, tender, thin, medium-long, smooth and 4-5 ridged pods with optimum seed setting ability.
- Pods should be free from conspicuous hairs.
- To breed early-maturing hybrids with a prolonged harvesting period.
- To evolve hybrids resistant to yellow vein mosaic virus and also to fungal disease like fusarium wilt, cercospora leaf spot, powdery mildew, alternaria leaf spot, fruit rot, anthracnose, aschochyta pod spots *etc.*
- To combine resistance to yellow vein mosaic virus with resistance to fruit and shoot-borer, white fly, jassid and root-knot nematodes.

- To develop multiple disease and pest resistant hybrids.
- To develop most suitable ideotype short plants and more number of nodes with short internodes length would be more productive than tall plant with long internodal length.
- Plants and fruits should be devoid of conspicuous hairs. Fruits should snap easily from the stalk for facilitating easy and economic harvest.
- To evolve hybrids tolerant to abiotic stresses especially tolerance to low temperatures, excessive rains, saline and alkaline soils, fungicides and other environmental pollutants.
- To develop hybrids suitable for processing industry and export markets.

Botany and Floral Biology

Plant is an erect, herbaceous annual and 1-2 m tall. Stem is green or with reddish tinge. Leaves are alternate, broadly cordate, palmately 3-7 lobed, hirsute and serrate. Flowers are solitary, axillary with about 2 cm long peduncle. Epicalyx is up to 10 cm, narrow hairy bracteoles that fall before the fruit reaches maturity. The calyx splits longitudinally as flower open. Petals are 5 cm, yellow with crimson spot on claw and 5-7 cm long. Staminal column is united to the base of petals with numerous stamens. Ovary is superior, stigma 5-9 cm with deep red. Fruit is a capsule, light green, or sometimes red in colour pyramidal oblong, beaked, longitudinally narrowed, 10-30 cm long and dehiscencing longitudinally when ripe. Seeds are green to dark brown rounded.

The flower bud appears in axil of each leaf above 6th to 8th leaf depending upon the cultivar. The crown of the stem at this time bears 3-4 under-developed flowers but later on during the period of profuse flowering, there may be as many as 10 under-developed flowers on a single crown. As the stem elongates, lower-most flower buds open into flowers. There may be a period of 2-3 or more days between the times of development of each flower but never does more than one flower appear on a single stem. The buds measuring 2.82 cm x 1.55 cm open into flowers (Purewal and Randhawa, 1947). Flower-bud initiation takes place 22-26 days after sowing. The first flowering occurs 20-22 days after initiation. Flowering continues for 40-60 days. Edible fruit is formed within 3-6 days of flower opening. Stigma is most receptive on the day of flower opening, anther dehiscence and dusting of the stigma with pollen being almost simultaneous. Anthesis is observed between 9 and 10 hours and is preceded by maximum anther dehiscence between 8 and 9 hours. The stigma remains receptive on day of anthesis. Protogyny has been reported in okra, which is mainly responsible for allogamy of the crop. According to Srivastava (1982), second to fourth positioned flowers carry more anthers (63 anthers/flower), while only 26 in first flower and 10 in 10th flower. Anther dehiscence commences 15-20 minutes after anthesis and is completed in 10-15 minutes. The dehiscence on anthers is transverse (Purewall and Randhawa, 1947). Pollen output/anther also differs by flower position. The flower numbers two to four are the best producers (Srivastava, 1982).

Pollen-grains of okra are polysiphonous (Nair *et al.,* 1974). All the tubes except one stop elongation and that one is the functional tube. In other tubes, protoplasmic activity is arrested by the formation of coagulated protoplasmic plug (callose plugs) at the tips. The middle part of the staminal column from the flower collected in the middle phase of flowering possesses the most balanced pollen biomass and therefore, the most desired for breeding. (Srivastava, 1982). The pollen remains viable in storage at 50% RH for 55 days (Dubey and Singh, 1968). Pollen fertility is maximum between the period one hour before and one hour after opening of flower. It takes 2-6 hours for fertilization after pollination. Studies on influence of ovule position on pollen tube entry and seed set revealed those ovule positions 7 to 14 received the maximum number of pollen tubes, *i.e.* about 90 % (Chandra and Bhatnagar, 1975). It is basically a self-pollinated crop, however, cross-pollination through insects can be as high as 19 %, and hence, it is classified as often cross-pollinated vegetable crop.

Methods of Hybrid Seed Production

Hand Emasculation and Hand Pollination

For emasculation, buds likely to open the next day are given a slight ring cut at the base of the floral bud with the help of a blade. Emasculate the perfect flower of female line a day prior to anthesis and bag them. In case of male sterile line only tie the tip of male sterile flower with a thread. In male parent tie the bud a day prior to anthesis. Collect male flower in morning (7.30 A M) and complete the pollination latest by 10.30 a.m. Pollination may continue for 40-50 days. It requires 5 labours/acre/day. As the flower in okra is large and solitary it is comparatively easy to emasculate and cross-pollination or self them. A day prior to opening the flower buds are emasculated, preferably in afternoon, by scarping off the anthesis from the staminal column with the help of the blunt edge of scalpel. As the floral organs are mucilaginous, there are chances of anthesis adhering to the staminal column or stigma while removing them. Hence, great care is needed in emasculation, so that no anther is left sticking to the floral parts. The flowers of male and female parents are enclosed separately in butter paper bags when a large number of crosses are to be attempted instead of bagging. It is a common practice to tie the apex of the corolla with a sting or adhesive cellophone tape to prevent their opening. For pollination, the pollen from the freshly opened bud previously bagged from the female parent is dusted directly on the stigma of the emasculated flowers of the female parent. The pollen can also be applied with the help of a camel hair brush. For selfing, it is convenient to cover the individual flower in butter paper bags or to tie the corolla tip with a sting or cellophone tape. In order to obtain better seed setting is used to pollinate the flowers by hand using the pollen from the same flowers.

National Released Hybrids

AOH-262

This hybrid has been developed at GAU, Anand. Plants are 77 cm tall with average of 1.8 branches. Each plant bears 10.0 fruits. Fruits are of 10.6 cm x 5.5 cm in size. The fruit average weight is 11.1 g. It gives an average yield of 780 q/ha.

DVR-2 (Shitla Uphar)

It is a cross between PDIB-1 and PDIB-3. This is a yellow vein mosaic virus and enation of leaf curl virus resistant hybrid developed at IIVR, Varanasi. Plants are medium-tall (74.0 cm) with dark green stem. Fruits are smooth, tender and dark green, 14.53 cm x 1.51 cm in size and less tuberous. It gives an average yield of 860 q/ha.

AOH-263

This is yellow vein mosaic virus resistant hybrid developed at GAU, Anand. Plants are 91 cm tall with average of 2.0 branches. Each plant bears 12.6 fruits. Fruits are of 11.4 cm x 5.3 cm in size and an average fruit weight of 12.6 g. This hybrid gives an average yield of 894 q/ha.

DVR-1 (Shitla Jyoti)

It is cross of PDIB-12 and PDIB-1, developed at IIVR, Varanasi. Plants are 135 cm tall, green stem. Fruits are smooth, 13.53 cm x 1.63 sizes, tender, dark green and less fibrous. Suitable for both rainy and summer seasons cultivation. Fruiting period of about 45-95 days. It gives an average yield of 760 q/ha, which is 27.25 % increase over the standard check. This is yellow vein mosaic virus and enation of leaf curl virus resistant hybrid.

Varsha

This hybrid has been developed by Indo-American Hybrid Seeds, Bangalore. A vigorous tall growing (1.8-2 m), early maturing hybrid with deep cut leaves. The fruits are five-ridged, medium in size, smooth and shining green in colour. It starts yielding 6-7 weeks after sowing. It is highly tolerant to yellow vein mosaic virus.

Vijay

This hybrid has also been developed by Indo-American Hybrid Seeds, Bangalore. Plants are dwarf compared to Varsha and Vishal. Fruits are slender, five-ridged, smooth and green. It is highly tolerant to yellow vein mosaic virus.

Vishal

This hybrid has also been developed by Indo-American Hybrid Seeds, Bangalore.This is taller and with lesser branches compared to 'Varsha'. Fruits are medium-sized, five-ridged, smooth and green. It is highly tolerant to yellow vein mosaic virus.

Seed Certification Standards

I. Field Inspection

A minimum of three inspections should be made, the first before flowering, the second during peak flowering and fruiting stage, and the third at mature fruit stage and prior to harvesting.

II. Field Standards

(A) General Requirements

Isolation: Seed fields should be isolated from the contaminants shown in column 1 of the Table below by the distances specified in columns 2 and 3 of the said Table:

Contaminant	**Minimum distance (m)**	
	Foundation	**Certified**
1	**2**	**3**
Fields of other hybrids	400	200
Fields of the same hybrids not confirming of varietal purity requirements for certification of wild okra (*A. ficuleneus* (L.) Wt. & Arn. *A.manihot* (L.) Medic. and *A. moschcatus* (L.) Medic.).	400	200

(B) Specific Requirements

Factor	**Maximum permitted (%)***	
	Foundation	**Certified**
Off types	0.10	0.20
** Objectionable weed plants	None	None

* Maximum permitted at and after flowering.

** Objectionable weeds shall be: (*A. ficuleneus* (L.) Wt. & Arn. *A. manihot* (L.) Medic. and *A. moschatus* (L.) Medic.).

III. Seed Standards

Factors	**Standards for each class**	
	Foundation	**Certified**
Pure seed (minimum)	99.0 %	99.0 %
Inert matter (maximum)	1.0 %	1.0 %
Other crop seeds (maximum)	None	5/kg
Total weed seeds (maximum)	None	None
Objectionable weed seeds (maximum)	None	None
Other distinguishable varieties (maximum)	10/kg	20/kg
Germination (minimum)	65.0 %	65.0 %
Moisture (maximum)	10.0 %	8.0 %
For vapour-proof containers (maximum)	8.0 %	8.0 %

13

COLE CROPS

Cabbage

Genetic Resources

Historical evidences indicates that modern hard-headed cabbage cultivars are descended from wild non-heading *Brassica* originating in the Eastern Mediterranean and in Asia Minor. The ancient Greeks held these early forms of cabbage in high regard and believed that they were a gift from the God. The Clelts and, the Romans disseminated cabbage throughout Europe. In fact, the Latin name *Brassica* is derived from the Celtic word *bresi* meaning cabbage. Over a period of centuries, hard-headed cabbage types evolved in northern Europe, while loose-heading, heat resistant types developed further in South. Cabbage was first introduced USA when the French explorer Jacques Cartier planted seed in Canada on his voyage in 1541. Cabbage is grown for the thickened main bud called head. The Portuguese introduced cabbage much earlier than cauliflower. There are reports that it was also grown during the Mughal period. In India, much varieties of cabbage are not available. However, genetic resources of cabbage for high yield and resistance to biotic stress are given in Table1.

Table 1: Genetic resources of cabbage

Trait	Promising accession
High Yield	EC-490162, 490165, 490174, 49176, 490191, 490200, Pride of Asia, Autumn Victory and PBOG-54.
Solerotinia rot (*Sclerotinia sclerotionum resistant*)	MR-1.
Cabbage yellow	AC-238, Spitzkool, EC- 93559, Globelle and Badger inbred 12 and 13.
Downy mildew resistant	MR-1.
Black rot resistant	Greenland Hammer, Pusa Mukta, EC-10109 and Spitzkool.
Diamond-back moth resistant	Green Acre and PI-3243.
Cabbage aphids resistant	KK cross, NY-1381 and 101181.

Production Objectives

Following objectives is considered while developing cabbage hybrids:

- Hybrids having longer staying capacity in field after head formation.
- High-yielding hybrids.

- Hybrids having short stem and narrow, short and soft core.
- Hybrids that can grow under mild winter conditions.
- Hybrids resistant to black rot (*Xanthomonas campestris*) and cabbage yellows (*Fusarium oxysporum* f. *conglutinous*).
- Hybrids resistant to cabbage butterfly caterpillar (*Pierih brassicae*), aphids (*Brevicoryne brassicae*) and diamond back moth (*Plutella oxygostella*).

Botany and Floral Biology

Plants are polymorphous, annual or biennial, erect, 30-150 cm high and much branched in the upper part. Stem is obtusely quadrangular or subterete, pruinose and glabrous. Leaves are very variable at the base of stem 2-10 cm long, petioled, lyrate or obovate, subentire or undulate, more or less deeply and irregularly lobed, terminal to orbicular or ovate, apex rounded, higher leaves mostly sessile, semi-amplexicaul, ovate-oblong or linear-lanceolate, entire or subentire. All leaves are variable in shape, colour and size, green, yellowish green or red, fleshy, thick, glabrous and more or less distinctly dotted. Inflorescence is terminal, elongate, lax, racemiform or paniculiform, 10-50 cm long and racemes 20-40 flowered. Flowers are rather large, buds raised far above the expanded flowers and pedicels 0.5-0.8 cm long. Sepals are erect, obtuse, 1 cm long, the outer ones ovate-oblong and narrow, saccate at the base, the inner ones more ovate or broadly ovate oblong. Petals are oblong-spathulate, rounded at tips, light yellow and 1.5-2 cm long. Stamens are subequal, erect and 1-1.3 cm long. Pods on stalk are 1.5-2 cm long, more or less patent, thin, cylindrical or linear elongate, with a long beak, many seeded, 5-9.5 cm long and 0.3-0.5 cm thick. Seeds are seriate, globuse and smooth (Fig.1).

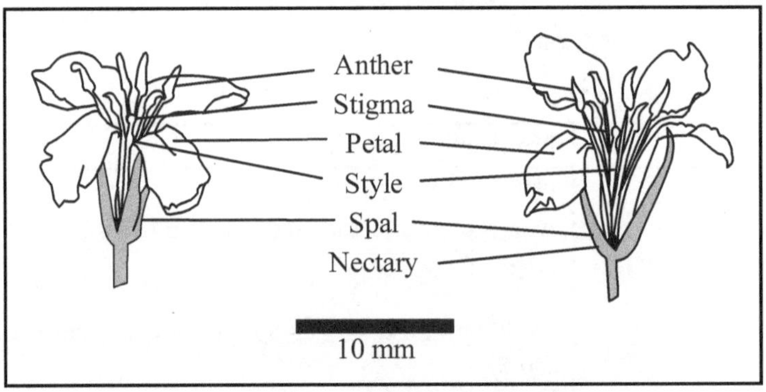

Figure 1: Floral parts of cabbage flower.

Methods of Hybrid Seed Production

Cabbage is a naturally cross-pollinated crop. Bees and flies normally pollinate stigma. The F_1 hybrids are produced by-hand pollination or by using free insect-pollination with the help of self-incompatible lines (Sampson, 1966). The use of male sterility and marker genes have also been recommended. In certain cases, bud-pollination is used when self-incompatible lines or male sterile lines is available for the parental lines. The hand pollination is not economical for commercial seed production in cabbage. Different methods proposed by various workers for production of F_1 hybrid seeds of cabbage are as follows:

(1) Hand pollination

- Not very convenient
- High cost of seed production

(2) Free Insect-pollination

(a) Using self-incompatibility

- Single-crosses (York *et al.*, 1957)
- Double-crosses (Odlland and Noll, 1950)
- Triple-crosses

(b) Using genetic male sterility (Rundfeldt, 1960)

(c) Using dominant marker gene (Swarup and Gill, 1964)

(d) Using cytoplasmic male sterility (Ogura, 1968).

Steps in Development of F_1 Hybrids by Using Self-incompatability

Development of Self-incompatible Parental Lines

Inbred lines are developed after 4-5 generations of inbreeding. Inbred lines can also be developed using pedigree method and single plant and family modifications, single seed descent (SSD) and anther culture. In parental lines at the time of their selection, self-incompatibility should be high. Each line should have one *S-allele*, which should be different from other *S-alleles* present in other lines to be crossed. Preferably, select dominant *S-lines*, which do not fulfil the requirement of their association with weak self-incompatibility, adequate seed quantity and quality, must be rejected unless used as pollinator.

Test Crossing and Identification of Superior F_1 Combinations

Crossing should be attempted amongst desired self-incompatible inbred lines using half-diallel or full-diallel or top-cross method. The selection of lines should be based more on intuitive or empirical approach, keeping into account differing maturity period, flowering synchrony, combination of desirable morphological and quality characters or disease resistance. Combinations should be assessed in replicated trials. Multilocational trials of identified promising combinations may prove more useful before their final recommendation. Identifications of superior combinations also followed consideration of their suitability for commercial seed production.

Type of F_1 Combinations

Following combinations are used:

• *Single-cross*

This method involves two cross compatible lines, each of them is highly self-incompatible. The lines A (S_{11}) and B (S_{22}) an individually propagated by bud-pollination or open-pollination at elevated temperature by selfing followed by pollination in isolation.

In hybrid seed production block, these lines are planted together in alternate row to cross-pollinate each other. The hybrid seed is harvested from both the parents. Scheme of producing single-cross hybrids using self-incompatible lines is given in Fig.2.

Line A: S_{11} Line B: S_{22}

S_{11} x S_{22}

↓

Commercial Hybrid

Figure 2: Method of producing single-cross hybrid using self-incompatibility.

In self-incompatible condition, selfing can be accomplished by bud-pollination. The flower buds are pollinated two days prior to their opening by their own pollen by applying fresh pollen from previously bagged flowers to the same plant. It has observed that seed setting is better if buds are pollinated two days before opening when there is at least 30% flowering on plant than bud-pollinating only a day prior to opening or just at the commencement of flowering. A special type of scissor is used (Fig. 3). The scissor which is a semi-circular opening in the cutting edge about 5 mm from the tip, can be used for this purpose (Wiering, 1958), as it facilitates removal of petals and accelerates bud-pollination, particularly when it is done on a large-scale. Emasculation is not necessary in bud-pollination. However, after pollination, the buds are enclosed in butter paper bag to avoid contamination from foreign pollen. It is more convenient to enclose one or two flowering branches in a muslin bag and use individual flower buds or flowers or bud-pollination or crossing as and when they become ready for this purpose. Each branch containing pollinated buds or flowers is tagged with a small tag label. Usually, about 8-10 flower buds or flowers are bud-pollinated or crossed in each inflorescence branch to ensure better seed setting and the unopened

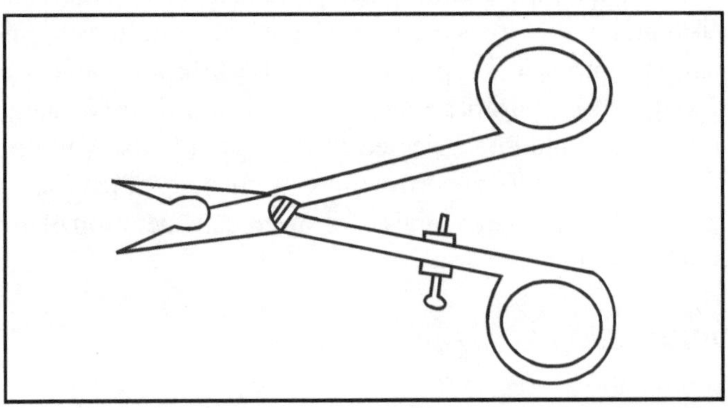

Figure 3: Special type of scissor used for bud-pollination in *Brassicas* vegetables.

young flower buds at the terminal end are removed. When a large quantity of crossed seed is required, the mature heads of female and male parents are planted in alternate rows spaced 60 cm apart. Later, about 3-4 days before opening of buds, the plants are covered under an insect-proof wire nets accommodating 25 plants, five plants/row and number of rows should be five. Twenty-five plants of cabbage grown under these cages produce about 675 g (1.5 lb) seeds.

Using single-cross method which follows a modified method of hybrid seed production in cabbage by using single-cross method to alleviate the problems encountered in maintaining parental inbred lines. The first step in this programme (Fig. 4) is the selection of parents expressing high combining ability and also possessing most desirable horticultural characters. Second step is the development of inbred lines. Cross-compatibility of lines 1 & 2 and line 3 & 4 in reciprocal crosses show that parental lines isolated for hybrid production have been ready. The prime reason for step 5 is to increase the inbred lines. About 28 g of seed can be produced from a single plant (5,000-7,000 seeds). Thus, an increase of several thousand times might be accomplished by step 5. The seed produced on both lines is F_1 hybrid. In this method, one feature is worth noting, *i.e.* the seed growers do not have to resort to tedious hand-pollination technique. In both, sexual increase of parental lines (step 5) and in commercial F_1 hybrid seed production (step 6) insect transfer the pollen and no manipulation of flower whatever is required. All rows produce practically identical seed.

A further modification of this method would be to grow line 'x' and line '3' together in isolated field as step 6. Line three would be primarily a pollen parent and only one-fourth or less of the field would be planted under this line. All the harvest would be F_1 hybrid seed. Utilization of three instead of four inbred lines is the only point of merit in this modification. This method of breeding can be applied with some modifications in production of hybrid seed in other cruciferous crops where self-incompatibility is prevalent.

• *Double-cross*

This method was first utilized by Thompson (1959) for the production of stem kale hybrid and later utilized in other cruciferous crops. This method is based on the maintenance of four homozygous lines. This method is relatively inexpensive. In this technique, A x B and C x D parents are needed for final hybrid seed production (Fig.5). This method requires obviously more preparatory breeding work both in respect of isolation of four incompatible lines as well as testing for estimating their combining ability. An additional disadvantage is the much less uniform character of final hybrid as compared to single-cross hybrid. Also, in many cases, the yield of double cross hybrids is lower than single-cross hybrids.

• *Triple-cross*

To reduce the total cost of producing parental inbred lines, which involves expensive bud-pollination, triple-cross method has been recommended in view of presence of sporophytic type of incompatibility. In this plan, presence of sporophytic type of incompatibility ensures presence of 100% hybrid seed. This otherwise is not possible with gametophytic type of self-incompatibility. Whereas, production of double-cross seed would

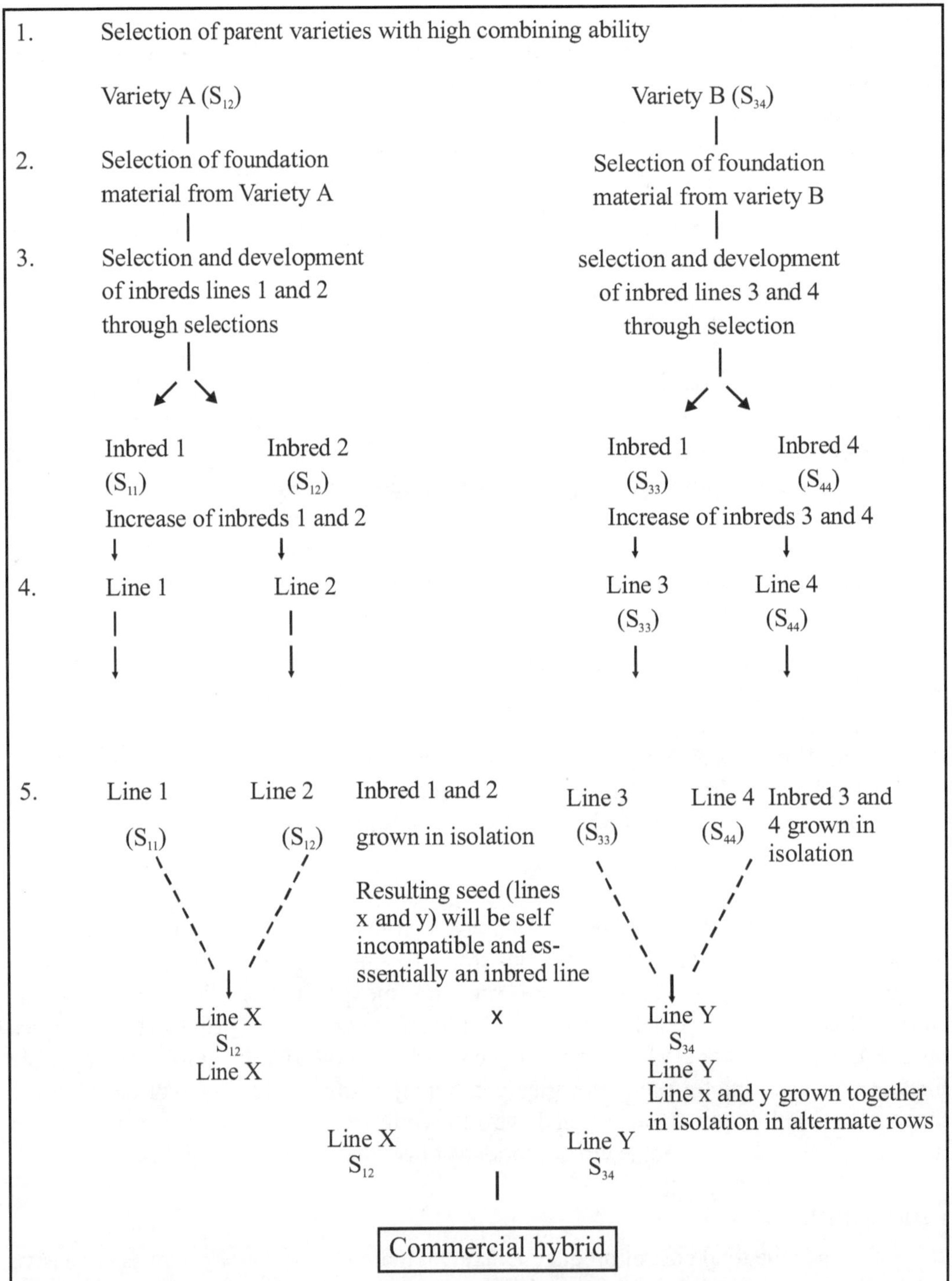

Figure 4: Modified method of single-cross hybrid seed production.
Source: Deshpandey and Reddy, 1996.

be possible, whether self-incompatibility is determined sporophytically or gametophytically. It is essential for the production of a triple-cross hybrid that the reaction of pollen and style are sporophytically controlled. However, existences of dominance relationship, which may occur between two *S-alleles* in pollen and style in inbred lines, complicate the multiplication of the triple cross hybrids (Fig.6). Therefore, for production of 100% triple-cross hybrid seed in final cross, *S-allele* 3 from inbred C must be active (dominant) in pollen and style with both *S-alleles* 1 and 2, then plants with two possible *S-alleles* S_{13} and S_{23} from three-way cross would be cross-compatible with each other when planted with other three-way cross to produce triple-cross hybrid seed.

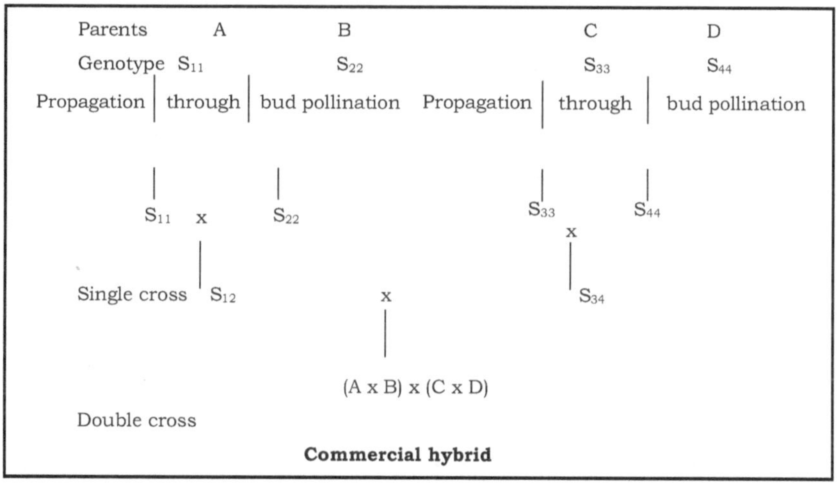

Figure 5: A Method of producing double-cross hybrid using self-incompatibility
Source: Deshpandey and Reddy 1996.

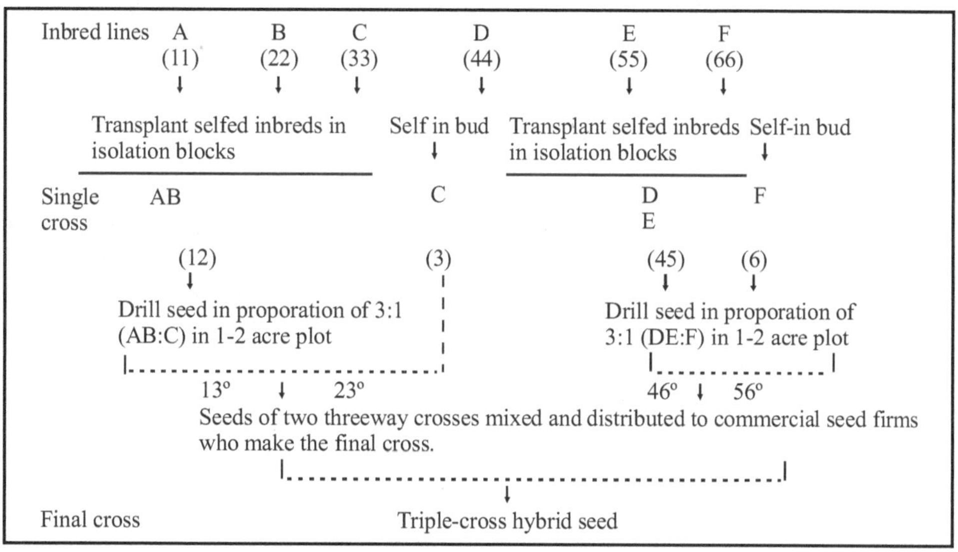

Fig. 6. Method for production of triple-cross hybrid seed (dot over *S-allele* denote dominance).
Source: Deshpandey and Reddy, 1996.

It is not essential for six basic inbreds to be homozygous for six different *S-alleles*. If *S-allele* 1 is dominant over *S-allele*-2 in pollen and style and similarly *S-allele* 4 is dominant over *S-allele* 5 in pollen and style. The same *S-allele* although not from the same inbred line can be used for inbreds B and E. Further, if *S-allele* three or six is dominant in pollen and style over *S-allele* 1 and 2 or 4 and 5, respectively. Inbreds A and B can be homozygous for the same two *S-alleles* as in the inbreds D and E.

Advantages of Triple-cross Hybrids

The advantages of triple-cross hybrids are:

- Since higher seed yield is expected from hybrids than inbreds, it reduces the extent of bud-pollination. The amount of bud-pollination required for triple-cross would be less than one-fifth of that necessary for producing hybrids by the double-cross method.
- Hybrid seed is produced at a rather low-cost as compared to single-crosses.
- It provides wider choice of productive and high combining lines.
- Triple cross hybrids have better winter survival and are more homeostatic because of wider genetic base.

Disadvantages

The disadvantages of triple-cross hybrids are:

- Overall yield is much less as compared to single-cross hybrids.
- The final hybrid is far from uniform.
- This method requires long preparatory breeding work with very little breeding flexibility after the original six breeding lines have been determined.
- In triple-cross method, restrictions on the arrangement of inbreds, are imposed because of the necessity for inbreds 'C' and 'F' to be homozygous for *S-allele* and use also high in the dominance series. Thus, although inbreds 'C' and 'F' might be interchangeable, it is less likely that inbreds 'A' and 'B' would replace 'C'. Further, it might not be possible to multiply highest yielding combination.

Genetic Male Sterility

Male sterility in *B. oleracea* is a recessive character governed by single gene *ms*, which is mutated from male fertile gene *ms* (Nieuwhof, 1971 and Rundfeld, 1960). The male sterile stock is maintained by crossing with heterozygous fertile plants. Half of the plants are sterile, while other half are heterozygous fertile which are removed as soon as flowering starts. In male sterile plants, flowers and anthers are smaller than male fertile ones.

Cytoplasmic Male Sterility

This trait has been found in Japanese radish by Ogura (1968). Sterility was induced by the interaction between homozygous recessive nuclear genes *ms* and the S-cytoplasm Bannerot *et al.* (1974) reported successful introduction of cabbage nucleus into cytoplasm of Ogura male sterile radish by means of repeated backcross. Unfortunately, all the *Oleracea* genotypes that are introduced into radish cytoplasm, exhibited mild to severe yellowing (chlorosis) of young leaves when seedlings are grown at 12°C or below (Bannerot *et al.*, 1977). The degree of yellowing varied with the choice of recurrent male parents. McCollum (1988) released two cytoplasmic male sterile cabbage germplasm by crossing *Raphanus sativus* (radish) and *B. oleracea* followed by repeated backcrossing with cabbage in the last seven generations. However, these two cabbage lines also showing mild chlorosis, a problem that is eliminated through protoplast fusion when radish chloroplasts are replaced with *Brassica* chloroplasts (Kagami *et al.*, 1990 and Robertson *et al.*, 1987). In India, Hybrid H-64 has been developed at IARI Regional Station, Katrain (Kullu Valley, Himachal Pradesh) by using cytoplasmic male sterility.

Use of Marker Genes

Earlier, Swarup and Gill (1964) suggested the use of a dominant marker gene in cabbage form purple stem pigmentation to facilitate identification of F_1 hybrid seedling before transplanting. Sampson (1966) reported that male sterility is closely linked with a seedling recessive marker gene, which blocks the synthesis of anthocyanin pigment. The marker gene, which produces anthocyanin-free plant with bright green hypocotyle, is obtained from culinary cury kale and variegated ornamental kale. With use of CC female lines and CC pollen lines in F_1 hybrid seed production, the sibs in inbred line can be easily identified at seedling stage. With the use of male sterility, sib-crossing instead of bud-pollination can maintain the inbred lines or by selfing if self-compatible plants are used. F_1 hybrid BRH-5 (Sel-8 x AC-204) has been developed in India by using dominant marker gene (purple pigmentation) in the pollen parent. The seedlings are produced by selfed seed of female parent if any, can easily be detected at nursery stage.

National Released Hybrids

BSS-50 (Bajrang)

It has been developed at Beejo Sheetal Seeds Pvt. Ltd., Jalna. Heads are dark green, smooth, highly compact, average head weight 1kg. Outer leaves are upright. It matures in 65 days but can stand in field for 45 days after maturity. It can tolerate high temperature up to 36°C. This hybrid is resistant to fusarium wilt. The female parent (BH-8-20) of this hybrid is characterized by early in maturity and has compact round head. However, male parent (NORT-14) is early in maturity with highly compact round head.

BSS-32 (Suvarna)

It has also been developed at Beejo Sheetal Seeds Pvt. Ltd., Jalna. Heads are round. Average head weight is 3 kg. Heads are strong with smooth outer leaves and mature in 110 days. The female (ET-2-2) parent of this hybrid is characterized by medium-late in maturity having blue leaves with vigorous growth. The heads are round and medium. However, male parent (BOL-13) is of late in maturity. The leaves are blue in colour. Heads are flattish-round and big.

H-113

This is a self-incompatible based hybrid between Sel-8 and 83-6 line, developed at IARI, Regional Station, Katrain. Plants bear roundish, compact head (13.5 cm x 16 cm), bluish green and waxy leaves, 0.5-0.7 cm stalk height, 50-60 cm frame size (13.5cm x 16 cm). It gives an average yield of 473.5 q/ha in 60-68 days of crop duration.

Sri Ganesh Gol

It has been developed at Maharashtra Hybrid Seed Co. Ltd., Jalna. Plants are vigorous. Heads are round, compact, attractive and bluish green. Head becomes ready for harvest 90-95 days after sowing. This hybrid is resistant to yellowing. It gives an average yield of 750-875 q/ha. It has good transport quality.

Seed Certification Standards

I. Field Inspection

A minimum of three inspections should be made, the first before the marketable stage, the second at marketable stage and the third at flowering stage.

II. Field Standards

(A) General Requirements

1. Isolation: Seed fields should be isolated from contaminants shown in columns 2 and 3 of the Table as given below.

Contaminant	**Minimum distance (m)**	
1	**2**	**3**
Fields of other varieties	1,600	1,000
Fields of the same	1,600	1,000

(B) Specific Requirements

Factor	**Maximum permitted (%)***	
Off types	0.10	0.20
**Plants affected by seed borne diseases	0.10	0.50

*Standards for off types shall be met at and after flowering and for seed-borne diseases at final inspection.

** Seed-borne diseases shall be: Black leg (*Leptosphaeria maculans* (Desm.) Cos.& de Not); black rot (*Xanthomonas campestris* pv. *campestris* (Pamm.) Dawson); soft rot (*Erwinia carotovora* L.R. Jones).

III. Seed Standards

Factors	**Standards for each class**	
	Foundation	**Certified**
1	**2**	**3**
Pure seed (minimum)	98.0%	98.0%
Inert matter (maximum)	2.0%	2.0%
Other crop seeds (maximum)	5/kg	5/kg
Weed seed (maximum)	10 /kg	10/kg
Germination (minimum)	65%	65%
Moisture (maximum)	7.0%	7.0%
For vapour-proof containers (maximum)	5.0%	5.0%

Cauliflower

Genetic Resources

Indian cauliflower is introduced as temperate crop, well-adapted to this soil to the extent that some of the genetic variant got developed here in nature by regular cultivation for several decades and becomes tropical in habit by producing seeds in plains. These strains are termed as early strain of cauliflower. In India, variability in Indian cauliflower widely exists in Bihar, Uttar Pradesh, Rajasthan, Nilgiri Hills, Assam, Meghalaya, Madhya Pradesh, Punjab and Haryana. The NBPGR, New Delhi, has organized several explorations in Bihar, Uttar Pradesh, Madhya Pradesh, Rajasthan and Karnataka. The IIVR, Varanasi, has initiated work on collection of local-land races available in Bihar and Uttar Pradesh in collaboration with NBPGR, New Delhi. More than 200 lines belonging to different maturity groups have been collected and maintained. At present, IIVR, Varanasi; IARI Regional Research Station, Katrain, KulluValley (Himachal Pradesh); Rajendra Agriculture University, Sabour; Central Horticulture Experiment Station, Namkum, Ranchi; Mahatma Phule Krishi Vidhyapeeth, Rahuri, and College of Horticulture, Dr. Y. S. Parmar University of Horticulture and Forestry, Solan, are actively engaged for collection, evaluation and characterization of cauliflower germplasms. On the basis of evaluation trials, following lines/accessions have been identified and are given below :

Table 2: Promising lines/accessions of cauliflower in India

Attribute	Lines/accession(s)
Self-incompatibility	CC-12, CC-14, CC-15, India Express, First Early, Kataki-3, Kataki-5, Aghani-1 and Aghani-26.
High-yielding Kunwari	Kunwari-2, Kunwari-13 and Kunwari-18-1, Kunwari-15, PDVR-8, JBI-22/79, Kunwari-23/39, Kunwari-23/42, Early Kunwari-2-1, Kunwari-14-1, JB-23/47, Kunwari-6, Kunwari-10 and Hazipur-1
Kataki	Kataki-21, Kataki-1, Kataki-6, Kataki-20, JBT-23/40, JBT-23/95, JBT-23/78, Katki-29, Katki-15, 81-5, 84-3, 91-1-92-2, 93-3, 94-1, Katki Early-23/40 Katki Early - 23/95, 2002-1, 95-3 and Kataki-13.
Aghani	Aghani-33, Aghani-10, Aghani-18 and Aghani-32, BSS-48, JBT-23/66, IIVR-12, IIVR-6, Aghani-4, Aghani Awasthi Seed, Aghani Prem Seed and Aghani-23/74.
Pusi	Pusi-4, Pusi-3, Pusi-5, Pusi-2, Pusi-1, Pusi-8, Pusi-13, Pusi Prem Seed and Pusi Chanderdev.
Maghi	Maghi-1, Maghi SS and SF Maghi.
Early (high-yielding)	91-1, 91-2, 81-5, 93-3, 92-1, 94-1, IIVRCE-2 and JBT/BS-23/95.
Mid (high-yielding)	81-5, 84-3, 91-1, IIVRCM-12, IIVRCM-11 and 78-3.
Late (high yielding)	Kathmandu Local.
High temperature (30-35°C)	Sel-23-7, 98-4 and 41-5.

Contd....

Table 2 Contd....

Sclerotinia rot (*Sclerotinia sclerotiorum* resistant)	EC-103576, EC-131592, Janavan, EWAWH and KN-81.
Downey mildew	
Resistant	BR-2, MGS-2-3, 3-5-1-1, CC, 12C and EWAWH.
Moderately resistant	Kunwari-17, Kunwari-8, Kunwari-4 and First Early.
Black rot resistant	MGS-2-3, Pua kea, Sn- 445, Br-2, EC-162587, Sel-2 and RBS-1, Pusa Subhra, Pusa Snowball, K-1 and Sel-8.
Black rot moderately resistant	Early India, Kataki-1 and Kunwari-2.
Diamond back moth resistant	PI-234559

Besides above lines/accessions, Pusa Kataki and Pusa Deepali, Pusa Synthetic, Pusa Shubra developed from IARI Regional Station Katrian, Punjab Kunwari at PAU, Ludhiana, Pant Gobhi-1, Pant Gobhi-2, Pant Gobhi-3 and Pant Gobhi-4 of G.B. Pant University, Pantnagar; NDC-1 of NDUAT, Faizabad, these varieties can also be used as one of the parent for developing hybrids for high yield and some extent resistant to diseases.

Production Objectives

Following objectives should be considered to develop cauliflower hybrids :

- Hybrids of tropical types, which can withstand heat and can form curds in August/ September through the use of incompatible lines.
- Hybrids for earliness, high yielding, better curd qualities like non-ricey, compact bract free protected curds with retentive cream/ white colour.
- Hybrids having high yield due to increase in curd size as well as compactness.
- Hybrids for earliness.
- Hybrids resistant to diseases like black rot, sclerotina rot, alternaria blight, downey mildew and erwinia rot.

Botany and Floral Biology

The stem of vegetative cauliflower plant is rather short, and with thickness of about the some extent that of cabbage. The leaves are large, generally oblong, younger ones being nearly always sessile. In contrast to situation in other crops, buds do not usually arise in leaf axils. During transition to generative phase, which, is in contrast to situation in most the other cole crops, is accomplished in many types of cauliflower in first year, the peduncles in axils of the bracts formed by the main growing point branch repeatedly, so that branches even of the fifth order can arise. At first the numerous peduncles do not grow length-wise but become thick and fleshy. Thus, a colourless, roughly spherical, terminal compact head arises of which the upper surface consists of vast numbers of naked apical meristems. The peduncles are composed of thin-walled parenchyma and vascular bundles, which do not become woody.

The young curd is at first entirely covered by the foliage, on becoming visible it already has a diameter of over 5 cm after sometimes the flower stems elongate and a number of the apices develop into normal flowers. The inflorescence may attain a length of 1-2 m, but the slender pedicels are only 1.5-2 cm long. There are 4 sepals, 4 petals, 6 stamens and 2 carpels. The carpel form a superior ovary with false septum and 2 rows of campyloto ropus ovules. The androecium is tetradynamous, *i.e.* there are 2 short and 4 long stamens. The pollen-grains are 30-40 μ in diameter and have three germinating pores. The bright yellow petals become 15-25 mm long about 10 mm wide. The sepals are erect. The buds open under the pressure of the rapidly growing petals. This process starts in afternoon, and usually the flowers become fully expanded during the following morning. The anthers open a few hours later, the flowers being slightly protogynous. Insects, particularly bees, which collect pollen and nectar, pollinate the flowers. Two nectarines situated between the basis of the short stamens and the ovary secretes the nectar. Situated outside, the basis of the pairs of long stamens is also two nectarines, but these are not active. Flowers are borne in racemes on the main stem and its branches. The fruits are glabrous siliques, 4-5 mm wide and sometimes over 10 cm long, with two rows of seeds lying along the edges of the replum (false septum, an outgrowth of the placentae). A silique contains from 10-30 seeds. Three to four weeks after the opening of the flower from which it is formed, the silique in two valves breaking away from below upwards, leaving the seeds attached to the placentas.

Anthers are emasculated by physically removing the anthers for controlled hybridization. Anthers rarely dehisce before the flowers are fully open and emasculation can safely be left until a day before anthesis, even though the stigma may be level with the anthers within the flower bud.

Methods of Hybrid Seed Production

Following techniques are used for hybrid seed production.

Hand-Emasculation and Hand Pollination

The flowers may be emasculated by removing six stamens using a pair of forceps. In self- compatible cauliflower (European types), the stamens are removed before the opening of the buds as the flowers is already fertile in bud stage; crossing can be done at the same time. In self-incompatible types, emasculation may be omitted. When pollination cages are available, insects such as honeybees can make crosses between self-incompatible types, bumble bees and flies. Initially this method was used for producing F_1 hybrids. Since, much of the labour is involved which results into lesser quantity of seed and increased production cost. Hence, it is not practiced on commercial scale.

Hand Pollination Without Emasculation

Where male sterility is available, the process of emasculation is eliminated. As male sterile lines are available in cauliflower, therefore, method of hand pollination without emasculation is practised. Some marker genes can also be used which are closely linked to male sterility.

Free Insect Pollination

It is feasible to produce F_1 hybrid under natural field conditions. But this method can be used when self-incompatible but cross-compatible lines are available. In this case, advantage of the occurrence of multiple allelic system of self-incompatibility. First self-incompatible lines are evolved which should be cross-compatible then only these can use for the production of F_1 hybrids. These lines are planted in alternate rows or two rows of each line and planted adjoining to each other. However, these should be sown as such that flowering time coincide for effective cross-pollination.The scheme of using self-incompatible line for production of F_1 hybrid is as follows:

$S_1 \times S_2$ or $S_1 \times S_2 + S_3S_2$ dominant to S_3

$S_1S_2F_1$ $S_1S_2\ F_1$

National Released Hybrids

Pusa Hybrid-2

This hybrid was developed at IARI, New Delhi, and identified in 1992 for its cultivation in West Bengal, Assam, Punjab, Uttranchal and Bihar. Plants are semi-erect with flush-green upright leaves. Curds are creamy white, highly compact. Average curd weight is 907 g. It matures from mid-November to mid-December in North Indian plains. It is resistant to downy mildew and gives an average yield of 250 q/ha.

Besides, Pusa Hybrid-3, Snow Crown, BSS-18, NS-60 and NS-90 are also popular among farmers and are widely cultivated.

Seed Certification Standards: Similar to those mentioned for cabbage (Page 312)

14

CUCURBITACEOUS CROPS

Bitter Gourd

Genetic Resources

Bitter gourd or balsam pear (*Momordica charantia* L.) is one of the most popular cucurbitaceous vegetable, commonly cultivated in many countries. Probably, India is the centre of origin and secondary centre of diversity of this crop. Bitter gourd is an annual monoecious herb, belongs to the genus *Momordica* that comprises 60 species. Out of these, only *M. charantia*, *M. cochinchinensis*, *M. dioica*, *M. balsamina* (balsam apple), *M. cymbalaria* (syn. *M. tuberosa*) are under cultivation. *Momordica charantia* is grown all over the country in tropical and subtropical climates and rich genetic diversity occurs in northern plains including tarai region, the northern-western plains, the central region/plateau and the western and eastern peninsular regions. Kakrol (*M. dioica*) is cultivated in West Bengal, Assam, parts of Bihar, while sweet gourd (*M. cochinchinensis*) is popular particularly in Tripura, Assam, Andaman Nicobar, Tamil Nadu, Uttar Pradesh, Bihar and West Bengal. *M. denundata* (Thw) Clarke (syn. *M. dioica* var. *denundatais)* is found abundantly in Kerala, *M. macrophylla* in Assam and *M. sabangulata* Blume is commonly found in Karnataka, Maharashtra, Meghalaya and Sikkim. In India, NBPGR , New Delhi, and IIVR, Varanasi, are actively engaged in the collection of bitter gourd germplasms.The NBPGR, New Delhi, has evaluated 33 collections for 37 characteristics. Variability has been reported in vine length (105-335 cm), number of primary branches (3-7), number of fruits/plant (1-25), fruit girth (8-27 cm), fruit length (9.5-20 cm) and fruit yield/plant (0.5-2.8 kg). Accessions, UB-22, UB-66, IC-74158, C-1621 and K-3374, have been found better yielder. At IIVR, Varanasi, 121 genotypes have been evaluated for yield and yield-contributing traits. VRB-40 & IC-44415A have been identified as promising lines. Accessions showing better charecters for different traits are given below (Ghosh and Kalloo, 2000):

For more number of fruits/plants

IC-45346A (20.2), VRBT-78 (24.0), IC-43261 (17.4), IC-44426D (18.6), IC-44428 (17.6), IC-45350 (18.4), NIC-72285 (20.0) and IC-85611 (18.0).

For high yield/plant

IC-44428 B, NIC-72285, IC-85604A, IC-85608BC, IC-85611, IC-85636, NIC-14034, NIC-13978, V-8910-103C, VRBT-29 and VRBT-50, VKB-127, VKB-120, VKB-129, VKB-130, VKB-139, VKB-113, VKB-125, VKB-127, JMC-14-2, JMC-14-3, IC-256226, VR-231, VR-235, VR-234, VR-232, VR-237, JMC-26, JMC-27, NDBG-57, NDBG-58, IC-92404 and EC-110596.

For CMV Resistance

IC-44428B, NIC-72285, IC-85604A, IC-85608B, IC-85611, IC-85636, V-890-103C, VRBT-29 and VRBT-50.

For Fruit Fly Resistance

IC-85641, MC-84 and V-890-103C.

Besides, several varieties, *viz.* Arka Harit, Coimbatore Long, Pusa Do Mausami, Punjab-14, Pusa Vishesh, NDB-1 and local materials like Pride of Surat and Pride of Gujarat can also be used for development of hybrids.

Production Objectives

Following objectives should be taken into consideration while developing hybrids:

- To develop hybrids of early fruiting with high female to male sex ratio.
- To develop hybrids having whitish green to glossy green fruit colour depending upon consumer's preference.
- To develop hybrids having less ridges on fruit surface and thick fruits particularly suitable for stuffing.
- To develop hybrids of different fruit size as per consumer's preference (small-7.5- 10 cm long, medium long, 10-15 cm long, 15-20 cm and extra long- 20-40 cm).
- To develop hybrids having immature seeds for longer period during green edible stage.
- To develop self-stalked hybrids, that is high yielding and early bearing.
- To develop hybrids resistant to major diseases and insect-pest.
- To develop hybrids suited for canning and dehydration.

Botany and Floral Biology

It is a slender, annual, tendril-bearing climber, somewhat fetid when bruised and 1-5 m long. Stem is thin, acutely quinquangular, with broad longitudinal furrows, light green and thinly white pubescent. Tendrils are inserted beside the leaves on long, robust stalks, simple or cleft, faintly 5-angular, yellowish green and 15-30 cm long. Leaves are alternate, long stalked, in outline orbicular or broadly ovate. Base is deeply cordate, apex shortly acutely acuminate, deeply pedately and 5-9 lobed. Lobes are more or less deeply lobed, all of them narrowed at the base, shortly mucronate, on both surfaces appressed pubescent on the nerves, afterwards becoming almost glabrous, above dark green, beneath light green or yellowish green, dull or feebly shining, 3-5 nerved or pedately 5-9 nerved, with transparent veins, densely pellucidely dotted and 5-17 cm diameter. Petioles are robust, half terate, furrowed above, thinly pubescent, light green, 3-10 cm long and stipule absent. Flowers are unisexual-monoecious and solitary in the leaf-axils. Pedicels are long, filiform, thinly pubescent or subglabrous and 5-15 cm long. Bract is below the middle of the pedicel, often near its base, sessile, large, orbicular-cordate or reniform with an obtuse or rounded, often emarginate or shortly cuspidate apex, entire, thinly pubescent, shortly ciliate and green 0.5-2 cm diameter.

Male flowers are much resembling the female ones but on shorter and thinner pedicels. Calyx is campanulate, deeply 5-fid and 1-1.25 cm long. Tube is short, thick, yellowish green, densely appressed pubescent and wartty. Segments are patent, elliptic or lanceolate, acute, light yellowish green, straight, densely shortly pubescent on surfaces, 0.4-0.85 cm long and 0.2-0.4 cm wide. Petals are 5 in numbers inserted in the throat of the calyx-tube, oblong lanceolate or elliptic with an obtuse or rounded, often-emarginated apex, entire, sulphureous, densely pubescent, 1.5-2 cm long and 0.75-1.2 cm wide. Stamens are three in number, filaments short, thick and white. Anthers are three, conduplicate and cohering. Ovary is fusiform, rostrate, warty, dark green and 2.5-3 cm long. Calyx is small, 0.75-1 cm in diameter. Sepals are five, oblong lanceolate, acute, yellowish green, shortly pubescent, 0.25-0.35 cm long and 0.1 cm wide. Petals are 5, patent, elliptic or oblong obovate, rounded at the apex, entire, citrine, densely shortly pubescent, 1.5-1.75 cm long and 1 cm wide. Staminodes are 5, tubercles-shaped; pale yellow, 4 of them approximate in pairs. Style is short and pale green. Stigmas are 5-6 and erect. Fruits are long peduncled, pendulous, and fusiform, beset with many irregular warts, very variable as to size, orange-coloured when ripe, 3-valvate at the top, 7-30 cm long and flesh blood-red. Seeds are numerous, flattened, ovate-oblong, warty, 1-1.5 cm long and 0.6-0.9 cm broad.

Bitter gourd is a monoecious crop, where male and female flowers are born on the same plant. Gynoecious sex form is also reported at IIVR, Varanasi. Varieties differs in days to flowering, time of anthesis and pollen viability (Deshpande *et al*., 1979). Anthesis occurs between 3.30 and 7.30 a.m. and anther dehiscence between midnight and complete opening of the flower (7.30 am). The stigma remained receptive for 24 hours before and after anthesis. Pollination is brought by bees. The appearance of first pistillate flower at a lower node number is an indication of the earliness. On the other hand, the higher the node number of the position of the first pistillate flower, the greater would be the production of pistillate flowers and a shorter interval between the appearance of first staminate flower and first pistillate flower indicates shorter plant life.

Method of Hybrid Seed Production

For hybridization, male flower buds from the chosen female parent can be nipped off as and when they appear and pollination resorted using pollen collected from male parents. This method can be used for large-scale F_1 seed production. For small-scale F_1 seed production and also generating variability through hybridization, following steps may be undertaken.

- Female mature buds are identified from female parent plants and covered using paper bag in the evening before pollination takes place.
- Matured male flower buds from male parent plants, which will open the next morning, are identified and covered with paper bags.
- Hybridization is done early the next morning (3.30-7.30 am). The paper bags are removed from female and male flowers. Pollen-grains are collected using a fine brush and transformed to the stigma of female flowers, which is then covered with the same paper bag. Pollination is also carried out by rubbing staminate flower anthers from

freshly opened flowers against the stigma of protected pistillate flowers. Tagging is done indicating parentage and date of pollination. In order to increase crossed fruit setting, other open- pollinated fruits already set, are removed to avoid inhibitory effect of these fruits on further fruit setting.

Bitter gourd has a strong male tendency. Studies *in vitro* on sex expression revealed that ethephon induced femaleness by increasing the size of pistillate buds. Foliar and seed treatment with ethephon (50-250 ppm) resulted in earlier and greater production of pistillate flowers (Mangal *et al.*, 1981; Sadhu and Das, 1978 and Verma *et al.*, 1984).

National Released Hybrids

BTH-7

This hybrid is a cross of VRBT-3A and VRBT-82-1. It bears long attractive green fruits of 20-25 cm length. Fruit bearing continue for longer period. It has field tolerance to powdery and downy mildews. It gives an average yield of 300 q/ha.

BTH-165

This hybrid is cross of KB-1 x VRBT-36. Fruits are long green with attractive colour and average length of fruit is 18 cm length. It is tolerant to powdery and downy mildews.

COBGH-1

It is a cross of MC-84 x MDU-1 and developed at TNAU, Coimbatore. Plants are with 14.38:1 male and female flower ratio. Fruits are light green to creamy white, 28 cm long, 23.45 cm in girth and 450.00 g an average fruit weight.

PUSA HYBRID-1

This hybrid has been developed from IARI, New Delhi. It is suitable for commercial cultivation in spring summer season of northern plains. Its vines are medium in growth. The leaves are broadly ovate, lobed, surface soft, medium-sized and dark in colour. The fruits are attractively green, medium-long and medium thick (fruit length 13.5 cm and breadth 15.0 cm) with irregular smooth tubercles on surface. On an average a fruit weight is 100 g. Fruit flesh is 7.3 mm thick. The fruits are suitable for making vegetable and also for dehydration purpose. It is early-maturing (50-55 days) and superior in fruit quality to Pusa Do Mausami and Pusa Vishesh. On an average, it gives 42 and 58% higher yield than Pusa Vishesh and Pusa Do Mausami, respectively. It gives an average yield of 218 q/ha.

Seed Certification Standards

Following guidelines should be followed for seed certification of bitter gourd hybrids.

I. Field Inspection

A minimum of three inspections should be made, the first before flowering, second during flowering and fruiting stage, and the third at the mature fruit stage and prior to harvesting.

II. Field Standards

(A) General Requirements

1. Isolation: Seed fields should be isolated from contaminants shown in column 1 of the table below by the distances specified in columns 2 and 3.

Contaminant	Minimum distance (m)	
	Foundation	Certified
Fields of other varieties	1000	500
Fields of the same variety not conforming to parietal purity requirements for certification and from balsam apple (*mokha): Momordica balsamina* L.; *M. cochinchinensis* Spreng.;	1000	500

(B) Specific Requirements

Factor	Maximum permitted (%)	
	Foundation	Certified
Off types	0.10	0.20
** Objectionable weed plants	None	None

* Maximum permitted at and after flowering.

** Objectionable weeds shall be: Balsam apple *(mokha) (Momordica balsamina* L.); (*M. cochinchinensis* Spreng); Jangli Karela *(M. dioica* Roxb. ex. Wild).

III. Seed Standards

Factor	Standards for each class	
	Foundation	Certified
Pure seed (minimum)	98.0 %	98.0 %
Inert matter (maximum)	2.0 %	2.0 %
Other crop seeds (maximum)	None	None
Total weed seeds (maximum)	None	None
Objectionable weed seeds (maximum)	None	None
Other distinguishable varieties (maximum)	5/kg	10/kg
Germination (minimum)	60 %	60 %
Moisture (maximum)	7.0 %	7.0 %
For vapour-proof containers (maximum)	6.0 %	6.0 %

Bottle Gourd

Genetic Resources

Bottle gourd is an important cucurbitaceous vegetable. It is a monoecious, annual having vine with long ribbed stem and strong tendrils. There are two subspecies of *Lagenaria siceraria*, one domesticated in India [ssp. *asiatica* (Kob.) Heiser] and the other in Africa and New World (ssp. *siceraria*). *Lagenaria leucantha* has a bitter character, which is governed by 'Bi' gene. Bottle gourd is distributed in northern plains including Tarai region. In Baster district of Chattisgarh, maximum variability has been recorded for its fruit shape and size. It appears that *Lagenaria* species occur in wild form of South America and India. Winter fruited type germplasm lines are very promising and able to withstand extreme winter. Germplasm collection areas for winter fruited types are eastern Uttar Pradesh, north Bihar, central West Bengal, Orissa, Rajasthan, Punjab, Haryana, Karnataka, Kerala and NEH region like Arunchal Pradesh, Tripura, Sikkim, Meghalaya, Nagaland and Mizoram.The NBPGR, New Delhi, is the nodal organization for collecting and maintaining the genetic resources of bottle gourd. This centre has collected more than 182 bottle gourd lines from north-western parts of Bastar district of Chattisgarh. The IIVR, Varanasi, has more than 68 collections; Jawaharlal Nehru Krishi Vishawavidyalaya, Jabalpur has, more than 77 lines, Vivekananda Parvatiya Krishi Anusandhan Shala, Almora, Uttaranchal has more than 30 land/ primitive races, IARI, New Delhi has more than 102 lines and these lines are being maintained as active collection centre. At NBPGR, New Delhi, bottle gourd germplasms have been evaluated during both summer and rainy seasons for 37 characters. Accessions U-10-315, U-11-10, U-8-217, U-8-46, U-8-198, BDJ-864 have been found high-yielding in summer crop, whereas U-11-216A, U-11-50, BDJ-53, BDJ-638, BDJ-783, BDJ-896, EC-18747 and BDJ-616 have been found high-yielding in *kharif* season. At IIVR, Varanasi 11 indigenous collection and 14 local races/varieties were evaluated for various traits. The promising germplosms for different traits are given below:

- *For high-yield:* U10-316, IC-92363, VRBG-2, VRBG-4, NDBG-10, NDBG-403, NDBG-512, NDBG-523, NDBG-524, NDBG-528, IABG-43, 42, 25, 26, VR-177, 178, 182, 184, 180, RHRBG-4,10, 33, 12, NDBG-601, NDBG-509-3-2, NDBG-619, NDBG-129, NDBG-140, NDBG-208, Big Round, Sel-1-1, IC-35286, IC-92429 and ND BG-202.
- *For 'off'- season cultivation:* NDBG-56, NDBG-10, NDBG-104, IC-92462, IC-92363A and U10-316.
- *For resistance to red pumpkin beetle:* U 10-316 and NIC-1225.
- *For resistance to leaf minor:* U-10-316, IC-92362 and IC-92418.
- *For resistance to downy mildew:* U-10-316, IC-92362, Laxmipur and Patna-1.

Besides, Pusa Summer Prolific Long, Pusa Summer Prolific Round, Arka Bahar, Pusa Naveen, NDBG-1, Punjab Komal and KBG-13 can also be used for developing hybrids. Some cultivars, White Surat, Hot Season, Long White, Garmiwali Safed Lambee, Doodhi Long and All Season Long, are very popular local land races in Madhya Pradesh. These germplasms can be utilized for developing hybrids. The lines PI- 271353 (resistant to viruses), Doodhi Long Green and Three Feet Long Green (moderately resistant to five isolates of *Sphaerotheca fuliginea)* and Taiwan variety Renshi (highly resistant to *Fusarium* wilt) have also been found good donors for development of disease resistant hybrids.

Production Objectives

Higher early and total yields, desirable fruit shape (straight long fruits having blunt end), size (40-50- cm in length), light green in colour, soft skin with tender flesh and resistance to common diseases and pests are the major objectives of heterosis breeding programme in bottle gourd.

Botany and Floral Biology

It is annual, tendril-bearing and disagreeably smelling climber. Stems are robust, obtusely quinquangular, deeply longitudinally furrowed, light green, densely clothed with soft, woody, gland-tipped hairs and 5-10 m long. Tendrils are inserted beside the leaves, on robust stalks of 3-6 cm long, cleft, densely clothed with soft, patent, white, gland-tipped hairs and 15-20 cm long. Leaves are distichous, long petioled, simple, in outline broadly cordate or reniform, base widely cordate, apex obtuse or rounded, suddenly caudate; faintly 5-7 lobed or undulate, patently acutely dentate, above dark green, beneath light green, on both sides dull, on both surfaces and along the edges densely clothed with soft, white, gland-tipped hairs, 5-7 nerved or pedately 9-15 nerved, with white transparent veins and 10-40 cm diameter. Petiole is robust, subfarctate or narrowly fistular, subterete, longitudinally furrowed, with a narrow groove on the upper side, densely clothed with patent, soft, white, gland-tipped hairs, light green, at the apex with two dark green, knob-shaped glands and 5-30 cm long. Stipule is absent.

Flowers are large, unisexual-monoecious, axillary, the two sexes on different branches. Flowers of both sexes are solitary and 7-10 cm in diameter. Male flowers are very long stalked, faintly fragrant and white. Peduncle is terete, yellowish green, densely clothed with soft, patent, gland-tipped hairs; longitudinally furrowed, obtusely quiquangular, jointed at the top, 15-25 cm long; receptacle campanulate or cupular, whitish green or yellowish green, longitudinally veined, on both sides densely clothed with white, gland-tipped hairs and 1.5-3 cm long. Sepals are 5, widely patent or obliquely erect narrowly triangular or lanceolate, narrowed towards the apex acute, green, 1-1.5 cm long and 0.2-0.3 cm wide. Petals are five, patent, broadly obovate, emarginate, cuspidate, undulate-crenulate, 5-nerved, densely wooly on both sides, white with 5 yellowish green, swollen nerves and 2.5-4 cm long. Filaments are 3, short and robust stalks. Stalk is terete, light green, densely clothed with wooly, white, gland-tipped hairs, not joined, lengthening after the anthesis and 2-3 cm long. Calyx and corolla are as in the male flowers. Ovary is inferior, ellipsoid, densely clothed with soft, wolly, gland-tipped hairs, light green or yellowish green and 2.5-3 cm long. Style short, stigma 3, thick, clavate or curved. Fruits are long peduncled, big, fleshy-watery, very variable as to shape and size, depressed globose, oblong-ellipsoid, pear-shaped or clavate, light green or whitish green, often with scattered grey or white blotches, yellowish when ripe, glabrous, many-seeded, 20-100 cm long and 10-15 cm diameter. Seeds are oblong-obovate or oblong-triangular, truncate or emarginate at the apex, white, and 0.7-2 cm long.

Flowers open at night. Being a monoecious crop, bottle gourd is strictly cross-pollinated. Bees are the pollinators. The time of anthesis in bottle gourd is in between 5 and 8 p.m. under Punjab conditions. Anthesis is preceded by anther dehiscence. Stigma is receptive 36 hours before anthesis. Both the male and female flowers open at the same time. Flowers remain open only for a few hours, after which the petals of all of flowers withered, thus the flowers are short-lived. In experimental studies on pollen germination, the maximum germination occurs in 7% sucrose solution. The pollen germination is accomplished in about half to 2 hours. After about 24 hours of pollination, the petals withered, the stigma lost its shine and become brownish and increased by about 3 mm in length, showing starting of fruit formation.

Method of Hybrid Seed Production

Like other cucurbits, in bottle gourd, hybrid seed production has been simplified by manipulation of sex mechanism and sex expression. Sex expression in cucurbits including bottle gourd is under genetic as well as environmental influences. Three of the primary variables of the plant environment, *i.e.* mineral nutrition, light and temperature regimes. Moreover, their effects are reasonably constant (Nitch *et al.,* 1952). In general: (i) high levels of nitrogen nutrition promote femaleness and depress maleness, (ii) short day treatment promotes femaleness and depresses maleness, low light intensity during growth may also inhibit staminal development, and (iii) low temperature promotes female sex expression and depress maleness. Techniques have been evaluated to reduce the cost of hybrid seed production, *viz*; use of male sterile lines, self-incompatible but cross-compatible lines, gynoecious and use of plant growth regulators. In case of monoecious cucurbits like bottle gourd, special attention is given to standardize techniques for the suppression of male flowers. Various techniques involved for hybrid seed production of bottle gourd are given below:

Pinching of Male Flowers

Production of hybrid seeds by this method is the most economical and practical way, which can easily be adopted by growers. F_1 hybrid seed production in bottle gourd can economically be done on large-scale by pinching all the male flowers before opening from the female parent and allowing the male parent to grow side-by-side of female parent for natural cross-pollination. For this, one row of female parent as suggested by Choudhury and Singh (1971) for producing F_1 hybrid seed on large-scale in bottle gourd. All the fruit setting in female parent would be necessarily through cross-pollination by insects. The precaution is taken that leave a single male bud in female parent. Because, it will promote self or sib-pollination within a female parent. The flowers of bottle gourd open in the afternoon, the pinching operation should, therefore, be done in the forenoon. There should not be any other variety except the two parental lines in a distance of about 400 m. The F_1

seeds can be extracted after harvesting the fruits from female parental line. Since, the male or staminate flowers in bottle gourd are quite big, showy, having long pedicels and less in number, the pinching operation can easily be performed. For maximum fruit setting and yield availability of pollinators are prerequisite. One medium-sized bee colony per hectare would be enough in seed production plots.

Chemical Sex Expression

There is another method of hybrid seed production by the use of chemicals. With the advent of growth regulating substances, very significant results have been obtained with regard to sex modification in bottle gourd. It has now been possible to prove that two true-leaf-stage is the most responsive for application of sex-modification. In bottle gourd, F_1 seeds can be produced by application of ethrel (2-chloroethyl-phosphonic acid) at the rate of 200-300 ppm at two true-leaf and four true-leaf stages. Another application is useful at flowering time. Ethrel helps in suppressing the staminate flowers and initiating pistillate flowers successfully in the first few flowering node on female parent. The row of male parent is grown by side of female parent and allows natural cross-pollination. In the absence of insect pollinators, hand pollination is possible when the two sexes are separate. The female flowers of the female parent and male flowers of the male parent are bagged separately prior to opening and hand-pollination brought about later. Precaution should be taken during natural cross-pollination, there should not be any other variety except the parents of hybrid. Four to five fruits set at initial nodes containing hybrid seed would be sufficient.

Use of Male Sterile Line

Dutta (1983) reported that in bottle gourd, genic male sterile line is controlled by ms-1. This can be used for the production of hybrid seed. Till date none has developed hybrids using male sterile lines in bottle gourd.

National Released Hybrids

Pusa Meghdut

This hybrid is cross between Pusa Summer Prolific Long x Selection-2. It has been released by IARI, New Delhi. It is an early, suited for both spring and summer seasons. Fruits are long, light green, tender and attractive and high-yielding.

Pusa Manjari

It is cross of Pusa Summer Prolific Round and Selection-11. It has released by IARI, New Delhi. Fruits are round, green, tender and attractive.

Pusa Hybrid-3

This hybrid as developed at IARI, New Delhi, its plants are medium long with medium internodes. Leaves are medium in size, green and softly pubescent. Fruits are attractively green, long and straight without any neck, weighing on an average 9 kg each. The fruits are highly suitable for easy packaging in cardboard boxes for distant market. This hybrid can successfully grow in both the seasons in northern plains. It is very early in maturity, becomes ready for picking in 60, and 50-55 days with an average yield of 425 and 470 q/ha yield in spring summer and *kharif* seasons, respectively.

Seed Certification Standards

Following guidelines should be followed for seed certification of hybrids.

I. Field Inspection

A minimum of three inspections should be done, the first before flowering and fruiting stage, and third at the mature fruit stage and prior to harvesting.

II. Field Standards

(A) General Requirements

1. Isolation: Seed fields should be isolated from the contaminants shown in column 1 of the table below by the distances specified in columns 2 and 3.

Contaminant	Minimum distance (m)	
	Foundation	Certified
1	2	3
Fields of other varieties	1000	500
Fields of the same variety not conforming to varietal purity requirements for certification	1000	500

(B) Specific Requirements

Factor	Maximum permitted (%)	
	Foundation	Certifed
Off types	0.10	0.20

*Maximum permitted at and after flowering.

III. Seed Standards

Factor	Standards for each class	
	Foundation	Certified
Pure seed (minimum)	98.0 %	98.0 %
Inert matter (maximum)	2.0 %	2.0 %
Other crop seeds (maximum)	None	None
Weed seeds (maximum)	None	None
Germination (minimum)	60 %	60 %
Moisture (maximum)	7.0 %	7.0 %
For vapour-proof containers (maximum)	6.0 %	6.0 %

Cucumber

Genetic Resources

Cucumber is cultivated all over South Asian region for commercial purpose. It is actually a warm season crop but these days it is being cultivated at higher altitudes under controlled conditions. It is native of India. *Cucumis hardwickii*, *C. anguria*, *C. callosus* and *C. metuliferus* are important wild relatives. In India, under All India Coordinated Vegetable Improvement Project, Vellanikara (Kerala), Katrain (Himachal Pradesh), Solan (Himachal Pradesh), Sabour (Bihar), IARI (New Delhi), Dapoli (Maharashtra), Ludhiana (Punjab) are the centres where cucumber improvement work is currently being undertaken. Kalyani (West Bengal) and Rahuri (Maharashtra) centres are engaged in collection and evaluation of cucumber germplasms. A large number of lines have been collected from these centres.The IARI, New Delhi obtained 11 lines from NBPGR, New Delhi, which possess male sterility genes (ms-1 and ms-2), white fruit (W) and Fusarium resistant genes (from 1 & 3). The promising germplasms of cucumber for different traits are given in Table1.

Cucumber apparently domesticated in India and wild species *C. hardwickii*, likely progenitor of cultivated cucumber, is found growing in natural habitants in foothill of the Himalayas. *C. sativus* var. *sativus* has 3-5 lobed leaves, obscurely trigonus or cylindric. This variety is cultivated all over India in tropical and subtropical conditions. However, *C. sativus* var. *sikkimensis* has 7-9 lobed leaves, ovary often 5 placentiferous, fruits ovoid, oblong, adapted in temperate and humid climate. *C. hardwickii* is adapted to cold climate and distributed in natural habitats in western Himalayas. The variability existing in cucumber is given in (Fig.1, see Plate-3).

Table 1: Promising germplasms of cucumber

Trait	Promising accession (s)
High yield	Chamba Local-11, EC-27080, VRC-7, Dhaneswar-2, Kheera Bharasati, VRC-11, FLCU-4, Patna-3, CH-124, CH-123, CH-136, HAC-71, HAC-31, HAC-26, UHF- Sel, LC-1, Ranch-9, VRC-7, Bihar-10, IC-203938, RHRC-2, RHRC-1 and Market Long.
Extended shelf life (Four months)	IC-203838 and 203839.
Early determinate	EC-398030.
Gynoecious line	EC-382737, 329300 and 382739.
Powdery mildew resistant	Poinsett, Yomaki, Sarton and Saladpi-197088.
Watermelon mosaic virus (WMV-1 and WMV-2) resistant	Tablegreen and Surinam.
Bacterial wilt resistant	PI-200815, PI-200818, 79-48 and 79-42-24
Multiple disease resistant	EC-320526.
Anthracnose resistant	PI-197087.
Angular leaf spot resistant	PI-200815, PI-169400, MSU-9402 and GY-14A.
Cucumber green mottle mosaic virus	*Cucumis anguria*.
Cucumber mosaic	TMG-1.

Production Objectives

Cucumber is a cross-pollinated crop. Bees are the main pollinating agents. The female flowers develop later than the more numerous male flowers and have a ring of nectiferous tissue surrounding the style. The main sex forms are gynoecious, hermaphrodite and monoecious, which have been greatly used in developing single and triple F_1 hybrids. The following traits should be considered for developing hybrids:

- To develop hybrids at early node which been first pistillate (or hermaphrodite) flower that gives an indication of early maturity.
- To develop hybrids having high female to male sex ratio especially in monoecious plant types.
- To develop hybrids having attractive green or dark green fruits with smooth fruit surface and without prominent spines or prickles.
- To develop uniform long cylindrical shape without crook neck.
- To develop fruits free from carpel separation showing hollow spots.
- To develop fruits free from bitterness.
- To develop hybrids, which do not produce mature seeds at edible maturity.
- To develop resistance to diseases and insect-pest.
- To develop hybrids for canning purposes.

Botany and Floral Biology

Plant is annual, tendril bearing, and creeping or climbing, sparingly branched and scabrous. Stem is robust, obtusely quinquangular, densely hispidulous, yellowish green and 50-250 cm long. Tendrils are inserted beside the leaves, robust, patent, peduncled, simple, hispid, yellowish green and 10-30 cm long. Leaves are distichous, long petioled, simple, in outline broadly ovate and angular. Base is deeply cordate. Apex is acuminate acute, distinctly 3-7 lobed, with unequal glandular teeth, hispid along the margins, above light or dark green, beneath pale green or yellowish green, dull on both surfaces, densely patently hispid, very scabrid, 5-7 nerved, the lowest main nerves branched on the outer side, lateral nerves white, transculent 7-18 cm long and 7-15 cm wide. Lobes are triangular, acuminate, acute and dentate. Petiole is robust, bractate, furrowed on upper side, clothed with numerous, patent, very stiff hairs, yellowish green and 3-20 cm long. Flowers are axillary, shortly stalked, unisexual monoecious, rather large, densely hispid, 3-4 cm diameter, for the greater part male, female ones solitary, and male ones in fascicles of 3-7. Stalk of male flowers is thin, lengthening after anthesis, clothed with patent hairs with much swollen and ovoid base, 0.3-0.5 cm long and thick. Receptacle is campanulate or tubular, 0.8-1.0 cm high and densely tomentose. Calyx is segmented on margin of receptacle, 5-partite more than half-way down, with widely patent segments, vitalline or sulphureous, paler on the outside, with numerous longitudinal ribs or nerves, which are clothed with scattered long, stiff hairs, between the nerves densely and very shortly hispidulous, wrinkled on inside, densely clothed with minute gland-tipped hairs and 2-3 cm diameter.

Segments are ovate-oblong, lanceolate or obovate, acute, veined, 1.5-2 cm long and 1-1.5 cm wide. Male flowers are broadly ovoid, yellowish white, 0.5-0.7 cm long column of anthers, which are supported by 2-3 very short, free and white filaments. Female flowers are with an ellipsoid or fusiform, pruinose ovary, clothed with scattered hairs with swollen, ovoid or wart-like bases. Style is simple and stigmas 3-5 and thick. Fruits are pendulous on robust stalks, very variable in shape and size, usually oblong, lanceolate or linear-cylindrical, small or large, straight or curved, at first green or whitish green, pruinose, with 5-10 more or less conspicuous longitudinal, white or yellowish white streaks, especially towards the apex and often with scattered spinous tubercles and warts, sordidly yellow or sordidly orange-coloured and glabrous when ripe, succulent, many-seeded, 10-20 cm long or longer and 3-7 cm diameter.

Flesh is pale green. Fruit-stalk is much lengthened, robust, 2-5 cm long and hispid. Fruits of hermaphroditic flowers, *i.e.* of perigynous flowers are more round than fruits of epigynous flowers. Staminate, pistillate & hermaphrodite flowers can easily be distinguished from each other visually. Their pedicels can recognize the staminate flowers. Since hermaphroditic flowers start afternoon for anthesis as well as pistil, their column at the top of the ovary is enlarged in comparison with pistillate flowers and can therefore, easily be distinguished from them. Seeds are flat, ovate-oblong, and rather acute at ends, sharply margined, 0.8-1 cm long and 0.3-0.5 cm broad. The whole developmental process from the initial bud stage to the stage when the flower is detached from the pedicel is divided into 8 stages (stage 1 to 9).

The sunrise and sunset mainly influence opening and closing of male flowers, which is by light and the time of the day. Anther in all varieties dehisced between temperature range of 20.5° C and 21.5^0 C. Pollen fertility is considerable up to noon and by afternoon (2 pm), fertility is greatly reduced and is found negligible by the evening. Stigmatic receptivity is of very short duration and pollination is carried out within two hours after anthesis. Rise in temperature causes early drying of stigmatic secretion. Different floral abnormalities like mixed inflorescence, hermaphroditism, fusion, dimorphic female flowers, reduction and increase in floral parts have been observed. A greater tendency has observed in exotic collections towards these abnormalities.

Methods of Hybrid Seed Production

Following methods may be used for hybrid seed production.

Artificial Pollination Alone

Since cucumber plants are monoecious, emasculation is not required. The female flowers of seed-parent (female parent) can 6^e artificially pollinated with pollen collected from male flowers of the pollen parent. A single pollination produces a large number of seeds and therefore, the cost of production of hybrid is not high. In this method, following techniques are used for hybrid seed production.

(1) Training and Pruning Operation for Hybrid Seed Production

Cucumber varieties can be classified into two groups based upon their fruit bearing and branching pattern as:

(a) Single Stem Type

- Train plants vertically using a suitable trellies system.
- Remove young fruits and branches on lower part of the plant (up to 5th node).
- Start pollinating the female flower of 8-10 nodes on the stem and continue the pollination operation up to 15th-16th nodes. Allow 5-6 fruits to set on the main stem.
- Remove all female flowers after 16th node.

(b) Branching Type

- Train the plants vertically using a suitable support.
- Remove all young fruits and branches on lower part of stem (up to 5th node).
- Start pollination the female flower on the branch emerging from 6th nodes onwards.
- Pinch off the tip of each branch after pollination is over.
- Pollinate female flower on the main stem of 9-10th nodes onwards.
- Set 6-7 fruits/plant and then stop pollination operation.
- Do not set fruit lower than 30 cm above the ground, because the extra early set fruits contain less seed, tend to touch the ground and rot. Fruit setting occurs on extremely upper nodes also produced poor seed.

(2) Pollination Operation

- Bag or tie the male and female flowers on male and female lines, respectively, a day prior to anthesis.
- Collect male flowers next day morning (5 am).
- Start the pollination work at 7.0 am and complete the operation latest by 9.30 am.
- Pollination may continue for a period of 10-15 days.
- Pollination operation requires 20-25/ labourers/acre/day, depending upon the plant population.

(3) Harvesting, Curing and Seed Extraction

Harvest the fruit 35-50 days after pollination, depending upon the variety. The success of controlled pollination may be enhanced by removal of any previously set fruit as first fertilized flower inhibits the development of subsequent fruits. Therefore, controlled pollination should be done as soon as possible after flowering begins. Cure the fruits under shade in a cool and dry place for a period of 7-10 days. Extract the seeds by fermenting the seed jelly for 24 hours. Wash the seeds with plenty of clean water. Dry the seeds under sunlight during the first day. Subsequently, seeds should be dried under shade for 3-4 days. Use nylon mesh wooden trays for drying the seeds.

(4) Seed Yield

About 300-350 kg/ha (1:4 male: female plant population, total plant population = 19,000 plants/ha).

Removal of Male Buds and Use of Insect Pollinators

In this method, male buds in monoecious parental line are removed before they open and only female flowers are left on the plant for pollination by insects in nature. The seeds is collected from female parental line only. Usually, the rows of female and pollen parent's lines are planted alternatively in an isolated field. In this method, insect and natural pollen vector proof greenhouses can also used for hybrid seed production. In this case, carefully dabbing anthers taken from fully opened staminate or hermaphrodite flowers to the stigma of an opening pistillate flower, using tweezers carry out crossings. An anther can also be carefully inserted into the opening in the middle of the stigma and left there. Tweezers need not be used if the petals of a staminate flower of male parents are removed, the flower held by the pedicel, the thumb and forefinger, and carefully to the stigma of the pistillate flower. If the greenhouse cannot be isolated to natural pollen vector or outdoor plantings have to be crossed, the pistillate flowers to be pollinated are closed with a small plastic bag or wire tie and thus protected against unwanted pollen transfer about 1-2 days prior to opening. The bag or wire is removed for pollination. However, it is advisable to replace it to protect the pollinated flower for another 3-4 days. The production of selfed seed in monoecious and hermaphrodite plant presents no difficulties. Although, self-pollination is possible in hermaphroditic flowers without artificial pollination, hand-pollination is recommended to increase seed yield after selfing.

Use of Gynomonoecious Lines

Peterson and Weigle (1958) advocated the use of gynomonoecious line for the production of hybrid seeds. They reported that Korean variety Shogoin produces same gynoecious plants. This variety was crossed and then backcrossed with a pickling variety (Wisconsin SMR 18) to develop gynomonoecious line, which produces a high percentage of gynoecious plant (Peterson, 1960). The percentages of gynoecious plants in different gynomonoecious lines have been found to be variable. The gynomonoecious and monoecious (pollen parent) lines are planted alternately and at the 10 node stage, all the monoecious and other intermediate-sex form type are removed from 'gynomonoecious' line and hybrid seed is collected from gynoecious plants which are left for seed production. The monoecious and other intermediate plants from 'gynomonoecious' line must be removed at an early stage, preferably before the pistillate buds on gynoecious plants open.

Use of Gynoecious Plants

Peterson and Andher (1960) and Peterson (1960) suggested using homozygous gynoecious line for production of hybrid seeds. The gynoecious and monoecious plants, as mentioned above, are produced in the progeny of 'gynomonoecious' line. Inducing formation of male flowers on gynoecious plants with the help of gibberellic acid can develop homozygous gynoecious line. According to Peterson and Andher (1960), spraying of GA_3 (2,000 ppm) at first true-leaf stage and again 14 days later, induced the production of about 10.9 male flowers / 100 nodes. The gynoecious plants can be selfed using the pollen from induced male flowers and thus maintained in a homozygous condition. The homozygous gynoecious line and pollen parent line can be planted alternatively, and hybrid seeds are collected from the gynoecious plants.

In this method, there are two stages involved in hybrid seed production of cucumber (Fig. 2). The first stage is breeding of female line having strong genetic female tendency. The genetic constitution of female or gynoecious line is M-, *st/st* (or *MMFF*). The Japanese cucumber variety 'Shogoin' is a highly gynoecious line and it has been crossed with other monoecious cultivars (*MMff* or *MMst⁺st⁺*). The female (gynoecious) segregants are backcrossed to suitable cultivars for a few generations and finally selfed to produce a homozygous gynoecious line. Treatment with GA_3 or GA_4 or GA_7 induces male flowers in female plants, which can be used for selfing. GA_4 or GA_7 is better than GA_3 (Pike and Peterson, 1969), and later silver nitrate (Tolla and Peterson, 1979) and silver thiosulphate @ 5,000 or 2,000 ppm (Nijs and Den Visser, 1979) has been found better than gibberellic acid (GA_3).

The second stage is hybrid seed production and multiplication of female lines. The hybrid seed is produced in an isolated field in which generally three rows of female (gynoecious) or seed-parent line is alternated with one row of pollinator line (monoecious). The hybrid seed is collected from female lines. The female line is propagated in another isolated field, in which one row out of three or four is sprayed with GA_3, silver nitrate or silver thiosulphate to induce male flowers. The natural sib and self-pollinated seeds are obtained for growing the female lines in hybrid seed production programme. The seeds of pollinator plants (monoecious) are obtained from hybrid seed production. In hybrid-seed production field, it is necessary to keep at least four beehives / ha to assure cross-pollination, because monoecious pollinator line has often poor pollen production.

Another method has been proposed by Kubicki (1974) for producing gynoecious seed parent. The gynoecious line (*MMFF*) is crossed with an andromonoecious line (*mmff*) and after several generations of backcrossing, hermaphrodite type (*mmff*) is stabilized. As a

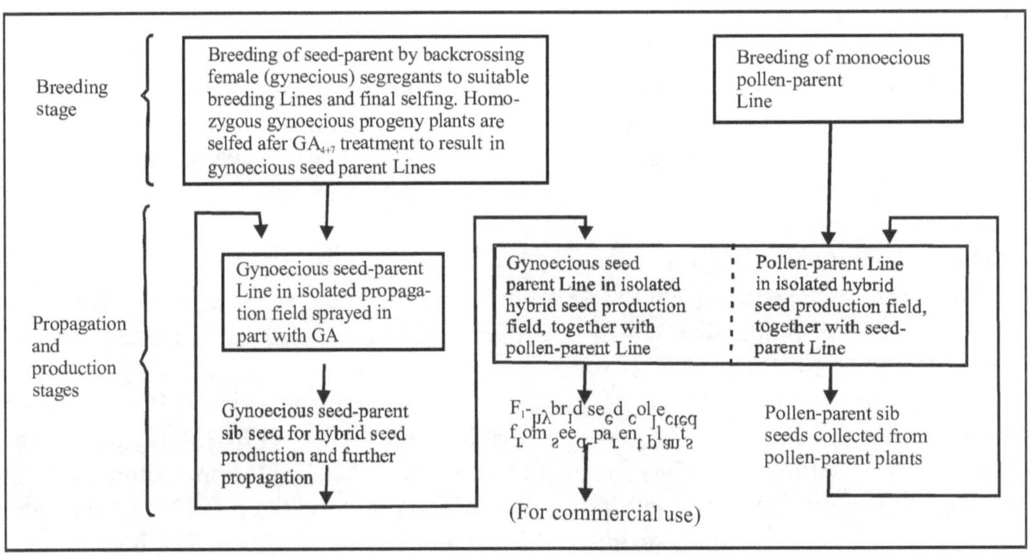

Figure 2: F_1 Hybrid seed production in cucumber based on gynoecious (female) line as seed parent (*Source*: Kalloo, 1994, p. 115).

result of backcrossing to recurrent gynoecious line, the hermaphrodite line developed will be similar (isogenic) to gynoecious line. On crossing, these two lines, gynoecious and hermaphrodite, 100 % gynoecious plants can be obtained to utilize for hybrid-seed production. This method is quite effective in production of female line without the use of gibberellic acid.

National Released Hybrids

KTCH-8

A cross of Poinsette and LC-3, it has been developed at IARI Regional Station, Katrain. Fruits are attractive, dark green, 25-30 cm long, sweet, and cylindrical with crispy flesh and white spines. Veins are medium-long. The average fruit weight is 250-300 g. It gives an average yield of 182 q/ha in 46 days of crop duration, which is 43.57 % more than Pusa Sanyog.

KTCH-11

This is a gynoecious line based prolific-bearing hybrid, developed at IARI Regional Station, Katrain. Fruits are attractive, 25-28 cm long, cylindrical, yellowish light green (local preference of Katrain region), smooth, sweet with brown spines and crispy flesh. It gives an average yield of 186 q/ha in 50-55 days of crop duration.

PH-9

This is a gynoecious line based hybrid. This is a cross between GYC-2 and Phule Shubhangi, developed at MPKV, Rahuri. Fruits are green and average weight of 175 g. Each plant bears 14 fruits. An average yield is 490 q/ha.

PH-18

This is gynoecious based hybrid. between GYC-4 and Poona Khira developed at MPKV, Rahuri. Fruits are yellow with average weight of 146 g. Each plant bears 11 fruits.

Phule Priyanka

A cross of RHRB-5 and RHRBG-4, it has been developed at MPKV, Rahuri. Fruits are dark green, highly prickled surface and 20 cm long. This hybrid is suitable for cultivation under rainy and summer seasons. This hybrid is tolerant to downy mildew. It gives an average yield of 282.7 q/ha.

Pusa Sanyog

This hybrid is cross between Japanese gynoecious line and Green Long Naples. It is an early-maturing (50 days) and high-yielding hybrid whose fruits are 28-30 cm long, cylindrical and dark green with yellow stripes, and crisp flesh. It was released by IARI Regional Station, Katrain (Gill *et al.,* 1973).

Seed Certification Standards

I. Field Inspection

A minimum of three inspections should be done, the first before flowering, the second during flowering and fruiting stage and the third at the mature fruit stage and prior to harvesting.

II. Field Standards

General Requirements

Isolation: Cucumber seed fields should be isolated from the contaminants shown in column 1 of the table below by the distances specified in columns 2 and 3.

Contaminant	Minimum distance (m)	
	Foundation	Certified
1	2	3
Fields of other varieties	1000	500
Fields of the same variety not conforming to varietal purity requirements for certification and from *Cucumis hardwickii* Royle	1000	500

Specific Requirement

Factor	Maximum permitted (%)	
	Foundation	Certifed
Off types	0.10	0.20
Objectionable weed plants**	None	None

*Maximum permitted at and after flowering.

** Objectionable weed shall be *Cucumis hardwickii* Royle.

III. Seed Standards

Factor	Standards for each class	
	Foundation	Certified
Pure seed (minimum)	98.0 %	98.0 %
Inert matter (maximum)	2.0 %	2.0 %
Other crop seeds (maximum)	5/kg	10/kg
Total weed seeds (maximum)	None	None
Objectionable Weed seeds (maximum)	None	None
Germination (minimum)	60 %	60 %
Moisture (maximum)	7.0 %	7.0 %
For vapour-proof containers (maximum)	6.0 %	6.0 %

Muskmelon

Genetic Resources

In muskmelon, there is a wide variability of desert and non-desert kinds (Fig. 3, see Plate-4), some of which are used like cucumber, *viz.* 'Vellarikkai' of Tamil Nadu, 'Nakkadosa Kaya' of Andhra Pradesh and 'Kakri' or long melon of north India. The muskmelon species are highly polymorphic. An African Botanist (1859) divided species into nine botanical varieties, *viz. agrestis*, *cultura*, *reticulatus* (cantaloupes of USA), *cantaloupensis*, *inodorus*, *flexous* (syn. *utilissimus* of India), *conomon*, *chito* and *dudain*. However, this trinomial nomenclature is not accepted by taxonomists because of their free inter-crossability. There is '*snapmelon*' called 'phoot' in north India (*C. melon* var. *momordica)* that has been found to resistant virus diseases like ZYMV (Pitrat and Lecoqe, 1984), WMV (Moyer *et al*., 1985) and aphids (Bohn *et al*., 1972). This snapmelon has been used for breeding resistance to downy mildew in muskmelon in Punjab. A wild melon (*Cucumis melo* var. *callosus,* 2n=24) found as weed in semi-arid region of Tamil Nadu, possesses desirable genes for resistance to fruit fly and leaf caterpillar and fusarium wilt. It is characterized by earliness and multiple fruiting habit. In Africa, there is a large distribution of wild species of *C. melo*, consisting of annuals and perennials, monoecious, dioecious and polyploids. These do not cross readily with *C. melo* and several disease resistant genes located in these wild species could not be yet transferred to *C. africanus* (green mottle virus and bean spider mite), *C. anguria* var. *longipes* (green mottle virus, bean spider mite and some nematodes), *C. zeyheri* (green mottle virus and nematodes), *C. asper* (green mottle virus and nematodes).

Some of the improved cultivar of muskmelon have been developed in India and can be used as parent for developing hybrids. A list of these cultivars is given in Table 2. There are some sources of resistance to diseases and insects that can be used as donor for development of hybrids for biotic stresses (Table 3).

Table 2: Muskmelon varieties used for development of hybrids

Variety	Characteristics
Arka Rajhans	Mid-season variety, large-fruited and tolerant to powdery mildew.
Arka Jeet	Fruits are flat and small but very sweet.
Pusa Sharbati	Moderately sweet and early variety.
Hara Madhu	Late variety, big juicy fruits and poor keeping quality.
Pusa Madhurus	Dark green stripes and rind, big fruit and poor shelf-life.
Punjab Sunehari	Early maturity and moderately sweet.
Durgapura Madhu	Small fruits and very sweet.

Table 3: Major donors of resistance to diseases and insects in muskmelon

Disease / insect	Donor
Powdery mildew	PMR-45, PMR-450. PMR-5, PMR-6, PI- 124111 and Seminole.
Downey mildew	MR I and PI- 414723.
Fusarium wilt	MR -1, PI -124111F and CM- 17187.
Cucumber mosaic virus	PI-161375.
Watermelon mosaic virus	PI-414723, MR-1, Punjab Rasila, Makua, Novogibirish & Silver World.
Fruitfly	*C. callosus*
Melon aphid	PI- 161375 and PI- 371795.
Multi disease resistance (downey mildew, powdery mildew and muskmelon mosaic)	MR-12.

Source: Ram (1997).

There are also some promising lines which can be used for development of hybrids as given in table-4.

Characters	Promising Lines
High yield	Australia-91-1, DMDR-1, T-7, Silver World IC-26372, IC-320154, IC-320167, Sarad Melon, and Australian-99-1
High TSS	PSDPM-1-7-50, IC-32057 and IC-267370

In addition, a large number of local varieties are grown in different parts of India. These varieties may be used for developing high-yielding hybrids. Some of them are Lucknow Safeda, Bhaput Melon of Merut, Jaunpuri Netted and Mau of Azamgarh (Uttar Pradesh), Tonk Melon of Rajasthan, Kharri of Hoshangabad, Goose Melon, Jam and Neil of north Maharashtra and Sharbat-e-Anar and Bathesa of Andhra Pradesh.

Production Objectives

Following objectives should be considered while developing hybrids.

- To develop early-maturing and high yielding hybrids.
- to develop attractive round/ spherical shape having small seed cavity, thick flesh, Netted and non-netted, and attractive colour.
- To develop sweet, juicy and flavoursome fruits. The sugar content in muskmelon should vary from 11 to 13 % or higher but not be less than 10 %.
- To develop hybrids resistant to diseases like powdery mildew, downey mildew and Mosaic viruses.
- To develop hybrids resistant to insects like red pumpkin beetle, fruit fly and aphids.

Botany and Floral Biology

Muskmelon belongs to genus *Cucumis*. The characteristics of genus *Cucumis* are fruit fleshy, many seeded pepo, flower solitary, lemon yellow to deep orange, leaves deep or shallowly lobed not pinnatified, corolla rotate deeply parled and small flowers. It is a polymorphic species where most of the cultivars are andromonoecious (hermaphrodite and staminate flowers), gynoecious (only pistillate flowers) or monoecious (pistillate and staminate flowers), monoecious and andromonoecious are most common. Stem is soft hairy to glabrous, striate or angled, leaves orbiculate to ovate to reinform, usually five-angled, sometimes shallowly 3 to 7 lobed, hairy or somewhat scabrous, 3-5 inches across. The staminate flowers are clustered. The pistillate flowers are solitary on stout pedicels (Whitaker and Davis, 1962). There are various horticultural forms within *C. melo* based on fruit characteristics, namely cantaloupes, nutmeg muskmelon, winter melons, white-skinned melons, snake melons, oriental pickling melons, mango melons, pomegranate melons which hybridize readily with each other. There is apparently very little sterility even among progenies from crosses involving variant types. Under open natural system, it is partly crossed and partly self-pollinated by bees but there are reports of obtaining vigorous lines by inbreeding up to 25 generations (Robinson *et al*., 1976). Muskmelon is highly cross-pollinated crop. Honeybees are pollinators. Nandpuri and Brar (1966) reported time of anthesis is 5.30-6.30 hours, dehiscence 5.00-6.00 hours, duration of pollen fertility 5.00-14.00 hours and stigma receptivity 2 hours before to 2-3 hours after anthesis.

Methods of Hybrid Seed Production

Following procedures may be adopted for producing F_1 hybrid seeds.

Artificial Pollination Alone

For developing hybrids through artificial pollination, following practices are done:

- Train the plants upright.
- Remove all the branches below the 8-10 nodes.
- Pollinate the flowers, which produced on branches emerging at 11th to 16th nodes on main stem. Retain 3-4 fruits / plant.
- Pinch off the tip of the branch after pollination.
- Pinch off the tip of the main stem after 25th node.
- Bag or tie the male and female flowers required for pollination, a day prior to anthesis and bag them.
- In case of female lines with andromonoecious sex expression, do the emasculation of the perfect flowers, a day prior to anthesis and bag them.
- Pollination continues for 15 days. It requires 25-30 labourers/acre / day, depending upon the plant population.
- Harvest the fruits at full slip stage. Those varieties, which do not show full slip, harvesting should be done when cracks develop at junction point of peduncle with the fruit. Such fruit requires curing for 7-10 days in a well-ventilated cool shade. Extract the seeds after fermenting the seed for 24 hours and washing them with clean water.

- Dry the seed in sunlight during first day and then under shade for 3-4 days using nylon mesh wooden trays.
- About 300-400 kg/ha (1:5 to 1:8 male: female plant population) can be obtained.

Removel Male Buds and Use Insect Pollination

In this method, male buds in monoecious parents are removed before they open and only female flowers are left on the plant for pollination by insects in nature. The seeds are collected from the female parental line only. Usually the rows of female and pollen parent lines are planted alternatively in a isolated field. Other procedures are the same as described in cucumber.

Use of Male Sterility

First recessive *ms* gene was reported by Bohn and Whitakar (1949), since then at least four additional male sterile recessive alleles, *viz.* ms-2 (Bhon and Principe, 1962), ms-3 (McCreight and Ebmstrom, 1984), ms-4 (Pilral, 1990) and ms-5 (Lecouviour *et al;* 1990) have been identified. The ms-1 line has been successfully utilized in India to develop first commercial hybrid Punjab Hybrid- 1 (Kumar *et al.,* 2000). The procedure for development of hybrid is same as described in tomato.

Use of Gynoecious Lines

Gynoecious lines can also be used for hybrid seed production of muskmelon. These lines are maintained by applying silver nitrate @ 100-200 ppm. Kubicki (1966) described a method in muskmelon for producing 100 % genetically pure gynoecious lines. Such lines are derived from a cross of hermaphrodite (*aagg*) with a monoecious (*AAGG*) line and isolating gynoecious (*Aagg*) hermaphrodite (*aagg*) lines. These genotypes when crossed would produce all gynoecious (*Aagg*) plants. But this approach would require a long breeding programme to complete desirable horticultural traits if necessary into these lines for commercial use.

National Released Hybrid

Punjab Hybrid-1

It has been developed at PAU, Ludhiana. Plant has long vines with medium green leaves, early in maturity, fruits set close to the base of the vine. The fruits are nearly round, sutured with medium netting and average weight of 800 g. Flesh is orange, flavoursome, juicy, sweet with 12 % TSS. Fruit develops 'full-slip' stage and rind turns light yellow at maturity. Fruit self-life is two days. It gives an average yield of 160 q/ha. It has poor transport quality. It is recommended for Punjab, Bihar, Uttar Pradesh, Karnataka, Tamil Nadu and Kerala. The female parent (MS-*1*) is characterized by medium-long vines. Leaves are dark green, deeply serrated with short petioles. Fruits are round, sutured and intensely netted. Flesh is orange, flavoursome, medium, juicy, medium sweet with 8-9 % TSS. Fruit develops 'full-slip' at maturity. Rind turns brown at maturity while, the plant of male parent (Hara Madhu) is having long vines. Leaves are light coloured with light petioles. Fruits are green, found tapering towards stalk end. Rind is smooth and turns light yellow on maturity, sutured and average weight of 1 kg. Flesh is green, juicy, flavoursome sweet with 13 % TSS. It gives an average yield of 125 q/ha.

MH-10

It has also been developed at PAU, Ludhiana. Plant has medium vine growth, early maturity and monoecious sex expression. Fruit setting takes place close to the base of the vine. Fruits are nearly round or oval, netted rind; non-sutured, thick brown on maturity and average fruit weight of 915 g. Flesh is thick, orange, medium juicy, flavoursome with 9-10% of TSS. Fruit develops 'full-slip' stage at maturity. It has good self-life and suitable for distant transportation. It gives an average yield of 240 q/ha. The female parent (W1-998) is characterized by dwarf and upright plant with stout stem and compact terminal growth. Plants bear entirely pistillate flowers. It takes 32 days to first female flowering. Fruits are round, flat, sutured rind, and sparsely netted having average weight of 495 g. It turns greenish yellow on maturity. Flesh is orange, medium, juicy, and moderately sweet with 6-7 % of TSS. Seeds are white and roundish flat. While, the plant of male parent (Punjab Sunehri) is of medium long vines and monoecious sex expression. It takes 35-40 days to first female flowering. Fruits take 75-80 days to maturity. Fruits are globular round, thick rind, non-sutured, intensely netted brownish at maturity. The fruit average weight is 700-800 g. Flesh is orange, thick and medium juicy. Fruit develops full-slip stage at maturity. It has an excellent keeping quality and suitable for distant transportation. It gives an average yield of 160 q/ha.

Seed Certification Standards

I. Field Inspection

A minimum of three inspections should be done, the first before flowering, the second during flowering and fruiting stages and the third at mature fruit stage and prior to harvesting.

II. Field Standards

(A) General requirements

Isolation: Seed fields should be isolated from the contaminants shown in column 1 of the Table below by the distances specified in columns 2 and 3.

Contaminant	**Minimum distance (m)**	
	Foundation	**Certified**
1	**2**	**3**
Fields of other varieties	1000	500
Fields of the same variety not conforming to varietal purity requirements for certification and from longmelon (Cucumis melo L. var. *utilissimus* Duth. & Full.), oriental picking melon (*Cucumis melo* L. var. *conomon* Makino); snapmelon (*Cucumis prophetarum;* weed melon (*Takmek)* (*Cucumis melo* L. var. *agrestis* Naud), and other non-dessert forms of *Cucumis melo* L., known to cross or suspected to being able to cross.	1000	500

(B) Specific Requirements

Factor	Maximum permitted (%)	
	Foundation	**Certifed**
Off types	0.10	0.20
**Objectionable weed plants	None	None
*** Plants effected by seed borne diseases	0.10	0.20

* Standards for off types and objectionable weeds shall be met at any inspection conducted at and after flowering and for seed-borne disease at final inspection.

** Objectionable weeds shall be: *Cucumis prophetarum*; weed melon (*Takmek): Cucumis melo* (L.) var. *agrestis* Naud., and other non-dessert forms of *Cucumis melo* (L.)

*** Seed borne disease shall be: Cucumber Mosaic Virus (CMV).

III. Seed Standards

Factor	Standards for each class	
	Foundation	**Certified**
Pure seed (minimum)	98.0 %	98.0 %
Inert matter (maximum)	2.0 %	2.0 %
Other crop seeds (maximum)	5/kg	10/kg
Total weed seeds (maximum)	None	None
Objectionable weed seeds (maximum)	None	None
Germination (minimum)	60 %	60 %
Moisture (maximum)	7.0 %	7.0 %
For vapour-proof containers (maximum)	6.0 %	6.0 %

Watermelon

Genetic Resources

Watermelon belongs to the genus *Citrullus*. It has two species of economic importance *viz*. watermelon (*Citrullus lanatus* (Thumb) and round gourd (*C. vulgaris* Schord var. *fistulosus*). Based on the evidence (DeCandole, 1982) and linguistic data (Filov and Vilenskaya, 1972), *Citrullus lanatus* originated in Africa and reached India in prehistoric times. Cultivation of large watermelon *C. lanatus* var. *citrodes*, is comparatively recent and Soviet grown today still have shape of their African ancestors (*C. lanatus* var. *coffer*). Shimotsuma (1963) reported that *C. vulgaris*, *C. colocynthis*, *C. ecirrhosus* and *C. nandinianus* are related and cross-compatible with each other. Whitaker (1933) considered *C. colocynthoides* is probably ancestor of watermelon and exhibit much variation in northwestern plains. It has similar characters with *vulgaris* but fruits are bitter with small seeds. Its cultivation is widespread in warmer parts, all over the country and particularly in river belts. Notable diversity has been observed in Indo-Gangetic plains like Rajasthan and Gujarat. The promising accessions of watermelon are given in table 5.

Table 5: Promising accessions of watermelon

Trait	Promising accession
Early and yellow fleshed	EC-380989 and 380992.
Red fleshed, round and small	EC-39240 and 393243.
Fusarium wilt resistant	Conqueror, Summit, Charleston Gray, Dixielee, EC-383809 and 378523.
Fusarium and gummy stem blight resistant	EC-217073 and 217074.
Anthracnose (*Colletrichum lagenarium*)	Congo, Fairfax, Charleston Gray, PI-189225, PI- 271775 and PI-271778.

Production Objectives

Following objectives should be considered while developing its hybrids:

- To develop hybrids bearing pistillate (or hermaphrodite) flower at early node, leads & earliness.
- To develop tough-skinned fruits, which are better suited for long distance areas.
- To develop fruits with 10% or higher TSS content so as to meet consumer acceptance.
- To develop fruits with fewer and smaller seeds and attractive deep red flesh.
- To develop high yielding hybrids in terms of weight and number of fruits of marketable quality.
- To develop biotic and abiotic resistant hybrids.

Botany and Floral Biology

The plant of watermelon is typically a trailing (vine) annual, but recently dwarf forms, which may be described as 'bush', have been located and are used to develop cultivars suitable for growth in a limited space. Dwarf may have a total diameter of as little as 60 cm, whereas more rampant vine cultivars may be 90 cm or more in diameter. Dwarfing is primarily related to shortened internodes (Mohr, 1956). In typical vine, there is a delay in the development of lateral branches, with a dominant single 'runner' reaching several feet in length before branching, is initiated and maintaining dominance for sometime thereafter. In some of the dwarf types, multiple branching occurs simultaneously from the crown of plant when it is quite young, which provides more potential bearing areas concentrated fruit set.

The root system of watermelon is very extensive but shallow, consisting of a tap root and many lateral roots within the top 60 cm of soil. Early destruction of the tap root (as in transplanting) may be advantageous in getting high yield. Fruit shape varies, ranging from long, cylindrical to spherical fruits with the melons of most cultivars intermediate in shape. Rind colour varies from white with various shades of green, with some cultivars having striped rind and some mottled rind. Rind ranges from brittle to tough and flexible, from very thin to excessively thick, and form very soft to very hard. Flesh texture varies from fine grained and 'melting' to firm coarse, and fibrous, seeds range in size from very small (size of tomato seed) to large size (size of pumpkin seed). Seeds usually range in colour from white to black, but red, green, brown and mottled colours also occur. The greater part of the fruit is the edible flesh, which is mostly derived from the placental, whereas cantaloupe and squash are largely pericarp. The pericarp of watermelon is its hard outer rind.

Honeybees generally pollinate watermelon. Monoecious and andromonoecious types of flowering habits are found, of which the monoecious is most common. The ratio of staminate to pistillate flowers under field conditions varies between 7:1 and 15:1. Staminate flowers usually appear earlier than pistillate ones. The first female and hermaphrodite flowers are formed on the main vine. The male flowers open first in the axil of 4th to 12th leaf and a few days later female and hermaphrodite flower open, those of early cultivars in the axil of 5th to 7th leaf and in late cultivars in the axil of 25th to 30th leaf. More female flowers are formed in spring than in autumn. The stigma of watermelon is of 'wet type' and consists of multicellular papillae invested by copious extracellular secretions (Sedgley, 1981 and 1982). Although, exudate production and secretion occur two days prior to anthesis, cells have the capacity to produce further exudates in response to pollination. Pollination of stigma at anthesis results in a rapid pollination. After 30 minutes, papillae and pollen-grains are submerged in the secretion. The anthesis of flower in watermelon can be observed from 6-7.30 am, with a peak between 6.30-7.30 am. The anther dehiscence starts one hour before anthesis and continues until 6.30 am. At high temperature stigmatic fluid starts drying and stigma becomes non-receptive (Sidhu *et al.*, 1980). Pollen from hermaphrodite flowers germinates rather slowly than that form female flowers, but is equally effective for fertilization. In watermelon crosses, higher seed set has been observed when pollination occurred in early morning immediately after flower opening. Seed setting has been found lower when pollination is carried out in the afternoon. Pollen kept at room conditions for 5 hours gave a lower seed set than freshly collected pollen and after 30 hours completely lost viability.

Methods of Hybrid Seed Production

Following procedures may be used for production of hybrids.

- Artificial pollination alone
- Removal of male buds and use of insect-pollination
- Use of marker genes
- Triploid watermelon

Artificial Pollination Alone

In this method following practices are adopted:

- Train the vines on the ground across the sowing channel and fix them on the ground 'U' shaped clips.
- Pinch off the 1st female flower as soon as it is noticed.
- Allow one or two fruits to develop in the middle portion of the plant *i.e.* between 12 to 25th nodes. Pinch off the growing tip after 28-30th nodes. For small-fruited varieties, allow four fruits to develop per plant.

- Bag the male and female flowers a day prior to anthesis and complete the pollination work next day morning latest by 9.30 am. Rebag the female flower after pollination. Do not touch the ovary at the time of pollination. Pollination continues for a period of 7-10 days. Pollination operation is done by 12-15 labourers/acre/day.
- Harvest the fruit when the tendril nearest to the set fruit is dry up the base. Ripened fruits are also judged by the dull sound produced when tagged with the knuckles. Usually, it takes 55-65 days after pollination, when the fruits are fully ripe.
- Seeds are extracted by the fermenting the pulp for 24 hrs and wash them with clean water. Dry the seeds in sunlight during the 1st day and under shade during the next 3-4 days. Use nylon mesh wooden tray for drying purpose.
- Hybrid seed yield (diploid watermelon) 100-250 kg/ha, if plant populations are maintained 8000 plant/ha (1:5 to 1:6 male to female population).
- For increasing fruit setting under controlled pollination, open pollinated fruit should be removed as there is inhibitory effect produced by the fruits already set.

Removal of Male Buds and Use of Insect-pollination

The procedures are same as for cucumber (see Page 331).

Use of Marker Genes

The non-lobing leaf character governed by a single recessive gene, which is expressed in the seedling stage, can be used as a marker gene for producing hybrid seeds as suggested by Mohr *et al.* (1955). The inbred lines non-lobed and lobed may be planted in alternate row at spacing of 1 m x 2 m and seeds collected from the non-lobed line only. The F_1 hybrid having lobed leaves can be easily distinguished in the seedling stage. Mohr *et al.* (1955) suggested sowing about 6-8 seeds per hill, as an average only one third of the plants will be F_1 hybrids. According to Whitaker and Davis (1962), this method may be quite easy and economical for production of hybrid seeds.

Triploid Watermelon

The seedless triploid watermelon is also F_1 hybrids of the cross made between tetraploid and diploid varieties. The credit for production of seedless watermelon goes to Japanese scientists who were the first to discover this possibility. Kihara (1951) gave an exceptionally clear account of various steps in production of triploid fruits and later publication by Kondo (1955) and Shimotsuma (1962) provide an analysis of several problems encountered in commercial exploitation. Japanese triploid were produced mainly from Fumin tetraploid stocks, 'Asahi 4x' and 'Ogon' stocks. Most of them are medium maturity and fruit flesh is crisp in texture in some and soft in others. The Vegetable Breeding Laboratory of USDA in Charleston (SC) in the USA has developed by Eigsti for commercial production of seeds of seedless watermelon. Tetra-2 line has also been used as cultivar 'Pusa Bedana' (Seshadri *et al.,* 1972). For production of seedless F_1 hybrids, following steps are involved (Fig. 4).

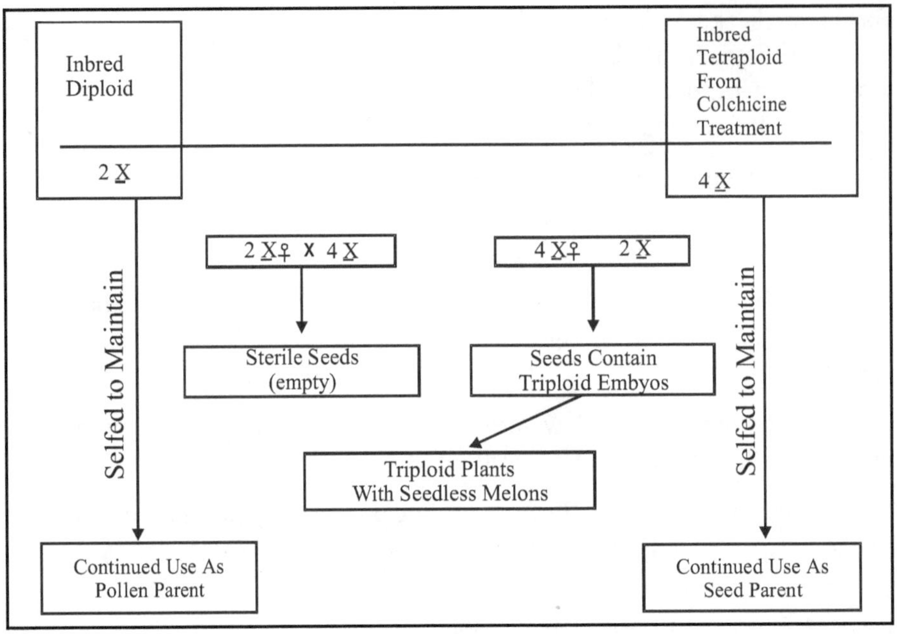

Figure 4: Procedure for producing seed of seedless triploid hybrid watermelon. The tetraploid line must be used as female parent, because the reciprocal cross using diploid as female results in empty seed. Seedless melons are the result of stimulative parthenocarpy (pollination with interplanted with seedlees hybrid. [*Source*: Breeding Vegetable Crops. Bassett M.J. (Ed.), p.46].

Production of Tetraploid Lines

Application of colchicine (0.2 % to 0.4 %) to growing points of young seedlings (at 1-2 true-leaf stages) for 2-3 days successively, gives satisfactory results. The treatment is done under controlled conditions avoiding direct sunlight. The growth of tetraploids (44 chromosomes) will be slow and terminal bud is carefully promoted to grow. Gigas characters like thick and broad leaves and bigger stomata characterize the 4x plants. When the plants start flowering, the pollen-grains are tested for fertility and size of pollen grains of 4x plants. The fruiting will be less in first generation with very few seeds, around 30-40 due to high pollen sterility. The seeds will be broad and bigger. The 4x plants are checked for chromosome number, both in somatic cells and in meiosis. There would not be 11 quadrivalents, but it will range from 4 to 10 with varying number of univalent, bivalents and trivalent.

Maintenance of tetraploid lines at stable level is important and continuous selection for improvement of quality and vigour in tetraploid lines is done for at least 4 or 5 generations. Numerous small seeds in the fruits can detect chance reversion to diploidy. Tetraploids fruits have characteristic large blossom scar. Higher fertility during continuous selection is possibly leads to the true breeding behaviour. Attainment of superior fruit quality (firm flesh, deep colour and stable TSS content of 12-13 %) in tetraploids is an important requirement for production of high quality seedless fruits. The need for caution in seed extraction of tetraploids is to be noted. Fermentation of the pulp should be avoided which will affect the seed germination. Similarly, at the time of sowing of seeds of tetraploids, overnight presoaking is not done and they will have to be germinated under controlled conditions, ensuring slow and continuous absorption of moisture.

Production of 3x Seeds

It can be done by: (i) hand-pollinating the female tetraploid by diploid male or (ii) in open fields with adjacent rows of tetraploid female and diploid male, pinching of male buds of tetraploid plants, leaving the pollination to insects or (iii) leaving the pollination and fruit setting to natural condition of insect pollination. This last method involves interference of self or sib-pollinated seeds in tetraploid lines among the cross-pollinated triploid seeds. In 1:1 planting, the percentage of 4x seeds (by sib/selfs) under open-pollinated condition comes to about 30 and the triploid and tetraploid seeds can not be distinguished. The production of 3x seeds is, therefore, expensive, because of low seed content of tetraploid lines and mixture of 4x seeds. Seedmens do not get more than 4.5-6.8 kg / acre of triploid seeds under open-pollinated conditions.

Germination of 3x Seeds

This is important because of high cost of production of 3x seeds. There is poor germination of 3x seeds because of hard seed coat. Seeds are shown in flat or peat cups with moisture, carefully maintained at a temperature of 24-29°C in first 72 hours. Naturally, germination of seedlings under controlled condition and their subsequent transplanting increase the cost of production of seedless fruits. Transplanting is done at two true leaf stage.

Production of 3x Fruits

There will be no fruit setting in triploid plants, if there are only 3x plants in the field. Even few tetraploid plants, occuring as a mixture would not be enough for good fruit setting. Stimulation of fruit development is brought about by 2x pollen through bee pollination. Hence, a pollinizer, a diploid cultivar is planted in the field. Another difficulty arises at the time of fruiting, when 2x, 3x and 4x fruits are distinguished. Hence at the time of choice of parents for crossing 4x plants with 2x male, contrasting fruit characters are chosen. If 4x plants produce fruits with light green rind and stripes and 2x male chosen, would produce fruits with light green rind, the 3x fruits will then have intermediate green rind with stripes. The 3x fruits are triangular configuration, in contrast to round fruits of 4x and 2x plants. The 3x plants are of long duration, fruits maturing in 120-130 days in contrast to 85-90 days in diploid cultivars. The firm texture of rind is attractive feature, but there may be few empty ovular cavities in fruits. Sweetness is uniformly distributed from the rind to the centre. Triploids are sterile both on male and female sides and unequal chromosomal distribution in meiosis causes pollen and ovular abortion. However, few ovules may develop into a hardened seed coat and these are called 'ex-embryonate' or 'false' seeds.

National Released Hybrids

Arka Jyoti

It has been developed at IIHR, Bangalore. Fruits are round to oval dark green with blue angular stripes and average weight of 5-6 kg. Flesh is bright crimson, granular texture with 12-13 % TSS. It has good keeping quality. First picking starts 90 days after sowing. An average fruit yield is 60-80 q/ha. The female parent (IIHR-20) is characterized by oval to round fruits, green round with deep blue angular stripes, red flesh with 11-12 % TSS. The crop duration is 90-100 days. However, male parent (Crimson Sweet) is of oval light green round fruits. A fruit has green to dark green stripes. The flesh is crimson. The crop duration is 110-115 days.

Madhu

A unique fusarium-resistant hybrid. It has been developed by Indo American Hybrid Seeds, Bangalore. This hybrid produces excellent uniformly-sized fruits. The fruits are oval in shape, medium green with green stripes outside and bright red, sweet flesh inside with small brown seeds.

Milan

Developed by Indo-American Hybrid Seeds, Bangalore. This is characterized by an early maturity and high-yielding. It produces oval, light grey green fruits with tough rind and smooth red, sweet flesh inside. The fruits stand well for shipping.

Seed Certification Standards

Following guidelines should be followed for seed certification.

I. Field Inspection

A minimum of three inspections should be done, the first before flowering, the second during flowering and fruiting stage and the third at mature fruit stage and prior to harvesting.

II. Field Standards

(A) General Requirements

Isolation: Seed fields should be isolated from the contaminants shown in column 1 of the Table below by the distances specified in columns 2 and 3.

Contaminant	Minimum distance (m)	
	Foundation	Certified
1	2	3
Fields of other varieties	1000	500
Fields of the same variety not conforming to varietal urity requirements for certification and from wild water melon (Indrayan): *Citrullus colocynthis* L.	1000	500

(B) Specific Requirements

Factor	Maximum permitted (%)	
	Foundation	Certified
Off types	0.10	0.20
** Objectionable weed plants	None	None

* Maximum permitted at and after flowering.

** Objectionable weed shall be: Wild watermelon (*Indrayan*): *Citrullus colocynthis* L.

III. Seed Standards

Factor	Standards for each class	
	Foundation	Certified
Pure seed (minimum)	98.0 %	98.0 %
Inert matter (maximum)	2.0 %	2.0 %
Other crop seeds (maximum)	None	None
Total weed seeds (maximum)	None	None
Objectionable Weed seeds (maximum)	None	None
Other distinguishable varieties (maximum)	5/kg	10/kg
Germination (minimum)	60 %	60 %
Moisture (maximum)	7.0 %	7.0 %
For vapour-proof containers (maximum)	6.0 %	6.0 %

Pumpkin and Squash

Genetic Resources

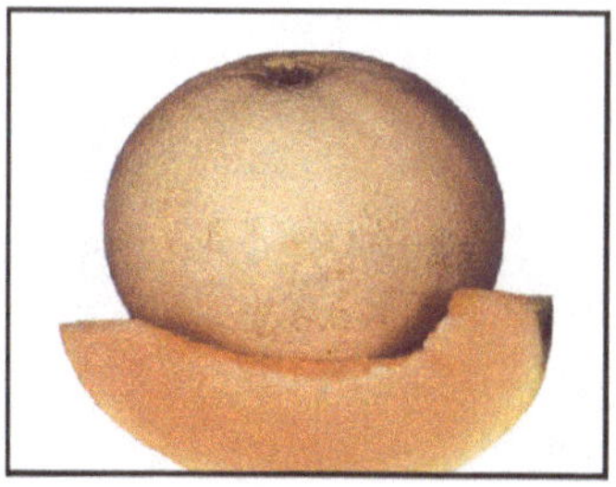

Pumpkin, also known as 'Kashiphal' or 'Sitaphal' or 'Lal Kaddu'. It is grown in different parts of India. It belongs to the genus *Cucurbita*. The genus has wide variability in respect to fruit size and colour (Fig. 5 a & b, see Plate-5). There are 26 species of *Cucurbita*, five of which are domesticated, but genetically isolated from each other. The wild species contain many eco-species, which are separated geographically but are genetically cross-compatible. The details of five cultivated species are given in table 6 (Whitaker, 1974, Whitaker and Bemis, 1976).

Table 6: Cultivated species of *cucurbita*

Species	Common Names/chief features	Distribution in pre-Colombian/times
C.pepo	Summer squash: Some varieties belong to winter squash, pumpkins, marrows, ornamental gourds, bush (bracchytic) plants, peduncle hard, sharply angular, grooved, foliage spiculate, stiff, upright, harsh and prickly to touch, leaves broad, triangular in outline, usually with deep lobes little or no expansion of peduncle at fruit atachment, although frost sensitive, has some resistance to cold.	North America from about Mexico city Northward to North-eastern US and Southern Canada, South Western US and North Western Mexico
C.mixta	Squash, cymlins, cushaws, used both as summer and winter squash, leaves large, shallowly to moderately lobed with or rarely without white blotches, non spiculate, fruits with a large, hard, corky peduncle.	From Southern Mexico to North Western Mexico and South-Western USA.
C.moschata	Winter or baking squash, field pumpkins, viny plants, tolerant to vine squash borer, foliage soft hairy, not harsh or prickly, leaves shallowly lobed, often with white spots along the veins, peduncle hard, smoothly grooved, flared at fruit attachment.	From South-Western USA through Mexico, Central America to Northern of South America (Peru, Colombia).
C.maxima	Winter or baking squash, large pumpkins, high quality, soft skinned winter squash with vigorous vines and, sometimes extremely large fruits, stem soft, round, leaves rounded in outline, not lobed, peduncle soft enlarged by soft cork, fruits globular, oblong cylindrical or flattened cylindrical.	In South America (Peru Bolivia, Chile, Northern Argentina).
C.ficifolia	Fig-leaf gourd plants, perennial, seeds black, long running vines, becoming somewhat woody, foliage prickly, leaves broad or circular ovate, usually lobed and fig like, peduncle hard, small, slightly expanded at the fruit attachment, fruit globular or cylindrical, 5-6 inches in diameter, 6-8 inches long, green with white stripes and blotches, rind hard, flesh white.	At high altitudes from Southern Mexico to Bolivia.

Source: Ram (1997).

The IIHR, Hessarghatta, Bangalore and NDUAT, Faizabad, have taken the responsibility of collection and evaluation of germplasms in all India co-ordinated trials. The NBPGR, New Delhi, evaluated 28 accessions in summer and 42 in *kharif* using Pusa Biswas as check parent. Collection, U-10-142, BDJ-1-507, UB-182, C-1750, U-9-59, UB-62 and UB-274, have been found promising. The promising lines of pumpkin for high yield are CM-50, CM-55, CM-67, CM-35, CM-34, CM-59, CM-64, CM-71, CM-74, MDPK-204, MDPK-205, MDPK-209, MDPK-210, MDPK-215, MDPK-227, MDPK-401, MDPK-402, MDPK-2409 and MDPK-2410.

Production Objectives

Following objectives should be kept in mind while developing hybrids.

- To develop early-maturing high-yielding hybrids with high female and male ratio.
- To develop hybrids with high carotenoids content.
- To develop hybrids with yellow or mottled and non-ridged fruits, which is more attractive.
- To develop hybrids with thick flesh and narrow seed cavity.
- To develop hybrids having a high flesh recovery per volume of fruit from yield point of view.
- Medium vine characteristics.
- To develop hybrids with mone no. 7 fruits or weight contributs towards higher yield expressed in terms of weight and number of fruits.
- To develop resistant hybrids against diseased pest.

Botany and Floral Biology

Plants are annual, robust, tendril-bearing climber, usually creeping over the ground, fetid when bruised. Stems are robust, thick, obtusely quiquangular or longitudinally furrowed, densely clothed with very stiff and rough hairs, densely gland-dotted, light green and 5-25 m long. Tendrils are inserted beside the leaves, on robust stalks, 3-4 fid, clothed with patent, unequal, white hairs, light or yellowish green, 10-40 cm long, inclusive the stalks of 1.5-7 cm long. Leaves are distichous, long stalked, simple, in outline orbicular-reniform or broadly ovate, base deeply and widely cordate, apex shortly acuminate, obtuse or rounded; 5-7 lobed, densely toothed; teeth obtuse, glandular at the base coarsely serrate or lobed, hispid along the margins, light or dark green and feebly shining above, light green and dull beneath, on both surfaces but especially above densely clothed with patent soft or rough hairs, beneath with scattered, pale yellow or greenish, knob-shaped glands, 3-nerved, two lateral main-nerves repeatedly branched on one side or pedately 5-9 nerved, with white transparent veins, 7-35 cm long and 10-35 cm wide. Petiole is very robust, fistular, numerous longitudinal faint grooves densely clothed with patent, very rough hairs, light or dark green, 4-30 cm long and 0.5-1 cm thick. Stipule is absent or replaced by a gland.

Flowers are peduncled, large, unisexual-monoecious, solitary in leaf-axils, for the greater part male, very hairy, dotted with many glands and 6-15 cm diameter. Peduncle is robust, light green, densely clothed with patent, unequal white, rough hairs and with gland-tipped hairs. Stalk of male flowers are longer and thinner than that of femalc ones, often subterete, 5-25 cm long and 0.5-0.7 cm thick. Stalk of female flowers is distinctly 5-angular and longitudinally furrowed, 2-7 cm long, 0.75-1 cm thick and much lengthened after the anthesis. Calyx-segments 5, obliquely patent, narrowly linear-lanceolate or spathulate, their tips foliaceous, acute or obtusely acuminate, irregularly lobed-dentate, light green, on both sides

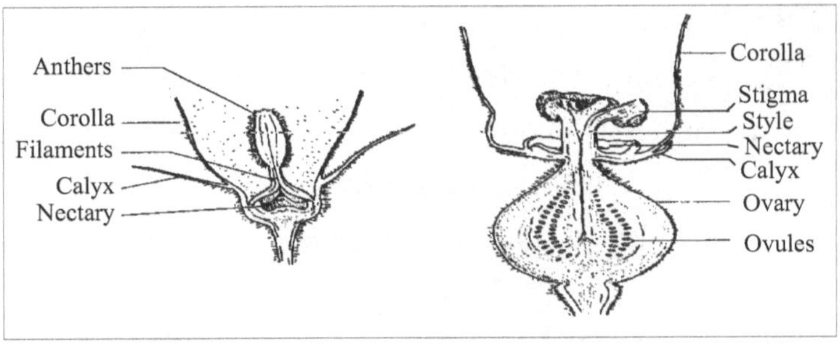

Figure 6: Typical staminate (left) and pistillate (right) flower of *cucurbita*.

clothed with patent, rough, white hairs and with gland-tipped hairs and somewhat fleshy. In male flowers, 2-5 cm long, 0.25-0.3 cm wide at the top and sometimes 0.5-1 cm wide. While, in female flowers 3-6 cm long, at the top 0.5-3 cm wide, Corolla is campanulate, 5-lobed less than half way down or about to the middle, on the outside light yellow with very prominent, yellowish green longitudinal ribs or veins, clothed with long, patent, white rough hairs, within light yellow in the tube, higher up orange-coloured or salmon pink, beset with scattered (or at the base crowded) patent, light yellow and orange-coloured long papillae or hairs. The female corolla is 7-15 cm long and 10-15 cm diameter. Lobes are broadly ovate or broadly rhomboid, acutely acuminate with light green tops, widely patent, 2.5-5 cm or in the female flowers 4-7.5 cm long and wide. Male flowers are with entirely connate stamens of 2-3 cm long. Staminal column is robust with a broad base, sulphureous and 0.5-1.25 cm long. Anthers are oblong, dorsally connate, ochreous and 1.5-2.5 cm long.

Ovary large, globose ovoid or oblong, light green or grayish green, densely irregularly globose or warty with white spots. Style is columnar, angular, at the apex 2-3 partite, yellowish white, glabrous and 1.5-2 cm long. Stigma is thick, deeply 2 lobed, ovoid or oblong, dark orange-coloured, densely papillose and 1.25-2 cm long. Receptacle and staminodes are curving upwards in the shape of a wide cup, faintly 3-lobed and yellowish white. Fruits are on long robust stalks, pendulous, very variable as to shape, colour and dimensions, rotate, ovoid or oblong lanceolate-ellipsoid, more or less deeply longitudinally furrowed or lobed, blotched or marbled, sordidly orange-coloured when quite ripe, flesh orange-coloured. Seeds are numerous and compressed.

Methods of Hybrid Seed Production

Manual Pollination Techniques

Male and female lines are planted in the field in ratio of 1:2 to 1:5. Cutting off the growing points, stops development of later male flowers after 2-3 fruit setting with a knife. Pinch off the main stem at 4th node and allow two strong branches to develop. Allow setting two fruits on each branch between 12th to 25th nodes. Pinch off the main branch at 30th node. For pollination, tie male and female flowers on male and female parental lines, respectively a day prior to anthesis. Collect male flowers at 5 am and complete the pollination latest by 8.30 am. Remove all male flowers at the bud stage in female parental line. Techniques for selfing or crossing in pumpkin and squash are simple. 'Surveyors' flags (small plastic strip attached to wire stakes) are a convenient way to indicate which plants or lines are to be pollinated. Flags of different colours can be used to indicate the pollen parent. Pistillate and staminate flowers that are to be used in pollination in following morning are located the afternoon prior to anthesis by appearance of a slight touch of yellow at the apex of corolla tube (Fig. 6). They are prevented from opening by tying the tips of corolla tube. Both pistillate and staminate flowers are secured to protect the flowers from insect pollination (Fig.7). The parentage of pollination can be indicated on a tag attached to the pedicel of pollinated flower, when selfing or making cross-pollination with the same male parent. It is convenient to indicate the pollen parent by plastic covered wire (twist tie) coded with a different colour for each source of pollen. It is desirable to pollinate the first pistillate flowers to develop on a plant, since fruit setting is usually better if the plant has not previously produced fruits. After pollination, the pistillate flowers are labelled and tied with or enclosed in a small bag to prevent the visit insects. A slender stakes 3-4 fits long is placed adjacent to pollinated fruits marking its location. Eight labourers/acre can complete the pollination in 10 days. Harvest the fruits 45-50 days after pollination. Cure the fruits under well-ventilated dry cool shade for a period of one month. Extract the seeds along with placenta after cutting the fruit. Mix rice bran in it and mesh it thoroughly. Wash it with clean water. Dry the seeds on a nylon mesh wooden tray in the sun. The average yield is about 350-500 kg/ha, if plant population is 5,000 plants/ha.

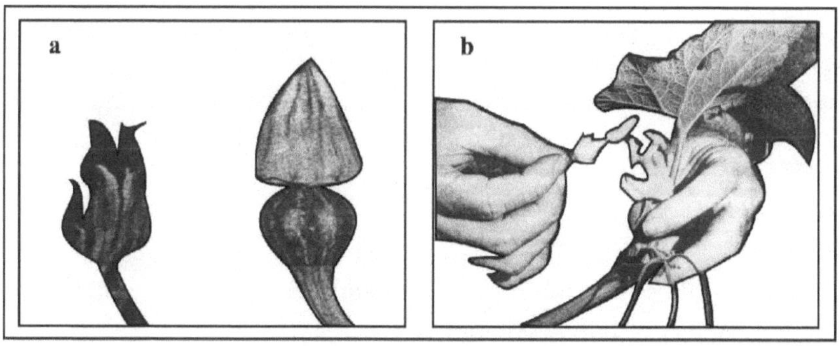

Figure 7: (a) Male (left) and female (right) flowers one-day before anthesis (b) Artificial pollination of *Cucurbita* flowers.

Removal of Male Buds and Use of Insect Pollination

Same as described in cucumber and watermelon. Curtis (1939) suggested planting the seed parent and pollen parent rows alternately and the removal of staminate buds from the seed parent line until one or two fruits have set on plants of the seed parent.

Use of Male Sterility

Male sterility in *Cucurbita pepo* has been reported by Shibriss (1945) and Eisa and Munger (1966) and in *C. maxima* by Scott and Riner (1945). A single recessive gene governs the male sterility. The male sterility reported by Eisa and Munger (1966), is derived from the male sterile cultivar 'Eskandarang,' which was introduced from Egypt in 1962. However, male sterility has not yet been used for commercial production of hybrid seed.

Use of Chemicals

The production of male flowers on monoecious plants in *Cucurbita pepo* can be temporarily suppressed for 2-3 weeks by repeated sprays of ethephon (2-chloroethyl phosphonic acid) at 250 ppm on young plants at first true-leaf stage (Robinson *et al.*, 1970). Two applications of ethephon at 600 ppm at two and four leaf stages result in complete male suppression during fruiting stage (Shannan and Robinson, 1979).

National Released Hybrids

Pusa Hybrid 1

This hybrid has been selected and tested in different agroclimatic conditions under All India Vegetable Improvement Project. It showed appreciable superiority in yield over 'Pusa Viswas'. It is a first hybrid of pumpkin available for commercial cultivation. Its vine growth is medium with an average length of 4.0 m. The leaves are lobed and surface of leaves is non-hirsute and soft in texture with white spots due to raised epidermal cells. The fruits are rounded-flat and medium in size. On an average, a fruit weight is 4.75 kg. The flesh is deep golden-yellow in colour. It gave 30 % higher yield than Pusa Viswas, a high-yielding commercial variety (Sirohi, 2000). It is ready for picking 100 days after sowing in spring-summer season. The average yield is 520 q/ha. It is suitable for growing in Punjab, Kerala, Delhi and other parts of northern plains.

Seed Certification Standards

Following specific standards should be followed for certification of pumpkin hybrid seeds.

I. Field Inspection

A minimum of three inspections should be made, the first before flowering, the second during flowering stage and the third at mature fruit stage and prior to harvesting.

II. Field Standards

(A) General Requirements

Isolation: Seed field should be isolated from the contaminants shown in column 1 of the table below by the distances specified in columns 2 and 3.

Contaminant	**Minimum distance (m)**	
	Foundation	**Certified**
1	**2**	**3**
Fields of other varieties	1000	500
Fields of the same variety not conforming to varietal purity requirements for certification and winter squash (*C. maxima Duch.*), summer squash (*C. pepo* Duch.) and cushaw (*C. mixta* Pang.)	1000	500

B. Specific Requirements

Factor	**Maximum permitted (%)**	
	Foundation	**Certified**
Off types	0.10	0.20

*Maximum permitted at and after flowering.

III. Seed Standards

Factor	Standards for each class	
	Foundation	Certified
Pure seed (minimum)	98.0 %	98.0 %
Inert matter (maximum)	2.0 %	2.0 %
Other crop seeds (maximum)	None	None
Weed seeds (maximum)	None	None
Germination (minimum)	60 %	60 %
Moisture (maximum)	7.0 %	7.0 %
For vapour-proof containers (maximum)	6.0 %	6.0 %

Ridged Gourd and Smooth (Sponge) Gourd

Genetic Resources

The genus *Luffa* comprises nine species. Of these, seven species, *i.e. Luffa acutangula, L. cylindrica, L. echinate, L. graveolens, L. hermaphordita, L. tuberosa* and *L. umbellate* are native to India. The Indian gene centre has rich diversity of genetic resources of ridged gourd (*L. acutangula*) and sponge gourd (*L. cylindrica*), especially in eastern peninsular tracts, Indo-gangetic plains, north eastern region and in tribal dominated belts of central India (Fig. 8, see Plate-6). In India, NBPGR, New Delhi, has collected 15 accessions of ridged gourd. Similarly, 13 accessions of sponge gourd have been collected from different parts of the country. Sri K.C. Mishra Research Institute of Wild Vegetable Crops, Bihar has a collection of *L. acutangula* (two land races from Godda-Bihar) and *L. cylindrica* (two land races from Bihar). At present, IIVR, Varanasi, is maintaining *L. cylindrica* (morethan 21 land races) and *L. acutangula* (morethan 11 land races). NBPGR, New Delhi, is having some gynoecious lines in its collection. In this organization, systematic attempts have been made for germplasm collection and 216 accessions of ridged gourd (*L. acutangula*), 144 of sponge gourd (*L. cylindrica*), and 16 of satputia (*L. hermaphordita*) have been collected. These collections were evaluated systematically for 15 agro- botanical and economic characters. The ridged gourd varied in plant habit (sometimes bushy to prostate), sex types (monoecious, andromonoecious, androecious, gynomonoecious, gynoecious and hermaphrodite), fruit shape (elliptical, elongated, globose/oblong, straight/curved, smooth to sharp angled); fruit skin colour (light green to green), fruit skin glossiness (dull mostly), vine length (1.0-30.0 cm), fruit / plant (5-27) days to first fruit harvesting (50-95) days to last fruit harvesting (86-132) and fruit weight / plant. Promising lines identified in ridged gourd are, IC-93399, U-324-138, U-20-202 A, U-24-134 and NIC-10216. In sponge gourd, IC-92779, 92745 (very early), D-24-163, NIC-10237 and in hermaphrodita (satputia) DCB-1363 are promising ones. In another evaluation characterization programme, accessions UB-217 and K-3847 in sponge gourd have been identified.

Production Objectives

For the development of hybrids in ridged gourd and smooth gourd, following objectives should be:

- To develop hybrids having lowest node number at which first pistillate flower appears.
- To develop hybrids having female to male ratio as it results in higher fruit yield.
- To develop hybrids possesing uniform thick cylindrical fruits free from bitterness.
- To develop hybrids and remains for long duration having fruits, tender and non-fibrous in at edible maturity.
- To develop hybrids and remains for long duration high fruit yield in terms of weight and number of fruits of marketable quality.
- To develop hybrids having Resistance against diseases and insect-pest.

Botany and Floral Biology

Ridged gourd and smooth gourd are climbers, stem acutely 5-angled, hairy, tendrils 3-bid or more leaves 5-7 angled or shallowly lobed. The sex form in ridged gourd is monoecious, androecious, gynoecious, gynomonoecious, andromonoecious and hermaphrodite. Male inflorescence in raceme of several flowers, peduncle 15-35 cm long, female flower solitary, borne axillary as male flowers. Flowers are 4-5 cm in diameter, fragrant, calyx 5 partite, glandular, petals yellow, obovate, 2-2.5 cm long, stamen 3, ovary inferior, filiform, with 10 longitudinal ribs on which are swollen glands, style short, stigma trifid flower open in evening, fruit club shaped, crowned by enlarged sepals, style is persistent, angled fruits in ridge gourd 10-ribbed, many seeded, seeds black and pilted. In smooth gourd, seeds are flattened and anthesis occurs early in the morning, while, in ridge gourd, it is in the afternoon. Pollen fertility is maximum on the day of anthesis and lasts till 2-3 days after anthesis in winter and one and a half days in rainy season. Stigma is receptive 6 hours before to 84 hours after anthesis (Singh, 1957). The main differences between ridged gourd and sponge gourd are as follows:

Ridged gourd	Smooth gourd
Six sex forms are seen	Generally, monoecious form is seen.
Leaves are shallowly lobbed and larger	Leaves are deeply lobbed and smaller.
No white patches on leaf	White patches on leaf.
Male flowers smaller	Male flowers larger.
Disc shaped and not tomentose ovary	Cylindrical and tomentose ovary.
Angled and 10-ribbed fruits	Smooth fruits.
Seed black and pitted	Seeds black, white flat and not pitted.
Anthesis takes place in afternoon	Anthesis takes place early in the morning.

Pathak and Singh (1949) have been found successfully in making reciprocal crosses between these two species. They observed that F_1 plants were generally intermediate between the parents. The F_1 showed various irregularities like univalent, rings, chain of four chromosomes, chromatin bridge and fragment at metaphase. The percentage of good pollen ranged from 18 to 40 %. Thus species are not easily crossable and F_1 appears to be of not much practical value.

Methods of Hybrid Seed Production

Following methods can be used for hybrid seed production of ridged gourd and smooth gourd.

Pinching of Male Flowers

Production of hybrid seeds in this method is most economical and practical, which can easily be adopted by growers. F_1 hybrid seed production in both crops can economically be done on large-scale by pinching all the flowers before opening from the female parent and allow the male parent to grow by side of female parent for natural cross-pollination. In both ridged and smooth gourds, where male flowers are produced in racemes. Pinching of male buds is not complete and very effective method. Interspecific crossing between both *Luffa cylindrica* and *L. acutangula* is not possible. For maximum fruit setting and seed yield, availibility of pollinators is a prerequisite. One medium-sized bee colony per hectare is enough in seed production plots. Since, both these crops have monoecious sex form, hybrid seed production can be resorted through bagging of unopened female flowers, and pollinating with male parents, pollination next day and again bagging to ensure controlled cross-pollination. This will give hybrid seed.

Chemical Sex Expression

There is yet another method of hybrid seed production by use of chemicals. With the use of plant growth regulators, very significant results have been obtained with regard to modification of sex in ridged and sponge gourds. Two true leaf stage is most responsive stage for application of plant growth regulators for sex modification. Ethrel (250 ppm) is used for suppressing staminate flowers and initiating pistillate flowers at flowering time. The row of male parent is grown by the side of female parent and allows natural cross-pollination. In the absence of insect pollinators, hand-pollination is possible when two sexes are separate. The female flowers of female parent and male flowers of male parent have to be bagged separately prior to opening and hand-pollination brought about later. Precautions should be taken that during natural cross-pollination, there should not be any other variety except the parents of hybrid. Four to five fruit set at initial stage, nodes containing hybrid seed would be a sufficient seed yield.

Use of Gynoecious Lines

In ridged gourd, a variety called 'Satputia' bears a hermaphrodite sex condition. It is observed that gynoecious sex forms are stable only under moderate regimes of temperature and photoperiodic conditions. When the temperature exceeds beyond 30^0 C, stability of gynoecious sex expression is affected. However, temperate gynoecious lines are found unstable for gynoecy under high temperature and long photoperiodic conditions because of their thermospecific response for gynoecious lines. Homozygous lines are planted in strict field isolation. The gynoecious lines are crossed with monoecious male parent to produce F_1 hybrids.

Seed Certification Standards

Following are the standards for certification of both ridged and sponge gourds hybrid seeds.

I. Field Inspection

A minimum of three inspections should be done, the first before flowering, the second during flowering and fruiting stage and the third at mature fruit stage and prior to harvesting.

II. Field Standards

(A) General Requirements

1. Isolation: Seed fields should be isolated from the contaminants shown in column 1 of the table below by the distances specified in columns 2 and 3.

Contaminant	Minimum distance (m)	
	Foundation	Certifed
1	2	3
Fields of other varieties of the same species	1000	500
Fields of the same variety not conforming to varietal purity requirements for certifcation and ridge gourd (*Luffa acutangula* Roxb.) in sponge gourd and ridged gourd.	1000	500

(B) Specific Requirements

Factor	Maximum permitted (%)*	
	Foundation	Certified
Off types	0.10	0.20

III. Seed Standards

Factor	Standards for each class	
	Foundation	Certified
Pure seed (minimum)	98.0 %	98.0 %
Inert mater (maximum)	2.0 %	2.0 %
Other crop seeds (maximum)	None	None
Weed seeds (maximum)	None	None
Other distinguishable varieties (maximum)	5/kg	10/kg
Germination (minimum)	60 %	60 %
Moisture (maximum)	7.0 %	7.0 %
For vapour-proof containers (maximum)	6.0 %	6.0 %

*Maximum permitted at and after flowering.

15

ROOT CROP

Carrot

Genetic Resources

Wild carrot (Queen Anne's Lace) grows in temperate region around the world but the extent of genetic variability in wild population has not yet been evaluated in *Daucus carota* or related species. Constance (1971), Heywood (1983) and Mathias (1971) have contributed much to our understanding of Umbelliferae systematics, but relatively little carrot germplasm has been collected for distribution. The cultivated carrot, *Daucus carota* L, intercrosses freely with almost all the wild forms. Although, no incompatibility has been observed but protandry is the main reason for cross-pollination. However, there is no barrier for selfing. Carrot variety Brasilia is resistant to nematode, *Meloidogyne incognita*, while species *Daucus carota* subsp. *hispanicus* has resistance to other nematode, *M. hapla* (Simon, 1993). The genetic material, resistant to disease and pest of carrot, which can be used for developing hybrids for abiotic stresses, are given in Table1.

Table 1: Genetic material of carrot used for hybrid seed production

Trait	Promising Lines
High yield Red colour	Jind Collection-1, HC-1, HC-131-1, Hisar Local-5, HC-100, HC-27, BC-3, Desi Red Kernal, HC-4-2, Hisar Local-1, HC-155, HC-153-3, HC-12, HC-191, HC-118-2, HC-256, HC-246 and HC-199.
Purple colour	HCP-118-1, HCP-261, HCP-227-1, HCP-226-1, HCP-160-1, HCP-4-2, HCPC-1, HCP-4-2-1 and HCP-160.
Orange colour	HCO-4 and HCO-1.
Black colour	HCB-170, HCB-22-1, HC-2B, HCB-4-2, HCB-22, HCB-165 and HCB-1.
Yellow colour	HCY-83, HCY-235, HCY-183, HCY-4-2-1 and HCY-4-2.
Cercospora carotae resistant	Inbred line WCRI.
Mottley virus resistant	Sweet Crop and Top Weight.
CMV resistant	'Virginia Savoy'.
Pythium spp. resistant	Nantes and Imperator.
Heterodera carotae (nematode) resistant	Vilmorium-66.
Tip burn (physiological disorder) tolerant	Scarlet and Nantes.

Production Objectives

The cultivated carrot, *Daucus carota* L. intercrosses freely with almost all the wild forms. Very little hybrids have been developed in our country. Major attributes on which the attention is needed for developing hybrids are:

- Hybrids with high carotene content by crossing Asiatic and European types.
- Hybrids with good colour, size, earliness, top root ratio and resistant to cracking.
- Hybrids that will suit to abiotic stresses.
- Hybrids resistant to diseases and insects.

Botany and Floral Biology

The carrot is an annual or biennial herb, with an erect to much branched stem, 30-120 cm high arising from a thick, fleshy taproot and 5-30 cm long. Usually, stem elongates and produces round, hispid branches during second year. Leaves are pinnately compound with long petioles extended into a stalk at the base. The edible portion of carrot root is actually enlarged fleshy taproot. It consists primarily of phloem or cortex and core or xylem. Good quality carrot has maximum cortex and minimum core. In so called 'coreless' cultivars, the core is small and deeply pigmented, so that cortex and core are evenly coloured.

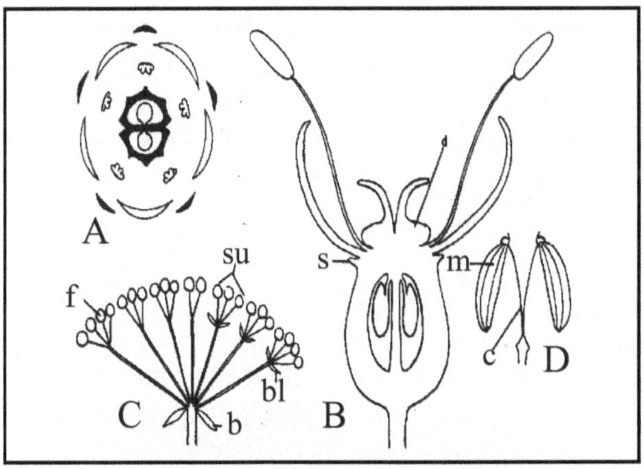

Figure 1: Flower parts of carrot.

The carrot umbel is a compound inflorescence. The development of umbel begins with a broadening of floral axis and internodes elongation (Borthwick *et al.*, 1931). A primary umbel can contain over 1,000 flowers at maturity, whereas secondary, tertiary and quaternary umbels bear successively fewer flowers. Usually, several floral stalks develop from a single plant. Flowers are arranged in umbellets and umbellets into an umbel. Within umbellets and umbels, floral development is centripetal and arrangement is spiral. Thus, first mature flowers are those on the outer edges of outer umbellets. The primary umbel consists mainly of bisexual flowers, but male flowers can occur frequently (between the edge and centre of an umbellet in subsequent umbels). The pollen from flowers at the centre of an umbellet is larger and more frequently fertile than that from peripheral flowers.

Carrot is potandrous. The petals separate and filaments begin unrolling to release the anthers at anthesis. After the filament is straightened, the pollen is shedded and the stamen is quickly abscised. The petals then open fully and the style elongates. The carrot has a split style, which separates when the flower is receptive to pollination. The petals of male fertile plants fall soon after the split stigma is receptive. It is interesting that petals of petaloid, but not brown anther, male sterile plants are persistent until the seed ripen. Carrot flowers are epigynous with five each small sepals, petals and stamens, and two carpels. The matured flowers and consequent matured fruits are approximately 2 mm long. Early in development, each carpel bears two ovule primordia, but only the lower one continues to grow. The carrot embryo sac is monosporic (developing from the chalazal macrospore) and 8-nucleate. Nectaries on ovary wall are important for insect attraction in pollination. The flowers are perfect with small petals, usually white or yellowish. The calyx is entire. There are five stamens. The ovary is inferior and consists of two locules, each with a single ovule. On the upper surface of ovary, there is a swollen nectar-secreting disk, which supports the style and stigma. The carrot fruit is a bilocular schizocarp, which dries and splits upon maturity to yield two mesocarp (achenes) with one seed each. The mericarp are covered with spines, which must be removed for easy handling. The carrot seed is actually a fruit, an indehiscent mericarp, containing a single seed. In fact, two mericarp pair to form the schizocarp, the true carrot fruit, which develop from a two-loculed ovary. The mericarp or seed is somewhat flattened on one side and opposite side has longitudinal ribs with bristly hairs, the fruit is oblong and about 3 mm long. It produces the first order umbel called king umbel, second order and tertiary umbels.

Pollen dehiscence may lasts 1-2 days per floret and 6-7 days after anthesis and may take about 4-6 weeks for complete blooming. The different order of umbels flower at intervals 8-12 days from each other. Anthesis in a single umbel is completed in 7-9 days. The studies on carrot variety Imperator revealed that umbels of the first two order set seeds in 100% umbels and third, fourth and onwards orders set in 82.7, 6.29 and 2.78% umbels, respectively. The setting of seed in late formed umbels is extremely poor. The stigma fertility starts when dehiscence is completed in whole umbel. First king umbel comes in flower and it gets pollen from secondary umbel. The secondary gets pollen from tertiary umbel. The temperate types start bolting by third week of April. The king umbels start flowering by the end of May. The secondary umbels or tertiary and other umbels flower at intervals of 8-12 days from each other. Anthesis in a single umbel is completed in 7-9 days. Flowering in first, second and third order umbels are almost completed by end of June. The fourth and onwards umbels flower during July and August. Over 95% of cross-pollination has been observed in carrot.

Methods of Hybrid Seed Production

Hand-Emasculation and Pollination

In hand-emasculation and pollination for selfing, king or main umbel is enclosed in muslin bag and a second order umbel of the same plant is also put the same bag or bagged separately. For a couple of days these two umbels are rubbed gently against each other by hand daily to ensure self-pollination. Sometimes, it is convenient to enclose two or three umbels of the same plant in a muslin bag before anthesis. After anthesis, the umbels are rubbed against each other daily for a few days for effective self-pollination. As soon as the first bud of an umbel opens, the whole umbels are emasculated, the remaining young flower buds are removed. The umbel of pollen parent previously kept covered in a muslin bag is cut off and its stalk placed in a glass bottle filled with water, which is fastened to a bamboo stake fixed in the soil close to female parent. The pollen parent umbel is enclosed in the same bag along with emasculated. Umbel of the-pollen parent continue to open and shed their pollen. The male umbel is gently rubbed over the emasculated umbel to ensure artificial cross-pollination in morning daily for a few days. A few common houseflies can also be introduced into bag for pollinating the flowers. Sometimes, when pollen parent possesses some dominant marker genes with the help of which hybrids can be distinguished at seedling stage or later at root stage. For this, it would be better to plant the carrot roots of male and female parents very close to each or about 20 cm apart, and enclose a few umbels each of male and female parents in the same clothes bag before anthesis. To ensure cross-pollination, umbel of male parent is gently rubbed against umbel of the female parent daily for a few days.

Use of Male Sterile Lines

Male sterility has been documented in the year 1885 (Kaul, 1988). Several other genetic male sterile mutants have been utilized for commercial seed production due to availability of more efficient *CMS* system in carrot. Two type of sterile cytoplasm have been reported: (i) petaloid and (ii) brown anther (Welch and Grimball, 1947; Morelock, 1974). Taki Seed Company of Japan developed first F_1 hybrid variety in 1982 using *CMS* (Pelletier *et al.* 1995). In USA, majority of hybrids are produced from one cytoplasm i.e. Cornell cytoplasm, Considering the risk of disease vulnerability of hybrid variety due to monolistic use of Cornell sterile cytoplasm, USDA has released a petaloid type new *CMS* line derived from sterile cytoplasm of 'Winconsin' (Morelock *et al*; 1996*)*. In this method, following steps are evolved for hybrid seed production:

- Study the location of cytoplasmic-genic male sterile lines of different origin.
- Development of maternal male sterile lines from the desired varieties.
- Development of good agronomic male fertile line with good general combining ability.
- Making the heterotic crosses and testing F_1 hybrids.
- Production of commercial (F_1) seed.

There are two types of male sterility in carrot, namely 'the brown anther type' in which the anthers are shrivelled and brown without any viable pollen and the 'petalloid type' in which the anther are modified into petalloid structure (Cross, 1970). The male sterility is due to interaction of a cytoplasmic factor and a number of genes. The inheritance of male sterility is, however, not fully understood. The male sterile line is maintained by crossing it with male fertile sister line. In heterosis breeding programme, three lines are used, namely male-sterile line, male-fertile sister line and pollinator line, which is male-fertile and has a good combining ability with male-sterile line. The male-sterile and pollinator lines are inbred for several generations for attaining uniformity even though there may be decline in vigour of these lines. However, loss in vigour is restored by hybridization. Usually, four rows of male-sterile seed parent are planted alternatively with two rows of pollinator line in an isolated field for production of F_1 hybrid seeds. The seeds are harvested from male-sterile line and pollinator plants are removed before collecting the seeds from male sterile female parent. The development of *ms* and maintainer lines is very laborious process due to dominant state of male sterility. Crossing, backcrossing, selfing and testing of progenies in following two generations, including isolation of positive progenies, are characteristic steps in breeding processes. The breeder is forced to eliminate male fertile plants or phenotypes with partial male fertility within the *ms* lines developed, and to develop new lines with low inbreeding depression. Such lines can be found but they rarely have good combining ability.

An excellent uniformity of hybrid varieties demands a high degree of homozygosity in *ms* and maintainer lines but selection of lines with low inbreeding depression reduce the probability of finding good parents for hybrid varieties. To overcome this problem, the development of three-way hybrids has been recommended. For this method, an original maintainer and a second line (exchange maintainer) are necessary. The universal maintainers, which can maintain all *ms* lines due to complete homozygous state of maintainer genes, are desirable for such a system. A universal maintainer must also possess specific characters of *ms* line and a good combining ability. Exchange maintainers have therefore been propagated which are selected following a diallel with all lines of one type. These lines must have the same genetic state as the specific *ms* line, a state that can be found relatively frequently. Thus, the best line for performance and uniformity can be selected. Recently, exchange maintainer lines for most important *ms* lines have become available and can be used for seed production of new lines. Pollinator lines, which are necessary for hybrid seed production, are derived from open-pollinated varieties or special breeding lines by using a top-cross system for the testing of general combining ability.

Achievements

So far development of carrot hybrid, one hybrid variety (Hybrid-1) developed at Maharashtra Hybrid Seed Ltd, Jalana, has been identified for cultivation in Humid West Himalayan Region and Semi-Arid Lava Plateau and Central High Lands.

Seed Certification Standards

Following seed standards should be followed for certification of carrot hybrid seed.

I. Field Inspection

Mother Root Production Stage

A minimum of two inspections are done as follows:

- The first inspection is done 20-30 days after sowing to determine isolation, volunteer plants, outcrosses, off types and other relevant factors.
- The second inspection is done after the mother roots have been lifted to verify the true characteristics of roots.

Seed Production Stage

A minimum of four inspections are done as follows:

- The first inspection is done before flowering to determine isolation, volunteer plants, outcrosses and another relevant factors

- The second and third inspections are done during flowering to check isolation, off types and other relevant factors
- The fourth inspection is done at maturity to verify the true nature of umbels.

II. Field Standards

(A) General Requirements

1.Isolation: Carrot seed fields should be isolated from the contaminants shown in column 1 of Table below by distances specified in columns 2 and 3.

Contaminant	Minimum distance (m)			
	Mother root production stage		Seed production stage	
	Foundation	Certified	Foundation	Certified
1	2	3	4	5
Fields of other vaireties	5	5	1000	800
Fields of same hybrid not conforming to varietal purity requirements for certification	5	5	1000	800

(B) Specific Requirements

Factor	Maximum permitted (%)*	
	Foundation	Certified
*Roots not confirming to varietal characteristics including forked roots	0.10 (by number)	0.20 (by number)
**Off types (plants)	0.10	0.20

*Maximum permitted at second inspection at mother root production stage.
** Maximum permitted at and after flowering at seed production stage.

III. Seed Standards

Factor	Standards for each class	
	Foundation	Certified
Pure seed (minimum)	95.0 %	95.0 %
Inert matter (maximum)	5.0 %	5.0 %
Other crop seeds (maximum)	5 /kg	10/ kg
Total weed seeds (maximum)	5 /kg	10 /kg
Other distinguishable varieties (maximum)	5 /kg	10 /kg
Germination (minimum)	60 %	60 %
Moisture (maximum)	8.0 %	8.0 %
For Vapour-proof containers (maximum)	7.0 %	7.0 %

BULB CROP

16

Onion

Genetic Resources

In onion, much variation exists in bulb characteristics for size, shape, colour of skin and flesh, number of hearts, skin retention, firmness, pungency, TSS, yield *etc*. Bulbs may be white, red, brown or yellow, round, globe, flat or spindle shaped. Diversity is also found in field tolerance to *Stemphyllium* blight, purple blotch, basal rot and thrips. Various workers have also reported tolerance to soil salinity, reduced bolting, keeping quality and high dry matter content. In India, NBPGR, New Delhi, is a nodal agency engaged in collection, evaluation and maintenance of onion germplasms both from indigenous and exotic collections. Accessions numbering more than 1,000 onion germplasm lines are being maintained at NBPGR, New Delhi. Other centres maintaining sizeable number of germplasms are IIHR, Bangalore, Mahatma Phule Krishi Vidyapeeth, Rahuri, NHRDF also NRC on Onion and Garlic, Nasik; IARI, New Delhi, PAU, Ludhiana and IIVR, Varanasi. Indigenous collections account for more than 95% of the germplasm lines maintained in onion, which indicates the existence of large amount of variability within the country for short-day onions. However, it also indicates the necessity to enrich the existing germplasm pool by importing more exotic collections. Promising germplasms for different attributes are given in Table1.

Table 1: Promising germplasms of onion

Trait	Promising accessions (s)
High-yielding and large-sized bulb	IC-48025, 48129, 48145, 48361, 48404, 48411, Collection No-381, W-418, W-414, W-415, W-430, W-438, W-441, ACC No.-397, ACC No.-343, and T-109/27.
Early-maturing and high-yielding	IC-34444, 42834, 47548, 48025 and 48051
Yellow colour and high yield	Early Stophen Yellow.
High TSS and attractive red bulb	IC-47954, 48145, 48544, 48764, 48807, 48916, W-427, W-425, W-406, W-395, Coll No. 595, Coll No. 606, White Onion, Phule Safed, No. 569, Early Stophen Red No. 472 No. 453, 608 and Udaipur 103 A.
Field tolerant to stemphylium blight and purple blotch	IC-42900, 47972, 47998, 48575, 48710 and 49812.
Non bolting salad onion	Barmuda Yellow and EC-109123.
High-yielding, large-sized, dark red and pungent types	Giza-6, Giza-20, EC-200936 and 200937, No. 799, Bangalore Rose, Jalgaon Vari, No. 5 and Barwad.
Short day, resistant to bolting and pink rot disease	Nu Mex Br-1 and EC-169372.

Production Objectives

Following objectives should be kept in mind while developing its hybrids:

- There is a need to develop high-yielding hybrids having improved quality of bulbs with longer storage life, suitable for different agroclimatic conditions.
- To develop musiple resistant hybrids against many of the diseases like purple blotch, basal rot, stemphylium blight and storage rots, besides its, insects-like thrips, which causes a heavy loss of onion crop both in the field and storehouses.
- To develop hybrids suitable for dehydration.
- Although, India is exporting large quantities of fresh onion bulbs, there exists a vast potential for increasing it by developing hybrids suitable for export.
- To develop hybrids which are resistant to abiotic stresses especially moisture stress, high temperatures, salinity and alkalinity.

Botany and Floral Biology

Onion is bulbous biennial or perennial herb, which give off a distinctive and pungent odour when the tissues are crushed. The onion smell and taste are important diagnostic features of the genus *Allium.* Other characteristics include the presence of bulb formed by the attachment of swollen leaf bases to the underground part of the stem, inflorescence in the form of umbels with numcrous small flowers, a single spathe which split into segments as the young flowers emerge and a superior ovary above the six stamens and petals like perianth. (Fig.1).

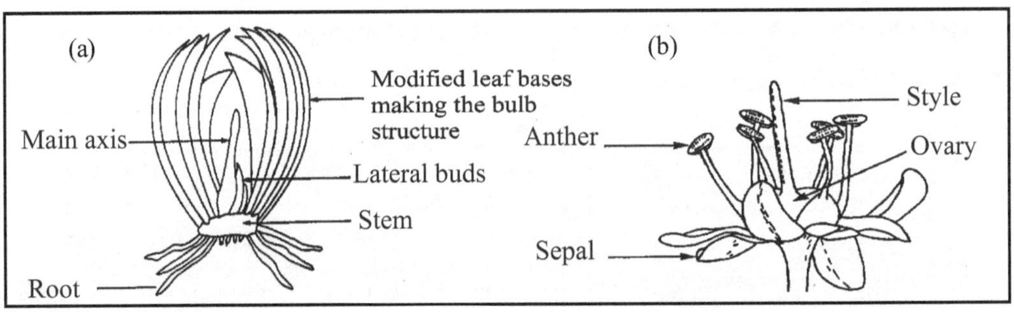

Figure 1: (a) Stem of onion (b) Individual onion flower.

Flowers are borne in simple umbels at the apex of floral stem, which is commonly hollow when mature, round in cross-section and somewhat swollen at the middle or near the base. The number of seed stems varies from 1 to 20 or more depending-before it is fully expanded, is enclosed in a papery spathe, consisting of 2-3 bracts, which split open by the pressure of developing flowering buds. The number of flowers may vary from 50 to 2000, depending upon species and cultivars, time of planting, size and storage conditions of mother bulbs. The umbels are aggregates of cymes of 5-10 flowers each and the flowers open in definite sequence. The flowers are white to bluish in colour. The perianth segments are six in two whorls spreading, reflexed, free and ovate. The stamens are also six in two whorls, anthers bilocular, ovary superior, three locules with nectary at base, each locule has two ovules (Fig.1b). The anthers of inner whorls dehisce first and usually it occurs between 9 am and 5 am. The style, which is approximately 1 mm long, when the flowers first open, is not receptive, until it elongates to a length of 5 mm. This requires a day or two days after anthers have all dehisced. An inflorescence may continue opening for two weeks or more and a plant may be in bloom for more than 30 days. The flowers are protandrous in nature and generally cross-pollinated. It is pollinated chiefly by honey bees, which visit the nectaries. Though, onion is cross-pollinated crop but selfing also takes place as the insect visit many flowers in a single umbel and also umbels of the same plant before leaving for other plant. In onion, anthesis start early in one morning (6 am to 7 am). An artificial pollination for breeding purpose, selfing is accompanied by bagging all the umbels of the same plant together and by shaking or by introducing houseflies.

Methods of Hybrid Seed Production

Hand Emasculation and Pollination

Manual emasculation and pollination of onion is tedious. Although, stamens are easily removed from a single flower, the size, number and continuous maturity of many small flowers/umbel, make hand-pollination unattractive. As a result, onion breeders have developed techniques to manipulate insect pollinators. Houseflies and blowflies are commonly used as pollinators. These techniques are:

- For selfing of onion flowers about 2 or 3 umbels of the same plants are enclosed in an ordinary large-sized muslin bag or kraft-paper bag before the buds open. It is, however, necessary to shake the bag frequently in the morning at least one or two times for a couple of days to ensure good pollination. If facilities for rearing houseflies exist, it would be helpful to introduce a few flies inside the bag to do the selfing.

  As soon as the first buds in an umbel open, the whole umbel of the female parent is bagged in a muslin bag. The flowers are removed daily for a few days until the peak flowering is reached after which no flowers are removed. The buds in this umbel are emasculated as they open and when sufficient buds have been emasculated. The remaining young flower buds are removed. The umbel of pollen parent previously kept covered in a muslin bags is cut off and its stalk placed in a glass bottle filled with water, which is fastened to a bamboo stake fixed in the soil close to female parents. The pollen parent umbel is enclosed in the same bag along with emasculated umbel of the female parent. For about a week or so, the buds in the enclosed umbel of the pollen parent continue to open and shed their pollen. Daily for a few days in the morning the male umbel is gently rubbed over the emasculated umbel to ensure artificial cross-pollination. A few common houseflies can also be introduced into the bag for pollinating the flowers.

- Sometimes when the pollen parent posses some dominant marker genes with the help of which the hybrid can be distinguished in seedling stage or later at the bulbs of male and female parents are planted alternatively in rows 60 cm apart with a plant spacing of 50 cm. It is more convenient to plant two bulbs each of the male and female parents at the same hill and later before anthesis encloses about 2 or 3 umbels each of the two parents together inside the same muslin bag. The bag must be shaken at least once or twice daily for a few days for affecting cross-pollination.

Use of Male Sterile Lines

In onion, the use of male sterile lines is common to produce seeds of F_1 hybrids. The male sterility in onion is due to interaction of cytoplasm and nuclear gene (Jones and Emsweller, 1936 and Jones and Clarke, 1943). The nuclear gene is single recessive *msms*, while its dominant allele *msms* is for male fertility. Similarly, there is a sterile cytoplasm *S* and the normal cytoplasm *N*. The genetic constitution of male sterile plant commonly known as female parent, is *Smsms* having both sterile, *S* cytoplasm and single recessive gene *msms* for sterility. The other genotypes like *Smsms, Smsms, Nmsms, Nmsms* and *Nmsms* are male fertile. These genotypes are male fertile because they have one sterility factor, either the sterile cytoplasm *S* or the sterile gene *msms,* but not both of them together.

In a hybrid breeding programme, it is necessary to have male sterile lines. Some of the male sterile lines developed in other countries like the USA, Japan and Holland may be introduced for use in our country. The male sterile lines can be obtained only from the research institutions or organizations because the private seed firms do not supply them. However, for introducing the male sterile lines into our country, the prevalent photoperiodic and temperature conditions, have to be taken in account. Another important consideration should be to choose those lines, which have good keeping or storage quality. Since, in our country storage type of onion is more commonly grown than the table types. Various steps involved for developing onion hybrid by using male sterile line, are described below:

Parental Lines

Three parental lines, namely A, B and C are used for hybrid seed production. The A line known as the female parent, is the male sterile having the genetic constitution *Smsms*, which produces hybrid seeds, the line B with the genotype *Nmsms* is the male fertile counterpart for the maintenance of male sterility of line A and the line C is the fertile pollen parent or inbred which is genetically diverse from the lines A and B. The lines A and B are genetically identical except that the former is male sterile while the latter is male fertile. The male sterile line A is prepared by crossing with line B, the male fertile counterpart.

Developing of the Parental Lines

Lines A and B

The lines A and B can be developed by four different methods, which are described below.

The simplest method is by backcrossing, the male sterile plants to male fertile plants of the genotype *Nmsms* of the cultivar in which the lines A and B are to be developed. The male sterility gene has been found to be widespread in several cultivars collected from many parts of the world (Little *et al*., 1946 and Davis, 1957). In many cultivars, plants of *Nmsms* type can be detected by a suitable test cross. For this purpose, select at random about 10 plants in the population of a particular cultivar and cross each one of them with male sterile plant. The progenies of some of these cases will be male fertile, while in some other cases they will all be male sterile. It would show thereby that the male fertile plants which are used for crossing with the male sterile plant and produces all male sterile progeny, has the genetic constitution of *Nmsms*. Such individual plants having *Nmsms* type, are selected and selfed for maintenance. By repeated backcrossing the male sterile plants to these *Nmsms* plants, it will be possible to develop a male sterile line of the desired cultivar. After 4 or 5 backcrosses, the lines A (*Smsms*) and B (*Nmsms*) will become genetically almost identical except for sterility in the line A, while other will be male fertile (Fig. 2).

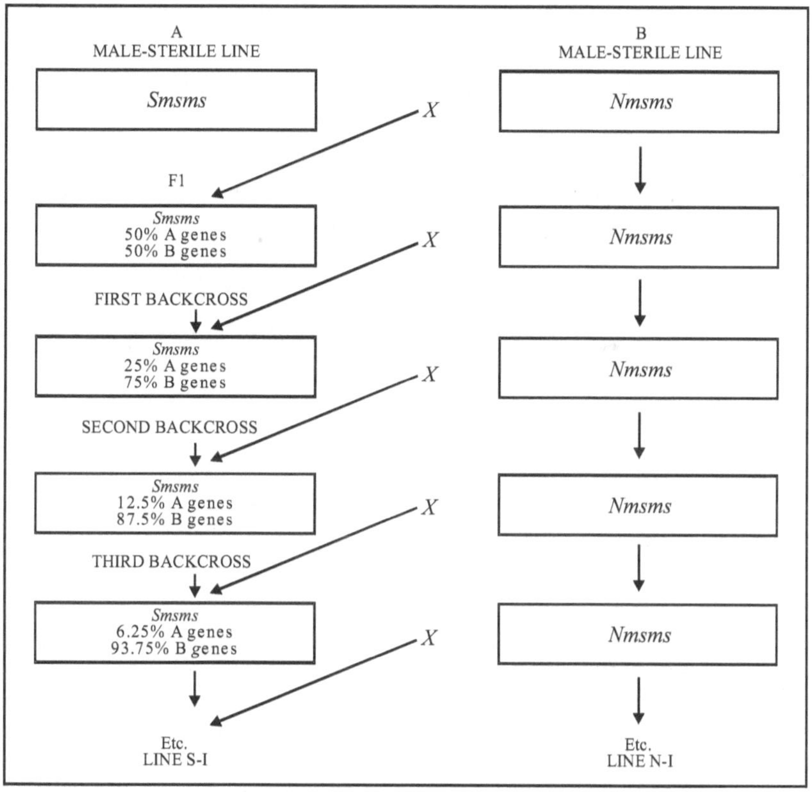

Figure 2: Incorporation of male sterility in an onion cultivar by backcrossing.
Source: Swarup, 1991, p. 34.

Following methods have been suggested for incorporating the recessive gene *ms* into those cultivars in which all plants are of the genetic constitution *Nmsms* and which do not carry the recessive gene *ms*. The plants of *Nmsms* genotype can be developed in the cultivar by the procedure outlined below:

- Cross the male sterile plants (*Smsms*) with male fertile plants (*Nmsms*). The F_1 plants will be male fertile, their genotype being *Smsms*.
- Cross male fertile F_1 plants (*Smsms*) with *Nmsms* using the former as the pollen parent and the latter as the seed parent. The progeny of this cross will be male fertile. The male fertile plants will be of two genotypes, namely *Nmsms* and *Nmsms*, which can be detected only by a progeny test. Cross about 10 such individual male fertile plants with male sterile plant and study their progenies. There will be two types of progenies, one in which all the plants will be male fertile, while in the other, half of the progeny

will be male fertile and half male sterile. In former case, the plants are of the *Nmsms* genotype and in the latter they are *Nmsms*.

- On selfing the *Nmsms* plants, three types of progenies are observed, *i.e. Nmsms*, *Nmsms* and *Nmsms*. All these progenies will be male fertile but the genotypes can be distinguished by a progeny test after crossing with the male sterile plant. The *Nmsms* plants can be easily detected as only these will produce all male sterile progeny when crossed with the male sterile plant (*Smsms*).
- After obtaining the desired B line (*Nmsms*) of new cultivar the back crossing programme can be followed as mentioned above. By four or five repeated back-crossing of the *Nmsms* plants to the male sterile plant, the male sterile A line (*Smsms*) can be developed, which will be otherwise genetically identical to the B line. The outline of the procedure outline are as follows:

a) *Smsms* x *Nmsms* $\rightarrow$ *Smsms* (male fertile)

b) *Nmsms* (female) x *Smsms* (male) $\rightarrow$ *N msms* + *N msms* (all male fertile)
S msms x *Nmsms* $\rightarrow$ *Msms* (male fertile)
Progeny test

c) *Smsms* x *Nmsms* $\rightarrow$ *Smsms* (male fertile)
Smsms x *Nmsms* $\rightarrow$ *msms* (male fertile) + *Smsms* (male sterile)

d) *NMsms* (selfed) $\rightarrow$ *msms* + *Nmsms* + *Nmsms* (all male fertile)

e) Progeny test:
Smsms x *Nmsms* $\rightarrow$ *Smsms* (male fertile)
Smsms x *Nmsms* $\rightarrow$ *Smsms* (male fertile) + *Smsms* (male sterile)
Smsms x *Nmsms* $\rightarrow$ *Smsms* (all male sterile)
Self *Nmsms* plants to obtain a new line B of the cultivar

f) Backcross *Smsms* to *Nmsms*, four or five times to develop the genetically identical A line.

- The naturally-occurring male sterile plants which are found to the extent of about 1-3% of the total plants in large planting of commercial seeds, can also be used for developing the lines A and B. The male sterile plants can be crossed with the male sterile plants of the same cultivar, in which, it was observed to test and develop the plants of the genotype *Nmsms*. After obtaining the male sterile plants of the genotype *Nmsms*, they can be backcrossed with male sterile plants (*Smsms*) in the same way as mentioned earlier to develop the lines A and B. However in this case, since the lines A and B selected from the same cultivar are genetically similar, fewer backcrossing will be necessary than when these are from different cultivars.

- Sometimes, lines A and B can be developed from seeds of a commercial F_1 hybrid. The seeds of F_1 hybrid obtained from abroad may be sown and later the fertility of the F_1 plants may be checked. If the F_1 plants are all male sterile it can be presumed that the F_1 hybrid is obtained by crossing the lines A (*Smsms*) and C (male parent, *Nmsms*).

When the F_1 plants are male fertile it would indicate that the F_1 hybrid is derived from the cross-made between the lines A (*Smsms*) and C (*Nmsms*) and the genetic constitution of the F_1 hybrid would be *Smsms*. On selfing (*Smsms*), it will segregate into a ratio of three male fertile plants (1 *Smsms* + 2 *Smsms*) and one male sterile plant (*Smsms*). The male sterile plant can be detected by progeny test and then used for developing the lines A and B by the method mentioned earlier.

Developing the Line C

The line C (pollen parent) can be developed by inbreeding in a promising cultivar for some generations to achieve uniformity. The line C is genetically diverse from the lines A and B. It is necessary to test the combining ability of the C line before producing the F_1 hybrid seeds.

Maintenance and Multiplication of Lines A, B and C

The first increase of seeds of three lines A, B and C must be in insect-proof cages followed by an increase in the open but not more than twice to obtain a high degree of purity in the stock seeds of parental lines used for production of F_1 hybrid seeds. The seeds of lines A and B can be increased in the same cage, while line C must be grown separately in another cage. A plastic or wire-net cage of 20 x 20 or 24 x 24 mesh measuring in 6 m x 3 m x 2 m (height) with a small door at one corner for entrance and exit is suitable for this purpose. This size of cage can accommodate four rows of bulbs spaced 0.6 m (2 feet) apart to produce about 1-2.5 kg of seeds. In the cage, plant the bulbs of lines A and B in alternate rows with equal number of bulbs of each as shown in Fig.3.

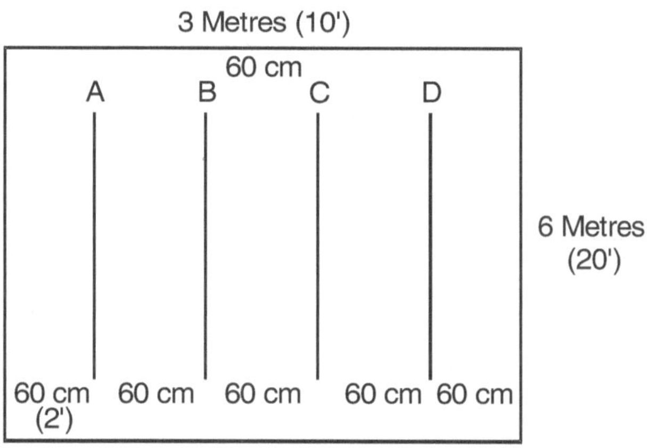

Figure 3: Insect-proof cage (*Source*: Swarup, 1991, p. 36).

Before flowering, the cage is put over the plants. The umbels should not be touching the cage, as this would allow the stigmas to rovide and get contaminated with foreign pollen carried by insects. Staking with bamboo and tying string can prevent this or standing on both sides of the plants to keep the seed stalks upright. About three days before the commencement of flowering, place a beehive containing a medium-sized colony of bees with a queen inside the cage and feed them with 10% sugar solution. It is necessary to rogue out the pollen-bearing plants from the line A and pollen sterile plants from the line B in the morning before dehiscence of anthers. The contamination of line B (*Nmsms*) with the *Ms* gene should be particularly avoided as it cannot be detected, but it will produce male fertile plants in line A in later generations. This will necessitate roguing in a seed field, which will ultimately reduce the seed yield. Other off type occurring in lines A and B should be rogued out. The line C can also be multiplied in the same way as mentioned above.

Increase in the Open

For planting large areas of commercial seed crop, it is necessary to increase the seeds of lines A, B and C in the open. The stock seed of line A can be produced by planting lines A and B in alternate rows in a large plot of 0.40 ha (1 acre) or more with several rows of B line, planted around and outside the field. If plot is smaller than 0.40 ha, it will have higher percentage of contamination from foreign pollen. The increase field should be at least 3.2 km (2 miles) away from other onion seed crops. The roguing of off type plants is done in the same way as in the cage. Usually, stock seed of line A produced in the open is used for production of hybrid seeds. The seeds of line C can also be increased in the open in a similar manner.

Production of Hybrid Seeds

The hybrid seed is produced in the open as in isolated field. The bulbs of line A (male sterile) and line C (pollen parent) are planted alternately in a ratio of four rows of line A and one row of line C. Sometime, eight rows of A line are planted with two rows of C line. The flowering in lines A and C must synchronize. If it does not, it can be accomplished by adjusting planting dates of lines A and C. In large fields, it is useful to keep about three or four beehives in 0.40 ha (1 acre), ensuring large population of honeybees for pollination. Daily roguing of pollen-bearing plants and other off types in line A in morning before the anthers dehisce is essential. Such off types or rogued plants must be pulled up and destroyed. At the time of harvesting much care is taken to avoid mixing of seeds of lines A and C and seed stalks of both lines are separated carefully before harvesting. When seed of line C is not being saved its flower stalks may be cut and destroyed as soon as the pollination is completed. The seeds collected from the rows of line A are F_1 hybrid seeds.

National Released Hybrids

Arka Kirtiman

This hybrid has been developed at IIHR, Bangalore. Bulbs are red with firm texture. Bulb has light flesh with 10% TSS. It takes 120-125 days to mature. Shelf-life of this hybrid is 5 months in ambient condition. It has good transport quality. Bulbs are free from split and bolting. Plants are resistant to biotic stress under field conditions. This hybrid is recommended for Andhra Pradesh, eastern parts of Madhya Pradesh and Orissa. The female parent (*MS 65*) of this hybrid is 50-60 cm tall. Bulbs are deep red, medium and flattish globe. It takes about 120-150 days to mature. However, male parent (Sel-13-1-1) is tall having deep red bulb. Bulbs are big and globe. It takes 120-125 days to mature.

Arka Lalima

This hybrid has also been developed at IIHR, Bangalore. Bulbs are deep red, globe with firm texture. Bulb has red flesh with 12% TSS. Shelf-life of this hybrid is five months at ambient condition. It has good transport quality. Plants are resistant to biotic stress under field conditions. It is free from split and bolting. It takes about 130-135 days to mature and gives 28% higher yield than the check N-53.

Seed Certification Standards

I. Field Inspection

A. Mother Bulb Production Stage

A minimum of two inspections are done as follows:

- The first inspection is done after transplanting of seedlings to determine isolation, volunteer plants, off types including bolters and other relevant factors.
- The second inspection is done after the bulbs have been lifted to verify true characteristic of bulbs.

B. Seed Production Stage

A minimum of four inspections are done as follows:

- The first inspection is done before flowering to determine isolation, volunteer plants, off types including bolters and other relevant factors.
- The second and third inspections is done during flowering to check isolation, off type and other relevant factors.
- The fourth inspection is done at maturity to verify true nature of plants and other relevant factors.

II. Field Standards

(A) General Requirements

1. Isolation: Onion seed fields should be isolated from the contaminants shown in column 1 of the Table below by the distances specified in columns 2, 3, 4 and 5 of the Table:

Contaminant	**Minimum distance (m)**			
	Mother root production stage		**Seed production stage**	
	Foundation	**Certified**	**Foundation**	**Certified**
1	**2**	**3**	**4**	**5**
Fields of other vaireties	5	5	1000	800
Fields of same hybrid not conforming to varietal purity requirements for certification	5	5	1000	800

(B) Specific Requirements

Factor	Maximum permitted (%)*	
	Foundation	Certified
*Roots not confirming to varietal characteristics including forked roots	0.10 (by number)	0.20 (by number)
**Off types (plants)	0.10	0.20

*Maximum permitted at second inspection at mother root production stage.
** Maximum permitted at and after flowering at seed production stage.

III. Seed Standards

Factor	Standards for each class	
	Foundation	Certified
Pure seed (minimum)	98.0 %	98.0 %
Inert matter (maximum)	2.0 %	2.0 %
Other crop seeds (maximum)	5 /kg	10/ kg
Weed seeds (maximum)	5 /kg	10 /kg
Germination (minimum)	70 %	70 %
Moisture (maximum)	8.0 %	8.0 %
For Vapour-proof containers (maximum)	6.0 %	6.0 %

17

HINTS FOR CULTIVATION

Tomato

Tomato, a warm season vegetable, requires a long season to produce a profitable crop. The optimum range of temperature for its cultivation should be 21–24°C. The mean temperatures below 16°C and above 27°C are not desirable. The minimum, optimum and maximum soil temperatures for seed germination are 10°, 30° and 35°C, respectively. Maximum seed germination occurs at 24°C. Maximum fruit setting occurs at night temperature of 15-20°C. Flowers fail to set fruit below 13°C and beyond 38°C. Formation of red colour due to lycopene is inhibited when temperature exceeds 32°C. Lycopene is highest at 21-24°C, while its production drops off rapidly above 27°C. Tomato plants are highly susceptible to high rainfall areas if proper drainage is not available.

It is grown in many types of soil from sandy to heavy clay. A well-drained fairly fertile loam soil with a fair moisture-holding capacity is ideal for growing its good crop. It prefers a soil having a pH of 6.0 - 7.0. Application of liming is beneficial if it is cultivated in acid soils.

It can be grown almost round the year. The planting seasons vary from place to place. In northern India, for spring summer crop, it is sown in November and transplanted in second fortnight of January. In area, where frost does occur, it is sown in July-August and transplanted in August-September. In hilly areas, it is sown in March-April and transplanted in April-May.

Tomato seeds are very light in weight. The seed rate varies from its size and weight. The seed requirement is 100-150 g/ha. Since, tomato is transplanted crop, hence, seeds are sown in nursery-bed for raising seedling. Generally, nursery-bed should be 7.5 m long, 1.2 m wide and 10 cm high from the ground. Well-decomposed farmyard manure is properly mixed into topsoil of the bed at the rate of about 3 kg/m^2. A fertilizer mixture of 0.5 kg NPK/ bed is also mixed in soil before sowing. For good seedlings, treating seeds with fungicide like captan/ bavistin is essential. Similarly, seedbeds are also treated with steam or formalin. A half-litre 40% formalin/m^2 soil is sufficient. Fumigation may be done with 10% formaldehyde in properly dug soil. After fumigation, beds are covered with polythene sheet for 24 hours. The seeds are sown 4-5 days after the removal of polythene sheet. The seeds are sown in bed either broadcast in rows. In broadcast method, incidence of damping off disease is less. A distance of 7.5 cm is kept between rows, this facilitates proper weeding and easy lifting of seedlings. The beds are covered with straw or polythene sheet until seeds germinate. The seedling should be protected against wind, exposure to sun and excessive rainfall. Fungicide, dithane M-45 or difolaton at 0.25%, if required should be sprayed weekly to reduce the post-emergence 'damping off'. Nursery can also be raised in polyhouse. Seedlings are ready for transplanting in 3-4 weeks, depending upon raising of nursery in open condition or polyhouse. In polyhouse, nursery becomes ready very early as compared to open condition. In areas, where, rainfall is a limiting factor for raising nursery, it should be raised in polyhouse. The seedling should be hardened before transplanting them in the field. This is done by withholding watering for 4-5 days before transplanting.

Tomato seedlings are also transplanted on flat beds or on sides of ridges. The seedlings are transplanted on side of ridge and later earthing-up is done to keep the plant in the centre of ridges. Planting distance varies with growth habit. For indeterminate hybrids, the row-to-row distance varies from 90-120 cm and plant to plant 60-75 cm, whereas determinate hybrids are planted at a spacing of 60-75 cm x 45-60 cm row-to-row and plant to plant, respectively. For good yield, 250-300 q/ha farmyard manure is incorporated in soil before transplanting. In general, nitrogen, phosphorous and potash are applied in the ratio of 200:100:100 kg/ha, respectively, depending upon fertility of soil. Half quantity of nitrogen and full of both phosphorus and potash are applied at the time of transplanting, while remaining quantity of nitrogen is applied either twice or thrice, depending upon soil condition at 30 and 45 days after transplanting.

Hoeing is done as soon as it is dry enough after every irrigation or shower. The weeds are removed in this process. Earthing-up is done after the topdressing. Weedicides can also be used for controlling weeds. Weedicides, oxyflurofen @ 0.25 a.i./ha (pre-emergence), goal @ 0.25 kg a.i./ha (pre-planting), pendimethalin @ 1.0 kg a.i./ha and basalin @ 1 kg a.i./ha (pre-planting) have been recommended. Staking is very essential operation in indeterminate group of hybrids for getting high yield and quality fruits. It may be done either by using bamboo stick or trained on wire. The beneficial effects of staking are: (i) improvement in plant yield and quality (ii) protects fruits from insects and diseases (iii) easy in harvesting and (iv) easy in spraying of any chemicals.

Tomato is a deep-rooted crop. It grows to a depth of 120–150 cm. Plants can withstand drought fairly well, especially on heavier or high humus soil. Its plants require adequate moisture throughout their growth period. Excess moisture is harmful. The crop requires 7-8 irrigations when 50 mm rainfall is received. The first irrigation should be done at the time of transplanting. Too much water after transplanting and before first fruit setting, causes the plants to run vines and drop the blossom off. In areas, where tomato cultivation is done through artificial irrigation source, its crop should be irrigated at 3-4 days intervals in summer and 10–15 days in winter, to maintain moderate moisture. Furrow irrigation is most widely used to irrigate tomato crop in India. In this method, it is essential to plant the seedling halfway down the ridges to maintain an optimum air water balance. Irrigation through drip system resulted 43.4% more yield and 78.4% water saving as compared furrow method of irrigation.

Apply fungicides and insecticides regularly for controlling diseases and insects, respectively. Various diseases and insects damaging tomato crops are given in Table 1 and 2, respectively.

Table 1: Major diseases of tomato

Disease and Causal Organism	Symptoms	Control Measures
Damping off and root rot (*Pythium* spp., *Rhizoctonia solani* and *Phytophthora parasitica*)	Pre-emergence 'damping off' consist of killing the seedlings from inital stage of seed germination to their emergence above soil. Post-emergence 'damping off' consists of rapid shrinking and darkening of cortical tissue of hypocotyle.	Soil treatment with formaldehyde two weeks before sowing and seed treatment with 1% mercuric chloride. Spraying of seedling with phytolan or captan is effective. Hot-water treatment of seeds at 50°C for 30 minutes is recommended along with five years crop rotation.
Early blight *(Alternaria solani)*	On seedlings, pale brown spots develop on cotyledons, stems and leaves. Cotyledons and young leaves may be killed and stem may be gridled. On established plant, dark brown spots with concentric rings develop first and old leaves spotted, leaves die prematurelly resulting in early defoliation. Fruits become sunscaled.	Seed treatment with thiram or captan or bavistin (2.5-3g/kg seeds). Foliar sprays of fungicides like difoltan (0.2 %) or dithane M-45 (0.2 %) or benlate (0.1 %) or (bavistin (0.1 %) have been recommended. Collection and burning of affected parts and residues of crop. Select seeds from healthy and unaffected plants.
Late blight *(Phytophthora infestans)*	Symptom appears on leaves and fruits. Irregular greasy greyish areas develop on leaves. These areas expand rapidly during the moist conditon and a white downy growth of fungus on affectged areas or surface of leaves. The foliage soon dries and shrivels. On fruits, olivaceous greasy spots are formed which gradually cover the entire fruit and the fruit soon disintegrates. This disease mostly occurs in cool, humid and cloudy weathers.	All the fungicides (except benlate and bavistin) recommended for the control of early and late blight disease can also use for controlling late blight.
Buckeye rot	The pathogen does not affect the foliage and thus the disease is distinct from late blight. The disease appears as brownish spot, often at the point of contact between fruit and spoil. As the spot enlarges, the surface of lesion assumes a pattern of narrow concentric ring of dark	Staking plants and removing foliage and fruits up to a height of 15-30 cm from ground level. Spray of difoltan (0.3 %)

Contd...

Table 1 Contd...

	brown and wide light brown bonds usally present within the affected area. When young green fruits are infected they usally become mummified and old affected fruits start roting.	at 10 days intervals.
Fusarium wilt *(Fusarium oxysporum)*	The disease is characterized by yellowing and wilting of leaves and finally the entire plants wilt are die prematurely. Stem tissue often is discoloured throughout the plant. Vascular browning takes place in the root.	In general, control of this disease is not possible. However, its infection can be minimized by crop rotation and planting of resistant hybrids. Wilt Incidence can be reduced significantly with the application of bavistin (0.1 %) or benlate (0.2 %) as soil drench.
Septoria spot *(Septoria lycopersici)*	The disease is characterized by numerous small grey circular leaf spots having dark border. A few black in point size pycnidia may be seen within the spot.	Seed treatment with dithane M-45 or dithane Z-78 (2 g/kg seed) is useful in checking seed borne infection. Above fungicides at @ 2.0 % can also used for effective control of disease in the field.
Bacterial wilt *(Pseudomonas solanacearum)*	Lower leaves may drop before wilting starts, the vascular system becomes brown and development of adventitious root along the stem may be enhanced. Slimy bacterial material may excude from the bundle when stem is cross-sectioned.	Use resistant hybrids for cultivation. Crop rotation such as cowpea-maize-cabbage, okra-cowpea-maize and maize-cowpea-maize Seedling treatment with streptocycline (1 g/40 litre) for 30 minute.
Tomato leaf curl virus	Severe stunting of the plants with downward rolling and crinkling of leaves. The newly emerging leaves exhibit slight yellow colouration and later they show curling. Older leaves become leathery and brittle. The nodes and internodes are reduced in size. The infected plants look pale and produce more lateral branches, giving a bushy appearance. The infected plants remain stunted. The disease is spread by vector (*Bemisia tabaci.).*	Grow resistant hybrids. Control insect vector *(Bemisia tabci)* by regular spraying insecticides, *viz.*, dimethoate (0.03%) or monocrotphos (0.04%) or metasystox (0.02%).

Table 2: Major insect-pests of tomato

Disease and Causal Organism	Symptoms	Control Measures
Tomato fruit-borer (*Heliothis armigera*)	The infection of this insect is more during July-November. The caterpillars are greenish. Immediately after the hatching the caterpillars crawl over the leaves and feeds vegetable parts. After that it entered in the fruit and make holes. Infested fruits are unfit for consumption.	Fort nightly sprays of any one of the insecticides like methomyl (0.2%), monocrotophos (0.05%), carbaryl (0.1%), endosulfan (0.07%) and quinalphos (0.05%) to control tomato fruit borer.
White fly (*Bamesia tabaci*)	This insect acts as vector, transmitting leaf curl virus causing severe loss of yield.	Foliar sprays of any one of insecticide *viz*; dimethoate (0.03%), phosphamidon (0.02%), monocrotophos (0.04%) and fenitrothion (0.1%) have been recommended for white fly control.
Root-knot nematode (*Meloidogyne incognita, M. javanica* and *M. arenaria*)	In sandy soil attack is more due to fast development of nematode populations and galls. Above ground symptoms consist of foliage discolourtion to pale than normal, unthrifty development, dwarfing, wilting in hot dry weather and in extreme cases plants die. When affected plants are uprooted, large and small roots contain swelling vary from spheroid galls to elongated spindles.	Use resistant hybrids for cultivation. Use nematicide like fensulfothion (2ml/l).

The stage of maturity at which tomato fruit is harvested, depends upon the purpose for which they are used and distance over which they are to be transported. Following stages of maturity for harvesting have been recognized.

Immature: Before the seeds have fully developed and the jelly like substance surrounding the seeds have formed.

Mature green: The fully grown fruit shows a brownish ring at stem scar, removal of calyx, light green colour at blossom end has change to yellowish green and seeds are surrounded by jelly like substances filling the seed cavity.

Turning: One-fourth of the surface at blossom end shows pink (breaker stage).

Pink: Three-fourth of the surface shows pink.

Hard ripe: Nearly all red or pink but flesh is firm.

Over ripe: Fully coloured and soft.

The yield depends upon various agroclimatic factors, region and hybrids. An average yield varies from 750 to 1250 q/ha.

Brinjal

Brinjal, a warm season crop, is susceptible to severe frost. It requires long warm season during growth and fruit maturation. The optimum growing temperature is 22-30^{0}C and growth stops at temperature below 17^{0}C. It is observed that pollen deformity increases at temperatures of 15-16^{0}C. It is observed that late round hybrids are more tolerant to frost than early long hybrids.

Brinjal can be grown in all soils from light sandy to heavy clay. The soil should be deep, fertile and well-drained. The soil pH should not less than 5.5-6.0 for its better growth and development.

In hills, the seeds are sown in April and the seedlings are transplanted in May. In the plains, crop can be grown in all the three seasons, *viz.* February-March, May-July and October-November.

About 125-200 g of seed is sufficient for raising seedlings for one hectare of land. Seedlings are raised as described in tomato. Seedlings of 8-10 cm height with 2-3 true leaves are ready for transplanting. The seedlings are hardened before lifting for transplanting. The planting is done on flat beds of convenient size. In high rainfall regions, brinjal plants are grown on raised bed. Spring summer crop is raised on ridges and furrow system for efficient use of water. Generally, long-fruited hybrids are transplanted at 75cm x 75cm while round-fruited at 90 cm x 60 cm.

Brinjal is a long duration crop occupying the land for nearly 6-8 months. Besides, it is a heavy nutrient feeding vegetable crop, removing large quantities of plant food in one cycle of plant growth. For good yield, 300-400 q/ha farmyard manure is incorporated in soil before transplanting. The nitrogen, phosphorous and potash are applied in the ratio of 200:100:100 kg/ha, respectively, depending upon soil conditions and type of hybrids. Half quantity of nitrogen and full of phosphorous and potash are applied at the time of transplanting. While, remaining half quantity of nitrogen is applied twice or thrice depending upon soil condition 30 and 45 days after transplanting.

Generally, manual weeding by khurpi and other small hand operated implements are used for removing weeds. Herbicides like fluchloralin @1.5 kg a.i./ha (pre-emergence) or pendimethalin @ 1.0 kg a.i./ha as pre-emergence can also be used for weed control.

Irrigation is essential for brinjal cultivation in the region where there is little or no rain during the growing season. Brinjal being shallow-rooted crop needs irrigation at frequent intervals. Timely irrigation is very important for high yield. It is recommended to irrigate brinjal after every third or fourth day in summer season and after 12-15 days in winter season. Irrigations are given as per the need of soil, stage of crop growth and weather conditions. When there is danger of frost, irrigation should be given at frequent intervals, so that, the soil is always kept moist. Brinjal is generally irrigated with furrow system of irrigation. However, drip irrigation has been found beneficial for reducing water use and weed growth as compared to conventional surface irrigation.

Brinjal is damaged by a number of diseases and insects, which are controlled by regular spraying of fungicides and insecticides as given in Tables 3 and 4, respectively.

Table 3: Major diseases of binjal

Disease and Causal Organism	Symptoms	Control Measures
Phomopsis blight (*Phomopsis vexans*)	The pathogen attacks all above ground plant parts. Leaves show well-defined grey to brown spots and the older spots exhibit numerous black pycnida. The affected leaves may turn yellow and die. Lessions may also develop on petioles and stems, causing blighting of affected portion of plant. On fruits, pale sunken spots are formed, which gradually invade the entire fruit, which subsequently turns into a black mummy.	Seed treatment by 0.1 % mercuric chloride or with other organomercural fungicides. Hot-water treatment of seeds at 50°C for 30 minutes. Spraying at regular intervals in the nursery and in the main field with 1% Bordeaux mixture (4:4:50). Foliar spraying of carbendazin @ 0.1%, starting first spray after a month of planting followed by subsequent spray at 10 days interval.
Damping off (*Pythium* spp. *Phytophthora* spp. and *Rhizoctonia* spp.	The disease is very severe in the nursery. The fungus attack usally starts on the germinating seeds spreading to the hypocotyls, basal stem and developing tap root. The affected seedlings are pale green and brownish lesion found at the basal portion of the stem. Lesion girdles the stem, later extending upwards and downwards. The affected tissue rot and seedling collapse.	Soil should be sterilized before sowing and the seed are treated with bavistin (0.1%) prior to sowing.
Fruit rot (*Phytophthora nicotianae* B.de Haan Var. *nicotiane*)	Small water-soaked lesions on fruits, which later enlarges in considerable size. Skin of infected fruits turns brown and develops white colony growth.	Staking the plants, removing foliage and fruits up to a height of 15-30 cm from ground. Spray difoltan (0.3%) twice at 10-15 days intervals.
Fusarium wilt (*Fusarium oxysporum* of. sp. *lycopersici*)	Yellowing and wilting of leaves characterize the disease and finally the entire plant wilts and dies prematurely. Stem tissue often discoloured through out the plant.	As mentioned in tomato.
Bacterial wilt (*Pseudomonas solanacearum* (E.F. Sm.)	The typical symptoms of disease are wilting, stunting and yellowing of foliage, followed by collapse of the entire plant. The pathogen is mostly confined to vascular region in advanced cases, it may be invade the cortex and pit and yellow brown.	As mentioned in tomato.

Table 4: Major insects of brinjal

Insect Name	Nature of Damage	Control Measures
Shoot-and fruit-borer (*Leucinodes orbonalis*)	Damage by this pest starts soon after transplanting and continues till harvesting of fruits. Eggs are laid singly on ventral surface of leaves, on flower buds and occasionally on young fruits. In young plants, the caterpillars bore into petioles, midribs of leaves, young shoots and feed within. As a result, the affected leaves dry and drop off the growing point is killed in case of shoot damage. At crop growth stage, caterpillars bore into flower buds and fruits entering from under the calyx. Affected fruits become partially unfit for consumption.	Spray insecticides, *viz.* cypermethrin @ 30 g a.i./ha at 10 days intervals or deltamethrin @ 10 g a.i./ha or endosulfan 700 g a.i./ha at 15 days intervals starting from 10 days after transplating or dimethoate @ 300 g a.i./ha (two sprays).
Jassids (*Amrasca biguttula*)	Both nymph and adults suck the sap from the lower surface of leaves. The infested leaves curl upward along the margins, which may turn yellowish and show burnt up patches.	Spray insecticides like cypermethrin (30 g a.i./ha or deltamethrin (10 g a.i./ha) or endosulfan (700 g a.i./ha) at 15 days interval, starting from 10 days after transplanting.
Aphids (*Aphis gossypii* Glov.)	Adults suck the sap of leaves	Application of granules of carbofuran @ 1000 g a.i. during transplanting followed by 2-3 sprays of cypermenthrin @ 30 g a.i./ha at 10-15 days interval.
Root knot Nematodes (*Meloidogyne* spp.)	The plants become stunted and leaves show chlorotic symptoms. Fruiting is adversely affected.	Use nematicides and resistant hybrids.

Fruits are harvested as soon as they attain good size and colour. A heavier crop is produced if fruits are removed before they reach full size. The fruit should not loose its bright and glossy appearance. When a fruit looks dull, is an indication of over maturity and loss of quality. Pressing the thumb against the side of the fruit can test the maturity of the fruit. If the pressed portion springs back to its original shape, the fruit is too immature. The average yield varies from 400 to 800 q/ha.

Chilli

- Chilli is grown in both tropical and subtropical areas up to 2,000 m above mean sea-level in India. A warm humid climate favours growth, while warm and dry weather enhances fruit maturity. Heavy rainfall leads to poor fruit setting and in association with high humidity leads to rotting of fruits. In general, a frost-free period of four months with maximum temperature of 20-30^{0}C and minimum not below 10^{0}C is optimum for raising its crop.
- It grows in a variety of soils, provided they are well-drained, well-aerated and rich in organic matter. It can be grown successfully in sandy loam soil provided assured irrigation and manuring. Black soils are also suitable for chilli as rainfed crop. An ideal soil is light loamy or sandy loam rich in lime and organic matter.
- In north India, its seeds are sown in nursery-beds during May-July, depending upon onset of monsoon. In north-eastern region, sowing is little earlier due to early monsoon arrival. In southern states, and rainfed areas, May-June and September-October are best seed sowing time.
- About 200-300 g seed is required to raise seedlings for one hectare. Nursery raising is similar as described in tomato.
- Chilli has a long-growing season and, therefore, requires a judicious use of manures and fertilizers. About 200-250 q/ha farmyard manure should be applied 10-15 days before planting. It is thoroughly mixed in the soil. For getting good yield, nitrogen, phosphorous and potash are applied in the ratio of 200:100:100 kg/ha, respectively. Whole quantities of both phosphorus and potash along with half nitrogen should be applied at the time of transplanting, while remaining half as top dressing in two equal splits at one month intervals after transplanting.
- Chilli is slow-growing crops, incapable of offering any competition to aggressive weeds. Infestation of weeds also harbours insect-pest that cause damage to crop. Hence, weeds are controlled either by hand weeding and hoeing or by the application of weedicides. Frequent shallow cultivation is done at regular intervals to keep the field free from weeds and to facilitate soil aeration and proper root development. Weedicides, pendimethalin (1kg/ha), oxadiazon (1kg/ha) and fluchloralin (1.25kg/ha), can also be used to control weeds.
- The first light irrigation is done just after transplanting for better establishment in the soil if there is no rain. Gap filling is done during the second irrigation 10 days after transplanting. Later on, field is irrigated as and when required. Generally, 8-9 irrigations are required depending on rainfall, soil type, humidity and temperature. Furrow irrigation is generally practised.
- Apply fungicides and insecticides regularly for controlling diseases and insects. The major diseases and insects of chilli, are described in Table 5 and 6, respectively.

Table 5: Major diseases of chilli

Disease and Causal Organism	Symptoms	Control Measures
Damping off (*Pythium aphanidermatum* and *Rhizoctonia solani*)	As described in tomato	As described in tomato
Anthracnose and fruit (*Colletotrichum capsici*)	Circular and sunken spots with black margins characterizes the disease. The spotted fruits drop off prematurely resulting in spots with black margins. In severe cases, it may also attack the fruit stalk and spread along with stem, causing die-back symptoms. The fungus is seed-borne and spread with wind blown rains.	Seed treatment with bavistin (2g /kg seed) Spraying of bavistin (0.1%) is also effective.
Fusarium wilt (*Fusarium annum*)	As described in tomato.	As described in tomato.
Powdery mildew (*Oidiopsis taurica*)	The disease is characterized by development of powdery mass on leaves and rapid defoliation of plants.	Three to four sprays of karathene (0.1%) at 10 days interval.
Phytophthora blight (*Phytophthora nicotianae*)	The pathogen can affect all parts of the plant. 'Damping off' can develop on seedlings. Symptoms on old plants include root rot, stem canker, leaf blight and fruit rot. Cool and wet weather condition promotes disease development.	Foliar applications of any one of the fungicide like blitox (0.3 %) or dithane Z-78 (0.2%) or Bordeaux mixture (1%) is effective.
Bacterial leaf spot (*Xanthomonas vesicatoria*)	Symptoms are characterized by yellowish-green spots on younger leaves and dark & water-soaked lesions developed on older leaves. Small blister like spots occurs on fruits, later becoming warty in appearance. The pathogen is soil-borne.	Chemical control methods for this disease are similar to tomato. Use resistant hybrids for cultivation.
Bacterial wilt *(Pseudomonas solanacearum)*	As described in tomato	As described in tomato.
Leaf curl virus	As described in tomato	As described in tomato. Use resistant hybrids like Hybrid No.1, 5, 6, and 9, Bijli, Agni, NS-2302 and NS-1820.

Table 6: Major insects of chilli

Insect Name	Nature of Damage	Control Measures
Thrips (*Scirtothrips dorsalis* Hood)	Thrips may cause loss from 25 to 50%. These are minute fragile insects with fringed wings. Nymphs and adults lacerate the tissues and suck sap from leaves, makes curl and dry at tips.	Spray malathion (0.1%) or carbaryl (0.2%) 50WP or nuvacron 0.1% or metasystox (0.1%) or monocrotophos (0.05%).
Fruit-borer (*Spodoptera litura* Fabricius)	The pest is polyphagous, breeds throughtout the year. The moths are brown with patterned forewings, tiny dark caterpillar, feed gregariously, scraping the leaves, later the caterpillars segregates and assume cutworm habits feeding voraciously on leaves in nurseries and fields.	Spray of carbaryl (0.2%) or quinalphos 25 EC (2 ml in 1 litre of water) or dimethoate (0.03%) or monocrotophos (0.03%).
Aphid (*Aphis gosypii* Glover, *Myzus persicae* Sulzer)	Aphids are the most common and widely distributed insects of chilli. They suck the sap from leaves. In addition to that they transmit mosaic disease.	Spray malathion (0.1%) or demeton methyl (0.05 to 0.2%).

- In chilli, flowering begins 40-60 days after transplanting, depending upon hybrids, climate and nutritional status of soil. Fruits start ripening about three months after transplanting and picking may go on for 2-3 months. Chilli hybrids can yield 250-300 q /ha green fruits.

Bell Pepper

All cultural practices as described for cultivation of chilli hybrid, are similar for bell pepper. However, some additional informations are given below:

- The spacing should be 60 cm x 30 cm.
- The yield varies from 200 to 400 q/ha of green fruits.

Okra

- Okra does best in long warm season. It is sensitive to frost and thus requires frost-free growing period. Optimum temperature requirement is 24-27 °C for its successful cultivation. Seeds do not germinate when temperature is below 20°C. Okra seeds germinate easily during rainy season but it is poor in north Indian plains, if sown in February. The plant growth is better in rainy season as compared to spring-summer and so also the yield. Temperature higher than 42°C may cause flower drop.
- Okra can be grown on any type of soil-loam to clay loams. Soil should contain medium to high organic matter. The soil should be well-drained as its roots are sensitive to water stagnation. The pH of soil should be 6.0-6.8.

- In north Indian plains, spring/summer crop is sown in February-March, whereas in eastern and western India, optimum sowing time is January-February. In West Bengal, its sowing starts in February and continues up to June at an interval of 3 or 4 weeks. In south India, summer season crop is sown in November. The rainy season crop is sown in June-July throughout India. In north Indian Hills, it is sown in April-June.
- The seed rate of okra hybrid ranges from 5 to 10 kg/ha, depending upon sowing time. The recommended spacing for summer season crop is 45 cm x 45 cm, whereas for rainy season, it should be 60 cm x 45 cm.
- For obtaining good yield, nitrogen, phosphorus and potash are applied in the ratio of 150:100:100 kg/ha, respectively. Besides NPK, farmyard manure @ 250 q/ha is applied 10-15 days before sowing. It is thoroughly mix in the soil. Half quantity of nitrogen and whole phosphorus and potash are given at the time of sowing. Remaining quantity of nitrogen is given twice 30 and 60 days after seed sowing.
- First weeding is done 15-20 days after seed sowing. Total 3-4 weedings are needed. Following herbicides can also be used for weed control.

Weedicide	Rate (kg/ha)	Time of Spraying
Basalin 48 EC (flurchloralin)	2.5 litres (1.2 kg)	Pre-plant incorporation
Lasso 50 EC (alachlor)	5 litres (2.5 kg)	Pre-emergence
Stamp 30 EC (pendimethalin)	2.5 litres (0.75 kg)	Pre-emergence incorporation

- There should be enough moisture in soil at the time of sowing. First irrigation is given after the seeds have germinated. Subsequent irrigations are given at 4-5 days intervals during summer. In rainy season crop, there is no requirement of extra irrigation if there is regular rainfall. Drip irrigation is better than furrow irrigation.
- Control disease and insects by regular spraying of fungicides and insecticides. Major diseases, insects with their control measures are given in Tables 7 and 8, respectively.

Table 7: Major diseases of okra

Disease and Causal Organism	Symptoms	Control Measures
Powdery mildew (*Erysiphe cichoracearum*)	The infection occurs on all plant parts, *i.e.* leaves, stem/and fruits. The pathogen produces blotches of white-greyish powdery coating at under surfaces of leaves. Severely affected leaves turn yellow and drop off.	The disease can be controlled by foliar spraying of any one of the fungicide like calixin (0.05 % five sprays) or sulfex (0.3% four sprays) or bengard (0.1% four sprays)
Wilt (*Fusarium oxysporum* var. *vasinfectum*)	Affected plants show yellowing and stunting followed by wilting and death. Dark streaks develop inside infected stems.	Seed-borne infection can be checked by soaking seeds in ziram (0.3 %).

Contd...

Table 7 Contd...

Leaf spot (*Cercospora abelmoschi*)	Spots are formed on leaves which often coalase and later defoliated. Affected leaves roll, wilt and fall.	Spraying of copper oxychlorides (0.3%) or maneb (0.2%) or zineb (0.2%) or captan (0.2%).
Yellow vein mosaic	The characteristics symptoms of this disease is homogenous interwoven network of yellow vein enclosing islands of small tissues within initially leaves exhibited only yellow colour veins. But in later stage, entire leaf turns completely yellow or becomes fully cream coloured. The chlorophyll content of leaf is partially or completely destroyed. Infected plant remains stunted and produced a very few small leaves. Most of the affected plant develops thickening of veins on their lower sides. Hardly one or two fruits, often deformed, small, pale in colour and tough in texture are produced on infected plants.	Spraying of copper oxychlorides (0.3%) or maneb (0.2%) or zineb (0.2%) or captan (0.2%). Use resistant hybrids for cultivation. Control insect vector by regular spraying of dimethoate (0.05%) or metasystox (0.02%) or nuvacron (0.05%).

Table 8: Major insects of okra

Insect Name	Nature of Damage	Control Measures
Shoot and fruit-borer (*Earias* spp.)	The caterpillar, which has black and brown spots, bores into the growing points of young plants. It also bores into flower buds and developing fruits. They also fill the holes with its excreta. Affected fruits become unfit for human consumption.	Apply any one of the systematic pyrethroids, *viz.* permenthrin (50 g) or cypermethrin (30 g) or fenvalerate (50 g) at 10 or 25 days interval.
Leafhopper (*Amrasca biguttula* Ishida)	Leafhopper attack causing the leaves to curl upwards along the tips and margins and develop necrotic areas, which extend over the entire leaf surface.	Apply carbofuran (1 kg a.i./ha) or phorate (1.5 kg a.i./ha). Besides above, monocrotophos (0.05%) or cypermethrin (0.05%) can also be used control leafhoppers.
Okra stem fly	Maggots of okra stem fly attack 15-30 days old plants, causing nodulated galls and swellings. In later stages, maggots of this fly damage petioles.	Seed treatment with imidacloprid @ 5g/ kg seed can control the insect.

- Its pods are ready for harvesting 45-60 days after sowing depending upon hybrid and season. Only tender fruits are harvested. Frequent pickings are necessary, so that pods do not grow too large. Harvesting is done at interval of four days to get the highest pod yield. The average fruit yield varies from 500 to 600 q/ha.

Cabbage

- Cabbage can grown easily under a wide range of environmental conditions but cool moist climate is most suitable. The optimum soil temperature for seed germination and growth should be 21.2-26.2 ^{0}C and 25.2-34.2 ^{0}C, whereas above 43.2 ^{0}C growth is arrested in most of the hybrids. Minimum temperature for growth is just above 0°C. After cold treatment at 4-10 ^{0}C for 7-9 weeks, the plant bolt sooner and flower more abundantly than those exposed to such temperatures for shorter period but before this temperature treatment, plants must have passed juvenile phase (Nieuwhof, 1969).
- Cabbage can be grown in a wide range of soils. Early maturing cultivars grow well in light soils, whereas late-maturing hybrids perform better in heavy soils. The optimum pH of soil for its cultivation is 6.0-6.5. Most of the cabbage varieties can tolerate salt to some extent.
- Seed sowing time in different parts of India varies. In hills, especially in high hills, it is sown in May-June for summer/autumn crop. In hilly areas receiving heavy rains, summer and autumn crop is rather limited and sowing is undertaken in autumn to harvest them in late spring in early summer by over watering them. In plains of northern India, sowing *in-situ* or in seed-bed starts from early-August and continues till November for late cultivars. In eastern India, sowing generally starts from mid to late September. Sowing is delayed further for areas in southern hills except in western and southern peninsular, where its crop can be grown almost round the year by selecting proper hybrids.
- About 150-300 g seed/ha is enough. Its seeds are generally sown in a seed-bed and 4-6 weeks old seedlings are transplanted to field. In plains, the seed-beds for early crop requires cover to save the small seedlings from rains, while in high hills, it may be better to grow them under glass flames or poly tunnels. However, seedlings of mid season or late cultivar may be raised in the open. The soil of nursery-bed should be well-prepared and free from pathogenic organisms. The sieved well-rotten FYM or compost @ 2-3 kg/m^2 is added in seed-bed. If necessary, NPK may be used. Before sowing, seeds are treated with suitable fungicide like thiram or cerasan @ 3 g/kg seed. The optimum spacing between rows in nursery-bed is kept 10 cm and depth of sowing should be 1.5-2.5 cm. One gram seed gives rise to about 1,000 seedlings. After sowing, seeds are properly covered with a thin layer of mixture of fine manure and soil. The beds are levelled gently. A regular and good moisture supply is needed for rapid germination of seed and optimum growth of seedlings. To keep the upper soil of seed-bed moist, a thin layer of dry grass is spread on beds and watering is done in the form of shower either with a watering cane or with the help of a sprinkler. Seed-bed cover is removed as soon as the emergence of young seedlings above the ground starts. If there is over crowding of seedling due to thick sowing, extra seedlings should be thinned out.
- Attention should also be paid for control of diseases and insects in nursery. 'Damping off' caused by different fungi, *viz. Pythium* spp, *Rhizoctonia* spp is common and severe disease problem in nursery-beds. The spray of blitox-50 @ 0.3% is effective against this disease. Generally, 4-6 weeks old seedlings are ready for transplanting.

But seedlings up to 8 weeks old can also be transplanted. However, older seedlings have adverse effect on the establishment and growth of young plants. Planting distance may be varied as per type of hybrids, planting season and soil conditions. The recommended planting distances on the basis of maturity are: 45 cm x 45 cm or 60 cm x 30 cm, for early-maturing hybrids, 60 cm x 45 cm for mid maturing and 60 cm x 60 cm for late-maturing ones. Planting is done on flat land, ridges or in furrows depending on climate and soil conditions. For early planting, ridge method is be more suitable, especially in areas where the rains occur at the time of planting. In saline soils, planting is done in furrows.

- About 200-250 q/ha farmyard manure is thoroughly mixed in soil 15-20 days before transplanting. Nitrogen, phosphorous and potash are applied in the ratio of 200:100:10 kg/ha, respectively. Half quantity of nitrogen and full of phosphorus and potash are applied at the time of transplanting. Remaining quantity of nitrogen is applied 30-45 days after transplanting.
- The regular intercultural operations are necessary for proper aeration of root system and control of weeds to promote healthy plant growth. Manual weedings 2-3 times are required. Weedicides, trifluralin @ 0.5 kg/ha (soil incorporation) and fluchloralin @ 0.5 kg/ha (soil incorporation), can also be used for weed control.
- Cabbage is very sensitive to soil moisture. Maximum growth and yield can only be obtained when a plenty of water is available to plants throughout growth. First irrigation is given just after transplanting of seedlings and thereafter irrigations may be done at 10-15 days interval, depending on season and soil conditions. But optimum soil moisture should be maintained regularly. Cabbage is usually irrigated by furrow method of irrigation. Excessive irrigations in early stages cause superficial rooting and washing down of nutrients. Heavy irrigation is avoided when heads are formed, as it results in splitting of heads.
- Apply fungicides and insecticides regularly for controlling diseases and insects. The major diseases and insects along with their control measures are given in Tables 9 and 10, respectively.

Table 9: Major diseases of cabbage

Disease and Causal Organism	Symptoms	Control Measures
Downey mildew (*Perenospora parasitica*)	Plants can be infected at any time during their growing period. Young plants infected early show a white mildew growth mostly on under side of leaves. Later, however, slight yellowing shows up the corresponding upper surface. The younger leaves may 'drop off'. Older leaves usually persist and the infected areas enlarge. When disease is severe, the whole leaf dies.	It can be controlled by spraying the seedling in nursery-bed with copper oxychloride (0.3%).

Contd...

Table 9 Contd...

Wire stem (*Rhizoctonia solani*)	Young plants show soft water-soaked spots on skin just above the ground, the cotyledons wither and plant eventually falls over and perishes.	Seed treatment and soil drenching with dithane M-45 (0.2%) or bavistin (0.1 %).
Black rot (*Alternaria brassicae*)	It is a destructive disease on seed ccrop, affecting its pods and seeds. The first symptom is a small dark spot on leaf surface. As the spot enlarges, concentric rings develop. The diseased spot enlarges progressively and defoliate plants.	The disease can be controlled by spraying of captan (0.2 %) or copper oxychloride (0.5%).
Cabbage yellow (*Fusarium oxysporum* fsp. *conglutinans*)	Affected plants have a sickly-dwarfed yellow appearance, lower leaves drop one by one and growth of plant becomes stunted.	Use resistant hybrids for cultivation. Field sanitation, crop rotation and using disease-free seeds are useful.
Black leg (*Phoma lingam*)	The seed carries the fungus and hence it may occur from the early stage. Stem of affected plant when split vertically, shows severe black discolouration of sap stream. Whole root systems decay from bottom upward. Frequently, the affected plants fall over in the field.	Spray copper oxychloride Use resistant hybrids for cultivation.
Black rot (*Xanthomonas campestris* pv. *campestris*)	The infected plants become stunted, often one side of cotyledons turn yellow to black bends down and shed off prematurely. Many lower leaves are also shed off early. The remaining leaves may turn yellow with blackened veins. Development of 'V' shaped chlorotic yellow lesions takes place from the leaf margin. The vein and veinlet turn black and leaf tissue necrotic and brittle. Lesions progress towards the midrib and from this systematic infection further spread into stem and root. Vascular tissues in affected parts become conspicuously black, followed by internal breakdown of fleshy tissues. Black sunken spots may develop on suculent seed stocks and siliquae.	Disease can be considerably reduced by seed treatment with bactercide, plantomycin (100 ppm) or agrimycin (100 ppm) or streptocyclin (100 ppm).

Table 10: Major insects of cabbage

Insect Name	Nature of Damage	Control Measures
Mustard sawfly (*Althalia proxima*)	The blackish caterpillar feeds on young seedling in early stages. The infested leaves become curl and fall on the ground when touched.	Dusting the plants with sevin @ 20-25 kg/ha or spraying the same @ 5g/ litre of water is effective.

Contd...

Table 10 Contd...

Mustard Aphids (*Lipaphis erysimi*)	The greenish white small insect whose attack on cabbage is more when there is cloudy weather. They suck the sap from the plants. The affected leaves get curled and plant wither and die.	Aphid can effectively be controlled by spraying of methyldemeton or dimethoate @ 1.5 ml/litre or phosphamidon @ 2.0 ml/litre
Diamond back moth (*Plutella xylostella*)	A small slender pale green caterpillar that feeds on leaves and makes holes in them. The infested crop becomes unsuitable for marketing.	Foliar spraying (twice or thrice) of permethrin (0.2 kg a.i.) or fanvalerate (0.1 kg a.i./ha) or prothiophos (0.75 kg a.i./ ha) is useful of mustard as trap crop proves is effective in restricting diamond back moth.
Cabbage butterfly (*Pieris brassicae*)	Green caterpillars with black tubercles feed gregariously on the plants causing extensive defoliation of leaves.	Insecticide like Fenvalerate 0.1 kg a.i./ha has been found most effective for controlling this insect.

- The heads are harvested when reach at marketable size. The early hybrids grown under comparatively warmer conditions, develop loose head initial stage, but become harder at maturity. In some cultivars, heads start cracking soon after the maturity. In such cases, quantity of heads is deteriorated fast if harvesting is delayed. Hence, harvesting is done at right stage for getting good quality heads. Its average yield ranges from 500 to 700 q /ha.

Cauliflower

- The cauliflower has a wide range in adaptation to diverse climatic conditions, prevailing in temperate, subtropical and tropical parts. Therefore, cultivation is widespread in many parts of India. The optimum temperature for growth of young plants is around 23°C but 17-20°C is more favourable in later stage. The tropical cultivars show growth even at 35°C. In temperate region, growth of young seedlings may be ceased when temperatures are slightly above 0°C, while, early cultivars grown in plains of north India and other tropical parts can grow even at 35°C or still higher temperature. Lower temperature ranging from 5°C to 28-30°C is needed from vegetative to curding phase. The temperature higher or lower than the optimum required for curd formation may cause physiological disorders, *viz*. riceyness, leafy curd and blindness.
- Cauliflower can be grown in all types of soils with good fertility and good moisture regime. In light soil, plants are more sensitive to drought and therefore, adequate moisture supply is important. For early crops, light soil is preferred, while loamy and clay loam soils are more suitable for mid-season and late-maturing types. Cauliflower is relatively more sensitive to deficiency of boron and molybdenum and it has high

requirement of magnesium. The deficiency of magnesium may quickly appear in acid soils. The optimum time of sowing in nursery depends upon climate, varieties and their temperature requirement for curd formation. The nursery for mid-season crop can be grown successfully but the protection of beds from heavy rain during July-August is essential. The seedlings for mid-late and late crop can be conveniently raised because of favourable temperature conditions. Depending upon curd maturity of hybrids, following sowing times have been recommended.

Table 11 :

Group	Time of Seeds Sowing	Time of Curd Maturity
Early group		
I (a)	May-June	September
I (b)	First fortnight of July	October to mid-November
Mid-season group II	July-August	Late-November to mid-December
Mid-late group III	September	Late-December to mid-January
Late group	October	Mid-January onwards

- About 200-300 g seed is needed for one hectare crops, though in early crops, its may be more (500 g/ha). For raising nursery, the nursery-beds are prepared as descried in cabbage. The planting distance may vary according to hybrid, climate and soil. Normally, seedlings of early hybrids are planted at spacing of 60 cm from row-to-row and 30-45 cm from plant-to-plant. The main season and late groups of hybrids are planted at a spacing of 60-75 cm between rows and 45-60 cm between plants. About 150-200 q/ha farmyard manure is applied in soil before 15-20 days of transplanting. It is mixed thoroughly in the field. Nitrogen, phosphorus and potash are applied in the ratio of 150:100:125 kg/ha, respectively. Half quantity of nitrogen and full of phosphorus and potash are applied at the time of transplanting. Remaining quantity of nitrogen is applied 30-35 days after transplanting.
- The intercultural operations are done regularly to keep its crop free from weeds and proper aeration of root system. Hoeing should not be deep to avoid injury to roots. Usually, in medium heavy and clay soils, there is crust formation soon after transplanting. This crust must be broken, otherwise, water and air penetration in root system is hindered which affect plant growth. During rainy season, the roots of plants in ridge planting may be exposed which requires adequate earthing. When earthing or ridging is done, side dressing of nitrogenous fertilizer may also be done for healthy growth. Weedicides, basalin @ 2-3 kg a.i./ha + 1-2 hand-weedings or tenoran 50 WP @ 1 kg a.i./ha can be used for weed control.
- For getting quality curds, blanching is an important operation to protect the curds from yellowing due to direct exposure to sun. The curds may also loose some of their flavour because of sun exposure. This problem generally occurs in such hybrids of early and mid-season maturity group, which have spreading and open plant type. In some hybrids, the curd remains naturally protected and surrounded by inner whorls of leaves. These are called self-blanching types. Drawing and tying the tips of leaves when curds are fully developed may do blanching. The other way to place a leaf over the cauliflower curd 4-5 days prior to harvesting.

- First irrigation is given just after transplanting. Further, irrigation depends upon weather, soil type and hybrids. However, regular maintenance of optimum moisture supply is essential during both growth and curd development stages. For early and mid-season crops, usually less irrigation is needed because of rains. In areas having rainfall, planting is done on ridges and furrow irrigation has been found beneficial.
- Apply fungicides and insecticides regularly for controlling diseases and insects. Most of the diseases and insects, which affect cauliflower, are similar to those of cabbage. Hence, follow same control measures for diseases and insects as described cabbage.
- The harvesting is done as soon as curds attain right maturity and become compact. The white colour of curds is maintained. If harvesting is delayed, curds become over-matured whose quality is deteriorated. Such curds may turn loose, leafy, ricey or fuzzy. The over-matured curds should be sorted out while sending them to markets., The curds are cut off stalk well below the curd with a sharp cutting knife, sickle or *khurpi*. On an average, yield of cauliflower hybrid ranges from 500 to 750 q/ha.

Bitter Gourd

- Bitter gourd is grown under tropical and subtropical climates but warm climate is considered best. It cannot tolerate high cold and frost. It requires 18^{0}C minimum temperatures for seed germination and 20-30^{0}C for growth and development. Short days help in increasing female flower production.
- It requires sandy to sandy loam soil for early and good crop. The soil pH should be 5.5-6.7. For higher yield, soil should be rich in organic matter with proper drainage facilities.
- The sowing time of bitter gourd are: (i) February-March for summer crop in northern plains but January-February in north-west India, (ii) West Bengal and south India, in October- November, (iii) April for rainy season crop in West Bengal, but June-July in the rest of India and (iv) April and May in hills of north India.
- The recommended seed rate is 2.0-2.5 kg/ha. In north India, sowing is done on raised beds or in furrow or trenches, whereas in south India, it is sown in pits. In West Bengal, pits of 60 cm x 60 cm size are dug and filled with farmyard manure and topsoil of pits. Two seeds per hill are sown in both sides of raised beds. The recommended spacing is 1.5-2.0 m between rows and 60-120 cm between plants.
- Farmyard manure @ 200-250 q/ha is applied 15-20 days before sowing in pits/furrows. The nitrogen, phosphorus and potash are applied in the ratio of 100:50:50 kg/ha, respectively. Half quantity of nitrogen and full of phosphorus and potash are applied in pits or furrows at the time of sowing. Remaining quantity of nitrogen may apply 30 days after sowing or just before flowering. Seed treatment with boron at 3-4 mg/litre may be useful for obtaining quality fruits with higher yield. For higher production, vines are trained on bamboo with rope or wire on bower system. This training/support has manifold advantages over ground method. They are:

- Expansion of fruit area, which result in increasing number of fruits and total yield.
- Fruits become bigger in size.
- Intercultural operations can be done easily.
- Better pollination and fruit setting.
- Fruit becomes very attractive due to proper exposure to sunlight.
- Less incidences of fruit fly or other insects, fungal, bacterial and viral diseases.
- Plant protection measures like spraying, dusting *etc.* can be done easily.
- Hundred per cent clean harvesting can be done easily.
- No scab or rotting of fruits.
- For preparation of bowers, bamboos, cemented/iron poles, wooden iron angles may be used. There will be another requirements like pole-connecting wire of 16 gauges, one heavy stone and for making net wire of 10 gauges are quite useful. Cemented/iron poles of 2 m length, which have base of 12.5 m x 12.5 m and upper end 10 cm x 10 cm size, are placed at the distance of 3 m x 3 m. From the surface of soil, poles are buried up to a depth of 0.45 m in soil. Poles at corners are supported by two extra poles in a banding manner, if required central pole may also be provided. Apart from this, if required a heavy stone may be kept in pits of 1m x 1m size. The stone is tied with the corner by connecting wire of 16 gauges. In place of cemented poles, angle iron poles can also be used leaving 30-35 cm upper ends. The long rod is placed in such a manner that at least 15 cm remains above the pole. The entire pole should be connected by 10-gauges single GI wire, then 16-gauge wire should be stretched over 10-gauges wire at the distance of 60 cm criss-cross. Total 1,550 cemented/iron poles, 150-200 kg supporting wire, 400 kg criss-cross netting wire and 100-120 kg jute/coconut or 25 kg plastic rope are required for preparation of bower for one hectare. In bower system of training, plant height should be 2.55 m, maintained a spacing of 1.5 m and 0.6 m from row-to-row and plant-to-plant, respectively. The hybrids Vevek and Tijarti are suited for bower system of training. The field should be free from weeds throughout cropping season. Avoid herbicides for controlling weeds.
- Since bitter gourd is sown both in rainy and summer seasons, rainy season crop normally does not require much watering except during dry spells. Summer crop is irrigated immediately after planting. Subsequently, light irrigation is given every fourth or fifth day until flowering. It is necessary to maintain adequate moisture during flowering and fruit development to achieve high yields. The crop is irrigated through basin or furrow.
- Apply fungicides and insecticides regularly for controlling disease and insects. Major diseases and insects with their control measures are given in Tables 12 and 13, respectively.

Table 12: Major diseases of bitter gourd.

Disease and Causal Organism	Symptoms	Control Measures
Downey mildew (*Pseudoperonospora cubensis*)	The symptom appears on upper surfaces of leaves as angular yellow spots. During favourable conditions, the undersides of these spots are covered with a greyish mouldy growth. As the spots enlarge, a general yellowing of leaf occurs followed by the death of tissue. The leaves subsequently wither and die.	The recommended fungicides for control of this disease are ridomil (0.2%) or blitox (0.2%) or dikar (0.3%) or daconil (2 kg/ha) or dithane Z-78 (0.3%).
Powdery mildew (*Erysiphe cichoracearum*)	White to dirty grey powdery growth appears on leaf, petiole and young stem. Severely infected leaves become yellow and eventually shrivelled. Stems may also be attacked and plants become chlorotic and stunted when disease severe, plants die.	For effective control measures of this disease spray fungicides like elosal (0.5%) or karathane (0.05%) or thiovit (5 kg/ha) or benomyl (300 g/ha).
Anthracnose (*Collectotrichum lagenariarum*)	The first symptom is noticed on older leaves as round brown spots. The centres of these spots fall out giving a 'shot hole' appearance. Light brown to black, elongated streaks develop on stems and petioles. Round sunken water spots appear on fruits.	Seed treatment with thiram or bavistin or dithane M-45 (2 g/kg seed) is effective. Spray difoltan (0.2%) or blitox (0.25%) or bavistin (0.2%) or benomyl (0.1%) or thiophanate (0.1%).
Cucurbit phyllody	The disease is caused by mycoplasma like organism. The characteristic symptoms of this disease are shortening of internodes and phyllody of normal flowers. The floral parts like corolla, androecium and gynoecium are transformed into green leaf like structures. Infected plants usually do not set fruits. If the fruits are formed, they are small, abnormal and distorted. The infected plants often show dull pale colour.	All the infected plants should be removed as soon as they are noticed. Spray systemic insecticides like rogor (0.05%) or metasystox (0.02%) or nuvacron (0.05%) at 10 days intervals to control insect vector population.
Bitter gourd mosaic	Virus causes the disease. The leaves in secondary branches produced at the apical end of the plant show symptoms of the disease. Scattered, small, irregular, yellowish patches are seen on leaves. Some leaves show vein clearing in one or two lobes of leaf, severely infected plants show reduction in their size and elongation and or suppression of one or two lobes. Young developing leaves are completely distorted, filiform and malformed with considerable reduction	Spray just after germination with nuvacron or dimecron (0.05%) at 10 days intervals to prevent aphid vectors. Mineral oil, *e.g.* krishi oil @ 1.5 % sprayed to minimize virus spread.

Contd...

Table 12 Contd...

	in their size. Some of the leaves show marked reduction in the development of lamina resulting in a shoe-string effect.	
Bitter gourd witches broom	Virus causes this disease. The characteristic symptoms of this disease are malformation and proliferation of axillary buds. Diseased plants show abnormally little leaves which fail to attain full size. The disease plants bear many flowers and blossom earlier than healthy plants. Internode buds sprout and give many buds like chlorotic leaves arising from internodes. Flowers on infected plants show characcteristic green, phyllody symptoms.	Application of furadon @ 1.5 kg a.i./ha at the time of sowing seeds followed by 5-6 foliar sprays of either dimecron or nuvacron or metasystox (0.05%) at 10 days intervals to keep the vector population under check. Spraying of oxytetra-cycline hydrochloride hydrochloride solution at 500 ppmat weekly intervals suppress its symptoms.

Table 13: Major insects of bitter gourd.

Insect Name	Nature of Damage	Control Measures
Red pumpkin Bettle (*Aulacophora foveicollis*)	This is elongated small beetle of orange red colour. The beetles attack seedling stage especially at cotyledonary leaf stage. They make holes in cotyledonary leaves.	The insect can be controlled by spraying of carbaryl (0.1-0.2%) or rogor (0.1%).
Aphids (*Aphis cucurbitae*)	These small green insects damage the plants by sucking leaf sap. In young stage, cotyledonary leaves crinkle and in severe cases, the plants wither. In grown-up vines leaves turn yellow and plant lose its vigour and yield.	Aphids can be controlled by spraying of malathion (0.1%) or metasystox (0.1-0.2%).
Fruit fly (*Daucus cucurbitae*)	Maggots of this fly cause severe damage to young developing fruits. Damaged fruit are not fit for marketing.	There is no direct control of maggots because they are inside the developing fruits. Using light traps in the night and poison baits can control the adult flies. Infestation can be controlled by baits containing sex attractants like protein hydrolysate 0.5 kg + 1.25 % malathion WP + 200 g of molasses. Spray of endosulfan or thiodan @ 6 ml/ 4.5 litres of water also partialy check its incidence.

- First picking is done 55-65 days after sowing. The picking is mainly done when fruits are still tender and green. The consumers generally like its tender fruits. The picking should be done at an interval of 2-3 days as bitter gourd fruits mature very fast and turn yellow. Irregular harvesting may delay the formation of successive fruits and affect their growth and development adversely. Further, this will also decreased the number of fruits/plant. The average yield of bitter gourd varies from 300 to 400 q/ha.

Bottle Gourd

- Bottle gourd is a warm season crop. It cannot tolerate high cold and frost. It is highly sensitive to photoperiod. Short days and humid climate promote femaleness. It requires 18°C minimum temperature for seed germination and 20-30°C for growth and development of plant.
- It can be grown in all types of soils but sandy to sandy loam soils with high organic-matter content are considered best for early and good crop. The soil pH should be 5. 5-6.7. Bottle gourd can be raised successfully in riverbeds.
- Sowing time of bottle gourd for different regions and conditions are: (i) for plain areas, *kharif* crop: 15 June-15 July; *rabi* crop: October-November; spring-summer crop: 15 February-15 March, (ii) for hilly areas: March-June and (iii) river-beds sowing: November-December.
- The recommended seed rate is 2-3 kg/ha. Bottle gourd is mainly sown directly in the field, on raised bed, in furrows or trenches or pits. Two seeds per hill are sown insides on raised bed or furrows. Pits size are dug of 60 cm x 60 cm x 45cm deep and filled with well-rotten farmyard manure or cattle manure and top soil of pit. The recommended spacing is 1.8-3 m from row-to-row and 0.6 -1.5 m plant to plant.
- For taking early crop, seeds are sown in plastic bags in December inspite of directly sowing to main field. For this, one part of FYM and three parts of soil are mixed and filled in polybags. After that, 1-2 seeds treated with captan (2 g/kg seed) are sown in each polybag. The polybags are kept under plastic house or should be covered with plastic sheet during night time to keep it warm and for early germination. The plants are transplanted in the main field when outside temperature is warm (February-end or early March).
- Farmyard manures @ 200-250 q/ha is applied 15-20 days before sowing of seeds in pits/furrow. The nitrogen, phosphorous and potash are applied in the ratio of 100:50:50 kg/ha, respectively. Half quantity of nitrogen and full of phosphorus and potash are applied in pits or furrows at the time of sowing. Remaining quantity may apply 30-45 days of sowing or just before flowering.
- For higher production and obtaining quality fruits, vines are trained on bamboo with rope or wire or in bower system. The methods of preparation of bower have been described in cultivation of bitter gourd. In bower system of training, plant height should be 2.25 m and planting distance is kept at 2.5 m x 0.6 m. Hybrids Warad, Swati, NS-381, 421, Kavari, and Shramik are suitable for bower system of training. Pandey and Ram (2001) reported that bottle gourd trained in bower system yielded 600 q/ha.

- The field should be free from weeds throughout cropping season. Pre-sowing incorporation of fluchloralin @ 3 litres/ha or butachlor @ 2.5 litres/ha can also be used as weedicides for weed control.
- The summer crop requires frequent irrigation after every third or fourth days, while, the crop sown in rainy season is irrigated, whenever required. Winter sown crop is irrigated sparingly once in 10-12 days. Soil moisture deficit during fruit development greatly reduces fruit size and yield. Over-watering results in fruit rot development. Basin and furrow irrigation is commonly adopted to raise bottle gourd. Irrigation through drip system increased 45-47% higher yield as compared to sprinkler and furrow.
- Spray fungicides and insecticides regularly for controlling diseases and insects. The disease and insects are same as described in bitter gourd. Hence, follow the same control measures.
- Crop becomes ready for harvesting 60-100 days after sowing depending upon the hybrid and season. The fruits take 12-15 days after fruit setting to reach marketable stage. Harvesting is done when fruits are tender, medium in size and dark green. The picking intervals should be 3-4 days. Tenderness and edible maturity are judged by pressing the skin and little pubescence persisting on skin. Seeds should be soft, if examined in transverse section. Hence, smaller fruits fetch better price in market. The yield depends on hybrids and cultural practices. It yields 500-600 q/ha.

Cucumber

- It is a warm season crop. It cannot tolerate sever cold and frost. High humidity and short day length promote female flower production. It requires 18^{0}C minimum temperatures for seed germination and 20-30^{0}C for growth and development of plants.
- Sandy to sandy loam soils are preferred for early and good crop, whereas heavy soils are good for high yield. It can be grown successfully on slightly acidic soils. The soil pH should be 5.5-6.7. For higher production, soil should be rich in organic matter with proper drainage facility.
- Sowing time varies from place-to-place. The sowing time in various parts of country is given below.

Region	Optimum Sowing Time
North India	
a. Plains	February -March
b. Hills	April-May
South and Central India	October-November
Western region	September-February
West Bengal	
a. Summer crop	March-April
b. Rainy season	June-July

- Four kilogram seeds are sufficient for sowing in one hectare. Sowing can be done on raised beds, in furrows or in pits according to system followed. Two seeds/hill are generally sown on both sides of bed. In pit system, prepare pits of 45 cm x 45 cm x 60 cm size and filled with farmyard manure, cattle manure and soil in equal proportion up to 30 cm above the ground level. Plant spacing should be kept 1.5-2 m from row-to-row and 60-90 cm plant-to-plant.

  For taking early crop, same methods are followed as mentioned in bottle gourd.

  The doses of nitrogen, phosphorus and potash varied from place-to-place. However, for getting good yield, nitrogen, phosphorus and potash are applied in the ratio of 150:80:60 kg/ha, respectively. In addition to it, 200-250 quintals farmyard manure is required for a hectare. Whole qualities of FYM are applied 10-15 days before sowing. Half quantity of nitrogen and full of phosphorus and potash are applied at the time of sowing. Remaining quantity of nitrogen is applied 30 days after seed sowing.

- For higher production and increasing the quality of fruits, vines are trained on bamboo with rope or wire. Cucumber can also be trained in bower system. It is helpful in increasing the production and caring the plants. It does not allow the fruit to touch the ground, hence, increases the quality. In bower system, height of bower is kept 1.45 m and planting spacing of 2.0 m and 1.0 m from row to row and plant-to-plant, respectively. The hybrids Priya and Pusa Sanyog are suitable for training in bower system.

- The field should be free from weeds throughout cropping season. First weeding may be given 15-20 days after sowing. Two more weedings are given 25-30 days interval. Weeds can also be controlled by application of any contact weedicides.

- Over-irrigation should be avoided, as cucumber cannot withstand under water-logging conditions. In dry weather, its crop is irrigated fourth or fifth day but during rainy season, the interval may be increased, depending on rainfall distribution. Irrigation from starting of flowering and at full bloom is particularly beneficial. Fruit enlargement also requires sufficient supply of water. Drought during flowering results in deformed and non-viable pollen-grains, leading to poor yield. While, irrigation care is taken to see that water from the basin or furrows do not overflow to the plants because if they stand in water for any length of time, the foliage becomes chlorotic or yellow and growth is retarded.

- Spray fungicides and insecticides regularly for controlling diseases and insects. Most of the diseases and insects are common as described for bitter gourd and follow the same control measures. However, specific diseases are described below:

Cucumber Mosaic Virus (CMV)

The first symptoms appear on young leaves, which develop small greenish-yellow areas, often not more than one mm or 2 mm in diameter, occasionally circular, but more often limited in outline by the smaller veins of leaf. The spots are slightly more translucent than remainder of leaf and are often scarcely visible except by transmitted light. The characteristic symptoms are a yellow mottle on all leaves developed after infection, some leaf distortion and stunting of plant. Affected leaves are wrinkled and savoyed in appearance and may be somewhat distorted and curled. All growth subsequent to infection is much dwarfed. The stem internodes are shortened. The stem end of young fruits first becomes mottled with yellowish-green and this gradually spreads over the entire fruit. As these progresses, the body of fruit ordinarily becomes a light yellowish green, intermingled with spots of a much green colour. These dark portions are usually raised above the surrounding surface in such a way as to form wart like projections and produce distortion of fruits. Spraying of plants with dimethoate (0.05%) or nuvacron (0.05%) or metasystox (0.02%) at weekly intervals controls aphid population and the disease spread.

Cucumber Green Mottle Mosaic

The disease is caused by a strain of tobacco mosaic virus (TMV). The characteristic symptoms are slight clearing of veins and crumpling of young leaves, followed by a light or dark green mottle, together with blistering and distortion of leaves and stunting of plant. The symptoms first appear on younger leaves. The mottle is independent of the season, but leaf distortion is more sever when the plant is growing slowly in winter. Occasional yellow flecks occur on leaves, showing green mottle and may on fully developed leaves, be a prominent symptom. Fruits are usually unmarked, though it may be slightly mottled. The disease can be controlled by: (i) use of clean seeds, (ii) seeds from apparently healthy plants should be sown, (iii) during field operations, the hands of workers as well as implements should be washed thoroughly, (iv) dry heat treatment of seeds for 3 days at 70 ^{0}C and (v) soil fumigation with methyl bromide.

The crop becomes ready to harvesting in about 60-70 days after sowing. The fruit takes about 7-10 days from setting to reach marketable stage. Fruits are harvested at tender and green stages. The over-matured fruits fetch less price in the market. The fruits are picked at an interval of two days. The fruit yield varies from 300-400 q/ha.

Muskmelon

- The crop requires hot and dry climate. Optimum temperature requirement for growth is 27-30^{0}C. The seeds do not germinate at temperature lower than 18°C. Short day length promotes female flower production. High temperature, low humidity and plenty of sunshine are essential for proper ripening and development of high sugar content. Plants are sensitive to low temperature and frost.
- It requires sandy soils for early crop, while loam soils are good for high yield potential. The soil pH should be 6.0-6.8. For higher yield, the soil should be rich in organic matter content with proper drainage facility.
- Seed sowing times are: (i) in northern plains, best sowing time is February-March, but if the crop is to be transplanted, sowing can be done in the end of January, (ii) in north-eastern region, sowing is done from November to March, (iii) in western region, sowing is done from September-October, (iv) in south India, best sowing time is October-November and (v) in north India, sowing is done in April-May.

- Two kilograms seed is sufficient to sow one hectare area. However, in transplanting technique, only 1 kg/ha seed would be sufficient. Seeds can be sown directly in the field or in the polythene bags. For raising of transplants of muskmelon, polythene bags of 15 cm x 10 cm size 100 gauges thickness punched at the base are filled with a mixture of field soil, silt and farmyard manure in equal proportions and placed on the eastern side of a boundary wall or other structure. Two seeds are sown in each bag and bags are watered daily with the help of sprinkler cane. The transplants are ready in 25-30 days and at the time of transplanting, polythene bags are cut with a razor and removed. In riverbeds, trenches about 0.5-1 m deep are dug at a recommended row spacing and sowing is done on spot specially fertilized with farmyard manure and fertilizers. In pit method of sowing, pit size is kept at 60 cm x 60 x 60 cm. They are dug at recommended row and plant spacings and filled with farmyard manure. Three or four seeds per pit are sown. The recommended spacings are 1.5-2.5 m between rows and 0.5-1.0 m between plants.

- About 300 q/ha farmyard manures has been recommended for its cultivation. It is mixed in soil at the time of last ploughing or applied 10-15 days before sowing. The nitrogen, phosphorous and potash are applied in the ratios of 150:100:75 kg/ha, respectively. Half quantity of nitrogen and full of phosphorus and potash are applied in pits or furrows at the time of sowing. Remaining quantity of nitrogen is applied 30 days after sowing or just before flowering.

- First weeding is done 15-20 days after sowing or transplanting. Two or four weddings are required before the vine covers the whole area. Weedicides like fluchloralin (pre-sowing) @1.2 kg a.i./ha or fluchloralin @ 0.48 kg/ha + nitrofen @ 0.5 kg/ha (pre-emergence) or nitrofen @ 1.25 kg/ha (pre-emergence) can be used for weed control in muskmelon.

- Pre-irrigation before sowing is good if moisture is not sufficient in the soil. Muskmelon plants require an abundance of moisture during the period when their vines are developing. This crop requires 11 irrigations, totaling of 550 mm water, applied at an interval of 4-9 days. Furrow and basin method of irrigation is commonly used for watering in muskmelon. In regions of extreme water scarcity, pitcher method of irrigation may also be used.

- Apply fungicides and insecticides regularly for controlling diseases and insects. Since, diseases and insects, which attack bitter gourd, also damages to muskmelon, the same control measures are followed for controlling them.

- The crop is ready for harvesting 70-90 days after seed sowing, depending upon the variety and season. Fruits take 28-30 days from fruit setting to reach to mature stage. The maturity in muskmelon can be determined from the change in outer colour to yellow, yellow-green or brown. The fruits also slip from the vine. This is due to development of an abscission layer resulting in automatic detachment of fruits from vine when fully mature. This is called slip stage. The fruit yield varies from 350 to 450 q/ha.

Watermelon

- It is a warm season crop. It cannot tolerate high cold and frost. Watermelon requires 18°C minimum temperature for seed germination and 24-27°C for growth and development. For high quality fruit production, there should be plenty of high temperature at the time of ripening.
- It requires sandy to sandy loam soil for early and good crop. The soil should be well drained. Alluvial riverbed soils are also good for production of watermelon. The soil pH should be 6.5-7.0.
- In north India, it is sown in February-March, whereas in north-east and west India, best time of sowing is from November to January. In West Bengal, it is sown in November. In south and central India, sowing is done in December-January. In Maharashtra, optimum-sowing time is December-January. In Rajasthan, rainy season crop is sown in August-September. In northern hills of India, its crop is sown in April-May.
- Two kilograms seeds are sufficient to sow one hectare area. The sowing may be done on raised beds, furrows, trenches or pits. The commercial crop is generally sown directly. Two seeds per hill are sown on both sides of raised bed or furrow and 3-4 seeds/hill in riverbed trenches. In pit method, pits of are dug 60 cm x 60 cm size at recommended spacing and filled with a mixture of farmyard manure and soil in equal proportions. Four seeds/pit are sown. The recommended spacing is 2.5-4 m x 0.6-1.2 m.
- The crop requires 250 q/ha farmyard manure, which is applied 15-20 days before sowing. The farmyard manure is mixed in soil thoroughly. The nitrogen, phosphorous and potash are applied in the ratio of 100:80:100 kg/ha, respectively. Half quantity of nitrogen and full of phosphorus and potash are applied in pits or furrow at the time of sowing. Remaining quantity of nitrogen is applied 30 days after sowing or just before flowering.
- For getting good yield and quality fruits, apical shoots are removed and side shoots allowed to grow. Only 3-4 side shoots are allowed. To keep crop free of weeds, start intercultural operations 20-25 days after sowing. Depending upon seasons, about 2-3 weedings are required. Herbicides like pre-planting application of bensulide @ 4 kg/ha or trifluralin @ 1.2 kg/ha can also be used for weed control.
- Watermelon planted in river-beds does not need any irrigation as it gets sufficient moisture through capillary action from subsoil of water. In other places its crop is irrigated soon after moisture stress during pre-flowering, fruiting or fruit development stages. Irrigation once in 7-10 days may be given depending on soil type and weather conditions. Watermelon is generally watered with furrow or basin irrigation in India. Pitcher irrigation can also be used.
- Spray fungicides and insecticides regularly for controlling diseases and insects. The major diseases and insects, which have been described in bitter gourd, also damage watermelon. Hence, the same control measures may be followed.

- Its fruits are ready for harvesting 75-100 days after sowing, depending upon cultivar and season. Fruits become fit for consumption 30-40 days after anthesis. Maturity in watermelon can be judged from withering of tendrils, change in belly colour or ground spot to yellow and thumping test on thumping the fruit, the mature fruit will give dull sound. There is another test, on pressing the fruit, the ripe fruit gives a crisp, cracking noise but this test should not be followed as it may damage the fruits. Fruits are separated from the vine with the help of a knife. A well-managed crop can yield 1,000 q/ha.

Pumpkin

- It can be grown in relatively cooler climate than other cucurbits. Optimum temperature requirement for its growth is 18-24°C. Short days, low night temperature and high relative humidity are best for its production.
- Clay and loam soils are good for yield. Soils should be well-drained and with ample organic-matter contents. Optimum pH requirement is 5.5-6.8. It can be grown on slightly acidic soils also.
- Summer crop is sown during February-March in north Indian plains and in December-January in rest of the country. The rainy season crop is sown in April-May in West Bengal but June-July in other parts. In Tamil Nadu, optimum sowing time of rainy season crop is July-August. The crop is sown in April-May in hills of north India.
- The recommended seed rate is 4-6 kg/ha. Seeds are directly sown in field on raised beds or in furrows or trenches or pits. Two seeds/hill are sown on both sides of raised bed. In pit method, pits are dug at spacing of 60 cm x 60 cm and filled with farmyard manure and topsoil of the pit. Four to five seeds/pit are sown. The recommended spacing is 3 m x 0.75-0.90 m.
- About 150-200 q/ha farmyard manures is applied 15-20 days before sowing or at the time of digging the pits or furrows. The nitrogen, phosphorous and potash are applied in the ratio of 100-150: 80-100: 50-80 kg/ha, respectively. Half quantity of nitrogen and full of phosphorus and potash are applied in pits or furrows at the time of sowing. Remaining quantity of nitrogen is applied 30-45 days after sowing.
- First weeding is done 15-20 days after sowing. Three weedings are required to keep the field weeds free. Herbicides like pre-plant incorporation of bensulide @ 4-6 kg/ha or @ 2.5 kg/ha as pre- emergence can also be used for weed control. For rainy season crop, normally irrigations are not required. The summer season sown crop is irrigated immediately after sowing. The second irrigation after a week or so. It is irrigated every fourth or fifth day in dry weather.
- Spray fungicides and insecticides regularly for controlling diseases and insects. Diseases and insects damaging the pumpkin are similar to other cucurbitaceous crops, hence follow the same control measures. However, some specific diseases, which damage its crop, are described below.

Pumpkin Mosaic Virus

There are many mosaic diseases reported on pumpkin, but pumpkin mosaic virus referred here is a strain of watermelon mosaic virus-1. This disease was first reported from Bangalore (Singh, 1979a and Singh, 1981b). The virus causes moderate to heavy losses in yield and quality of pumpkin fruits. If plants get infection in early stage of their growth (10-15 days old), its plant growth is arrested. The size and number of leaves are reduced considerably. The plants produce on an average 0.5 kg in comparison to 2.5-3 kg in healthy plants. The virus infection also reduces the vitamin A content in fruits (Singh, 1979 a). The infected leaves initially show a conspicuous vein-banding consisting of a narrow border of dark green parallel with each of the main veins of leaf, remaining portion of leaf shows unusual lighter green. The peculiar serration is noticed in some foliage in which tips of veins protrude beyond the edges of the leaf, sometimes giving toothed appearance. Diseased leaves may also have deformations, which vary from an elongation to a rather showing outlined cordate pattern in mild cases. The leaves become smaller than the normal. The fruits from infected plant exhibit various deformations and dark green lines. Fruits are blistered, small and deformed. Carotene content of the infected fruit is reduced.

Control Measures

(i) Spraying of nuvacron or dimecron (0.05%) at 10 days interval, (ii) selection of seeds from healthy plants, (iii) spray of mineral oil, *e.g.* krishi oil has been found effective in checking the disease spread in the field.

Yellow Vein Mosaic of Pumpkin

This disease was first reported Vasudeva and Lal (1943), and Varma (1955) from India. The characteristic symptoms are yellowing of veins and mosaic symptoms. Initially, virus causes vein clearing on young leaves. Later on, mosaic symptoms are produced on leaves. Leaves become smaller as compared to healthy ones. Fruits are small and deformed.

Control Measures

(i) The disease can be kept under control by insecticide treatment, (ii) straw mulching and colour mulching delay the onset of infection.

- The pumpkin crop reach maturity 75-180 days after sowing, depending upon hybrid, season and other conditions. The mature fruit have a brown colour but in some hybrids, mature fruits may not have brown colour. Fruit yield varies from 400 to700 q/ha.

Ridged Gourd and Smooth Gourd

- Both crops require long and warm climate. Humid and warm climate is best. It grows best during rainy season. The optimum temperature for growth is 24-27^{0}C. Long day promotes femaleness.
- Loamy soil is considered best if there is good organic-matter content. The optimum soil pH is 5.5-6.7.

- Summer crop is sown during February-March in northern plains but January-March in eastern and western regions. In West Bengal, sowing is done in December-January. In south India, crop is sown in December-January. The rainy season crop is sown during June-July. In hills of north India, sowing is done in April-May.
- The recommended seed rate is 1.0-1.5 kg/ha. Sowing can be done either on raised bed or in pits. The recommended spacing is 1.6-1.8 m between rows and 60-70 cm between plants.
- About 250 q/ha farmyard manures is applied 10-15 days before sowing the seeds in raised bed or pits. Nitrogen, phosphorous and potash are applied in the ratio of 80:60:40 kg/ha, respectively. Half quantity of nitrogen and full of phosphorus and potash are applied at the time of sowing in raised bed/pits. Remaining quantity of nitrogen is applied 30-45 days after sowing.
- First weeding is done 15-20 days after seed sowing, while second weeding may be done 30-35 days after first weeding. Avoid chemical weedicides for control of weeds.
- First irrigation is given soon after sowing if there is insufficient moisture in soil. Later, frequent irrigation once in 4-5 days is given during summer season. Flowering and fruit development are very susceptible to soil moisture stress. Rainy season crop hardly requires any irrigation if there is good rainfall distribution. The crop is normally watered in basin or furrow system of irrigation. Irrigation through drip is not much more beneficial than sprinkler and furrow irrigations.
- Spray fungicides and insecticides regularly for controlling diseases and insects. Most of the disease and insects, which attack bitter gourd, also damages the ridged and sponge gourds. Hence, same control measures, which have given in bitter gourd, should also follow in ridged and sponge gourds.
- The crop is ready for harvesting 60-90 days after sowing, depending upon season and hybrids. The fruits attain marketable maturity 5-7 days after anthesis. Harvesting is done at an interval of 3-4 days. Only immature fruits are harvested. The average fruit yield varies from 300 to 400 q/ha.

Carrot

- Carrot is a cool season crop, but some Asiatic types can tolerate higher temperature. The optimum temperature for seed germination is 7.2-23.9^{0}C. Carrot produces largest roots between 10 and 15.6^{0}C with poor root colour and between 15.6 and 21.1^{0}C with good root colour. Plants grown continuously at a temperature of 21.1– 26.7^{0}C fail to develop flower primordia. That is why the seeds of European types of carrots are produced only in hills of India, where, winter is severe. At higher temperature, roots become shorter, thicker and non-juicy. Also brightness of colour is reduced. At lower temperature (10-15^{0}C), roots become longer and develop poor colour.
- Carrots grow well in many soil types, preferably on well-drained, deep loose and neutral soil (pH 5.5-7.0). Sandy loam soils rich in humus free of clods and rocks are

preferred for smooth straight roots. Soil compaction affects root growth and length. Very loose or highly compact soils are detrimental to growth of roots.

- In regions where climatic conditions are mild, carrot can be grown almost all the year round. Several successive sowings may be done every fortnight to obtain a continuous supply of roots. In northern plains of India, carrot is sown from mid-August to early-December. It is grown from March to August in hills.
- The seed rate varies from 3 to 4 kg/ha. For better germination, carrot seeds are soaked for 12 hours in 50 ppm IBA solution. The seeds are usually sown either by broadcasting or by drilling in rows 15-20 cm apart. The seeds are sown on ridges or in flat beds. In flat bed method of planting, the spacing is kept 20-25 cm x 10-15 cm.
- Well-rotten farmyard manure is applied @ 200 q/ha at 10-15 days before sowing. The nitrogen, phosphorous and potash are applied in the ratio of 120:80:80 kg/ha, respectively. Half quantity of nitrogen and full of phosphorus and potash are applied at the time of sowing. The remaining quantity of nitrogen is given 30-45 days after sowing.
- Within a week of seed sowing, the seedlings start emerging from the ground. In line-sown crop, these are thinned 10-15 cm apart as soon as they are well-established. In broadcast sown crop, thinning is done twice, once earlier to remove surplus plants from the clump, leaving only one seedling at a hill. The second thinning is done at a later stage by removing the larger roots for table use, leaving smaller one to develop. This operation should be done when soil is adequately moist.
- Weeds are a serious problem in carrot fields and timely control of weeds is essential to avoid heavy loss in yield of top quality due to weed competition. Weedicides like tenoran (2.5 kg/ha pre-emergence) linoran (0.5-1.0 kg/ha pre-emergence) or propazine (1-1.2 kg/ha post-emergence) can be used to control weeds. During weeding and hoeing, crown of root is often exposed to light and this causes greening and lowering the quality of roots. Hence, for preventing discolouration of roots, earthing-up is necessary.
- Insufficient moisture for carrot increases roughness, decreases root size and yield. Low moisture grown carrot has strong and pungent flavour. Final stage of root development is very sensitive to moisture stress. Dryness of soil followed by rain or irrigation is likely to cause cracking in carrot roots. First irrigation is done soon after sowing followed by another in winter at fortnight, when the weather is dry. During spring and summer seasons, it is necessary to irrigate once in 3-8 days, depending on soil type, weather conditions and stage of development. Carrot is irrigated before any wilting of leaves appears, since they may be able to obtain enough water to keep the leaves erect without having enough moisture to make the desired growth. It is generally irrigated by furrow system in India. Basin irrigation is also followed occasionally.
- Apply fungicides and insecticides regularly for controlling diseases and insects in carrot. The major diseases and insects are given in Tables 14 and 15.

Table 14: Major diseases of carrot

Disease and Causal Organism	Symptoms	Control Measures
Cercospora leaf spot (*Cercospora carotae*)	Diseased lesions are formed on leaves, petioles, stems and floral parts. On leaves, brown to black spots appear and latter may shrivel and break. Flora parts when attacked also shrivel.	Seed treatment with hot water at 50° C for a minute is effective to control the disease.
Alternaria blight (*Alternaria* spp.)	The disease appears on mostly older and weaker leaves. Brown to black irregular spots appeare on leaves.	The control measures adopted for *Cercospora* leaf blight can also be used for the control of this disease.
Black rot (*Alternaria radicina*)	The symptoms on foliage are just like those caused by Alternaria blight. On roots, black, sunken areas irregular to circular in outline may develop. The decayed tissue is green black to jet back due to presence of mosses of black spores.	Store carrots at proper temperature. Treat washed carrots with thiabendazole.
Soft rot (*Erwinia carotovora* and *E. atroseptica*)	Infected roots become soft, discoloured and unpleasant smelling in the centre, although the outside may look normal. High temperature is favourable for this disease.	Remove and destroy infected roots, so that bacteria may not infect healthy roots. Possibly the incidence of soft rot may be due to excessive manuring which should be avoided. Avoid bruising during transportation storage and handling. Clean the roots properly.
Bacterial blight and roots scab (*Xanthomonas carotae*)	Affected leaves show small, irregular yellow areas, which turn dark brown and water-soaked. On petioles and stems dark brown streaks are formed. Brown to maron lessions are formed on roots.	Follow long-term crop rotation. Treat the seed with hot water at 50°C for 15 minutes. Plant the crop at relatively wider spacing.
Carrot yellows	This is viral disease transmitted by six-spotted leaf hopper (*Macrosteles divisus*). First symptoms apear on leaves, which become yellow, sometimes accompanied by vein clearing. Dormant buds in the crown grow into chlorotic shoots that give a "Witches broom" appearance on the top. Older leaves are reddish, twisted and may eventually	The most effective method of control of this disease is to spray insecticides.

Contd...

Table 14 Contd...

	break off. Size and quality of roots are reduced. The internal colour, texture and flavour show marked changes, causing reduction in value of carrot for fresh market as well as for processing. Roots of infected plants have a bitter and astringent flavour.	

Table 15: Major insects of carrot

Insect Name	Nature of Damage	Control Measures
Semi-loopers (*Plusia* spp.)	The larvae burrow into the roots aften causing it to become misshoped and decay, rendering them unfit for marketing. If roots are injured severely, leaves become rusty or even dried.	Application of organophosphate or organocarbamate insecticides in the seed furrow as drenches also controls the insect up to 100 days after seed sowing.
	It is a serious pest of seed crops. The pest damages the embryo of the seed. Sometimes the seed is without any embryo.	Seed treatments with insecticides may control this pest.

- Carrots should be harvested at proper stage of maturity. Otherwise, they become puffy and unfit for consumption. The Asiatic hybrids attain marketable stage of maturity at 2.5-4 cm diameter at upper end. Processing carrots are left in the ground for a longer period. Both dry matter and colour increase with maturity. In India, harvesting is done manually, while mechanical harvesting is the rule in advanced countires. Usually light irrigation is given 2-5 days before harvesting. The carrots grown on ridges can be easily pulled out. Then their tops are cut. In flat beds, tops are first cut and roots are then dug out with *kudali* or a wooden plough. Generally, Asiatic hybrids produce higher yield of 400-600 q/ha, while European hybrids produce around 300 q/ha.

Onion

- Onion can be grown under wide range of climatic conditions. However, it grows well under mild climate without extreme heat or cold or excessive rainfall. In early stages of growth before bulbing, fairly cool temperature ranging from 12.8 to 23°C and for bulb development 15.5 to 21°C are required. Very low temperature in the begining favour bolting. The requirement of day length varies with type of hybrids. At least 10 hours light for growth, development and bulbing with favourable temperature is essential. Extension of day length with incandescent light is more effective in promoting bulbing. It can be grown as a rainfed crop even at elevations of 1,500-2,100 m between April and August. Regions receiving 75-100 cm or more annual rainfall are not good for growing onion as a rainfed crop.

- For successful onion production, the soil should be deep friable loam and alluvial, where, there should be free drainage, absence of persistent weeds and presence of organic matter. Avoid always too heavy soils for cultivation. For obtaining higher yields, care should be taken that onion is planted in soil having high acidity and alkalinity. The seed sowing in nursery and seedling transplanting in the field vary from region to region-in-different parts of the country as given in Table16.

Table 16: Sowing and transplanting times of onion in different regions.

Region	Sowing time of Nursery	Transplanting time of seedlings
North India	October to mid-November (*rabi*) May to June (*kharif*)	Mid-December to January
Pune and Nifad area in Maharashtra	October	July to August November
Madras and Andhra Pradesh	June to July (monsoon crop) October to November (winter crop) January-February (summer crop)	August to September December to January February to March
Madhya Pradesh and Vidarbha	Mid-December	January-end
West Bengal	September to November	October to December

- For nursery raising, sandy loam soil is preferred. The soil should be well-prepared for sowing. Seeds are generally sown in raised bed in nursery. Seeds can also be sown in flat beds under proper drainage. Raising seedlings in 500 m^2 area are sufficient for planting one hectare crop. The nursery-bed is kept 10 -15 cm raised bed of about 3-6 m length and 1 m width, are prepared. About 70 cm distance is kept between two beds for irrigation and intercultural operation. For *kharif* season, 15-20 cm raised beds above the ground level and 30 cm wide channels are prepared all around the beds. It serves as drainage channel during rains and irrigation channel during long dry spell. Nursery can also be raised in polyhouse. The top surface up to 2-3 cm should be enriched with fine sieved well-decomposed farmyard manure or compost before sowing. If seeds are not treated, use difolton or monosan @ 2.5 g/kg seed for checking pre-emergence death of seedlings caused by damping off, which is very much prevalent during rainy season before sowing. To check post-emergence damping off, drenching of nursery with 0.1 % brassicol/copper oxychloride or 0.2 % capton is done. Seeds are sown in lines at the distance of 4-5 cm in *rabi* and 5-7 cm apart in *kharif*. Seeds should not be sown beyond 2-3 cm deep in soil. After sowing seeds are mulched with dry grass or straw or any other material to maintain and preserve the required soil moisture till the germination is completed. If there is lack of moisture in bed, it should be watered by a watering cane. On the other hand, if mulching is not used, light irrigation with sprinkling cane at every third or fourth day is required.

- The seedlings are ready for transplanting in 6-7 weeks time for *kharif* planting and in 8 weeks for *rabi* planting. Seedlings of 0.8-0.9 cm in diameter and about 20-35 cm in height are ready for transplanting. Over-aged seedlings result in bolting and take longer time to start new growth, whereas under-aged seedlings do not establish well after transplanting and withstand adverse weather conditions. The spacing of 15 cm between rows and 8-10 cm between plants is most conducive to high yield in most of the onion-growing belts of the country. About 4-5 kg seed in *rabi* and 6-7 kg seed in *kharif* are required to raise seedlings for planting one hectare crop. However in *kharif,* more onion seeds are required as there is a heavy incidence of damping off. In bulb method of planting, bulbs are dibbled 15 cm on side of 45 cm wide ridges or in beds.

- About farmyard manure 300-400 q/ha is throughly mixed in soil 10-15 days before transplanting. Apply 75 kg/ha nitrogen, 90 kg/ha phosphorus and 90 kg/ha potash at the time of transplanting. Whole quantity of nitrogen is applied 30-45 days after transplanting.

- Generally, hand-weeding is in practice for controlling weeds in onion. But there are some herbicides like basalin @ 2 kg a.i./ha along with one hand-weeding 45 days after transplanting or fluchloralin 1.0 kg (PPI) + one hand-weeding 45 days of transplanting are very effective to control weeds.

- Onion roots are shallow and most of the roots being limited to upper 30 cm of soil. Just after transplanting, field is irrigated as soon as possible after being set if soil is dry. The irrigation schedule should be 13 days intervals during November-December, 10 days interval during January and 7 days interval during February-March. The most common method of applying to the onion crop to irrigate with basin or border strip flooding or furrow irrigation. Irrigation through drip can also be used for getting good yield. Onion field is always kept moist. Drastic soil moisture fluctuation results in setbacks and in onion that starts new growth after being retarded the inner scales may split. The outer ones causing splits and doubles which greatly reduces marketability. Bulbs grown under low moisture are to dry out earlier and loss more weight during storage than those grown with adequate moisture. Over-irrigation as well as under-irrigation may result into lower yields. Irrigations heavily within 2-3 weeks of harvesting may cause immature and thick-skinned bulbs. When plants start to mature, irrigation is discontinued and the soil allowed drying out as much as possible which facilitates bulbs to ripen sooner, otherwise a secondary root growth, which is difficult to stop, may start and complicate the process of proper curing of onion.

- Apply regularly fungicides and insecticides for controlling diseases and insects. The major diseases and insects are described in Tables 17 and 18.

Table 17: Major diseases of onion

Disease and Causal Organism	Symptoms	Control Measures
Downey mildew (*Peronospora destructor*)	The first sign of disease is formation of elongated patches on leaves varying in size under favourable conditions. These areas become covered with white to purplish fruiting bodies. Often leaves fold over the affected areas and the leaf tips wither. Bulbs of such plants do not attain full growth and are often soft and immature. The disease damages seed crop, when the flower stem is destroyed the seed ripen.	The disease can be controlled by spraying of difoltan (0.1%) or cuman (0.1%) or ridomil (0.1%).
Purple blotch (*Alternaria porri*)	Above ground parts of the plant and bulb are affected. The first blotches usally appear on oldest leaves. In beginning, they are small, elongated, sunken and whitish, generally with purple centre. Under favourable conditions, with a blackfruiting bodies of affected harvest when the fungus entire injuries or at the neck causing storage rot.	Seed treatment with captan or thiram or bavistin (2.5 g/kg seed) and follow crop rotation. Foliar spraying of dithane M-45 (0.25%) or dithane Z-78 (0.25%) or difoltan (0.2%) or cuman (0.25%)
Black mould of onion (*Aspergillus* spp.)	The infection starts from the top, *i.e.* cut portion above the neck and any injury present in the bulb. The affected tissue becomes water-soaked and at first a white mould develops between the scales which follows development of black spores on stalks, can be seen with naked eyes and the mosses remains exterior of scales, and can easily rubbed off.	Avoid bulb injury during harvesting, curing, transportation and storage. Sort out all the bulbs showing any wound, green and thick neck, doubles *etc*; and allow only perfect bulb for storage. Clean the stores properly and chek sufficient ventilation. Temperature should be below 15°C.
Stemphylium blight (*Stemphylium vesicarium*)	The infection occurs on leaves in the form of small yellow to pale orange streaks in the middle of leaves. These son develop into elongate spindle-shaped ovate spots, which later turn brown in colour and coalesce. Leaves dry premature and affect bulb development.	Dithane M-45 (0.25%) is effecctive for its control.
Bulb rot (*Fusarium oxysporum*)	The soil-borne pathogen attack onion bulb which later is covered with a whitish mycleium. The first symptoms of disease are yellowing and withering of foliage. The plants can be easily pulled out showing decay of almost all their roots.	Benlate (0.25 %) is effective in inhibiting fungus.

Contd...

Table 17 Contd...

Bacterial stalk rot of of onion (*Pseudomonas gladili*)	The disease is characterized by internal brown discolouration and dry rotting in basal portion of flower stalk resulting partial or complete withering of inflorescence prior to seed formation. Such roting may extend internally up to the seed bulb. Outwardly, plants appear normal till either inflorescence gets dried up or the stalks tople down. Yellowing of outer leaves from tip downwards followed by drying is seen in bulb crop. During storage, inner scale of affected bulbs become water-soaked, followed by browning and drying which results in the hollowing of bulbs.	Soaking of bulbs with any bactericides is effective. Use resistant hybrids for cultivation.

Table 18: Major insects of onion

Insect Name	Nature of Damage	Control Measures
Onion thrips *(Thrips tabaci)*	Affected leaves show silvery white blotches, which later become brownish. Spring-summer crop is affected much by thrips. A long spell dry weather is favourable for its multiplication. Larvae attacks on leaves, flowers and flower stalks.	In the areas where thrips infestation is severe, grow resistant hybrids. Apply malathion (0.1%) or nuvacron (0.1%) or phorate (0.1%) or G @ 1 kg/ha).
Borer (*Heliothis armigera*)	Larvae attack on leaves, flowers and flower stalks.	Collect and destroy affected flowers and flower stalks. Apply endosulfan at 0.1 %.
Mites (*Rhizoglypus* spp.)	These are very small insects. They remain mostly on the under surface of leaves. They suck the sap of leaves. Affected plants become pale yellow, giving sickly appearance.	Expose infested bulbs to sun for about two days. Dust the crop with sulphur at the rate of 20-25 kg/ha.

- Crop is ready for harvesting in five months for dry onion. However, for marketing green onion, the crop becomes ready three months after transplanting. Plants are pulled out by hand or with the help of *khurpi* and immediately lifted to shade of curing. The harvested bulbs should never be left in the field in heat of May and June; otherwise, rotting of bulbs in the store will take place. In *kharif*, after harvesting, crop should be left in field for little shade for drying of leaves. In both seasons of crop after curing, roots and tops are removed by leaving about 2-2.5 cm above the bulbs. The yield of onion depends on cultivars, soil types, planting season, and management practices. However, average yield of bulbs is 400-600 q/ha in *rabi* season and 300-450 q/ha in *kharif* season.

GLOSSARY

Abortion of pollen-grains: Shedding of pollen-grains before attaining full maturity.

Adaptation: The process by which individuals (or parts of individuals), populations or species change in form of function in such a way to survive better under given environmental conditions.

Additive gene effect: Gene action in which the effects on a genetic trait are enhanced by each additional gene, either an allele at the same locus or genes at different loci.

Allele: An allele is an alternative form of a gene.

Allogamy: When pollen-grains from flowers of one plant pollinate the flowers of another plants.

Anther culture: The culturing of anthers *in vitro* for generating haploid plantlets.

Anthesis: Full flower expansion including anther extrusion.

Antibody: Substances in a tissue or fluid of the body that acts in antagonism to a foreign substance (antigen).

Antigen: A substance, usually a protein, introduced into a living organism that elicits an antibody formation.

Apical dominance: In plants, the inhibition of lateral buds by high level of auxins, produced in the lead shoot or apical meristem.

Apomixis: Reproduction from an unfertilized egg or from somatic cells associated with the egg.

Artificial selection: The practice of choosing individuals from a population for reproduction, usually because these individuals possess one or more desirable traits.

Asexual reproduction: Any process of reproduction that does not involve the formation and union of gametes from different sexes or mating types.

Asnapsis: The failure or partial failure in pairing of homologous chromosomes during meiotic prophase.

Assortive mating: Mating in which partners are chosen because they are phenotypically similar.

Autopolyploid: A polyploid that has multiple and identical or nearly identical sets of chromosomes (genomes). A polyploid species with genomes derived from the same original species.

Autoradiograph: A record or photograph prepared by labelling a substance such as DNA with a radioactive material such as tritiated thymidine and allowing the image produced by decay radiations to develop on a film over a period of time.

Autozygote: A diploid individual in which two genes of a locus are identical by descent from an ancestral gene.

Auxotroph: A mutant microorganism (*e.g.* bacterium or yeast) that does not grow on a minimal medium but that requires the addition of some compound such as an amino acid or a vitamin.

B-line: The fertile counterpart of a line. This line does not have fertility restorer genes and used as the male parents to maintain A-line.

Backcross: Crossing F_1 hybrids with either of the parents. This may be done to test the genotypic ratio of F_1 or to transfer specific gene complex from one species to another.

Backcross method: It is method of breeding in which desirable characters of a non-recurrent (donor) parent are added to the genetic background of recurrent (recipient) parent through subsequent generations of backcrossing and selection.

Balanced heterosis: It is a type of true heterosis, which occurs from the balanced combinations of genes in a hybrid.

Basic number: The number of chromosomes in ancestral diploid ancestors of polyploids, represented by x.

Biometrics: The science dealing with the application of statistical methods to biological problems.

Biotechnology: The application of recombinant DNA, cell and tissue culture, and other methods used to develop new and improved plants and plant products.

Biotype: Distinct physiological race or strain within morphological species. A population of individuals with identical genetic constitution. A biotype may be made up of homozygotes, of which only the former would be expected to breed true.

Bisexuality. A system of producing both stamen and carpel in the same flower on a plant.

Bract: A more or less modified subtending a flower or belonging to an inflorescence.

Breeders' seed: Seed increased by the originating, or sponsoring, plant breeder or institution, used as the source for production of foundation seed.

Breeders' rights: Legislation giving the equivalent of patent rights for plant cultivars. The effect is to give the developer of a cultivar sole legal possession of that cultivar and a legal basis for compensation for its use by others. Breeders' rights may also be referred to as '*Plant cultivar protection*'

Bud-pollination: Pollination of flowers before they attain maturity (open and shed pollen-grains) with the pollen-grains collected from the same plant/genotype.

Certified seed: The progeny of a foundation or registered seed, which maintains the satisfactory genetic identity and purity and has been certified and approved by the certifying agency.

Chromosomes: Darkly staining nucleoprotein bodies that are observed in cells during division. Each chromosome carries a linear array of genes.

Clone: A group of individuals derived by a single original plant propagated by vegetative means.

Closed anther mutant: A mutated anther type which through produces viable pollen-grains but does not shed them due to non-rupturing and causes bareness in plants particularly where self-pollination is the rule. Closed anther mutants can be maintained by hand pollination with pollen collected from such flowers by force opening at full maturity.

Codominant genes: Alleles, each of which produce an independent effect in F_1.

Colchicine: An alkaloids extracted from seeds or corms of *Colchicum autumnale,* which induces polyploid by arresting, spindle formation during mitosis.

Combinational heterosis: The heterosis in quantitative characters resulting from the overall combination of favourable cumulative effects of a number of componential characters.

Compatible: Capable of fertilization.

Complementary genes: Those genes which have more or less similar phenotypic expression individually but when they come together they interact to produce a new character expression. If two such genes are complementary for a dominant effect, a 9:7 ratio results in F_2; if two are complementary for a recessive effect, a 15:1 ratio results in F_2.

Composites: The advanced generations seed mixture of an interval or interracial cross.

Convergent improvement: A system of double backcrossing for improving each of the two inbred lines without greatly modifying the yield of their F_1 cross.

Cross: The product of mating between two or more parents of dissimilar genetic constitution. Various types of crosses utilized in heterosis breeding programmes are single, three-way, double-crosses, top-cross, multiple-crosses *etc.*

Cross-pollinated crops: An assembly of genetically heterozygous individuals under commercial cultivation which share a common gene pool and in which each individual takes new genotype generation after generation.

Cross-pollination: Transfer of pollen from the anther of one plant to the stigma of another plant. It affects the union of genetically dissimilar gametes.

Crossing: Artificial mating of two or more parents of unlike genetical constitutions.

Cultivar: A subdivision of species as a group of plants within a species which are uniform in characters.

Cytogenetic/cytonuclear male sterility: Male sterility is determined by interaction of genes and cytoplasm but none of them can singly control sterility.

Cytoplasmic sterility: Transmission of male sterility by cytoplasm.

Cytoplasmic: Concerned with the cytoplasm in cell.

Dehiscence: Splitting open of a fruiting structure or anther.

Diallele cross: A diallele cross can be defined as all possible combinations, *i.e.* n (n-1)/2 of single-crosses among *n* parents.

Dichogamy: The maturing of male and female gametes at different times when the male gametes matures earlier than the female, it is called protandry, but when female matures first, it is called protogyny. This situation is called dichogamy.

Dicliny: A situation, where plants produce unisexual flowers.

Dioecious: The condition in which male and female sexes are produced on different individuals.

Dioecy: Where male and female flowers are borne singly and on different plants.

DNA: Deoxyribonucleic acid; the information-carrying genetic material that comprises genes.

Dominance: The phenomenon in which the dominant gene has an overriding effect on its allele in such a way that heterozygote (Aa) is phenotypically indistinguishable from the dominant homozygote (AA).

Dominant gene: When two parents with contrasting characters are crossed, the character of one parent appears in F_1 generation to the exclusion of the character of the other parent even though both the genes are present in the hybrid. The gene, which expresses itself, is said to be dominant over other recessive.

Double-cross hybrid: A hybrid obtained by crossing two single-crosses, *i.e.*

(A x B) x (C x D).

Double fertilization: A phenomenon in angiosperms whereby one male nucleus unites with the egg nucleus to form a zygote (2n) which develops into the embryo and the second male nucleus unites with two polar nuclei in embryo sac to form the endosperm (3n).

Double top-cross: A cross-obtained from crossing single-cross with an open-pollinated variety.

Electrophoresis: The migration of suspended particles in an electric field.

Electroporation: A process whereby cell membranes are made permeable to DNA by applying an intense electric current.

Emasculation: In bisexual flowers, the removal of stamens before they burst and shed pollen-grains. Its purpose is to check self-pollination and it is done before effecting cross-pollination.

Embryo: The portion of seed, which contains the dormant, miniature and rudimentary plant. It arises from the zygote.

Embryo culture: The cutting of an immature embryo on a sterile nutrient medium.

Endosperm: The nutritive tissue formed inside the embryo sac in the seed. It arises from triple fusion of sperm nuclei with polar nuclei of embryo sac.

Enzyme: A protein that accelerates a specific chemical reaction in a living system.

Epistasis: The phenomenon in which a non-allelic or gene combination exert a dominant effect over another gene or combination of genes (non-allelic interaction).

Evolution: The process of origin of the organisms (varieties, species, genera, families, etc).

F_1 : An abbreviation to designate the first hybrid generation. It is composed of progenies raised by sowing the seeds obtained from cross between two genetically unlike parents.

Fertility: The ability of an organism to form viable off-spring.

Fertility restorer gene: Usually a gene when put into a cytoplasmic male sterile background is able to bring back the production of normal functional pollen-grains.

Foundation seed: The seed stocks that are so handed as to most nearly maintain the specific genetic identity and purity of original stock and provide the source for the production of called certified and registered seed.

Functional male sterility: Where plants produce normal pollen-grains but anthers remain closed and cause male sterility.

Gamete: A matured sex cell, capable of fusing with another to form a zygote.

Gamete selection: It is a type of selection for detecting and combining of desirable gemete from a genetically-variable heterozygous population into the background of an inbred line of known performance and combining ability.

Gametocide: A chemical substance destructive to gametes.

Gametophyte: A phase of life-cycle of plants, which has haploid nuclei; during it the sex cells are produced. It arises from a spore produced by meiosis from a sporophyte (diploid).

Gene: A heredity determinant of a specific biological function; a unit of inheritance (DNA) located in a fixed place on the chromosome.

Gene expression: The hereditary properties of an organism as represented by the gene and are expressed in generation under a set of environmental factors.

Gene flow: The spread of genes from one breeding population to another by migration, possibly leading to allele frequency changes.

Gene frequency: Proportion of a gene or allele or its series present in a population or a sample thereof. It is usually expressed as number between 0 to1.

Gene interaction: The modification of gene action by a non-allelic gene or genes.

Gene pool: Sum total of all genes in a breeding population.

General combining ability: The comparative ability of a line or a genetic stock to combine with a tester or a group of testers.

Genetic advance: The expected gain in the mean of population for a particular quantitative character by one generation of selection of a specified per cent of highest ranking plants.

Genetic architecture: A term used to denote the general genetic structure of a species.

Genetic breakdown: A term used to indicate the loss of vigour and often the early death of F_2 plants which lack the necessary adaptive complexes of either or both the original parents.

Genetic diversity: The genetic differences as observed between individuals, or genetic stocks with respect to individual trait or an array of traits.

Genetic engineering: Genetic manipulation that use recombinant DNA methods (gene splicing) to change the genetic make-up of an organism.

Genetic equilibrium: In a random mating population, the stage in which genotype frequencies do not change from one generation to another.

Genetic erosion: Gradual disappearance of various forms of a cultivated species and of its wild relatives.

Genetic homeostasis: The ability of a random mating population to equilibrate its genetic composition so as to resist sudden environmental changes.

Genetic male sterility: Sterility of male gametes in a flower governed by genes and is heritable.

Genetics: The science of heredity and variation.

Genotype: The hereditary properties of an organism as represented by the gene constituents. They may be expressed or latent.

Genotype X environmental interaction: The interplay in effect of the genetic and non-genetic factors on the development of an organism.

Genus *(plural genera)***:** A taxonomic category that includes groups of closely related species.

Geographical diversity: The diversity of biological population produced by the presence of a geographical barrier such as mountains, rivers, canyons *etc*.

Germplasm complexes: The advanced generation of mixed seeds obtained by a purposeful intermixing of a number of genotypes or hybrids among them.

Germplasm: In plant breeding sense, germplasm is the sum total of genetic stocks of a particular crop species.

Heredity: Resemblance among individuals related by descent; transmission of traits from parents to off-spring.

Heritability: Degree to which a given trait is controlled by inheritance.

Hermaphrodite: An individual with both male and female reproductive organs.

Heterobeltiosis: Heterosis expressed over the better parent of the cross.

Heterochromatin: Chromatin staining darkly even during interphase, often containing repetitive DNA with few genes.

Heterosis: Hybrid vigour such that an F_1 hybrid falls outside the range of parents with respect to same characters. Usually applied to size, rate of growth or general anthocyanin absent fitness.

Heterozygosity: The phenomenon in which homologous chromosomes of an organism possess different genes of the same allelic series.

Heterozygote: An organism with one or more heterozygous pairs of genes or unlike alleles at one or more corresponding loci. As a result of heterozygosity, the organism will not breed true.

Homologus chromosomes: Chromosomes that cover in pairs and are generally similar in size and shape, one having come from the male parent and the other from the female parent. Such chromosomes contain the same array of genes.

Homozygote: An organism with identical genes at corresponding loci on homologous chromosome.

Horticultural traits: Characters of economic importance in horticultural crops.

Hybrid. The progeny of a cross between two or more individual plants of unlike genetic constitution.

Hybrid vigour. Increase in vigour, growth, yield or function of a hybrid over the parents that result from the crossing of genetically unlike organisms.

Hybridization: A method of crop improvement in which two or more plants of unlike genetical constitutions differing in one or more characters are crossed together.

or

The process of making hybrids.

In situ: From the Latin, meaning in natural place. Refers to experimental treatments performed on cells or tissues rather than on extracts from them.

In vitro: From the Latin, meaning within glass; biological processes made to occur experimentally outside the organism in a test tube or other container.

Inbred line: A line produced by continued inbreeding. In crop breeding, a nearly homozygote line usually originating from continued or selfing, accompanied by selection.

Inbreeding depression: The loss of vigour as a consequence of inbreeding. It is primarily due to the breakdown of specific gene system governing the expression of vigour, a particular trait or traits and is often accompanied by reduction in yield, size, fecundity *etc*.

Inbreeding: The mating of closely-related organisms.

Incompatibility: Failure to obtain fertilization and seed formation after self-pollination, usually due to failure of pollen tube to penetrate the stigma, or to reduced growth of the pollen tube in the styler tissue.

Inflorescence: (1) A flower clusters (2) the arrangement and mode of development of the flowers on a flower axis.

Introgressive hybridization: A type of hybridization in which a number of genes or gene block of one species are added to the genetic background of another species by crossing and often by backcrossing.

Irradiation: Exposure of plants or plant parts to X-rays or other radiations to increase mutation rate.

Isogenic lines: Lines differing from each other genetically at one locus only, i.e. lines identical in all traits but one.

Isolation: The condition in which individuals of common ancestry is separated into two or more mating groups that mating between or among groups is prevented.

Landraces: Traditional cultivars with sufficient genetic integrity to be morphologically identifiable and differing in adaptation and to cultural practices, but genetically variable.

Line: This term refers to a group of individuals obtained from a common ancestry. In maize programme, it is synonym to inbred line.

Line breeding: A system of breeding in which a number of genotypes which have been progeny tested in respect to some characters or group of characters are composite to form a variety.

Line X tester analysis: A system of lines for combining ability in genetic background of a number of proven testers.

Linkage: Association of two or more non-allelomorphic genes, so that they tend to be passed from generation-to-generation as an inseperable unit and fail to show independent assortment. The parental types appear in greater frequency than expected in F_2. This is due to locating of linked factors on the same chromosome.

Lipase: An enzyme that joins the ends of two strands of nucleic acid.

Luxuriance: It refers to the phenomenon in which the crossing of two parental forms brings in an excessive, accidental, unadaptable and often unbalanced expression of a number of an attribute.

Maintainer line: A genotype used to maintain the male sterility of cytoplasmic-genic system.

Male sterility: An inherent inability of plant to produce viable pollen.

Marker genes: Common gene differences which assort independently from all other readily testable gene loci are called marker genes, *e.g.* potato leaf in tomato brown seed in onion, anthocyanin pigment in cotyledons of chilli/brinjal *etc*.

Mating system: The system in which individuals are arranged in pairs leading to sexual reproduction.

Meristem culture: Culture of apical meristems, particularly shoot apical meristem, for production of shoots and plantlets.

Momeostatic: A genotype giving better performance even in adverse environments because of wide genetic base.

Monoecious: The condition in which both male and female sex organs are produced separately but on the same plant.

Monoecy: Where male and female flowers are unisexual but produced on the same plant.

Mutant. Any plant, which has originated or acquired a heritable variation as a result of mutation.

Mutational heterosis: It is a type of true heterosis, which results from the occurrence of the balanced types of mutations.

Non-recurrent parent: The donor parent in the backcrossing programme. The desired character of this parent is added in the genetic background of recurrent parent.

Nucleic acid: A macromolecule composed of phosphoric acid, pentose, sugar, and organic bases; DNA and RNA.

Nucleotide: A unit of DNA and RNA molecules containing a phosphate, a sugar, and an organic base.

Nucleus: The part of a eukaryotic cell that contains the chromosomes, separated from cytoplasm by a membrane.

Obtuse: Blunt pointed.

Open-pollinated: The progeny of individuals where pollination takes place automatically in flowers in nature.

Orbicular: Circular, round.

Out breeding: Mating of unrelated individuals.

Out cross: A type of natural cross obtained from the crosses of a number of unknown genotypes.

Ovate: Egg-shaped; broadest below centre (contrast ovate and elliptic).

Over-dominance: An effect the heterozygote (Aa) that is greater than the effect of homozygous dominant (AA).

Ovule: The term generally applied to whole seed forming apparatus inside an ovary, i.e. nucellus and the integument.

Parthenocarpy: The development of fruits without fertilization and formation of normal seeds.

Parthenogenesis: The development of an individual from the female gamete without fertilization.

Pedicel: The flower stalks, which grow from the upper branches of peduncle.

Peduncle: The main flower stalk, often branched once or twice (or more) in its upper part and bearing pedicels near the ends of branches.

Pentagonal: In which the lobes are rather broader than long, giving the corolla a pentagonal or 5-pointed appearance.

Pericarp: The outer wall of fruit.

Petiole: The stalk of a leaflet.

Petiolulate: With petioles.

Phenotype: The visible manifestation of genotype produced as a consequence of growth and development.

Phenotypic assortative mating: Mating of individuals with similar phenotypes. *e.g.* crossing between tall individuals with Aa x AA or Aa or Aa x AA and aa x Aa genotypes.

Phenotypic disassortative mating: Mating of individuals with contrasting phenotypic characters. *e.g.* crossing between tall x dwarf plants.

Phenotypic stability: The stability of a genotype (or genotypes) over a spectrum of environmental conditions. It is also known as developmental homeostasis.

Pinnate: A leaf divided into terminal and lateral leaflets.

Pleiotrophy: Phenomenon of a single gene affecting two or more different characters.

Polycross: An isolated group of plants or clones arranged in same fashion to facilitate random interpollination.

Polymerase chain reaction (PCR): A procedure involving multiple cycles of denaturation, and polynucleotide synthesis that amplifies a particular DNA sequence.

Polymerization: Chemical union of two or more molecules of the same kind to form a new compound having the same elements in the same proportions but a higher molecular weight and different physical properties.

Polymorphism: Two or more kinds of individuals maintained in a breeding population.

Polynucleotide: A linear sequence of joined nucleotides in DNA or RNA.

Polypeptide: A linear molecule with two or more amino acids and one or more peptide groups. They are called dipeptides, tripeptides, and so on, according to the number of amino acids present.

Polyploid: An organism with more than two sets of chromosomes (2n diploid) or genomes (*e.g.* triploid (3n), tetraploid (4n) and so on.

Positional sterility: The condition of gynoecium being infertile and incapable of reproducing due to unsuitable position of androecium.

Post-harvest: After harvesting.

Progeny testing: The practice of ascertaining the genotype of an individual by mating it to an individual of known genotype and examining the progeny.

Progeny: The off-springs of a particular mating.

Prokaryote: A number of a large group of organisms (including bacteria and blue green algae) that lack true nuclei in their cells and that do not undergo meiosis.

Pseudo-dominance: The phenomenon of apparent dominance of a recessive gene in area opposite a chromosome deficiency.

Pseudo-incompatibility: Incompatibility due to physiological and physical reasons only and is not heritable, *e.g.* physiological factors like temperature and light, and physical factors (heterostyly, protoandry, closed anther, *etc*).

Pseudo-heterosis: *See luxuriance*.

Pure line: The descendents obtained from self-fertilization of a single homozygous or an inbred homogenic strain.

Qualitative characters: The characters, which show discrete variation and easily identified by visual observations, *e.g.* flower colour, fruit shape *etc*.

Quantitative characters: The characters, which show continuous variation such that visual identification of individual genes segregation is not possible. These are usually governed by the cumulative effect of polygenes and are highly influenced by environmental conditions *e.g.* are yield / plant, fruits / plant *etc*.

Random mating: Arrangements of pairs by chance, *i.e.* each individual has equal chances to mate with another *e.g.* cross-pollinated crops.

Recessive gene: When parents with contrasting characters are crossed, the character of one parent does not appear in F_1 even though genes for both the factors are present in the hybrid. The gene, which does not appear, is said to be recessive.

Reciprocal hybrids: Two hybrids produced by crossing the same parents but the male of first is used as female in the another and similarly the female of first is used as male in another such as (A x B) and (B x A).

Registered seed: The progeny of foundation seed that is so handled as to most nearly maintain the satisfactory genetic purity and has been approved by a certifying agency.

Relative heterosis: Usually refers to heterosis expressed over the mid-parental value of a cross.

Restriction enzyme: An endonuclease that recognizes a specific short sequence in DNA and cleaves the DNA molecule.

Restriction map: A linear or circular diagram of a DNA molecule showing the sites that are cleaved by different restriction enzyme.

Reverse transcriptase: An enzyme that catalyzes the synthesis of DNA using an RNA template.

RFLP (Restriction fragment length polymorphism): A genetic difference among individuals that is decided by comparing DNA fragments released by digestion with one or more restriction enzyme.

Ribosome: Cytoplasmic organelle on which proteins are synthesized.

RNA: Ribonucleic acid. The information-carrying material in some viruses. More generally, a molecule derived from DNA by transcription that may carry information (messenger or mRNA), provide subcellular structure (ribosomal or rRNA), transport amino acids (transfer or TRNA), or facilitate the biochemical modification of itself or other RNA molecules.

Roguing: The act of removing undesirable individuals from varietal mixtures or seed production plots.

***S-allele*:** A set of alleles controlling sporophytic self-compatibility.

Seed: The matured ovule having all the essential structures to produce a new plant.

Segregating generations: The F_2 and onward generation of a hybrid where separation of parental from material chromosomes at meiosis and consequent separation of genes takes place leading to possibility of recombination in off-spring.

Selection: The process in which a number of individuals with certain desirable characteristics are favoured for further reproduction.

Self-pollinated crops: An assembly of homozygous plants. These crops often have one single genotype and reproduce it as such from generation to generation.

Self-incompatibility: Inability of viable pollen to fertilize an ovule carrying an identical *S-allele* and inhibiting self or sib-pollination.

Self-sterile: The organism which fails to fertilize and set seeds after self-pollination.

Self-incompatibility: The inability of a functional gamete or the plant to set seed when self-pollinated, even though it forms normal zygotes when cross-pollinated.

Selfing: The process of putting the pollen of the same flower on its stigma, *i.e.* enforcing self-pollination. It is extensively practised in maize and other cross-pollinated crops to obtain inbred lines for utilization in the hybrid programme.

Self-pollination: The transfer of pollen from the anther to stigma of the same flower or the plant. It is also the union of genetically similar gametes.

Semi-sterility: A state of only partial fertility in plant zygotes usually associated with chromosomal translocations.

Semigamy: Abnormal fertilization in which the male gamete fertilizes the egg, but does not fuse with the egg nucleus.

Shuttle vector: A plasmid capable of replicating in two different organisms such as yeast and *E. coli*.

Sib-mating: Sibbing or crossing at random the two or more individual obtained from the same parentage. It is form of inbreeding and refers to brother-sister mating.

Sib-pollination: Pollination between closely related biotypes.

Silencer: A DNA sequence that helps to reduce or shut off the expression of a nearby gene.

Single-cross: A cross between two genetically dissimilar parents.

Single-cross hybrid: A hybrid involving two cross-compatible lines.

Somatopalstic sterility: A type of sterility, which results as a consequence of the collapse of fertilized ovules or zygote during the embryonic or early development stages due to disturbance in embryo endosperm relations.

Southern blot: The transfer of DNA fragments from an electrophoretic gel to a cellulose or nylon membrane by capillary action.

Species: A unit of taxonomic classification containing group of individuals enough alike so that it may be reasonably assumed that they have arisen from a common ancestor.

Specific combining ability: The deviation in performance expected on the basis of general combining ability.

Sporophytic incompatibility: Failure of fertilization due to genetic abnormalities.

Stamens. Organ of flower, which produces pollen.

Standard heterosis: Heterosis expressed over the standard or check variety. It is also known as useful heterosis.

Sterility: The phenomenon in which an organism is infertile and is incapable of reproducing.

Strain: The mating group within a variety or species with distinct morphological or physiological features.

Synthetic varieties: These are the open-pollinated advanced generation population of a number of hybrids obtained by crossing a number of tested lines grown in isolation.

Taxa (Singular Taxon): A general term for taxonomic classification, irrespective of rank.

Test cross: The cross of the F_1 with recessive homozygous parent, *i.e.* Aa x aa.

Three way cross: The cross between a single-cross and a inbred line with 3 lines (A, B, C), it could be represented as (A x B) x C. It has been genetically used in the maize-breeding programme for the production of commercial sweet maize hybrids.

Tissue culture: The development of an entire organism from plant cells and tissues *in vitro* or artificial media.

Top-cross: An out cross of selections clones or inbred, to a common pollen parent. In corn, a commonly an inbred-variety cross.

Transformation (bacteria): Genetic alteration of bacteria brought about by the incorporation of foreign DNA in the bacterial cells.

Transgenic: A term applied to organisms that have been altered by introducing DNA molecules into them.

Transgressive segregation: The appearance of individuals showing an extreme development of a character than either of the parent in F_2 or later generations. This usually occurs because of cumulative and complementary effect of genes contributed by the parents of the cross

Transition: A mutation caused by the substitution of one purine by another purine or one pyrimidine by another pyrimidine in DNA or RNA.

Triple-cross: The cross between two three way crosses with six lines (A, B, C, D, E, F) it could be represented as (A x B) x C) x (D x E) x F).

Variation: Differences amongst the individuals arising as a consequence of differences in the genetic make up; the effect of the environment or interplay of both. It is the chief characteristics of living organisms and is the basis of evolution.

Varietal blend: Mechanical mixtures of seed of two or more varieties

Variety: An agricultural variety is a population of similar individuals having common identifiable plant fruit or seed characteristics along with a good agronomic base. This is generally utilized for commercial cultivation.

Vector: An animal or insect, which transmits parasites *e.g.* white fly, is the vector of viral diseases of tomato okra, chilli, *etc.*

Vertical resistance: Resistance conferred by few genes, specific to few known races.

Vestige: An organ, the size, structure and function of which have diminished and simplified to the course of evolution until only a trace remains. Such vestiges may still have important function in organizing embryogenic development, even if their former adult function has disappeared.

Viability: The capability to live and develop normally.

Xenia: The immediate effect of pollen (*i.e.* male gamete) on the endosperm. It is frequently observed in open-pollinated maize with respect to kernel colour.

Zygote: The cell formed by the fusion of male and female gametes.

APPENDICES

Table 1: List of genes used as markers in tomato

Symbol	Characters
aa	Anthocyanin absent
alc	Alcobaca
Aps [1]	Acid phosphatase
B	Beta carotene
bs	Brown seed
c	Potato leaf
d	Dwarf
h	Hair absent
hp	High pigmented
j	Jointless
ls	Lateral suppression
ms	Male sterility
nor	Non-ripening
Nr	Never ripe
rin	Ripening inhibitor
sp	Self-pruning
tmf	Terminating flower
u	Uniform light-green colour of fruit
Cf, I, I-1, I-2, Mi, h_2, Ph, Py Tm-1, Tm-2, $Tm\text{-}2^2$	Genes conferring resistance to diseases

Table 2: List of genes used as markers in bell pepper

Symbol	Characters
A	Anthocyanin-basic gene for purple colour in foliage, stem, flower and fruit, and incompletely dominant
B	Beta-carotene-high in mature fruit, interacts with *t* for higher level in mature fruit
t	High beta carotene-complementary with B
Y	Yellow or orange mature fruit colour
Ys	Yellow spot on corolla of *C. pendulum*
Yt_1, Yt_2	Yellow top young expanding leaves are yellow and gradually turn green
Dw_1, dw_2	Dwarf plant
fa	Fasciculate-flowers and fruit borne in cluster or bunches,

Contd...

Table 2 Contd...

	compounded nodes, bushy plants with determinate tendency
fi_1	Filiform- thread like leaves, blossom irregularities and female sterile
fr	Frilly-leaf margins undulated
H_1	Hairless-smooth stem in *C. annuum* var. *minimum* (Blanco) dominant over fruit shape dominant over elongate
O	Oblate or round fruit shape dominant over elongate
P	Pointed fruit apex incompletely dominant over blunt
rl	Round leaf-leaf length reduced changing the length: width ratio from 1.50 to 1.24
ru	Rugose or savoyed mature leaves.
up	Upright or erect pedicel and fruit orientation, intermediate expression in some genetic background
fs_1	Female-sterile mutant in *C. annuum*, mature plants larger, otherwise normal, male fertile, sets parthenocarpic, seed less fruits
ms	Male sterile
C	Capsaicin-pungent fruit, modifiers increase or decrease capsaicin to produce a bimodal distribution of pungency in F_2
Bs_1	Bacterial spot resistance to *Xanthomonas campestris* pv. *vesticatria* race-1 in *C. chacoence* PI-260435
Bs_2	Bacterial spot resistance to race-2
	Root-knot nematode resistance to *M. incognita*

Table 3: List of genes used as markers in cabbage

Symbol	Characters
c	Anthocyaninless
A	Basic anthocyanin colour factor
A^{rc}	Coloured lamina
B	Light red midrib
M	Magenta plant colour (reddish purple)
As	Asparagoides, bizarre
Er	Entire leaf margin
fc	Fused cotyledons.
gl	Glossy foliage, dominant types produce wax on stems
Hr-1	Hairy first leaf hairs on margins
sm	Smooth leaves
Pet	Petiolate
W	Wide *vs* narrow leaf
K	Dominant factor for heading
Ax	Axil sprouts
T	Tall
dw	Dwarf, short internodes and round leaves
wp	White petal
cp	Crinkly petal

Contd...

Table 3 Contd...

ms-1	Broccoli (male sterile)
2	Brussels sprouts
4	Purple cauliflower
5	Cauliflower
6	Broccoli
S	Self-incompatibility multi-allelic
f	Major gene for resistance to black rot

Table 4: List of genes used as markers in watermelon (Robinson *et al.*, 1976).

Symbol	Synonym	Characters
A		Andromonoecious, recessive to monoecious
Ar	(B) Ge	Anthracnose resistance, resistance to *Giomerella cingulata* var. *orbiculare*
C		Canary yellow flesh, dominant to pink
d		Dotted seed coat, black dotted seed when dominant for r, t and w
dw-1		Dwarf-1, short internodes, due to fewer shorter cells than normal
dw-2		Dwarf-2, short internodes, due to fewer cells
e	(t)	Explosive rind, thin, tender rind, bursting when cut furrowed fruit surface, recessive to smooth
g	(D)	Light green skin, light green fruit, recessive to dark green
g^s	(d^s)	Stripped green skin, recessive to dark green but dominant to light green skin
go*	(c)	Golden, yellow colour of older leaves and mature fruit
gms	(msg)	Glabrous male sterile line, foliage lacking trichomes, male sterile
1		Long seed, long recessive to medium length of seed, interact with s
m		Mottled skin, greenish white mottling of fruit skin
n*		Non-lobed leaves, leaves lack lobbing, dominance incomplete
O		Oval fruit, incompletely dominant to spherical
P		Penciled lines of skin, inconspicuous stripes, recessive to netted fruit
Pm		Powdery mildew susceptibility, susceptibility to *Sphaerotheca fuliginea*
r		Red seed coat, interact with w and t
s		Short seeds, epistatic to 1

Contd...

Table 4 Contd...

Su*	(suB_i)	Suppressor of bitterness, non bitter fruit
t	(bt)	Tan seed coat, interacts with r and w
w		White seed coat, interact with r and t
Wf*	(W)	White flesh, dominant to red
y	r	Yellow flesh, recessive to red

*Proposed new symbols.

Table 5. List of genes used as markers in muskmelon (Robinson *et al.*, 1976)

Symbol	Synonymous	Characters
a	(M)	Andromonoecious, mostly staminate, fewer perfect flowers; interacts with g
ab		Abrachiate, lacking lateral branches; ab, a plant produce only staminate flowers
Af*		*Aulacophora foveicollis* resistance, resistant to red pumpkin beetle
Ag		*Aphis gossypii* tolerance free from leaf curling following aphid infestation
b		Bush, short internodes, compact plant habit
Fom-1*	(Fom 1)	*Fusarium oxysporum* f. *melon* is resistance resistance to race-1
Fom-2*	Fom 1, 2	*Fusarium oxysporum* f. melon is resistance resistance to race-1 and 2 of fusarium wilt
g		Gynomonoecious, mostly pistillate, fewer perfect flowers; a g plants produce perfect flowers exclusively
gf*		Green flesh colour, recessive to salmon
gl		Glabrous, trichomes lacking
gp*		Green petals, corolla leaf-like in colour and venation
h		Halo cotyledons, yellow cotyledons, later becoming green
Mc		*Mycosphaerella citrullina* resistance, high degree of resistance to gummy stem blight
Mc-2*	(Mc^1)	*Mycosphaerella citrullina* resistance, high degree of resistance to gummy stem blight
ms-1*	(ms_1)	Male sterile-1, indehiscent anthers with empty pollen walls in tetrads
ms-2*	(ms_2)	Male sterile-2, anthers indehiscent, containing mostly empty pollen walls and growth rate reduced
n		Nectarless, nectaries lacking in all flowers
O		Oval fruit shape, dominant to round; associated with a
p*		Pentamerous, five carpels and stamens; recessive to trimerous

Contd...

Table 5 Contd...

Pm-1*	(Pm_1)	Powdery mildew resistance, resistance to race-1 of *Sphaerotheca fuligenea*
Pm-2*	(Pm_2)	Powdery mildew resistance 2, race-2 resistance, incompletely dominant
Pm-3*	(Pm_3)	Powdery mildew resistance 3, resistance derived from PI-124111
Pm-4*	(Pm_4)	Powdery mildew resistance 4, resistance derived from cv. *seminole*
Pm-5*	(Pm_5)	Powdery mildew resistance 5, resistance derived from cv. *seminole*
r		Red stem, red pigment under epidermis of stems, especially at nodes
s		Sutures presence of vein tracts ("sutures"; recessive to ribless)
So*		Sour taste, dominant to sweet
sp*		Spherical fruit shape, recessive to obtuse; dominance incomplete
st*		Striped epicarp, recessive to non-striped fruit.
v		Virescent chlorotic foliage
w*		White colour of mature fruit, recessive to dark green fruit skin
wf*		White flesh, recessive to orange
wi*		White colour immature fruit, recessive to dark green fruit skin
wt*		White testa, dominant to yellow or tan seed coat colour
Y		Yellow epicarp, dominant to white fruit skin
yg*	(y)	Yellow green leaves, reduced chlorophyll content

* Proposed new symbols

Table 6. List of genes used as markers in cucumber (Robinson *et al.*, 1976)

Symbol	Characters
a	Androecious, products only male flower if recessive at Acr/acr locus
Acr (F, acr, D)	Female, high degree of female sex expression
bi	Bitter free, lacking cucurbitacin
cp	Compact, reduced internode length, poorly developed tendrils, small flowers
de(I)	Determinate, short vines with stem terminating in flowers
df	Delayed flowering under long photoperiod
dvl (dl)	Divided leaf

Contd...

Table 6 Contd...

gy	Gynoecious, recessive gene for high degree of female sex expression
m	Male, high degree of male expression, m plants are andromonoecious if recessive for F and hermaphrodite if dominant for F
m-2	Perfect flowers-2, bisexual flowers with normal ovaries
ms-1	Male sterile-1, male flowers abort before anthesis
ms-2	Male sterile-2, male flowers abort
n	Negative geotropic peduncle response, pistillate flowers upright
ns	Numerous spines on fruits
Pc (P)	Parthenocarpy, fruit setting without pollination
sp	Short petiole, 20% of normal size at first node
T	Tall plant, incompletely dominant to short
Tr	Trimonoecious, producing male, bisexual and female flowers in that sequence during plant development
ul	Umbrella leaf, leaf margins turn down at low relative humidity giving cup shaped leaves
B	Black or brown spines dominant to white spines on fruits
B-2	Black spine-2, interact with B to produce an F_2 of 15 black: 1 white
B-3	Black spine- 3, interact with B-4 to produce an F_2 of 9 black: 7 white
B-4	Black spine -4, interacts conversely with B-3
c	Cream mature fruit colour, interaction with R (red mature fruit) giving an F_2 ratio of 9 red (R+): 3 orange (Rc) : 3 yellow (+ +): cream (+ c)
D	Dull skin colour , dull skin colour of American cultivars dominant to glossy skin of most European cultivars
gl	Glabrous, foliage lacking trichomes, fruits without spines.
H	Heavy netting of fruit, dominant to no netting
R	Red mature fruit, interact with c, linked or pleiotropic with B and H
s	Spine size and frequency, many small fruit spines, characteristics of European cultivars, recessive to the few large spines of most American cultivars
te	Tender skin of fruit, thin tender skin of some European cultivars recessive to thick, tough skin of most American cultivars
Tu	Tuberculate fruit, warty fruit, characteristic of American cultivars dominant to smooth, non-warty fruits of European cultivars
u	Uniform immature fruit colour recessive to mottled or stippled colour
w	White immature fruit colour, recessive to green.
wf	White flesh
yf	Yellow flesh
Ar	Anthracnose-PI-175111, PI-175120, PI-179676 and PI-182445
Ccu	Scab-Davis Perfect
dm	Downey mildew-Poinsett
pm-1, pm-2 & pm-3	Powdery mildew-Natsufushnari, PI200815, PI-200818
Foc	Fusarium wilt-Wiscisson-248
Cmv	Cucumber mosaic virus-Wiscisson-SMR-12, Wiscission-SMR-15 and Wiscission-SMR-18

Table 7: List of genes used as markers in *Cucurbita* (Robinson *et al.*, 1976)

Symbol	Characters
a	Androceous produces only male flowers
B	Bicolor fruit, yellow fruit pigmentation developing before anthesis
Bi	Bitter fruit, high cucurbitacin content in fruit
bl	Blue fruit colour, incompletely recessive to green
Bu	Bush habit, short internodes; dominant to vine habit in young plant stage but recessive at maturity
C	Coloured fruit, green fruit; epistatic to r
cu	Cucurbitacin content, reduced content of cucurbitacin B
D	Dark green stem, dominant to light green stem
Di	Disc fruit shape, dominant to spherical
Est	Esterase, electrophoretic form of α-naphthyl acetate esterase
Hr	Hard rind, dominant to soft fruit rind
L	Light fruit colour, uniform light intensity of fruit pigmentation
Lap	Leucine aminopeptidase, electrophoretic form of leucine aminopeptidase
lt	Leaf tendril, tendrils with laminae
ly	Light yellow corolla, recessive to orange yellow
M	Mottled leaves, silver gray areas in axils of leaf veins
ms-1 (ms1)	Male sterility-1 male flowers abort before anthesis
ms-2 (ms2)	Male sterility-2, male flowers abort
n	Naked seed, lacking a lignified seed coat.
Pm	Powdery mildew resistance, resistance to *Sphaerotheca fuliginea*
r	Recessive white, white fruit colour, hypostatic to C
Rd	Red skin, red external fruit colour; dominant to green white, yellow and gray
ro	Rosettee leaf, lower lobes of leaves slightly spiralled
s	Sterile, male flowers small, without pollen; female flower sterile
St (I^{st})	Striped fruit, longitudinal stripes on fruit conspicuous if I but inconspicuous if I+
W	White fruit dominant to green mature fruit, partially epistatic to Y
Wf	White flesh, dominant to cream flesh colour
Wt	Warty fruit, dominant to smooth
Y	Yellow fruit colour, dominant to green
ys	Yellow seedling, lacking chlorophyll; lethal

Table 8: List of genes used as markers in onion

Symbol	Characters
a/a	Albino seedlings
y^1/y^1	Yellow seedling linked with glossy
y^2/y^2	Yellow seedling not linked with glossy
pg/pg	Pale green seedling
v/v	Virecent seedling
gl/gl	Glossy foliage
ea/ea	Exposed anther
ya/ya	Yellow anther
pr/pr	Pink root resistance
ms/ms	Male sterility
i/i C/C R/R	Homozygous red
i/i C/c R/R	Heterozygous red
i/i C/C R/r	Heterozygous red
i/i C/c R/r	Heterozygous red
i/i C/c r/r	Heterozygous yellow
i/i c/c R/R	Homozygous recessive white
i/i c/c R/r	Heterozygous recessive white
i/i c/c r/r	Homozygous recessive white
I/I—	Homozygous dominant white
I/i - -	Heterozygous dominant white (buff)

Table 9: List of genes used as markers in cowpea (adapted from Rery, 1985 a)

Symbol	Characters
A	Alfaalfa like pod shape
ax	Axillary buds, active buds of cotyledons
B	Blue seed coat
Bc-1	Bacterial cancker resistance-1
Bc-2	Bacterial cancker resistance-2
bcm	Blackeye cowpea mosaic virus resistance, likely a synonym of blc
Bcy	Brown calyx colour, dominant to green
Bk	Black pod, dominant to white pod
Bl	Black seed coat, also conditions anthocyanin production in the pod tip, calyx and peduncle, heterozygote produces mottled seeds
blc	Black eye cowpea mosaic virus resistance, likely a synonym of bcm
Bp	Brown pod, dominant to straw colour
Bpl-1	Bacterial pustule resistance-1

Contd...

Table 9 Contd...

Bpl-2	Bacterial pustule resistance-2
bpl-3	Bacterial pustule resistance-3
bpl-4	Bacterial pustule resistance-4
bpl-5	Bacterial pustule resistance-5
by	Bean yellow mosaic virus resistance
CC	Cowpea chlorotic mottle virus resistance, the recessive allele at the Mvi locus is likely a synonym
Ci	Compound inflorescence
Cls-1	Cercospora leaf spot resistance-1
cls	Cercospora leaf spot resistance-2
Cm	Cucumber mosaic virus resistance.
crpt	Crumpled petal
Cy	Cylindrical-length pod
D	Dark flower colour
df	Dwarf (Slow growth, dark green leaves, short internodes
Ef-1	Early-flowering-1
Ef-2	Early-flowering-2
Er	Erect pod attachment, dominant to drooping pod attachment
Gp	Green pod, dominant to cream pod
Gr	Green pod, dominant to white pod
La	Lanceolate leaf
lg	Light green pod
Llf	Long leaf
ls	Leaf size, small leaf recessive to large leaf
ms-1	Male sterile-1
ms-2	Male sterile-2
ms-3	Male sterile-3
ms-4	Male sterile-4
Nlf	Narrow leaf, dominant to broad leaf
Nv	Necrotic synergistic reaction associated with cowpea stunt
O	Hilum ring seed coat pattern
P	Purple pod, dominant to green, also causes anthocyanin production in the calyx and peduncle
pa-1	Pod apperarance-1 wrinkled dry pod recessive to smooth appearance
pa-2	Pod apperarance-2 wrinkled dry pod recessive to smooth appearance
Pb	Purple petiole base
Pbr	Purple branch base
Pf	Purple flower
Pg	Pale green plant
Pn	Penducle length, long penduncle dominant to short
Pu	Purple pod, stem and petiole and completely purple

Contd...

Table 9 Contd...

R	Red seed coat
rh	Beetle resistance
Rk	Root-knot resistance, allelic to rk^i
rk^i	Root-knot resistance- intermediate, allelic to Rk
S	Spotting pattern, patches of black pigment on certain types of seed coat
Sbm	Southern bean mosaic virus resistance
Sh	Spindly growth habit, marked elongation of the main stem few side branches
shp	Shrunken pericarp
Sr	Stem rot resistance
St	Standard petal exhibit full expression of colour
stx	*Sesquipedalis*- like texture of pod (soft)
Tr	Stem root resistance
un	Unifoliate leaf, petiole, all stipellate, and the two lateral leaflets with their petioles and missing
V	Seed coat mottling
Vw	Verticillium wilt resistance
Ymr	Cowpea yellow mosaic virus resistance

Table 10: List of genes used as markers in French bean (adapted from Varnell, 1965)

Symbol	Characters
Are	Resistant to anthracnose
Da	Straight pod
Dw1 dw2	Duplicate gene causing dwarf plant
Ext	Interspecific gene for external stigma in *Phaseolus coccineus*
Fa	Basic gene for pod membrane
Fin (fin^+)	Indeterminate *vs* fin determinat plant growth
gas	Causes both male and female sterility
in	Determinate growth = (fin)
itor	With ram triple branched inflorescence
L	Long vs1 short internodes
mo	Conditions resistance to bean virus
neu^+	Short day flowering response
ri	Confers resistance to bean virus 1
St	Stringy pods
te	Twining habits *t* non-twining
To	Cell wall fibre
tri	Produces three cotyledons
Uni	Unifoliate leaves

Table 11: List of Genes used as markers in carrot. (Peterson and Simon, 1986).

Symbol	Characters
A	Alpha carotene synthesis
Ce	Cercospora leaf spot resistance
Cr	Cracking roots
Eh	Downey mildew (*Erysiphe heracle*) resistance
g	Green petiole
gls	Glabrous seed stalk
io	Intense orange xylem
L	Lycopene synthesis
ms-1, ms-2, ms-3, ms-4 and ms-5	Male sterility
O	Orange xylem
P-1 and P-2	Purple root
Rs	Reducing sugar in roots
Y	Yellow xylem
Y-1 and Y-2	Differential xylem / phloem carotene levels

COLOUR PLATES

1-15

PLATE - 1

TOMATO

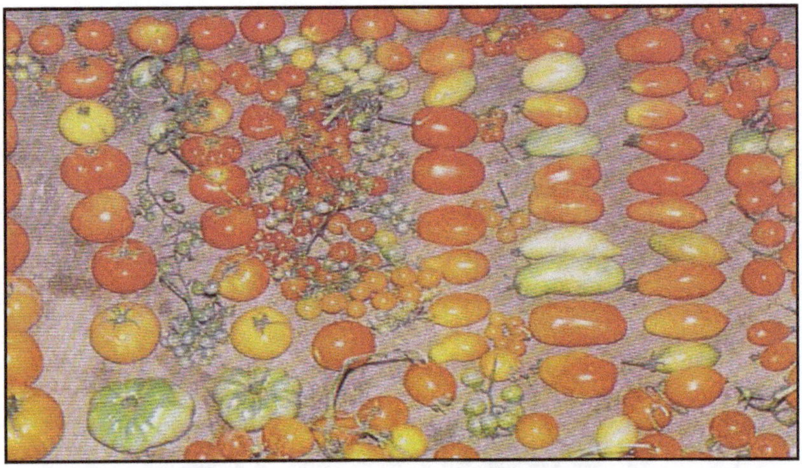

Figure 1: Variability in fruit shape, size and colour in tomato

BRINJAL

Figure 5: Variability in fruit shape, size and colour in brinjal

PLATE - 2

CHILLI

Figure 7: Variability in fruit shape, size and colour in chilli

BELL PEPPER

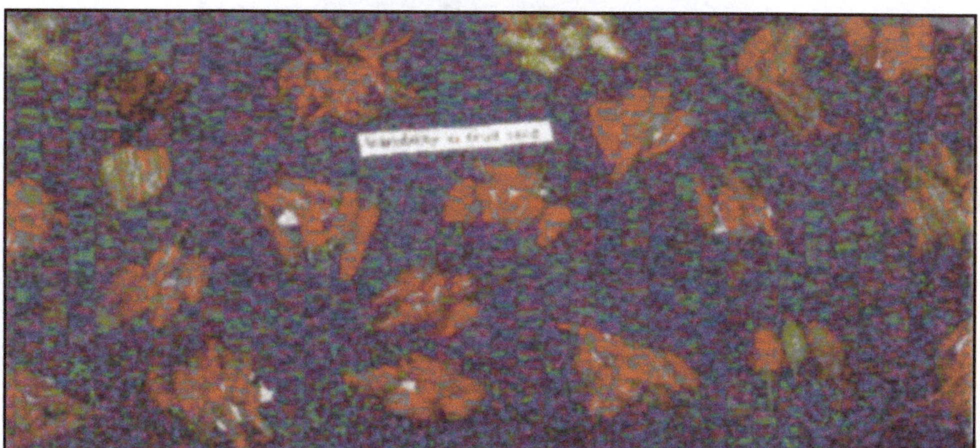

Figure 8: Genetic diversity in fruit shape and size in capsicum

PLATE - 3

OKRA

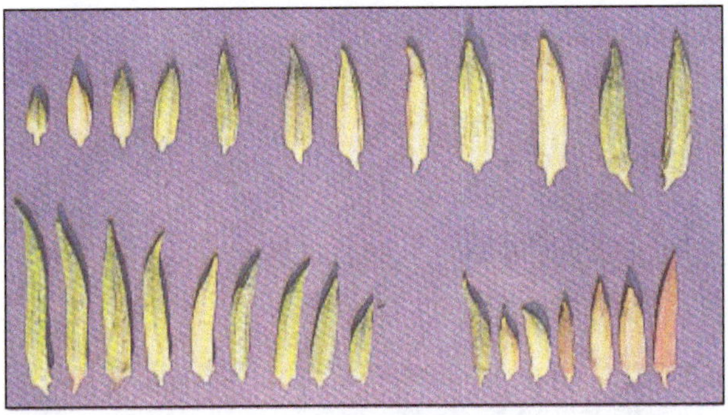

Figure 1: Variability in fruit shape, size and colour in okra

CUCUMBER

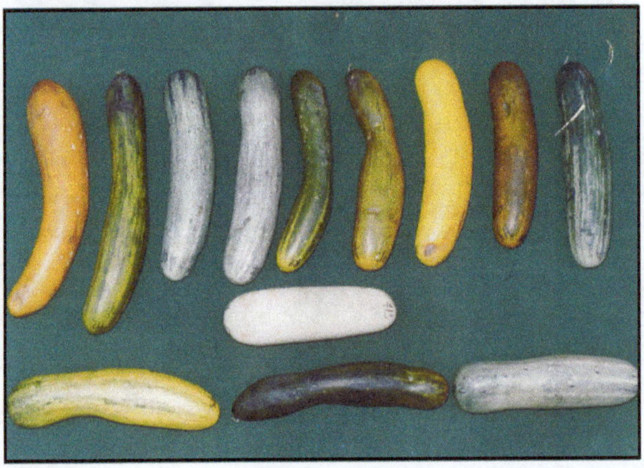

Figure 1: Variability in shape, size and colour in cucumber

PLATE - 4

MUSKMELON

Figure 3: Genetic variability in shape and size in muskmelon

PLATE - 5

PUMPKIN

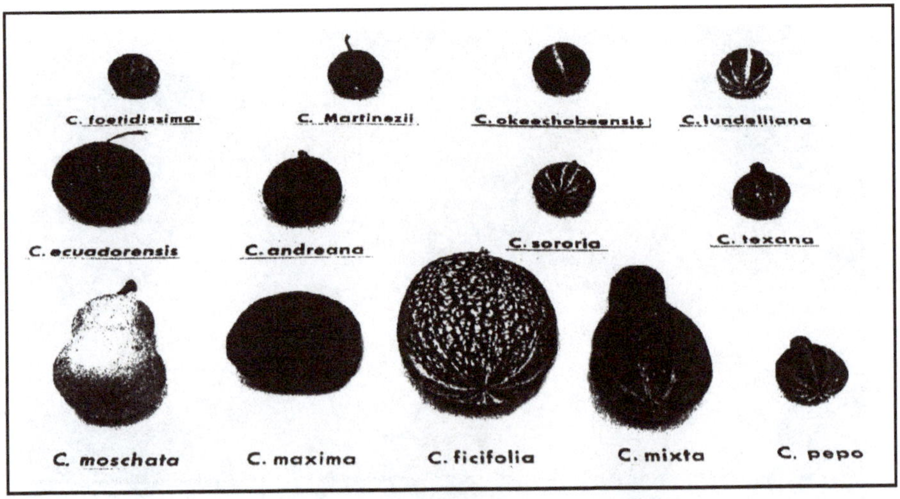

Figure 5a: Variability in *cucurbita* spp

Figure 5b: Genetic Variability in pumpkin

PLATE - 6

RIDGE GOURD AND SMOOTH (SPONGE) GOURD

Figure 8: Genetic variability in shape and size in ridge gourd

PLATE - 7

TOMATO HYBRIDS

Figure: 1. ARKA VISHAL

Figure: 2. ARTH-3

Figure: 3. ARTH-4

Figure: 4. MEENAKSHI

PLATE - 8

TOMATO HYBRIDS

Figure: 5. MADHURI

Figure: 6. MEGHA

Figure: 7. CTH 1

Figure: 8. MANEESHA

PLATE - 9

TOMATO HYBRIDS

Figure: 9. GULMOHAR

Figure: 10. PUSA HYBRID-1

Figure: 11. RHRTH - 57

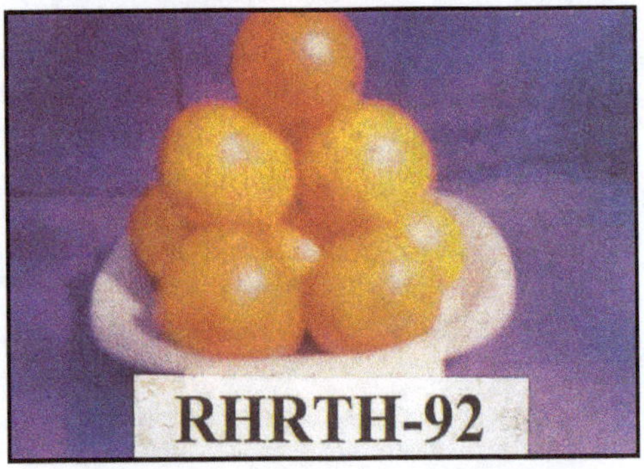

Figure: 12. RHRTH-92

Figure: 13. ARKA VARDAN

Figure: 14. PUSA HYBRID-4

PLATE - 10

TOMATO HYBRIDS

Figure: 15. TH-802

Figure: 16. THL 2312

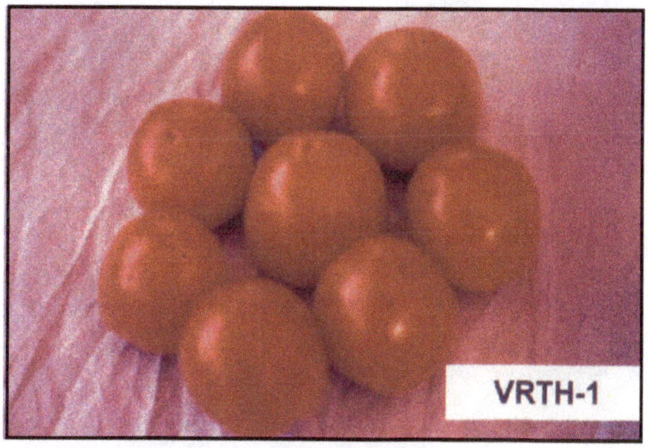

Figure: 17. VRTH-1

Figure: 18. VRTH-2

PLATE - 11

BRINJAL HYBRIDS

Figure: 19. BH-201

Figure: 20. ARKA NAVNEET

Figure: 21. BHL-2

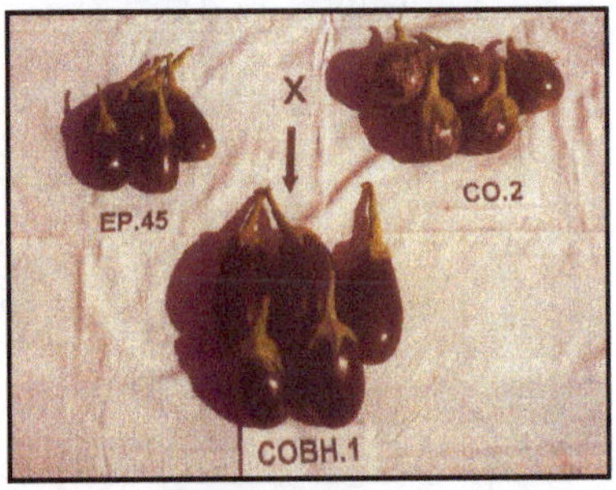

Figure: 22. COBH-1

Figure: 23. MHB-39

Figure: 24. NARDENA HYBRID-2

PLATE - 12

BRINJAL HYBRIDS

Figure: 25. NDBH-1

Figure: 26. PUSA HYBRID-5

Figure: 27. RHR-3x12

Figure: 28. PUSA HYBRID-9

Figure: 29. RHR-3x21

Figure: 30. VRBHR-1

Figure: 31. RHR-3x31

PLATE - 13

CHILLI HYBRIDS

Figure: 34. ARCH-236

Figure: 35. CH-3

Figure: 36. BSS-141

Figure: 37. BSS-138

Figure: 38. CH-1

PLATE - 14

OKRA HYBRIDS

Figure: 39. OKRA JYOTI

Figure: 40. PVR-2 OKRA

PLATE - 15

CABBAGE HYBRIDS

Figure: 41. BSS-115

Figure: 42. BSS-32

Figure: 43. BSS-44

Figure: 44. BSS-50

Figure: 45. SRI GANESH GOL

REFERENCES

Aarts M G M, Dirkse W G, Steikema W J and Pereira A (1993). Transposon tagging of a male sterlity gene in *Arabidopsis*. *Nature* **363**: 715-17.

Absaleha and Dutta OP (1995). Studies on heterosis in bottle gourd. *Indian Journal of Horticulture* **52**: 222-25.

Agble F (1972). Seed size heterosis in cowpea [*Vigna unguiculata* (L) Walp.]. *Ghana Journal of Science* **72**: 30-33.

Aghora TS and Pathak C S (1991). Heterosis and combining ability in a line 'X' tester cross of onions (*Allium cepa* L.). *Vegetable Science* **18**: 53-56.

Agrawal PK (1998). *Transgenic Plants in India: Implications and Issues*, pp 12-14. Agro India.

Ahmed N, Hakin M A and Gandroo M Y (1999). Exploitation of hybrid vigour in okra (*Abelmoschus esculentus* (L.) Moench.). *Indian Journal of Horticulture* **56**: 247-51.

Ahmed N, Muzafar Hurra, Hurra M (2000). Heterosis studies for fruit yield and some economic characters in sweet pepper (*Capsicum annum* L.). *Capsicum and Eggplant Newsletter* **19**: 74-77.

Ahmed N, Narayan RJ (2005) Hybrid vegetable seed production in Kashmir valley: A highly remunerative enterprise. *Proceedings held at 24th Entrepreneurship Development Programme under Agriclinics and Agribusiness Centre Scheme,* SKUAST (K), Srinagar, India.

Ai Y, Singh A, Coleman C E, Ioeger TR, Kheyr-Pour A and Kao TH (1990). Self-incompatibility in *Petunia inflata*: Isolation and characterization of cDNAs encoding three *S-allele* associated proteins. *Plant Reproduction* **3**: 130-38.

Alexandrove SN (1952). Using hybrid seed to increase the yield of indoor cucumbers. *Plodoov Khoz* **10**: 43-45.

Alipieva M and Dimitrov S (1984). Possibilities of producing F_1 hybrids of cabbage resistant to *Peronospora parasitica* Pers. Fr. *Genetic Science*, **17**: 95.
American Orchid Society Bulletin **33**: 473-78.

Anand H and Dubey S (1980). Investigation of hybrid effects in peas (*Pisum sativum* L.). *Archiv Fur Zuchtungs Forschung* **10**: 55-62.

Anbu S, Muthukrishnan CR and Irulappan I (1976). Line 'X' tester analysis in tomato. I. Combining ability. *South Indian Horticulture* **28**: 128.

Anbu S, Muthukrishnan CR and Irulappan I (1981). Line 'X' Tester analysis in tomato. *South Indian Horticulture* **29**: 49-53.

Anderson MA, Cornish EC, Mau SL, Williams EG, Hoggart RM, Atkinson RJ, Roche P, Haley J, Penschow J, Niall H, Tregear G, Coghlan J, Crawford R and Clarke AE (1986). Molecular cloning of a cDNA of a stylar glycoprotein associated with the expression of self-incompatibility gene of *Nicotiana alata. Nature* **321**: 38-44.

Anderson MA, McFaddon GI, Bernatzsky R, Atkinson A, Orpin T, Dedman H, Tregear G, Fernley R and Clarke AE (1989). Sequence variability of three alleles of the self-incompatibility gene of *Nicotiana alata. Plant Cell* **1**: 483-89.

Anonymous (1962). Studies on hybrid vigour. *Agricultural Research* (ICAR) **2**: 5.

Anonymous (1977-78). *Annual Report.* Vegetable Crops Breeding Station, Avignan.

Arasu NT (1968). Self-incompatibility in Angiosperms: A review. *Genetica* **39**: 1-24.

Arce TP and Lopez CB (1994). *Capsicum and Eggplant Newsletter* **13**: 68-71.

Arora SK (1993). Diallel analysis for combining ability studies in okra (*Abelmoschus esculentus* (L.) Moench). *Punjab Horticulture Journal* **33**:116-22.

Arumughamn R, Chelliah S and Muthukrishnan CR (1975), *Abelmoschus manihot* - A source of resistance to bhindi yellow vein mosaic. *Madras Agricultural Journal* **62**: 310-12.

Aruna S and Veergavathatham D (1996). Performance of parents and hybrids for yield and component traits in tomato (*Lycopersicon esculentum* Mill.). *South Indian Horticulture* **44**: 153-56.

Arus P, Shields C R and Orton T J (1985). Application of isozyme electrophoresis for purity testing and cultivar identification of F_1 hybrids of *Brassica oleracea. Euphytica* **34**:651.

Arus, P (1983). Genetic purity of commercial seed lots. *In: Isoenzymes in Plant Genetics and Breeding*, pp. 415-25. Tanksley, BD and Orton J.J. (Eds). Elsevier Science Publication.

Arvind H and Dubey S (1980). Investigation of hybrid effects in peas (*Pisum sativum* L). *Archiv fur Zuchtungs forschung* **10**: 55-62.

Arya PS and Saini SS (1976). Genetic variability and correlation studies in bell peppers. *Indian Journal of Agricultural Research* **10**: 223-28.

Arya PS and Saini SS (1977 a). Variability studies in salad type peppers. *Progressive Horticulture* **9**: 37-42.

Arya PS and Saini SS (1977 b). Variability studies in pepper (*Capsicum* spp.). *Indian Journal of Horticulture* **34**: 415-22.

Asao H, Arai S, Sato T and Hirai M (1994). Characteristics of a somatic hybrid between *Solanum melongena* L. and *Solanum sanitwongsei. Carabbian Breeding Science* **44**: 301-305.

Ascher PD and Peloquin SL (1966). Effect of floral aging on the growth of compatible and incompatible pollen tubes in *Lilium longiflorum. American Journal of Botany* **53**: 99-102.

Atanassova B (1999). Functional male sterility in tomato (L. *esculentum*) and its application in breeding and hybrid seed production. *Euphytica* **107**: 13-21.

Atanassova B and Georgiev H (1986). Investigation on tomato male sterile lines in relation to hybrid seed production. *Acta Horticulturae***22**: 219-22.

Attavar M and Bhat NK (1973). High yielding Indo American *Capsicum* hybrid. *Lal Bagh* **18**: 28-32.

Avecherkova ZG (1975). The quality of hybrid seed of white cabbage in relation to the place of their formation on the plant. Daklody TSKHA 211: 112.

Axelsan F (1956). Cucumber news from Weibullsholm Weibulls llehanda. **2**: 13-15.

Babu S and Thirmurugan T (2000). Effect of heterosis in brinjal (*Solanum melongena). Journal of Economic Taxonomy and Environment Monograph* **10**: 63-66.

Bain MS and Kang US (1963). Inheritance of some flower and fruit characters in muskmelon. *Indian Journal of Genetics* **23**: 101-6.

Bairagi SK, Singh DK and Ram Harihar (2002). Studies on heterosis for yield attributes in cucumber (*Cucmis sativus* L.). *Vegetable Science* **29**: 75-77.

Baishya KC, Syamal MM and Singh KP (2001). Heterotic studies in tomato. *Vegetable Science* **28**: 168-69.

Bajaj YPS and Singh H (1980). *In vitro* induction of androgenesis in mung bean (*Phaseolus aureus* L*). Indian Journal of Experimental Biology* **18**: 13-16.

Baker B, Coupland G, Federoff N, Starlinger P and Schell J (1987). Phenotypic assay for excision of the maize controlling element in tobacco. *EMBO Journal* **6**:1547-54.

Baker B, Schell J, Lorz H and Federoff N (1986). Transposition of the maize-controlling element 'Activator' in tobacco. *Proceedings of National Academic Science* USA

Baker LR (1978a). *Research Report*, Michigan State Agriculture University Experiment.

Baker LR (1978b). *Research Report*, Michigan State Agriculture University Experiment.

Baker LR (1978c). *Research Report*, Michigan State Agriculture University Experiment.

Baker LR and Valk M (1978). *Research Report*, Michigan State Agriculture University Experiment.

Ballvora A, Hesselbach J, Niewohner J, Leister D, Salamini F and Gebhardh (1995). Marker enrichment and high resolution map of the segment of potato chromosome VII harbouring the nematode resistance gene Grol. *Mol Gen Genetics* **249**: 82-90.

Balmohan TN, Subbiah R and Shanmugavelu KG (1983). Studies on heterosis in brinjal. *In:Proceedings of Scientific Meetings of Genetics and Improvement of Heterotic Systems*, held at Tamil Nadu Agricultural University, Coimbatore.

Bannerot HO, Boulidard L and Chupeau Y (1977). Unexpected difficulties met with the radish cytoplasm in *Brassica oleracea. Eucarpia crucifereae Newsletter* **2**:16.

Bannerot HO, Boulidard L, Cauderson Y and Temple T (1974). Cytoplasmic male sterility transfer from *Raphanus* to *Brassica* in *Eucarpia* cruciferae. Scottish Horticulture Research Institute, Scotland, 52.

Barabas Z (1991). Hybrid seed production using nutritional mutants. *Euphytica* **53**: 67-72.

Baranski P (2000). Application of a rapid glucose-6-phosphate isomerase assay in white cabbage breeding. *Physiology of Plants* **22**: 45-51.

Barnabas B (1985). Effect of water loss germination ability of maize (*Zea mays* L.) pollen. *Annual Botany* **48**: 861-64.

Barnes WC (1961). A male sterile cucumber. *Proceedings of American Society of Horticulture Science* **77**: 415-6..

Barringer SA, Yasseen MY, Schloupt, RM and Splittstoesser WE (1996). *Journal of Vegetable Crop Production* **2**: 27-33.

Basavaraja N (1986). Studies on hybrid seed production in brinjal (*Solanum melongena* L). *Abstract, Mysore Journal of Agricultural Sciences* **20**: 249.

Batnaru H (1979). Prospects for the utilization of large style for producing heterotic tomato seed. *Lucrari Stiintifice, Institutul Bergh.* Pergaman Press, N.Y.

Beachy RN, Loesh Fries S and Tumer NE (1990). Coat protein mediated resistance against virus infection. *Annual Review of Phytopathology* **28**: 451-74.

Beal WJ (1880). *Indian Corn Report*. Michigan State Board Agriculture **19**: 279-89.

Belivanis T and Dore C (1986). Interspecific hybridization of *Phaseolus vulgaris* L. and *Phaseolus angustissimus* using *in vitro* embryo culture. *Plant Cell Reporter* **5**:329-31.

Bennett MD and Hughes WG (1972). Additional mitosis in wheat pollen induced by ethrel. *Nature* **240**: 566-68.

Berger (1974). Estrase of Drosophilla. II. Biochemical studies of esterase in *D. pseudoobscura. Genetics* **78**: 1157-72.

Bhadula SK and Sawhney VK (1991). Protein analysis during the ontogeny of normal and male sterile stamenless-2 mutant stamens of tomato (*Lycopersicon esculentum* Mill.). *Biochemical Genetics* **29**: 29-41.

Bhagchandani P, Singh N and Thakur PC (1980). Combining ability in summer squash (*Cucurbita pepo* L.). *Indian Journal of Horticulture* **37**: 62-65.

Bhagchandani P and Choudhury B (1970). Heterosis and inbreeding expression in carrot. *Progressive Horticulture* **2**: 65-73.

Bhagyalakshmi PV, Shankar CR, Subrahmanyam D and Babu VG (1991). Heterosis and combining ability studies in chillies. *Indian Journal of Genetics and Plant Breeding* **51**: 420-23.

Bhansali RR and Ramavat P (1993). *Journal of Horticultural Science* **68**: 25-30.

Bhaskaraih KB, Shivashankar G and Virupakshappa (1980). Hybrid vigour in cowpea. *Indian Journal of Genetics and Plant Breeding* **40**: 334-37.

Bhor TJ, Kute NS, Dumbre AD and Sarode ND (1997). Heterosis and inbreeding depression in cowpea. *Indian Journal of Agricultural Research* **31**: 122-26.

Bhuller GS, Singh KB, Dhaliwal HS and Malhotra RS (1978). Heterosis and combining ability in pea. *Genetica Agraria* **30**: 35-46.

Bhushana HO, Viswanathan KP, Arunachalam P and Halesh GK (2000). Heterosis in cowpea (*Vigna unguiculata* (L) Walp.) for seed yield and its attributes. *Crop Research* **19**: 277-80.

Bhutani RD (1980). Diallel analysis for yield and quality traits in tomato (L. *esculentum* Mill.). Ph. D. Thesis, CCS Haryana Agricultural University, Hisar.

Bhutani RD (1981). Screening and diallel study for yield and quality characters in tomato (*Lycopersicon esculentum*). *Thesis Abstract* **7**: 170-1.

Bhutani RD, Kalloo G and Pandita ML (1983). Genetic variability and scope of improvement of quality traits in tomato *(Lycopersicon esculentum* Mill.). *Haryana Agriculture University Journal of Research* **8**: 408-12.

Bhutani RD, Kalloo G, Pandita ML and Singh GP (1973). Studies on heterosis in tomato. *Haryana Journal of Horticultural Science* **2**:156-163.

Bhutani RD, Kalloo G, Singh GP and Sidhu AS (1980). Heterosis and combining ability in brinjal. *Haryana Agriculture University Journal of Research* **10**: 476-84.

Binzel ML, Sankhla N, Joshi, S and Sankhla D (1996). *Plant Cell Repoter* **15**: 536-40.

Birat, RBS (1964). Variability of resistance in bhindi varieties to root galls. *Science and Culture* **30**: 244.

Birchler JA and Auger DL (2003). Biological consequences of dosage dependent gene regulatory mechanism in multicelluar eucaryotes. *In: Biology of Dominance*, Veitia R. (Ed.).

Bird CR, Ray JA, Fletcher JD, Boniwell JM, Bird AS, Teulieres C, Blain I, Bramley PM and Schuch W (1991). Using antisense RNA to study gene function: Inhibition of carotenoid biosynthesis in transgenic tomatoes. *Biotechnology* **14**: 875-79.

Blestsos FA, Roupakias DG, Tsaktsira ML, Scaltsoyjannes AB and Thanassoulopoulos CC (1998). Interspecific hybrids between three eggplant (*Solanum melongena* L.) cultivars and two wild species (*Solanum torvum* Sw. and *Solanum sisymbriifolium* Lam.). *Plant Breeding* **117**: 159-64.

Boaxi Z, Sanwen H, Guimei Y and Jaizhen G (2000). Two RAPD markers linked to a major fertility restorer gene in pepper. *Euphytica* **113**: 155-69.

Bockish T, Saranga Y, Altman A and Ziv, M (1997). *Acta Horiculture* **447**: 241-42.

Bohn GW and Davis GN (1957). Earliness in F_1 hybrid muskmelon and their parent varieties. *Hilgradia* **26**: 435-71.

Bohn GW and Whitaker TA (1949). A gene for male sterility in the muskmelon. *Proceedings of American Society of Horticulture Science* **53**: 309-14.

Bohn GW, Kishaba AN and Toba HH (1972). Mechanisms of resistance to melon aphid in a muskmelon line. *HortScience* **7**: 281.

Bohn, GW and Principe JA (1964). A second male sterility gene in the muskmelon. *Journal of Heredity* **55**: 211-15.

Bondarenko LD (1983). Cabbage varieties and hybrids for mechanical harvesting. *Kartofel Ovashchi* **8**: 27.

Bonnet A (1978). *Biuletin Warzywniczy* **22**:147-50.

Bonnierbale M, Plaisted RL and Tanksley SD (1988). RFLP maps of potato and tomato based on a common set of clones reveal modes of chromosomal evolution. *Genetics* **120**:1095-03.

Bora GC, Hazarika MH, Shadeque A and Verma MM (1993). Heterosis for yield and its component in tomato. *Processings of Seminar on Heterosis Breeding in Crop Plants - Theory and Application*, held during 23-24 February, pp. 10-11.

Borovec V (1988). *Bull.Vyzkmny Slech.University* **15**: 31-45.

Borthwick HA, Phillips M and Robbins, WW (1931). Floral Development in carrot (*Dacus carota L.*). *Euphytica* **7**: 131-39.

Brain AD (1957). Differences in the flowers visited by four species of bumble-bees and their causes. *Journal of Animal Ecology* **26**: 71-98.

Branca C, Torelli A and Bassi M (1990). *Plant Cell, Tissue and Organ Culture*, **21**: 17-19.

Brar, JS and Padda, DS (1972). A study of heterosis for root and leaf characters in radish (*Raphanus sativus* L.). *Journal of Research.* Punjab Agricultural University, Ludhiana. **9**: 27-31.

Brar JS and Sukhija BS (1978). Hybrid vigour in intervarietal crosses in watermelon (*Citrullus lanatus* (Thumb.) Mansf). *Indian Journal of Horticulture* **34**:277.

Bray EA (1991). Regulation of gene expression of endogenous ABA during drought stress.: Davies *WJ and Jones HG* (Eds). *In Abscisic Acid: Physiology and Biochemistry.* BIOS Scientific Publishers, Oxford, pp. 81-99.

Bredemerijer GM and Blaas J. (1981). S-specific proteins in styles of self-incompatible *Nicotiana alata. Theoretical and Applied Genetics* **59**: 185-90.

Brewbaker TL (1957). Pollen cytology and self-incompatibility systems in plants. *Journal of Heredity* **48**: 271-77.

Brezhnev DD (1976). New aspects of heterosis in tomato selection. *Eucarpia* **7**:59-62.

Broglie R and Broglie K (1993). Production of disease resistant transgenic plants-Current Opinion. *Biotechnology* **4**: 148-51.

Brommonschenkel SH and Tanksley SD (1997). Map based cloning of the tomato genomic region that spans the 100-seed weight-5 tospovirus resistance gene in tomato. *Molecular Genetics* **256**: 121-6.

Brunetti A, Tavazza M, Noris E, Tavazza R, Caciagli P, Ancora G, Crespi S and Accotto GP (1997). High expression of truncated viral REP protein confers resistance to tomato yellow leaf curl virus in transgenic tomato plants. *Molecular Plant-Microbe Interactions* **10**: 571-79.

Bucio Alanis L, Perkins JM, Jinks JL (1969). Environmental and genotype-environmental component of variability. V. Segregating generations. *Heredity* **24**: 115-27.

Bunin MS and Yoshikawa H (1989). Heterosis breeding in carrot in USSR and Japan. *Japanese Journal of Breeding* **39**: 235-41.

Burza W and Malepszy S (1995). *Plant Breeding* **114**: 41-45.

Butnaru H (1979). Prospects for the utilization of longstyly for producing heterotic tomato seed. *Lucrari stiintifice, institutal Agronomic Timisoare Agronomie* **16**: 197-200.

Camargo LE, Savides L, Jung G, Nienhuis J and Osborn TC (1997). Location of the self-incompatibility locus in an RFLP and RAPD map of *Brassica oleracea. Journal of Heredity* **88**: 57-60.

Carlsson G (1952). Experiments with hybrid seed of the greenhouse cucumber. *Meddn. Gullakers Vaxtfordl. Anst.* (9-10): 213-18.

Carmi Y, Salts Y, Shabtai S, Dedicovz B, Pilowsky M and Barg R (1997). Transgenic parthenocarpy due to specific oversensitization of the ovary to auxins. *Acta Horticulturae***447**: 579-81.

Cassells AC and Bajaj YPS (1991). In: Biotechnology in Agriculture and Forestry Bajaj YPS (Ed.), Springer-Verlag.

Cassells AC, Coleman M, Farrell G, Long R, Goetz EM and Boyton V (1986). Screening for virus resistance in tissue culture adventitious regenerants and their progeny. *Genetic Manipulation in Plant Breeding*, pp. 535-45. Horn W *et.al.* (Eds). W de Grujter, Berlino.

Celine VA and Sirohi PS (1996). Heterosis in bitter gourd (*Momordica charantia* L.). *Vegetable Science* **23**: 180-85.

Chadha KL (1997). Status and future of F_1 hybrids of vegetable crops in India. *In: Summer School on Hybrid and Hybrid Seed Production Technology of Vegetables*, Chakrabarty AK, Katarai AS, Mishra JP and Sharma VK (Eds), IARI, New Delhi, pp1-11.

Chadha ML and Sidhu AS (1982). Studies on hybrid in brinjal. *Indian Journal of Horticulture* **39**: 233-38.

Chadha ML, Hedge RK and Bajaj KL (1988). Heterosis and combining ability studies of pigmentation in brinjal (*Solanum melongena* L.). *Vegetable Science* **15**: 64-71.

Chadha ML, Joshi AK and Ghai TR (1990). Fifth gene for male sterility in *Cucumis melo. Genet Coop Rep.* **13**: 34.

Chadha ML, Joshi AK and Ghai TR (1990). Heterosis breeding in brinjal. *Indian Journal of Horticulture* **47**: 417-23.

Chandra Deo, Singh KP and Panda PK (2000). Screening of okra parental lines and their F_1's for resistance against yellow vein mosaic virus. *Vegetable Science* **27**: 78-79.

Chandra S and Bhatnagar SR (1975). Reproductive biology of *Abelmoschus esculentus* (L.). Reproductive behaviour, floral morphology, anthesis, pollination mechanism. *Acta Botanica Indica* **3**: 104-13.

Chaubey AK, Singh BP and Ram H (2002). Heterosis studies in early cauliflower. *Haryana Journal of Horticultural Science* **31**: 76-78.

Chaubey, Arun Kumar and Ram, Harihar (2004). Heterosis for fruit yield and its components in bitter gourd (*Momordica charantia* L.). *Vegetable Science* **31**: 51-53.

Chauhan Sunil and Singh Yudhvir (2002). Heterosis studies in okra (*Abelmoschus esculentus* (L.) Moench). *Vegetable Science* **29**: 116-18.

Chen LZ and Adachi T (1998). Protoplast fusion between *Lycopersicon esculentum* and *L. peruvianum* complex: Somatic embryogenesis, plant regeneration and morphology. *Plant Cell Reporter* **17**:508-14.

Cheverud JM, and Routman EJ (1995). Epistasis and its contribution to genetic variance components. *Genetics* **139**: 1455-61.

Chilton MD, Saiki RK, Yadav N, Gordon MP and Quetier F (1994). T-DNA from Agrobacterium Ti plasmid is in the nuclear DNA of crown gall tumour cells. *Proceedings of National Academic of Science* USA **77**:4060-64.

Choer E, Augustin E, Antunes IF, Silva JB, Peters JA and Silva JB (1999). Cultivar identification and genetic purity analysis of beans through isoenzymatic patterns. *Seed Science and Technology* **27**: 517-23.

Choudhary RC and Khanna KR (1972). Exploitation of heterosis in tomato yield and components. *South Indian Horticulture* **20**: 59-65.

Choudhary SM (1987). Studies on heterosis, combining ability and correlation in bitter gourd. Ph D Thesis, Mahatma Phule Agricultural University, Rahuri, Maharashtra, India.

Choudhury B and Pathak SC (1959). Studies on floral biology in cucumber (*Cucumis sativus* L.). *Indian Journal of Horticulture* **18**: 212-21.

Choudhury B and Roy HK (1964). Indian Agricultural Research Institute, New Delhi, *Agriculture Research* **4**: 213.

Choudhury B and Singh B (1971). Pusa Meghdut and Pusa Manjari, two high-yielding bottle gourd hybrids. *Indian Horticulture* **16**: 15-16.

Choudhury B, Punia RS and Sangha HS (1965). Manifestation of hybrid vigour in F_1 and its retention in F_2 generation of tomato (*L. esculentum* Mill.). *Indian Journal of Horticulture* **22**:52-59.

Choudhury SM and Kale PN (1991). Combining ability studies in bitter gourd. *Journal of Maharashtra Agricultural University* **16**:34.

Chunwongse J, Bunn TB, Crossman C, Jiang J and Tanksley SD (1994). Chromosomal location and molecular marker tagging of the powdery mildew resistance gene (Lv) in tomato. *Theoritical and Applied Genetics* **89**:76-79.

Chuong PV, Beversdorf WD, Powell AD and Pauls, KP (1988). The use of haploid protoplast fusion to combine cytoplasmic atrazine resistance and cytoplasmic male sterility in *Brassica napus*. *Currrent Plant Science-Biotechnology, Agriculture* 191-194.

Churchill GA and Doerge RW (1994). Empirical threshold values for quantitative trait mapping. *Techanical Report on Biometric Unit Cornell University, Ithica.*

Cizov ST (1945). Yields of hybrid cucumbers in glasshouses. *Proceedings of Scientific Conference. Timirjazev Agriculture Academy* **1**: 42-43.

Clements FE and Long FL (1923). *Experimental Pollination: An Outline of the Ecology of Flowers and Insects.* Washington: Carnegie Institution.

Cohen Y, Eyal H and Cohen A (1993). Gylona gynoecious muskmelon. *HortScience* **28**: 855.

Constance L (1971). History of classification of Umbelliferae (Apiaceae). Biology and Chemistry of Umbelliferae. Heywood, V.H. (Eds). Academic Press, New York.

Corelissen BJC Does MP and Melchers LS (1996). Strategies for molecular resistance breeding and transgenic plants. *In : Taxonomy, Molecular Biology, Ecology, Pathology and Control,* Sneh, B., Jabaji, Hau S., Neale, S. and Deist, G (Eds). Kluwer, pp. 529-36.

Courret JP, Delourme R and Renard M (1992). Expression of *ogu* cytoplasmic male sterility in cybrids of *Brassica napus*. *Theoretical and Applied Genetics* **83**: 549-56.

Crane MB (1915). Heredity of types of inflorescence and fruits in tomato. *Journal of Genetics* **5**: 1-12.

Crockett PA, Singh MB, Lee CK and Bhalla PL (2002). Genetic purity analysis of hybrid broccoli (*Brassica oleracea* var. *italica*) seeds using RAPD-PCR. *Australian Journal of Agricultural Research* **53**: 51-54.

Cross JW and Ladyman JAR (1991). Chemical agents that inhibit pollen development: Tools for research. *Sexual Plant Reproduction* **4**: 235-43.

Cross RE (1970). Hybrid carrots. *The Asgrow Farmer* **23**: 7-8.

Crossway A, Oakes JV, Irvine JM, Ward B, Knauf VC and Shewmaker CK (1986). Intergation of foreign DNA following micro-injection of tobacco mesophyll protplasts. *Molecular General Genetics* **202**: 179-85.

Crowe J H, Hoekstra F A, Crowe L M, Achordoguy T J and Drobnis E (1989b). Lipid phase transitions measured in intact cells with Fourier transform infrared spectroscopy. *Cryobiology* **26**: 76-83.

Crowe JH, Hoekstra FA and Crowe LM (1989a). Membrane phase transitions are responsible for imbibitional damage in dry pollen. *Proceedings of National Academy of Sciences* **86**: 520-23.

Curtis LC (1939). Heterosis in summer squash (*Cucurbita pepo*) and the possibility of producing F_1 hybrid seed for commercial planting. *Proceedings of American Society of Horticultural Science* **37**: 827-28.

Dahiya MS, Dhankar BS and Kalloo G (1984). Hybrid performance in eggplant (*Solanum melongena* L.). *Haryana Journal of Horticulture* **13**: 147-49.

Danam Shashi Bhushan and Chaudari, BP (2000). Heterosis studies in cowpea. *Annals of Agricultural Research* **21**: 248-52.

Daskalov and Poulos JM (1994). Updated *Capsicum* gene list. *Capsicum and Eggplant Newsletter* **13**: 15-26.

Daskalov H and Popova D (1958). Heterosis in melons (*Citrullus edulis*) and the prospects of utilizing it in practice. *Selskostop Misal* **3**: 785.

Davenport CB (1908). Degeneration, albinos and inbreeding. *Science* **28**: 454-55.

Davis EW (1966). An Improved method of producing hybrid onion seed. *Journal of Heredity* **57**: 55-7.

Davis GH (1957). The distribution of the male sterility gene in onion. *Proceedings of American Society of Horticulture Science* **70**: 316-18.

Dayasagar P (1994). Studies on heterosis in bhindi (*Abelmoshus esculentus* (L) Monech.). *Annals of Agricultural Research* **15**: 321-26.

DeBlock M, Botterman J, Van de Wiele M, Docky J Thoen C, Gossele V, Rao Mova N, Thompson C, van Montagu M and Leemans J (1987). Engineering herbicide resistance in plants by expression of a detoxifying enzyme. *EMBO Journal* **6**:2513-18.

DeCandole (1882). *Origin of Cultivated Plants*, New York.

DeNettancourt D, Saccardo F, Laneri U, Capaccio E, Westerhof Mand Ecochard R (1974). Self-incompatibility in a spontaneous tetraploid of *Lycopersicon peruvianum* Hull, in polyploidy and induced mutations in plant breeding, IAEA, Vienna, 77.

DeVilliers RP, Van RJ, Janse VR and Van SJ (1993). *Journal of South African Society of Horticultural Science* **3**: 24-27.

Deepali Tiwari and Ram Harihar (1999). Heterosis in bitter gourd (*Momordica charantia* L.). *Vegetable Science* **26**: 27-29.

Deepali Tewari, Ram Harihar and Jaiswal HR (2001). Studies on heterosis and combining ability in indigenous bitter gourd (*Momordica charantia* L.) for fruit yield. *Vegetable Science* **28**: 106-8.

Delannay DE and Lower LR (1987). Generation Mean analysis of plant characters in crosses between two determinate cucumber lines and *Cucumis sativus* var. *hardwickii* R line. *Journal of American Society of Horticulture Science* **112**: 701-11.

Delannay X , La Vallee BJ, Proksh RK , Fuchs RL, Sims SR Greenplate JT, Marrone PG, Doson RB, Augustine JJ , Layton JG and Fischoff DA (1989). Field performance of transgenic tomato plants expressing the *Bacillus thuringinensis* var Kurstaki insect control protein. *Biotechnology* **7**: 1265.

Deshpandey AA and Reddy Pratap VV (1996). Hybrid seed production in cabbage and cauliflower. *In: Compendium on Vegetable Hybrids and their Seed Production, Training Sponsored by Food and Agriculture Organization, IARI, New Delhi, during 15 March to 14 June,* pp.55-60.

Deshpandey AA, Bankapus VM and Nalawadi UG (1978). Some aspects of blossom biology in brinjal varieties (*Solanum melongena* L.). *Current Research,* **7**:174.

Deshpandey AA, Venkata Suffaiah K, and Nalwadi VG (1979). Studies on floral biology of bitter gourd (*Momordica charantia* L.). *Mysore Journal of Agricultural Science* **29**: 572.

Deshpandey RB (1933). Studies in Indian chilli III. The inheritance of some characters in *C. annum* L. *Indian Journal of Agricultural Science* **3**: 219-300.

Dev H, Rattan RS and Thakur MC (1994). Heterosis in tomato. *Horticulture Journal* **7**: 125-32.

Dhaliwal MS (1989). Techniques of developing Punjab Hybrid muskmelon seed. *Seeds and Farms* **15**:25.

Dhankar BS and Dhankhar SK (2001). Heterosis and combining ability studies for some economic characters in okra. *Haryana Journal of Horticultural Sciences* **30**: 230-33.

Dhankar SK, Sharna BS and Dhankar BS (1996a). Heterosis studies in okra (*Abelmoschus esculentus* (L) Monech.). *Haryana Journal of Horticultural Sciences* **25**: 81-87.

Dhankar SK, Sharma BS and Dhankar BS (1996). Combining ability studies in okra (*Abelmoschus esculentus* (L.) Moench). *Haryana Journal of Horticultural Sciences* **25**: 54-57.

Dhankhar SK, Dhankhar BS and Tewadia AS (1998). A note on heterosis and combining ability in okra (*Abelmoschus esculentus* (L.) Monech.). *Haryana Journal of Horticultural Sciences* **27**: 211-14.

Dharmati PR, Madalageri BB, Patil RV and Gasti VD (1996). Heterosis studies in tomato. *Karnataka Journal of Agricultural Sciences* **9**: 642-48.

Dhawan NL (1965). Breeding methodology for yield and other characters in maize. *Proceedings of 2nd International Asian Corn Improvement Workshop* pp. 4-14.

Dhillon NPS (1994). F_1 hybrid seed production in muskmelon and management of male sterile population. *Seed Science and Technology* **22**: 601-5.

Dhillon TS and Sharma BR (1982). Interspecific hybridization in okra (*Abelmoschus* sp.). *Genet. Agrar* **36**: 247-56.

Dickson MJ, Jones DA and Jones JD (1993). Close linkage between the *cf-2/cf-5* and *Mi* resistance loci in tomato. *Molecular Plant-Microbe Interaction* **6**: 341-347.

Dickson MH (1970). A temperature sensitive male sterile gene in Broccoli, (*Brassica oleracea* L. var. *Italica). Journal of American Society of Horticultural Science* **95**: 13-14.

Diojode SD (1981). Genetic architecture of yield and its components in pumpkin. *Genetica Agraria* **36**: 87-94.

Diojode SD and Sulladmath UV (1984). Preliminary studies on heterosis in pumpkin (*Cucurbits moschata* Poir.) *Mysore Journal of Agricultural Science* **18**: 30-34.

Diojode SD, Nath Prem and Sulladmath UV (1982). Hybrid vigour in pumpkin (*C. moschata*). *Genetica Agrarian* **36**: 87-94.

Direlewanger E Issac PG, Ranade S, Belajouza M, Cousin R and Vienne D De (1994). Restriction fragment length polymorphism analysis of loci associated with disease resistance genes and developmental traits in *Pisum sativum* L. *Theoretical and Applied Genetics* **88**: 17-27.

Dixit J and Gautam NC (1987). Studies on hybrid vigour in eggplant (*Solanum melongena* L.). *Indian Journal of Horticulture* **44**: 74-77.

Dixit J and Kalloo G (1983). Heterosis in muskmelon (*Cucumis melo* L.). *Haryana Agricultural University Journal of Research* **13**: 549.

Dixit J, Bhutani RD and Dudi BS (1982). Heterosis and combining ability in eggplant. *Indian Journal of Agriculture* **52**: 444-47.

Dixit J, Kalloo G, Bhutani RD and Sidhu AS (1980). Line 'X' tester analysis for the study of heterosis and combining ability in tomato. *Haryana Journal of Horticultural Sciences* **9**: 56-61.

Dixit SK, Singh BP and Harihar Ram (2004). Heterosis and combining ability in Indian cauliflower (*Brassica oleracea* var. *botrytis* L.). *Vegetable Science* **31**: 164-67.

Dixon MS, Jones DA Hatzixanthis K, Ganal MW, Tanksley SD and Jones JD (1995). High resolution mapping of the physical location of the tomato *cf-2* gene. *Molecular Plant Microbe Interaction* **8**: 200-6.

Dobzhansky T (1952). Nature and origin of heterosis. *In: Heterosis*, pp. 330-35. Iowa State College Press, Iowa USA.

Dod VN, Kale PB, Wankhade RV (1995). Heterosis and combining ability in tomato (*Lycopersicon esculentum* Mill.). *PKV Research Journal* **19**: 125-29.

Dod VN, Kale PB, Wankhade RV and Jadhao BJ (1992). Heterosis in the intervarietal crosses of tomato (*Lycopersocon esculentum* Mill.). *Crop Research* **5**: 134-39.

Does MP and Corenelissen BJC (1998). Emerging strategies to control fungal diseases using transgenic plants. Chopra VL, Singh RB and Varma Anupam (Eds), *In: Proceedings of 2nd International Crop Science.* Oxford IBH Publication Co. Pvt Ltd, New Delhi, pp 233-244.

Dogra BS, Rastogi KB and Ashok Kumar (1997). Exploitation of hybrid vigour in cucumber (*Cucumis sativus* L.*). Indian Journal of Horticulture* **54**: 261-64.

Dogra BS, Rastogi KB, Korla BN and Sharma RH (1997). Heterosis and combining ability studies for yield and quality characters in cucumber (*Cucmis sativus* L.). *Journal of Hill Research* **10**: 39-42.

Dore C (1973). Androgense *in vitro* par culture danthes de, *Asparagus officinalis*. I. Etat actuel des recherches. *In*: *4th Reunion sur la Selection de I, Asperge, Eucarpia Sect. Horticole*, Thevenin, L; (Ed). Versaillies, France.

Dore C (1974). Production de plantes homozygotes males at females a partird, anthers d asperge, cultivees *in vitro* (*Asparagus officinalis* L). *CR Academic Science Ser D*, 278, 2135.

Dorosiev L (1970). Possible use of stamenless type male sterile tomato mutants in heterotic seed production . *Genetics and Plant Breeding* **3**: 171-79.

Driscoll CJ and Barlow KK (1976). Male sterility in plants: induction, isolation and utilization. *In: Induced Mutations in Cross Breed*, pp. 123-31. Vienna: FAO/IAEA Publ.

Dryanovska OA and Ilieva IM (1983). *In vitro* anther cultures in muskmelon, *C.R. Academic Bulgarian Science* **36**: 1107.

Dubey RS and Singh SR (1968). Longevity of the pollen of okra and comparison of four qualitative tests for pollen viability. *Indian Journal of Horticulture* **25**: 210-13.

Duck N and Evola S (1997). *In: Advances in Insect Control: The Role of Transgenic Plants*. Carozzi N and Koziel M (Eds), pp. 1-20. Taylor and Francis.

Dumas de Vaulx R, Chambonnet D and Pochard E (1981). Culture *in vitro* d'anthers de piment (*Capsicum annum* L.): Amelioration de taux d' obtention des Plantes Chez differents genotypes par des traitements a +35 °C *Agron. Tropocal* (Paris) **1**: 858.

Dundi KB (1991). Development of F_1 hybrid in tomato (*Lycopersicon esculentum* Mill.). *M.Sc. (Agriculture) Thesis*, University of Agricultural Sciences, Dharwad.

Dundi KB and Madalageri BB (1991). Heterosis for shelf life and its component in tomato (*Lycopersicon esculentum* Mill.). *South Indian Horticulture* **39**: 353-55.

Dutta OP (1971). Effect of gamma irradiation on seed germination, plant growth, floral biology and fruit production in *Abelmoschus esculentus* (L.). *Third International Symposium on Tropical Horticulture*, held at IIHR, Bangalore.

Dutta OP (1983). Male sterility in okra [*Abelmoschus esculentus* (L) Monech] and bottle gourd [*Lagenaria siceraria* (Mol.) Standl] and its utilization in hybrid seed production. *Thesis Abstract* **9**: 341.

Dutta OP (1984). Breeding in okra for resistance to yellow vein mosaic virus and enation leaf curl virus. *Annual Report*, IIHR, Bangalore.

Dutta OP (1992). *Indian Horticulture* **37**: cover page and p. 55.

East EM (1908). Inbreeding in corn. *Report*, Connecticut Agricultural Experiment Station for 1907, pp 419-28.

East EM (1909). The distinction between development and heredity in inbreeding. *Ann. Nat.* **43**: 173-81.

Eileen Kabelka, Wenai Yang and David M Francis (2004). Improved tomato fruit colour within an inbred backcross line derived from *Lycopersicon esculentum and L. hirsutum* involves the interaction of loci. *Journal of American Society of Horticultural Science* **129**: 250-400.

Eisa H and Munger HN (1966). Male sterility in *Cucurbita pepo*. *Proceedings of XVII International Horticulture Congress* **1**: 47-8.

El-Kharbotly A, Leonards- Schippers C, Huigen DJ, Jacobsen E, Pereira A, Stiekema WJ, Salamini F and Gabhart C (1994). Segregation analysis and RFLP maping of the R_1 and R_3 alleles conferming race specific resistance to Phytophthora infestans in the progeny of dihaploid potato parents. *Molecular General Genetics* **242**: 749-54.

El-Maksoud MA, Hetal RM and Mohamad MH (1986). Studies on an intervarietal cross and hybrid vigour in Okra. *Annals of Agricultural Sciences* **29**: 43-78.

Elangovan M, Muthukrishnan CR and Irulappan I (1981). Hybrid vigour in bhindi (*Abelmoshus esculentus* (L) Monech.) for some economic characters. *South Indian Horticulture* **29**: 4-14.

Elleman CJ and Dickinson HG (1994). Pollen stigma interaction during sporophytic self-incompatibility in *Brassica oleracea* L. *In: Genetic Control of Self-incompatibility and Reproductive Development*, pp. 67-68. Willams EG and Knox RB (Eds).

Elleman CJ, Franklin-Tong Vand Dickinson HG (1992). Pollination in species with dry stigmas: The nature of the stigmatic response and the pathway taken by pollen tubes. *New Phytology* **121**: 413-24.

Emerson S (1948). A physiological basis for some suppressor mutations and possibly for one gene heterosis. *Proceedings of National Academy of Science* **34**: 72-74.

Erickson EH, Garment MB and Peterson C (1982). Structure of cytoplasmic male-sterile and fertile carrot flowers. *Journal of American Society of Horticultural Science* **107**: 698.

Eshbaugh WH (1975). Genetic and Biochemical systematic studies of chilli pepper. *Bull Torrey Bot. Club.* **102**: 396.

Ezura H, Nishimiya S and Kasumi M (1993). *Plant Cell Reporter* **12**: 676-80.

Fabre J and Dereuddre (1990). Encaspsulation-dehydration, a new approach to cryopreservation of solanum shoot tips. *Cryoletter* **11**: 413-26.

Fageria MS, Kohli UK and Dhaka RS (2001). Studies on heterobeltiosis for fruit yield and yield attributing traits in tomato. *Haryana Journal of Horticultural Sciences* **30**: 131-33.

Fang Z, Sun P, Liu Y, Wang X, Hou A and Bcan C (1997). A male sterile line with dominant gene (ms) in cabbage (*Brassica oleracea* var. *capitata*) and its utilization for hybrid seed production. *Euphytica* **97**:265-68.

Farash EM, Abdulla HI, Taghian AS and Ahmed MH (1993). *Australian Journal of Agricultural Sciences* **24**: 3-14.

Farnam MW (1998). Doubled haploid broccoli production using anther culture: Effect of anther source on seed set characteristics of derived lines. *Journal of American Society of Horticultural Science* **123**: 73-77.

Fatokun CA and Singh BB (1987). Interspecific hybridization between *Vigna pubescens* and *Vigna unguculata* Walp. Through embryo rescue. *Plant Cell Tissue Org Culture* **9**: 299-33.

Fatokun CA Menancio- Hautea DI, Danesh D and Young ND (1992). Evidence of orthologous seed weight genes in cowpea and mungbean based on RFLP mapping. *Genome* **132**: 841-46.

Federoff NV, Futerk DB and Nelson OE (1984). Cloning of the bronze locus in maize by a simple and generaliziable procedure using the transposable controlling element activator (Ac). *Proceedings of National Academy Science*, USA, **81**: 3825-29.

Feher T (1985). Use of the non-dissected leaf marker in hybrid breeding of watermelon. *Zoldsegtermesztesi Kutato Intezet Bulletin* **18**: 55.

Fery RL (1985). The genetics of cowpea- A review of the world literature. Singh SR and Rachie KO, (Eds). *In: Cowpea: Research Production and Utilization*, John Willey & Sons, pp. 25-62.

Filov AI and Vilenskaya GM (1972). Cultivated cucurbits in different languages of the world. *Trudy po Prikadnoi Botanike, Genetike Selektsii* **47**: 138-50.

Fischhoff DA, Bowdish KS, Perlak, FJ, Marrone PG, Mc Cormick SM, Niedermeyer JG, Dean DA, Kusano-Kretzmer K, Mayer EJ, Rochester DE, Rogers SG and Fraley RT (1987). Insect tolerant transgenic tomato plants. *Biotechnology* **5**: 807-13.

Folkertsma RT, Rouppe van der Voort JN, De Groot KE, van Zandvoort PM, Schots A, Gommers FJ, Helder J and Baker J (1996). Gene pool similarities of potato cyst nematode population assessed by AFLP analysis. *Molecular Plant-Microbe Interaction* **9**: 47-54.

Foolad MR and Chen FQ (1997). RAPD markers associated with salt tolerance in an interspecific cross of tomato (*Lycopersicon esculentum* x *L. pinnellii*). *Plant Cell Reporter* **17**: 306- 12.

Foster RE (1968). F_1 hybrid muskmelon I. Superior performance of selected hybrids. *Proceedings of American Society of Horticultural Science* **91**: 205-7.

Frang G and Grumet R (1993). Genetic engineering of potyvirus resistance using constructs derived from the zucchini yellow mosaic virus coat protein gene. *Molecular Plant Microbe Interaction* **6**: 358-67.

Frankel R and Galun E (1977). Pollination Mechanism, Reproduction and Plant Breeding. Springer-Verlag, New York.

Frost HB (1923). Heterosis and dominance of size factors in *Raphanus. Genetics, Princeton* **8**: 116-53.

Fujimaki H, Hiraiwa S, Kushibuchi K and Tanaka S (1977). Artificially induced male sterile mutants and their usages in rice breeding. *Japanese Journal of Breeding* **27**: 70-77.

Gallais A (1988). Heterosis: Its genetic basis and utilization in plant breeding. *Euphytica* **39**: 95-104.

Galun E and Aviv D (1983). Manipulation of protoplasts, organelles and genes: applicability to plant breeding. *In*: *Proceedings of the 10th Congress of the European Association for Release on Plant Breeding Eucarpia*, Lange W, Zeven AC and Hogenboom NG, (Eds). Wageningen, Netherlands, pp. 228-37.

Gamburg KZ, Semenova LA and Ziv M (1997). *Acta Horticlture* **447**: 147-48.

Ganal MW, Broun P and Tanksley SD (1992). Genetic mapping of randomly repeated telomeric DNA sequences in tomato (*Lycopersicon esculentum*). *Genomics* **14**: 444-48.

Ganal MW, Czihal R, Hannappel U, Kloos DU, Polley A and Ling HQ (1998). Sequencing of cDNA clones from the gentic map of tomato (*Lycopersicon esculentum*). *Genome Research* **8**: 842-47.

Ganal MW, Simon R Brommonschenkel S, Arndt M, Phillps, MS Tanksley, SD, Kumar A (1995). Genetic mapping of a wide spectrum nematode resistance gene (Hero) against *Globodera rostochiensis* in tomato. *Molecular Plant-Microbe Interaction* **8**: 886-91.

Gandhi SD, Navale PA, Venkatakrishna Kishore and Kishore V (2000). *Crop Research* **19**: 493-99.

Gangadhar Rao, Murthy SN and Reddy NS (1981). Heterosis and inbreeding depression in chilli. *Andhra Agriculture Journal* **28**:191-199.

Gangopadhyay KK, Gill HS and Sharma SR (1997). Heterosis and combining ability studies in early group of Indian cauliflower involving self-incompatible lines. *Vegetable Science* **24**: 26-28.

Garret RE, Mehlschau JJ, Smith NE and Redenbaugh MK (1991). Encapsulation of tomato seeds. *American Society of Agricultural Engineering* **7**: 25-31.

Gatehouse AMR and Gatehouse JA (1998). Identyfying proteins with insecticidal activty: Use of encoding genes to produce insect- resistant transgenic crops. *Pesticides Science* **52**: 165-75.

Gatehouse AMR, Boulter D and Hilder VA (1992). *In: Plant Genetic Manipulation for Crop Protection.* Gatehouse AMR, Hilder VA and Boulter D (Eds). CAB International, pp, 155-81.

Gaungchu Y, Xuezhen L, Zhen X, Lz C, Xhiyin Z and Fengynn B (1980). Anther culture in *Glycine max. Kexue Tongbao* **25**: 976.

Gavrilova OM (1967). Sel, *Skohoz Biology* **2**:559-62.

Geetha N, Venkatachalam P, Reddy PS and Rajasekharan G (1998). *Advances in Plant Science* **11**: 253-57.

Gemes JA and Martinovich L (1998). *Zoldsegtermesztesi-Kutato-Intezet-Bulletinje* **28**: 39-45.

Georgiev H (1991). Heterosis in tomato breeding. In: Kalloo G (ed), In: *Genetic Improvement of Tomato*, pp. 83-98. Springer Verlag, Berlin.

Ghosh B and Sen S (1994). Plant regeneration from alginate encapsulated somatic embryos of *Asperagus cooperi* Baker. *Plant Cell Reporter* **13**: 381-85.

Ghosh PK, Syamal MM and Rath S (1997). Heterosis studies in tomato (*Lycopersicon esculentum* Mill.). *Haryana Journal of Horticultral Sciences* **26**: 114-17.

Ghosh SP and Kalloo G (2000). Genetic resources of indigenous vegetables and their use in South Asia. *Technical Bulletin 4,* IIVR, Varanasi, pp. 36-39.

Gielen JL, Haan DP Kool JA, Peters D, Grinsven MQJM and Goldbach RW (1991). Engineered resistance to tomato spotted wilt virus, a negative strand RNA virus. *Biotechnology* **9**: 1363-67.

Gill BS and Kumar JC (1988). Combining ability analysis in watermelon [*Citrullus lanatus* (Thumb.) Mansf]. *Indian Journal of Horticulture* **45**: 104-9.

Gill HS, Arora RS, Tiwari RN and Sharma RK (1976). Exploitation of heterosis in tomato (*L. esculentum* Mill.). *Indian Journal of Horiculture* **33**:176.

Gill HS, Singh JP and Pachauri DC (1973). Pusa Sanyog out yielded variety. *Indian Horticulture* **18**: 11, 13, 30.

Gill HS, Singh JP and Singh R (1972). Pusa Alankar : An early, high-yielding variety. *Indian Horticulture* **17**: 15-16, 30.

Givanmoni JJ, Noensie EN, Ruezinsky DM, LuXH, Tracy SL, Ganal MW, Martin GB, Pillen K Alpert K and Tanksley SD (1995). Molecular and Genetic analysis of the ripening inhibitor and non- ripening loci of tomato: A first step in genetic map-based cloning of fruit ripening genes. *Molecular and General Genetics* **248**: 195-206.

Globerson D (1967). *Diss. Abstract* **28** : 67-95.

Goldberg RB, Beals TP and Sanders PM (1993). Anther development: Basic principle and practical applications. *Plant Cell* **5**: 1217-29.

Gonalves D, Chee P, Provident R, Seem R and Slightom JL (1992). Comparison of coat protein- mediated and genetically-derived resistance in cucumbers to infection by cucumber mosaic virus under field conditions with natural challenge inoculations by vectors. *Biotechnology* **10**: 1562-70.

Gonsalves C, XueB Yefes M, Fuchs M, Ling K, Namba S, Chee P Slightom JL and Gonsalves D (1994). Transferring cucumber mosaic virus white leaf strain coat protein gene into *Cucumis melo* L. and evaluating transgenic plants for protection against infection. *Journal of American Society of Horticultural Science* **119**: 345-55.

Gonslaves D Chee P, Providenti R, and Slightom JL (1992). Comparison of coat protein mediated and genetically derived resistance in cucumbers to infection by cucumber mosaic under field conditions with natural challenge inoculations by vectors, *Biotechnology* **10**: 1562-70.

Good Night (1999). *Genetics and Exploitation of Heterosis in Crops*. pp. 59-68. Coors J.G. and Pandey S (Eds). Crop Science Society of America.

Gopalakrishnan TR (1986). Diallel analysis in bitter gourd (*Momordica charantia* L.). M.Sc. (Horticulture) Thesis, Tamil Nadu Agricultural University, Coimbatore.

Gopalakrishnan TR (1985). Inheritance of clusterness, destalkness and deep colour in chilli. *Ph. D. Thesis*, Kerala Agricultural University, Vellanikara.

Gopalakrishnan TR, Gopalakrishnan PK and Peter KV (1987). Heterosis and combining ability analysis in chilli. *Indian Journal of Genetics* **47**: 205-9.

Gorman SW, Banasiak D, Fairley C and Mc Cormick S (1996). A 610 kb YAC clone harbors 7 cM of tomato (*Lycopersicon esculentum* Mill.) DNA that includes the male sterile 14 gene and hot spot for recombination. *Molecular General Genetics* **251**: 52-59.

Gourret JP, Delourme R and Renard M (1992). Expression of *Ogu* cytoplasmic male sterility in cybrids of *Brassica napus*. *Theoretical and Applied Genetics* **83**: 549-556.

Govindrasu P, Mithukrishnan CR and Irulappan, I (1981). Combining ability for yield and its component in tomato. *Scientia Horticulturae* **14**: 125-30.

Gowen JW (1952). Hybrid vigour in *Drosophila*. *In*: Heterosis Gowen JW (Ed.). Iowa State Coll Press, Ames Iowa, pp. 474-53.

Graham RD (1976). Physiological aspects of time of application of copper to wheat plants. *Journal of Experimental Biology* **7**: 717-724.

Gresshoff PM and Doy CH (1972). Development and differentiation of haploid of tomato seedling frozen in liquid nitrogen, *Carybiology* **15**: 478-483.

Gresshoff PM and Doy CH (1972). Development and differentiation of haploid in *Lycopersicon esculentum* (tomato). *Planta* **107**: 161-170

Grover RK and Singh G (1970). Pathology of wilt of okra caused by *Fusarium oxysporum* f. *vasinfectum* (Abk) Synder and Hassan, its host range and histopathology. *Indian Journal of Agricultural Science* **40**: 989-96.

Guha S and Maheshwari SC (1964). *In vitro* production of embryos from anthers of *Datura. Nature* **204**: 479.

Guitierrez JA and Singh SP (1985). Heterosis and inbreeding depression in dry bush bean (*Phaseolus vulgaris* L.). *Canadian Journal of Plant Sciences* **65**: 243-49.

Gupta and Rai (1990). *Indian Horticulture* **35**: cover page.

Gupta CG, Lakshmi N and Srivalli T (1998). *Capsicum and Eggplant Newsletter,* **17**: 42-45.

Gupta PK (2001). *Elements of Biotechnology*. Rastogi Publications, Merrut. India.

Hafen, L and Stevenson, E C (1956 c). Male sterile plants speed up hybrid seed production. *MKT Grow. Journal* **85**: 26.

Hagedorn DJ (1951). The relation of perfection-type peas to Wisconsion Bean virus of isolates from. *Phytopathology* **41**: 494.

Hagedorn DJ (1953). *Research Bulletin Wisconsion Agriculture Experiment Station,* 504-8.

Hageman RH, Leng ER and Dudley JW (1967). A biochemical approach to corn breeding. *Advances in Agronomy* **19**: 45-83.

Haley CS and Knot SA (1992). A simple regression method for mapping quantitative trait loci in line crosses using flanking markers. *Heredity* **69**: 315-24.

Hamilton AJ, Lycett GW and Grierson D (1990). Antisense gene that inhibits synthesis of the hormone ethylene in transgenic plants. *Nature* **336**: 284-87.

Hanning E (1904). Zur Physiologie pflanzenlicher Embryomen. I. Uber die kultue Von Cruciferen-Embryonen auferhalf des Embryosack. Z. Bot. **62**: 45-80.

Hanson MR (1991).Plant mitochondrial mutations and male sterility. *Annual Review of Genetics* **25**:461-68.

Hanson MR and Folkerts O (1992). Structure and function of the higher plant mitochondrial genome. *International Review of Cytology* **141**: 129-72.

Haque MS, Wada T and Hattori K (1997). *Plant Cell Tissue and Organ Culture* **50**: 83-89.

Haring MA, Gao J, Volbeda T, Rommens CMT, Nijkamp HJJ and Hille J (1989). A comparative study of tam 3 and AC transposition in transgenic tobacco and petunia plants. *Plant Molecular Biology* **13**:189-201.

Harn C, Kim MZ, Chov KT and Lee YI (1975). Production of haploid callus and embroid from the cultured anther of *Capsicum annuum*. *SABRAO Journal* **7**: 71-77.

Harrison BD, Mayo MA and Baulcombe DC (1987). Virus resistance in transgenic plants that express cucumber mosaic virus satellite RNA. *Nature* **328**: 799-802.

Harska S (1978). The relationship between the number of chloroplasts and number of lamellae in the grana of chloroplast of winter wheat hybrids. *Polnohospodarstvo* **24**: 22-27.

Haruta T (1966). Studies on the genetics of self and cross-incompability in Cruciferous vegetables. Minneapolis Minn. Northrop King and Co., pp 67.

Hassan AA, Nassar HH, Barakat MA and Talba MS (1999). Tomato breeding for salinity tolerance. III. Genetics of tolerance. *Egyptian Journal of Horticulture* **26**: 391-403.

Hayes HK (1963). *A Professor's Story of Hybrid Corn*. Burgess Publication Company, Minneapolis, 237 pp.

Hayes HK and Garber RJ (1919). Synthetic production of high protein corn in relation to breeding. *Journal of American Society of Agronomy* **11**: 309-19.

Hayes HK and Jones DF (1916 a). First generation crosses in cucumber. *Republic Conference Agriculture Experiment Station* **5**: 319-22.

Hayes HK, Immer FR and Smith DC (1955). *Methods of Plant Breeding*, 2nd edn., McGraw Hill Book Co. Inc.

Hazra P (1991). Genetic divergence, yield components and gene action in cowpea [*Vigna unguiculata* (L) Walp.]. Thesis, Bidhan Chandra Krishi Vishwavidyalaya, Mohanpur.

Hazra P, Som MG and Das PK (1993). Studies on heterosis and combining ability for forage and seed yield characters in cowpea [*Vigna unguiculata* (L) Walp.]. Ph.D.Thesis, Rajasthan Agriculture University, Udaipur.

Hedau N (2002). Genetical studies in ridge gourd (*Luffa acutangula* Roxb. L.). *Ph.D. Thesis*, IARI, New Delhi.

Hedric U and Booth NO (1908). Medelian characters in tomatoes. *Proceedings of American Society of Horticultural Science* **5**: 19-24.

Hemenwey C, Fang RX, Kaniewsk WK, Chua NH and Turner NE (1988). Analysis of the mechanism of protection in transgenic plants expressing the potato virus X coat protein or its antisense RNA. *EMBO Journal* **7**: 1273.

Hernandez PR, Prado O, Gill DV, Perez PJ and Ruiz M (1994). *Centro-Agricola* **21**: 369-74.

Herrera Estrella L and Simpson J (1995). Genetically engineered resistance to bacterial and fungal pathogens. *World Journal of Microbiology and Botany* **11**: 383-92.

Hesemann CV (1980). Haploid cells in calli from anther culture *Vicia faba. Z. pflanzenzuecht:* 18-84.

Heslop-Haarrison J (1979). An interpretation of the hydrodynamics of pollen. *American Journal of Botany* **66**: 737-43.

Heslop-Haarrison J and Heslop-Haarrison Y (1970). Evaluation of pollen viability by enzymatically induced fluorescence; intracellular hydrolysis of fluoresce in diacetate. *Stain Tech.* **45**: 115-20.

Hilder VA, Gatehouse AMR, Sheerman SE, Barker RF and Boulter D (1987). A novel mechanism of insect resistance engineered in tobacco. *Nature* **330**: 160-63.

Ho JY, Weide R, Ma HM, van Wordragen MF, Lambert KN, Koornneef M, Zabel P and Williamson VM (1992). The root knot nematode resistance gene (Mi) in tomato: Construction of a molecular linkage map and identification of dominant cDNA markers in resistant genotypes. *Plant Journal* **2**: 971-82.

Hodgkin T, Lyon GD and Dickinson HD (1988). Recognition in flowering plants: A comparison of the *Brassica* self-incompability system and plant pathogen interaction. *New Phytology* **110**: 557.

Hoekstra FA aand Van der Wall EW (1988). Initial moisture content and temperature of imbibition determine extent of imbibitional injury in pollen. *Plant Physiology,* **133**: 257-62.

Hogenboom NG (1972). Breaking breeding barriers in *Lycopersicon* II. Breakdown of self-incompatibility in L. *peruvianum* Mill. *Euphytica* **21**:228.

Hojby, HR (1958). Utilization of hybrid vigour in tomatoes: Possibilities and limitations. *J.S. MKr. Agric. Sci;***1**: 249-61

Hojby, HR (1960). *Pretoria Horticultural Research Station Newsletter* **11**:12.

Honne G, Melchers LS, Vleehouwers VGGAA, Van Roekel JSC and de Wit PJGM (1995). Production of the AVR 9 elicitor from the fungal pathogen *Eladosporium fulvum* in transgenic tobacco and tomato plants. *Plant Molecular Biology* **29**: 909-20.

Hormuzdi SG and More TA (1989). Heterosis studies in cucumber (*Cucumis sativus* L.). *Indian Journal of Horticulture* **46**: 73-79.

Howard HW (1942). Self-incompatibility in polyploid froms of *Brassica* and *Raphanus Nature* **149**: 302.

Hsu FC and Kleier DA (1990). Phloem mobility of xenobiotics.III. Sensitivity of unified model to plant parameters and application to patented chemical hybridizing agents. *Weed Science* **38**: 315-23.

Hu J, Sadowski J, Osborn TC, Landry BS and Quiros CF (1998). Linkage group alignment from four independent *Brassica oleracea* RFLP maps. *Genome* **41**: 226-35.

Hundal JS (1995). Progress of heterosis breeding in chilli at PAU, Ludhiana. *Punjab Vegetable Growers* **30**:58-61.

Hundal JS and Khurana DS (1988). Heterosis potential in chillies (*Capsicum annum* L.). *Proceedings of National Seminar on Chilli, Ginger and Turmeric,* held on 11-12 January at Hyderabad, pp. 28-32.

Hutchins AE (1939). Some examples of heterosis in the cucumber (*C.sativus* L.). *Proceedings of American Society of Horticultural Science* **36**: 660-4.

Hutchins AE and Croston FE (1941). Productivity of F_1 hybrids in the squash (*Cucumis maxima*). *Proceedings of American Society of Horticultural Sciences* **39**: 332-6.

Ichimura K and Oda M (1995). *Journal of Japanese Society Horticultural Science* **64**: 135-41.

Ignatov B and Popova D (1977). Intervarietal *Capsicum* hybrids highly resistant to stolbur virus. *B'lgarski Plodove Zelenchutesii Konservi* **8**: 31-2.

Indira P, Rajan S and Peter KV (1996). *Indian Horticulture* **40**: cover page and p. 23.

Ingale BV, Patil SJ (1997). Heterosis breeding in brinjal (*Solanum melongena*). *PKV Research Journal* **21**: 25-29.

Ioerger TR, Gohlke JR, Xu B and Kao TH (1991). Primary structural features of the self-incompatibility proteins in the solanaceae. *Sex Plant Reproduction* **4**: 81-87.

Iqabaljit Singh, Sharma JR and Kumar JC (1996). Heterosis studies in long fruited genotypes of bottle gourd. *Indian Journal of Horticulture* **53**: 64-67.

Irrulappan I (1992). *Indian Horticulture* **36**: cover page and p. 35.

Irulappan, I (1992). Genetic Diversity in cucurbits-ridge gourd (*Luffa acutangula* Roxb. L.). *Indian Horticulture* **37**: 1-2.

Irwin MR (1952). Specificity of gene effects. *In: Heterosis.* Iowa State College Press, pp. 236-55.

Irwin MR and Cole LJ (1936). Immunogenetic studies of species and species hybrids from the cross of *Columbiaq livia* and *streptopelia risoria*. *Journal of Experimental Zoology* **73**: 309-18.

Islam AK, Joarder OI, Rahman SM and Hossain M (1992). *Indian Journal of Horticulture* **49**: 249-52.

Isouard G Raquin C and Demarly Y (1979). Intention de plantes haploids et haploids et diploids par culture in vitro d, anthers d, Aubergine (*Solanum melongena* L.). *G.R Academic Science* **288**: 987.

Itoh K, Iwabuchi M and Shinaonoto K (1991). *In Situ* hybridization with Species-Specific DNA probes gives evidence for asymmetric nature of *Brassica* hybrids obtained by X-ray fusion. *Theoretical and Applied Genetics* **81**: 356-62.

Ivanoof SS and Albritton RC (1962). Several hybrid watermelons in testing stage. *Miss. Farm Research* **25**: 1.

Iwanami Y (1972 a). Viability of pollen grains in organic solvents. *Botanique* **3**: 61-68.

Iwanami Y (1972 b). Retaining the viability of *Camellia japonica* pollen in various organic solvents. *Plant Cell Physiology* **13**: 1139-41.

Iwanami Y and Naakaamura N (1972). Storage in an organic solvent as a means for preserving viability of pollen grains. *Stain Technology* **47**: 137-39.

Jagadish M (1995). The heterosis and combining ability studies in chillies (*Capsicum annuum* L.) using line 'X' tester analysis. *M.Sc. (Hort.) Thesis*, UAS, Bangalore.

Jain S and Jain S (1982). Studies on heterosis and combining ability of forage and seed yield characters in cowpea (*Vigna unguiculata* (L) Walp.). *Ph D. Thesis*, Rajasthan Agricultural University, Udaipur.

Jaiswal RS, Rattan RS and Saini SS (1984). Hybrid vigour and combining ability in tomato (*L. esculentum* Mill.). *Indian Journal of Horticulture* **33**: 176.

Jaiswal RS, Rattan RS and Saini SS (1984). Hybrid vigour and combining ability in tomato. *South Indian Horticulture* **32**: 69-74.

Jalani BS and Graham KM (1973). A study of heterosis in cross among local and American varieties of okra. *Malasian (Agriculture) Research* **2**: 7-14.

Jambhale ND and Nerkar YS (1983). Interspecific transfer to yellow vein mosaic disease in okra. *Journal of Maharashtra Agriculture University* **8**: 197.

Jambhale ND and Nerkar YS (1983). Introduction of polyploidy in the *Abelmoschus* spp. hybrids. *Progressive Horticulture* **15**: 157-8.

James C (1998). Movement towards global food security accelerates. *Asian Seed* **5**: 18-21.

Jamwal RS, Sharma PP, Sant Parkash and Prakash S (1991). Heterosis and combining ability for curd compactness, curd size index and yield in cauliflower (*Brassica oleracea var. botrytis* L.). *Himachal Journal of Agricultural Research* **17**: 37-42.

Janakiram T and Sirohi PS (1989). Heterosis studies in round-fruited bottle gourd. *Madras Agriculture Journal* **76**: 339-42.

Janakiram T and Sirohi PS (1992). Studies on heterosis for quantitative characters in bottle gourd. *Journal of Maharashtra Agriculture University* **17**: 204-6.

Janick J, Kim YH, Kitto S and Saranga Y (1993). Desiccated synthetic seeds. Synseeds: *In: Applications of Synthetic Seeds to Crop Improvement*. Redenbaugh, K. (Ed.) CRC Press, pp. 11-33.

Jansen RC and Stam P (1994). High resolution of quantitative traits into multiple loci via interval mapping. *Genetics* **136**: 1447-55.

Jawahar M, Mohamed SV and Jayabalan N (1998). *Advances in Plant Sciences* **11**: 199-204.

Jennersten O, Morse DH and O'Neil P (1991). Movement of male and worker bumble bees on and between flowers. *Oikos* **62**: 319-24.

Jeong HB and Cho MA, Ha SH and Kang KY (1998). *RDA-Journal Hotriculture Science* **40**: 78-82.

Jeong HB and Park HG (1997). *Journal of Korean Society of Horticulture Science* **38**: 123-28.

Jinks JL (1955). A survey of genetic basis of heterosis in a variety of diallele cross. *Heredity* **9**: 223-38.

Jinks JL (1978). Unambiguous test for linkage of genes displaying-non-allelic interaction for a metrical trait. *Heredity* **40**: 171-73.

Jinks JL (1983). Biometrical genetics of heterosis. *In*: *Heterosis: Reappraisal of Theory and Practice, Monograph on Theoretical and Applied Genetics* Frankel R (Ed). Springer, Berlin Heidelberg, pp 1-16.

Jinks JL and Jones RM (1958). Estimation of the components of heterosis. *Genetics* **43**: 223-34.

Jogedjik EH, Tigelaar JSC van Roekel SA, Bres-Vloemans, Dekker I, van den Elzen PJM, Cornelissen BJC and Melchers LS (1995). Synergistic activity of chittinase and β-1, 3 glucanase enhances fungal resistance in transgenic fruit plants. *Euphytica* **85**: 173-80.

Johnson AG (1971). Factors affecting the degree of self-incompatibility in inbred lines of Brussels Sprouts. *Euphytica* **20**: 561.

Johnson KW (1957). Inheritance studies in *Pisum sativum* L. *Diss Abstr* **17**, paper (22) 369, 1862.

Jones DF (1917). Dominance of linked factors as a means of accounting for heterosis, *Genetics* **2**: 466-79.

Jones DF (1918). The effects of inbreeding and cross breeding upon development. *Connecticut Agriculture Experiment Station Bulletin* **207**: p 100.

Jones HA (1932). Vegetable breeding at California. *Proceedings of American Society of Horticultural Science* **29**: 572.

Jones HA and Emsweller SL (1936). A male sterile onion. *Proceedings of American Society of Horticultural Science* **34**: 582-5.

Jones HE and Clarke AE (1943). The story of hybrid onions. *Yb US Dep. Agriculture,* pp. 320-6.

Jones ST and Isbell CL (1956). Selection of varieties for use as parents in the breeding of southern pea. *Proceedings of American Society of Horticultural Science* **67**:412.

Josekutty PC, Swati S and Prathapasenan G (1993). *Journal of Horticulture Science* **68**: 31-35.

Joshi AB, Singh HB and Gupta PS (1958). Studies in hybrid vigour-III, Bhindi, *Indian Journal of Genetics and Plant Breeding* **18**: 57-58.

Joshi Arun, Thakur MC and Kohli UK (2005). Heterosis and combining ability for shell-life, whole fruit firmness and related traits in tomato. *Indian Journal of Horticulture*, **62**: 33-36.

Joshi HC and Tandon TP (1976). Heterosis for yield and its genetic basis in onion. *Indian Journal of Agricultural Sciences* **46**: 88-92.

Joshi KM and Shukla PT (2000). Expression of heterosis in chilli (*Capsicum annum* L.). *Capsicum and Eggplant Newsletter* **19**: 66-69.

Joshi S (1986). Results of heterosis breeding on sweet pepper (*Capsicum annum* L.). *Capsicum and Eggplant Newsletter* **5**: 33-34.

Joshi S and Singh B (1980). A note on hybrid vigour in sweet pepper (*Capsicum annum* L.). *Haryana Journal of Horticultural Sciences* **9**: 90-92.

Joshi S and Singh B (1987). Results of combining ability studies in sweet pepper (*Capsicum annum* L.). *Capsicum Newsletter* **6**: 49-50.

Joshi S, Thakur PC and Verma TS (1995). Hybrid vigour in bell shaped paprika (*Capsicum annum* L.) *Vegetable Science* **22**: 105-8.

Joshi, S. (1986). Results of heterosis breeding on sweet pepper (*Capsicum annum* L.). *Capsicum Newsletter* **5**: 33-34.

Jourdan PS, Earle ED and Mutschler MA (1989). Atrazine resistant cauliflower obtained by somatic hybridization between *Brassica oleracea* and *B. napus*. *Theoretical and Applied Genetics* **78**: 271-79.

Jtoh K, Iwabuchi M and Shinaonoto K (1991). *In situ* hybridization with species-specific DNA probes gives evidence for asymmetric nature of *Brassica* hybrids obtained by X-ray fusion. *Theoretical and Applied Genetics* **81**: 356-62.

Kadam PY, Desai UT and Kale PN (1995). Heterosis studies in ridge gourd. *Journal of Maharashtra Agricultural University* **20**: 119-120.

Kagami T, Akamatsu T and Shiga T (1990). Effective transfer of *Ogura* type cytoplasmic male sterile traits from *B. compestris* or *Raphanus sativus* to *Brassica oleracea* by protoplasm fusion. *In Proceedings of the 6th Crucifer Genetics Workshop.* McFerson JR, Kresovich, S and Dwyer SG, (Eds), Cornell University.

Kagimiki Y, Nakamura K, Hida KI, Yosikawa H (1997). Varietal differences in embryogenic and regenerative ability in microspore culture of Chinese cabbage (*Brassica vespa* L. ssp. *perkinensis*). *Breeding Science* **47**: 341-46.

Kakihara F, Masahiro K and Tokumasu S (1988). Relationship between pollen degeneration and amino acids, especially proline, in male-sterile Japanese radish (*Raphanus sativus* L. var. *longipinnatus* Bailey). *Sci. Hort.* **36**: 17-23.

Kakizaki (1931). Hybrid vigour in eggplant and its practical utilization. *Genetics, Princeton* **16**: 1-25.

Kale PB and Seshadri VS (1981). Study of heterosis in watermelon (*Citrullus lanatus* L.). *Vegetable Science* **8**:15-24

Kale PN, Chatterjee SS and Swarup V (1979). Heterosis and combining ability in Indian cauliflower (*Brassica oleracea* var. *botrytis* L.). *Vegetable Science* **6**: 11-20.

Kale PS and Dod SU (1975). Heterosis and combining ability analysis in okra. *Agriculture Research Journal*, **29**: 49.

Kalia Pritam (2004). Fruits of biotechnology in Vegetables: A boon to farmers, consumers, processors and environment. *Indian Horticulture* **49**: 26-34.

Kalloo G (1986). Tomato. *In: Breeding Habits Selection Bases and Breeding Methods*. Allied Publisher Private Limited, p. 92.

Kalloo G (1988). *Vegetable Breeding,* Vol. 1. CRC Press, Inc.Florida. p. 16.

Kalloo G (1991). *In: Genetic Improvement of Tomato*. Springer-Verlag, Berlin Heidelber, p. 46.

Kalloo G (1994). *Vegetable Breeding* (combined edn.) Vol. I, II & III, Panima Educational Book Agency, New Delhi, p. 115.

Kalloo G and Singh Kirti (2001). *In : Souviner : Emerging Scenerio in Vegetable Research and Development*. Research Periodical and Book Publishing House, New Delhi; pp.104-117.

Kalloo G, Singh RK and Bhutani DS (1974). Combining ability studies in tomato (*Lycopersicon esculentum* Mill.). *Theoretical and Applied Genetics* **44**: 358-63.

Kalloo G, Dixit J and Sidhu, AS (1982). Genetic divergence in muskmelon (*Cucumis melo* L.), *Genetica Agrar* **36**:1.

Kalloo G, Singh RK and Bhutani RD (1974). Combining ability studies in tomato (*L. esculentum* Mill.). *Theoretical and Applied Genetics* **44**:358-63.

Kalloo G, Singh VP, Dudi BS and Pratap PS (1980). Analysis of variation and genetic divergence in garden peas (*Pisum sativum* L.). *Haryana Agricultural University Journal of Research* **10**: 540-46.

Kaloshian I, Yaghoobi J, Liharska T, Hontelez J,Hanson D, Hogan P, Jesse T, Wijbrandi J, Simons G, Vos p, zabel P and Williamson VM (1998). Genetic and physical localization of root-knot nematode resistance locus Mi in tomato. *Molecular and General Genetics* **257**: 376-85.

Kamada H (1985). Synthetic Seeds. *In: Practical Technology on the Mass Production of Clonal Plants*. Tanaka R, (Ed) CMC Publication, Tokyo, p. 48.

Kamboj R K and Jackson J K (1986). Self-incompatibility alleles control a low molecular weight basic protein in pistils of *Petunia hybrida. Theoretical and Appllied Genetics* **71**: 815-19.

Kameya T, miyazawa N and Toki S (19900. Production of somatic hybrids between *Solanum melongena* L. and *S. integrifolium. POIR Japanese Journal of Breeding* **40**: 429-34.

Kandaswamy P and Singh N, Kalda TS, Sirohi PS and Choudhry B (1983). Heterosis and combining ability in eggplant. *Indian Journal of Agricultural Science* **53**: 201-206.

Kang BC, Nam SH, Huh JH, You HS and Kim BD (1998). Construction of a molecular linkage map and marker selection of hot pepper. *Plant and Animal Genome VIC Conference,* 18-22, January, San Diego, CA.

Kaniewski W, Liardi V, Tomassoli L, Mitsky T, Layton J and Barba M (1999). Extreme resistance to cucumber mosaic virus (CMV) in transgenic tomato expressing one or two viral coat proteins. *Molecular Breeding* **5**: 111-19.

Kanno T and Hinata K (1969). An electron microscopic study of the barrier against pollen tube growth in self-incompatible Cruciferae. *Plant Cell Physiology* **10**: 213-16.

Kanoh K, Hayashi M, Serizawa Y and Konishi T (1988). Production of interspecific hybrids between *Lilium longiflorum* and *L. elegance* by ovary slice culture, *Japanese Journal of Breeding* **38**: 278-82.

Kanthaswamy V and Balakrisanan R (1989). *South Indian Horticulture* **37**: 77-83.

Kartha KK, Leung NL and Gamborg OL (1979). Freeze preservation of pea meristems in liquid nitrogen and subsequent plant regeneration. *Plant Science Letter* **15**: 7-15.

Kartha, KK and Gamborg OL (1978). Meristem culture techniques in the production of disease-free plants and freeze preservation of germplasm of tropical tuber crops and grain legumes. *In: Disease of Tropical Food Crops*, Maraite H and Mayer JA (Eds.) Universite Catholique, Louvain-La-Neuve, pp 267-283.

Kathal R, Bhatnagar SP and Bhojwani SS (1994). *Plant Science Limerick* **96**: 137-42.

Katsumata H and Yasui H (1965). *Engeishikanjo Hokou/Bulletin Horticulture Research Station Sericulture, D. Kurume* **3**: 79-112.

Kaufmann H, Salamini F and Thompson RD (1991). Sequence variability and gene structure at the self-incompatibility locus in *Solanum tuberosum. Molecular and General Geneics* **226**: 457-66.

Kaul BL and Sharma PP (1988). Heterosis and combining ability studies for some character in bell pepper. (*Capsicum annum* L.). *Vegetable Science* **15**:171-80.

Kaul MLH (1988). *Male Sterility in Higher Plants*. Berlin, Spinger-Verlag.

Kawaka Y, Sakiyama F and Tamaoki H (1990). Identification of two essential histidine residues of ribonuclease T_2 from *Aspergillus oryzae. European Journal of Biochemistry* **176**: 683-697.

Kawchuk LM, Lynch DR, Hachey J, Bains PS and Kulcsar F (1994). Identification of a codominant amplified polymorplic DNA marker linked to the verticillium wilt resistance gene in tomato. *Theoretical and Applied Genetics* **89**: 661-64.

Kearns CA aand Inouye DW (1992). *Techniques for Pollination Biologists.* Niwot, Co: University Press of Colorado.

Kendall DA and Smith BD (1975). The pollinating efficiency of honeybee and bumblebee visits to field bean flowers (*Vicia faba* L.) *Journal of Applied Ecology* **12**: 709-17.

Keppler E (1941). Inbreeding performance and the effect of hybridization in radish (*R. sativus*). *Z. Pflanzena* **23**: 661-84.

Kesavan PK and More TA (1991). Use of monoecious lines in heterosis breeding in muskmelon *(Cucumis melo* L). *Vegetable Science* **18**: 59-64.

Kesavan PK, More TA and Sharma JC (1987). Use of monoecious sex form in muskmelon. *In: National Symposium on Heterosis Exploitation: Accomplishments and Prospects*, held at Parbhani, 15-17 Oct., p 54.

Kesseli RV, Paran I, Michelmore RW (1994). Analysis of a detailed genetic linkage map of *Lactuca sativa* (lettuce) constructed from RFLP and RAPD markers. *Genetics* **136**: 1435-46.

Khan R and Rao GR (1976). Some results interspecific pollination with *Solanum melongena*-a popular crops. *In*: *Indo-Soviet Symposium. Embryol. Crop Plants Abstract,* New Delhi.

Khanna KR, Choudhury RC and Prabha C (1972). A high yielding hybrid tomato (*L. esculentum* Mill.) under low temperature condition. *Haryana Journal of Horticultural Sciences* **17**: 261-64.

Khantaswamy V and Balakrishnan R (1989). Studies on hybrid vigour in F_1 and its retention in F_2 generation of tomato (*Lycopersicum esculentum* Mill.). *South Indian Horticulture* **37**: 77-83.

Khattra AS, Singh NJ and Thakur JC (1994) . Studies of combining ability in bitter gourd. *Vegetable Science* **21**: 158-162.

Khattra AS, Singh NJ and Thakur JC (2000). Combining ability studies in bitter gourd in relation to line 'X' tester crossing system. *Vegetable Science* **27**: 148-51.

Kheyr-Pour A, Bintrim SB, Ioerger TR,Remy R, Hammond SA and kao T-h (1990). Sequence diversity of pistil S-proteins associated with gametophytic self-incompatibility in *Nicotiana alata. Sex Plant Reproduction* **3**: 88-97.

Khimic PE (1935). Heterosis in cabbage. Seljkhozgiz, Moscow 163-172.

Kho YO, Nits APM den Franken, J, Rodriques JG. Resistance of cucurbits to two-spotted spider mite. *Environmental Entomology* **4**: 507.

Kieffer M, Fuller MP and Jelings AJ (1995). *Plant Science Limeric,* **107**: 229-235.

Kihara H (1951). Triploids watermelons. *Proceedings of American Society of Horticultural Science* **58**: 217-30.

Kik C, Verbeek WHJ, Wietsma WA and Van Ooijen (1995). Towards an AFLP linkage map of onion (*Allium cepa*). *Plant Genome IV Conference,* January 1995, San Diego, CA.

Kikuchi T (1965). Role of interspecific dominance-sub-ordination relationship on the appearance of flower-visiting insects. *Sci. Rep. Tohoku Univ. Ser. IV (Biol.)* **31**: 275-96.

King JR (1960). Peroxidase reaction as an indicator of pollen viability. *Stain Technol.* **35**: 225-27.

King TH, Davis DW, Shehata MA, Pfleger FL and Minnesota (1981). All pea germplasms. *HortScience* **16**: 100.

Kirch HH, Uhrig H, Lottspeich FG, Salamini F and Thompson (1989). Characterization of proteins associated with self-incompatibility in *Solanum tuberosum*. *Theoretical and Applied Genetics* **78**: 581-88.

Kitto S and Janick J (1982). Production of synthetic seeds by encapsulating asexual embryos of carrot. *Journal of American Society of Horticultural Science* **110**: 277.

Klee H and Estelle M (1991). Molecular genetic approaches to plant hormone biology. *Annual Review of Plant Physiology and Molecular Biology* **42**: 529-51.

Knapp SJ, Bridges WC and Brikes D (1990). Mapping quantitative traits loci using marker linkage map. *Theoretical and Applied Genetics* **79**: 583-92.

Koelrecuter KG (1766). Vorlaufigen nachrict von einigen da geschlect der pflanzen betreffenden. *Versuchen und beobachtungen, Leipzig*, pp. 266.

Kofer W, Glimelius K and Bonnett HT (1990). Modification of floral development in tobacco induced by fusion of protoplasts of different male-sterile cultivars. *Theoretical and Applied Genetics* **79**: 97-102.

Kohle AK (1972). Exploitation of hybrid vigour in cucurbits. *Indian Horticulture* **29**: 77-81.

Kohle VK, Dua IS and Sainy SS (1981). Gibberellic acid as an androcide for bell pepper. *Scientia Horticulturae* **15**: 17-22.

Komochi S (1966). Genes controlling early maturation of F_1 hybrids of eggplant. *Japanese Journal of Breeding* **16**: 31-38.

Kondo N (1955). Studies on the Triploid watermelon. The Institute for Breeding Research, Tokyo Agricultural University.

Korla B and Sharma PP (1987). A note on genetics of yield in okra (*Abelmoshus esculentus* (L) Monech.). *Haryana Journal of Horticultural Sciences* **16**: 304-7.

Korzeniewska A, Golecka T, Neimirowicz Szezytt K (1995). Effect of ethephon (2-chloro ethylphosphonic acid) on monoecious muskmelon F_1 hybrid seed. Production. *Folia-Horticulture* **7**: 25-34.

Kraft JM and Roberts DD (1970). Resistance in peas to *Fusarium* and *Pythium* root-rot. *Phytopathology* **67**: 1057.

Kruger HU (1985). Studies on anther culture in asparagus. *Arch. Zuechtugsforsch* : 15-53.

Kubicki B (1962). Inheritance of some characters in muskmelon. *Genet Polonica* **3**: 265-274.

Kubicki B (1966). Genetic basis for obtaining gynoecious muskmelon lines and the possibility of their use for hybrid seed production. *Genet Polonica* **7**: 27-29.

Kubicki B (1974). New sex types in cucumber and their uses in breeding work. *Proceedings of 19th International Horticultural Congress,* III, pp. 475-485.

Kubicki B (1979). Recessive femaleness in cucumber and the possibility of making use of it inbreeding heterotic hybrids. *Hodowla Rosl. Aklem. Nasiem* **23**: 357-70.

Kugimiki Y, Nakumura K, Hida KI, Yosikawa H (1997). Varietal differences in embryogenic and regenerative ability in microscope culture of Chinese cabbage (*Brassica vespa* L. ssp. *Pekinensis*). *Breeding Science* **47**: 341-46.

Kumar JC and Dhaliwal MS (1991). Techniques of Developing Hybrids in Vegetable Crops. Agro-Botanical Publishers.

Kumar PA, Ajin Mandaokar, Sreenivasan Karra, Chakabarti SK, Suman Bisaria, Sharma SR, Sarvjeet Kaur and Sharma RP (1998). Insect resistant transgenic tomato plants. *Molecular Breeding* **4**: 33-37.

Kumar R, Prasad KK and Singh (1999). Heterosis in brinjal (*Solanum melongena* L.). *Journal of Research* **11**: 217-21.

Kumar R, Singh SP, Singh SB and Singh NK (2001). Environmental effect on the standard heterosis in bottle gourd. *In: Diamond Jubilee Symposium on Hundred Years of Post Mendelian Genetics and Plant Breeding. Retrospect and Prospects,* 6-9 November, 2001, New Delhi, Organised by Indian Society of Genetics and Plant Breeding, pp. 156-57.

Kumar S, Banerjee MK and Pratap PS (1995). Studies on heterosis for various characters in tomato. *Haryana Journal of Horticultural Sciences* **24**: 54-60.

Kumar S, Singh M, Kalloo G and Banerjee MK (2001). *In: National Conference on Role of Biotechnology in Indian Agriculture*, held at Vasant Rao Naik Mahavidyalaya, Aurangabad, 7-8 Oct. 2001 (Abstr.).

Kumar S, Singh M, Ram D and Banerjee MK (2002). *In: International Conference on Vegetables,* PNASF, Bangalore (India), 11-14 November (Abstr).

Kumar S, Singh SP and Singh NK (1998). Line 'X' tester analysis for combining ability and heterosis in bottle gourd (*Lagenaria siceraria* (Molina) Standl.). *Vegetable Science* **25**: 78-80.

Kumar S, Singh, SP and Jaiswal RC (1999). Heterosis over mid parent and top parent under line 'X' tester fashion in bottle gourd [*Lagenaria siceraria* (Mol.) Standl]. *Vegetable Science* **26**: 30-32.

Kumar Sanjay, Singh KP and Panda PK (2000). Heterosis for green pod yield and its components in garden pea. *Haryana Journal of Horticultural Sciences* **29**: 99-101.

Kumar Sanjeet, Banerjee MK and Kalloo G (2000). Male sterility mechanism and current status on identification, characterization and utilization in vegetables. *Vegetable Science* **27**: 1-24.

Kumar Sanjeev, Singh, Major and Banerjee, MK (2003). Commercial micropropagation of vegetable crops. *In: Comprehensive Micropropagation of Horticultural Crops*. Chandra R.and Maneesh M. International Book Distribution, Lucknow, pp. 308-326.

Kumar Vinod and Pathania NK (2004). Heterosis in bacterial wilt resistant lines of brinjal (*Solanum melongena* L.). *Vegetable Science* **13**: 158-60.

Kumar Vinod, Chandel KS and Pathania NK (2002). Heterosis in Asiatic and European cultivars in radish (*Raphanus sativus* L.). *Vegetable Science* **29**: 34-36.

Kumar Sanjeet , Banerjee MK, Kalloo G and Kumar S (2000). Morphocytological feature of a heterotic sweet pepper x hot pepper hybrid. Promising for pickle type cultivation and recombinant inbred lines (RILS) development. *Capsium* and *Egg Plant New letter* **19**: 54-57.

Kumbhani RP, Godhani PR and Fovgai RS (1993). Hybrid vigour in eight-parent diallel cross in okra (*Abelmochus esculentus* (L) Monech.). *Gujarat Agricultural University Research Journal* **18**: 13-18.

Kunik T, Salomon R, Zamir D, Navot N, Zeidan M, Michelson I, Gafni Y and Czosnak H (1994). Transgenic tomato plants expression the tomato yellow leaf curl virus capsid protein are resistant to the virus. *Biotechnology* **12**: 500-4.

Kurian, A. (1990). Evaluation of processing characteristics in tomato and their expression in a bacterial wilt resistant background. Ph. D. Thesis, Kerala Agricultural University, Vellanikara.

Ladeinde TA and Bliss FA (1977). A preliminary study on the production of plantlets from anthers of cowpea, *Tropical Grain Legume Bulletin* 8-13.

Lagercrantz U and Lydiate DJ (1996). Comparative genome mapping in *Brassica*. *Genetics* **144**: 1903- 10.

Lakshmi N, Prakash NS and Harini J (1988). Genetic and breeding behaviour of male sterile mutant in *Capsicum*. *Proceedings of National Seminar on Chilli, Ginger and Turmeric*, held on 11-12 January, at Hyderabad, pp. 28-32.

Lal MS and Dhaliwal MS (1992). *Indian Horticulture* **37**: cover page and p. 55.

Lal SD, Seth JN and Solanki SS (1976). Note on the heterosis in bitter gourd (*Momordica charantia*). *Indian Journal of Agricultural Research* **10**: 195-97.

Lal T and Dhaliwal MS (1993). Performance of selected muskmelon hybrids over environments. *In : Symposium on Heterosis Breeding in Crop Plants: Theory and Application*, Ludhiana, India, pp. 34-35.

Lal T and Dhaliwal MS (1996). Genetics of some important characters using line 'X' tester analysis in muskmelon. *Indian Journal of Genetics* **56**: 206-7.

Lal SD, Seth JN and Solanki SS (1976). Note on the heterosis in bitter gourd (*Momordica charantia* L.). *Indian Journal of Agricultural Research* **10**: 195.

Lalitha Anand (2001). Genetic transformation in vegetable crops: Present status and future prospects *In: Emerging Scenario in Vegetable Research and Development*, Kalloo G and Singh, Kirti, (Eds). Research Periodical and Book Publishing House, pp.86-95.

Larson RE and Paur S (1948). The description and inheritance of a functional sterile flower mutant in tomato and its possible value in hybrid tomato seed production. *Proceedings of American Society of Horticultural Science* **52**: 355-64.

Laughnan JR and Gabay- Laughnan S (1983). Cytoplasmic male sterility in maize. *Annual Review of Genetics* **17**:27-48.

Laurie DA and Bennett MD (1989). Genetic variation in sorghum for the inhibition of maize pollen tube growth. *Annals of Botany* **64**: 675-81.

Lawande KE (2004). Heterosis in onion: Prospects and challenges. *In: National Symposium on Harnessing Heterosis in Crop Plants*, held at IIVR, Varanasi, 13-15 March, pp.119-120.

Lawande KE and Patil AV (1989). Studies on heterosis as influenced by combining ability in bitter gourd. *Vegetable Science* **16**: 49.

Leal NR and Arume AK (1974). The productivity of cabbage cultivars and F_1 hybrids in the hilly region of the state of Rio de Janeiro. *Revta Ceres* **21**: 515-517.

Lecouvior M, Pitrat M and Risser G (1990). A fifth gene for male sterility in muskmelon. *Cucurbit Genetic Crop Report* **13**: 10-13.

Lee HS, Huang Sand Kao TH (1994). S-proteins control rejection of incompatible pollen in *Petunia inflata. Nature* **367**: 560-63.

Leonardi A, Damerval C, Herbert Y, Gallais A and de Vienna D (1991) Association of protein amount polymorphism (PAP) among maize lines with performances of their hybrids. *Theoretical and Applied Genetics* **82**: 552-60.

Leonardi A, Damerval C, Vienne D de (1987). Inheritance of protein amounts: Composition of two-dimensional electrophoresis patterns of leaf sheath of two maize lines *(Zea mays* L.) and their hybrids. *Genetic Research* **50**: 1-5.

Lesely, MM and Lesely JW (1958). Male sterile mutants in the tomato from seed treated with the radioactive isotope P^{32}. *Proceedings of American Society of Horticultural Science* **71**: 339-343.

Levin DA and Berube DE (1972). Phlox and Colias: The efficiency of a pollination system. *Evolution* **26**: 242-252.

Levings CS and Pring DR (1979). Molecular basis of cytoplasmic male sterility in maize. *In: Physiological Genetics*, Scandalios JG, (Ed.) Academic Press, pp. 171-93.

Lewis D (1949). Incompatability in flowering plants. *Biological Review* **24**: 477-96.

Lillo C and Nansen M (1987). Anther cultutre of cabbage. Influence on growth temperature of donor plants and media composition on embryo yield and plant regeneration. *Norwegian Journal of Agricultural Science* **1**: 105-9.

Ling Hongquing Pich A Scholz G and Ganal MW (1996). Genetic analysis of two tomato mutant affected in the regulation of iron metabilism. *Molecular and General Genetics* **252**: 87-92.

Linganagouda, Mulge Ravindra and Madalageri MB (2003). *Capsicum* x chilli crosses: Heterosis and combining ability for growth parameters. *Indian Journal of Horticulture* **60**: 262-67.

Linskens HF and Pfahler PL (1973). Biochemical composition of maize (Zea mays L.) pollen. III. Effects of allele X storage interactions at the waxy (wx), sugary (su), and shrunken (sh_2) loci, on the amino acid content. *Theoretical and Applied Genetics* **43**: 49-53.

Lippert LF and Legg PD (1972). Appearance and quality characters in hot pepper fruit evaluated by a ten cultivars diallel cross. *Journal of American Society of Horticultural Science* **97**: 84.

Little JM, Jones HA and Clarke AE (1946). The distribution of male sterile gene in varieties of onion. *Hereditas* **111**: 310-12.

Litvinova MK (1979). Tsitoplazmatich, Muzhsk, pp. 200-3.

Litz RE and Grey DJ (1995). Somatic embryogenesis for agricultural improvement. *World Journal of Microbiology and Biotechnology* **11**: 416-25.

Liu GQ, Li SJ and Zhang XP (1996). *Journal of Nanjing Agricultural University* **19**: 31-36.

Liu W, Ma L and Cao Z (1985). Studies on the viability of pollen of 11 species stored in some organic solvents. *Acta Hort Sinica* **12**: 69-71.

Livneh O, Nagler Y, Taj Y, Harush SB, Gatric JJ, Beckman J. and Sela, I (1990). RFLP analysis of a hybrid cultivar of pepper (*Capsicum annum* L.) and its use in distinguishing between parental lines and in hybrid identification. *Seed Science and Technology* **18**: 209-14.

Lockwood JL and Ballard JC (1960). *Bulletin Mich Station University Agriculture Experiment Station* **42**: 704.

Lohithaswa HC (1997). Genetics of yield and other qualitative characters in chillies (*Capsicum annum* L.). *Ph. D. Thesis*, UAS, Bangalore.

Lonnquist JH (1960). Modifying double cross hybrid performance through convergent improvement. *Agronomy Journal* **52**: 226-28.

Lou H and Kako S (1994). *HortScience* **29**: 906-9.

Loy JB, Natti TA, Zack CD and Frittrs SK (1979). Chemical regulation of sex expression in a gynoecious line of muskmelon. *American Society of Horticultural Science* **104**: 100-101.

Lukoschus F (1957). Quantative Untersuchungen uber den Pollen-Transport im Haarkleid der Honigbiene. *Z. Beinenforsch* **4**: 3-21.

Luo LJ, Mei Lizk HW,Shu QY, Tabien R, Zhong DB, Ying CS, Stansel JW, Khush GS and Paterson AH (2001). Overdominant epistasis loci are the primary genetic basis of inbreeding depression and heterosis in rice II. Grain yield component. *Genetics* **158**: 1755-71.

Ma WJ, Mi SE and Li B (1991). *Gansu-Nongye-Daxue-Xuebao* **26**: 409-16.

Maheswari K and Natrajan S (1999). Studies on heterosis for yield and quality in tomato *(Lycopersicon esculentum* Mill.). *South Indian Horticulture* **47**: 216-17.

Mahajan Vijay, Gill HS, Ram Singh and Sharma SR (1999). Heterosis studies in Indian cauliflower (group - II). *Annals of Agricultural Research* **20**: 292-97.

Maisonneuve B, Bellec Y, Anderson P and Michelmore RW (1994). Rapid mapping of two genes for resistance to downy mildew from *Lactuca serriola* to existing clusters of resistance genes. *Theoretical and Applied Genetics* **89**: 96-104.

Mak C and Yap TC (1977). Heterosis and combining ability of seed protein yield and yield components in long bean. *Crop Science* **17**: 339-41.

Makesem K, Leister D, Peleman J, Zabeau M, Salamini F and Gebhardt C (1995). A high resolution map of the vicinity of the R_1 locus on chromosome V of potato base on RFLP and AFLP markers. *Molecular and General Genetics* **249**: 74-81.

Malavika Dadlani, Priyadarshani S and Arun Raju AL (2003). Modern approaches for testing: Genetic purity of vegetable seeds. *In: Compendium of Winter School on Advances in Production and Evaluation of Quality Seeds*, held at IIVR, Varanasi, from 2-22 December, 2003, pp.47-51.

Malik VS (1989). Biotechnology : The Golden Age. *Advances in Applied Microbiology* **34**: 263-306.

Mamidwar RB, Nerkar YS and Jambhale ND (1979). Cytogenetics of interspecific hybrids in the genus *Abelmoschus. Indian Journal of Heredity* **11**: 35-39.

Mandal AK, Pandit MK and Maity TK (1994). Heterosis in brinjal (*Solanum melongena*). *Crop Research* **8**: 291-95.

Mandal AR, Hazara P and Som MG (1989). Studies on heterobeltiosis for fruit yield and quality of tomato (*Lycopersicon esculentum* Mill.). *Haryana Journal of Horticultural Sciences* **18**: 272-79.

Mandal N and Dana I (1993). Heritability and heterosis study in brinjal *Solanum melongena* L. *Environment and Ecology* **11**: 295-98.

Mandal N and Dana ND (1993). Heterosis and inbreeding depression in okra. *Environment and Ecology* **11**: 649-52.

Mangal JL, Pandita ML and Singh GR (1981). Effect of various chemicals on growth, flowering and yield of bitter gourd. *Indian Journal of Agricultural Research* **15**: 185.

Mariani C, De Beuckeller M, Truettner J Leemans J and Goldberg RB (1990). Induction of male sterility in plant by a chimaeric ribonuclease gene. *Nature* **347**: 737-41.

Mariani C, Gossele V, de Beuckeleer M, De Block M, Goldberge RB, De Greef W and Leemand J (1992). A chimeric ribonuclease inhibitor gene restores fertility to male sterile plants. *Nature* **357**: 384-87.

Markelt CL, and Mollar F (1959). Multiple forms of enzymes tissue ontogenic and specific patterns. *Proceedings of National Academy of Sciences* **45**: 753-63.

Martin GB, Ganal MW and Tanksley SD (1992). Construction of a yeast artificial chromosome library of tomato and identification of cloned segments linked to two disease resistance loci. *Molecular General Breeding Science* **43**: 327-33.

Maruyama E and Kinoshita I (1997). High plant recovery from double layered synthetic seeds. *Silvae Genetica* **46**: 17-23.

Masuda K, Hatakeyama E. Ito A, Takahashi S and Inoue M (1994). *Bull. Akita Prefectural College Agriculture* **20**: 43-48.

Mather K (1949). *Biometrical Genetics*, I[st] edn. Mathuex, London, pp. 345.

Mather K and De Winton D (1941). Adaptation and counter adaptation of the breeding system in *Primula. Annual Botany* **15**: 297-311.

Mather K and Jinks JL (1971). *Biometrical Genetics*, 2nd edn. Chapman & Hall. London, pp. 382.

Mather K and Jinks JL (1982). *Biometrical Genetics*, 3rd edn. Chapman & Hall, London, pp. 396.

Mathias ME (1971). Systematic Survey of New World Umbelliferae. *In: Biology and Chemistry of Umbelliferae*, Hetword, V. H. (eds). Academic Press.

Matsubara S (1980). Overcoming the self-incompatibility of *Raphanus sativus* L. with high temperature. *Journal of American Society of Horticultural Science,* **105**: 842.

Matsubara S (1984). Overcoming the self-incompatibility of *Raphanus sativus* L. by application of plant harmones, amino acid and vitamins and by temperature treatment of pollen, *Euphytica* **33**: 113.

Matsumoto A, Imanishi S, Hossain M, Esclante A and Egashira H (1997). Fertile somatic hybrids between F_1 (*Lycopersicon esculentum* x *L. peruvianum* var *humifusum* and *Solanum lycopersicoides*). *Breed Science* **43**: 25-32.

Mau SL, Williams EG, Atkinson MA, Cornish EC, Grego B, Simpson RJ, Kyeur-Pour A and Clarke AE (1986). Style proteins of a wild tomato (*Lycopersicon peruvianum*) associated with the expression of self-incompatibility. *Planta* **169**: 184-91.

Maurya IB, Singh SP and Singh NK (1993). Heterosis and combining ability in bottle gourd (*Lagenaria siceraria* (Molina) Standl). *Vegetable Science* **20**: 75-81.

May, Wang HL, Jhang CJ and Kang YQ (1994). *Plant Cell Reporter* **14**: 65-68.

McClure BA, Gray JE, Anderson MA and Clarke AE (1990). Self-incompatibility in *Nicotiana alata* involves degradation of pollen rRNA. *Nature* **347**: 757-60.

McClure BA, Gray JE, Anderson MA, Simpson RJ, Sakiyama F and Clarke AE (1989). Style self-incompatability gene products in *Nicotiana alata* are ribonucllease, *Nature* **342**: 955-57.

McCollum GD (1988). CMS (ESG) 508 and CMS (ESG) 512 cytoplasmic male sterile cabbage germplasm (with radish cytoplasm). *HortScience* **23**: 227.

McCreight JD and Elmstron GW (1984). A third male sterile gene in muskmelon. *HortScience* **19**: 268-70.

McDaniel RD and Sarkissan IV (1966). Heterosis complementation of mitochondria. *Science* **152**: 1640-42.

McDonald MB (1998). Seed quality assessment. *Seed Science Research* **8**: 265-75.

McGarvey PB, Montasser MS and Kaker JM (1992). Transgenic tomato plants expressing at RNA are tolerant to some strains of cucumber mosaic virus. *Journal of American Society of Horticultural Science,* **119**: 642-47.

McGurl B, Orozo-Cardenas M, Pearce G and Ryan CA (1994). Over expression of the prosystemin gene in transgenic tomato plants generates a systemic signal that constitutively induces proteinase inhibitor synthesis. *Proceedings of National Academy of Sciences,* USA, **91**: 9799-9802.

McRae DH (1985). Advances in chemical hybridization. *Plant Breeding Review* **3**: 169-91.

Meer, OP Vander and Bennekom JL Van (1982). Gibberellins as gametocides for the common onion (*Allium cepa* L.), GA_{4+7} and pollen viability. *Euphytica* **31**: 503-6.

Meksem K, Leister D, Peleman J, Zabeau M, Salamini F and Gebhardt C (1995). A high resolution map of the vicinity of the R_1 locus on chromosome V of potato base on RFLP and AFLP markers. *Molecular and General Genetics* **249**: 74-81.

Melchers G, Sacristan MD and Holder AA (1978). Somatic hybrid plants of potato and tomato regenerated from fused protoplasts. *Carlsberg Research Communication* **43**: 203-18.

Melotto M, Afanador L and Kelly JD (1996). Development of a SCAR marker linked to the I gene in common bean. *Genome* **39**: 1216-19.

Meng XD, Hong M, Zhang WH, Hong M, Zhang WH and Wang DS (1998). A fats procedure for genetic purity determination of head Chinese cabbage hybrid seed based on RAPD markers. *Seed Science and Technology* **26**: 829-33.

Meredith WR and Bridge RR (1972). Heterosis and gene action in cotton. *Gossypium hirsutum*. *Crop Science* **12**: 304-10.

Metwally EI, Moustafa SA, El- Sawy BI, Haroun SA and Shalaby TA (1998 a). Production of haploid plants from *in vitro* culture of unpollinated ovules of *Cucurbita pepo*. *Plant Cell Tissue Org Culture* **52**: 117-21.

Metwally EI, Moustafa SA, EL-Sawy BI, Haroun SA and Shalaby TA (1998 b). Haploid plantlets derived by anther culture of *Cucurbita pepo*. *Plant Cell Tissue Org Culture* **52**: 171-76.

Meyer RC, Milbourne D, Hackett CA, Bradshaw JE, McNichol JW, McNichoj W and Waugh R (1998). Linkage analysis in teteraploid potato and association of markers with quantitative resistance to late blight (*Phytophthora infestans*). *Molecular and General Genetics* **259**: 150-60.

Michigan (1978). USA Michigan State University Agricultural Experiment Station. *Spartan Hybrid Carrot Series 1968-76. Research Report*, Agricultural Experiment Station, Michigan State University 359:9

Milborrow BV (1998). A biochemical mechanism of hybrid vigour. *Journal of Experimental Botany* **49**:1063-71.

Milbourne D, Meyer RC, Collins AJ, Ramsay LD, Gebhardt C and Waugh R (1998). Isolation, characterization and mapping of simple sequence repeat loci in potato. *Molecular and General Genetics:* 259-233.

Mirghis E, Mirghis R, Lacatus V and Gomez ML (1995). *Acta Horticulturae* **412**: 111-16.

Mirza MN and Narkhede MN (1996). *Annals of Plant Physiology* **10**: 148-152.

Mishra GM (1969). Investigation on hybrid vigour in brinjal (*Solanum melongena* L.). *Indian Journal of Horticulture* **18**: 305-17.

Mishra GM, Nigam K and Singh RR (1982). Studies on investigation of hybrid vigour in bhindi. *Proceedings of Bihar Academy of Agricultural Sciences* **2**: 221-24.

Mishra JP and Sheshadri VS (1980). Studies on heterosis in muskmelon. *Vegetable Science* **7**: 27-40.

Mishra RS, Lotha RE, Mishra SN (1989). *South Indian Horticulture* **37**: 179-180.

Mishra RS, Lotha RE, Mishra SN, Paul PK and Mishra HN (1988). *Capsicum Newsletter* **7**: 49-50.

Mishra SN and Mishra RS (1990). Diallel analysis for combining ability in brinjal. *Indian Journal of Horticulture* **47**: 239-43.

Mishra SP, Singh HN and Singh A (1976). A note on heterosis in chilli (*Capsicum annum* L.). *Haryana Journal of Horticultural Sciences* **10**: 116-21.

Mishra VP and Sheshadri VS (1988). Male sterility in muskmelon. I. Studies on combining ability. *Indian Journal of Genetics and Plant Breeding* **48**: 347-354.

Mishra JP (1981). Studies on male sterility in muskmelon. *Ph.D. Thesis*, Kanpur University.

Mishra, JP and Sheshadri VS (1985). Male sterility in muskmelon. II. Studies on heterosis. *Genet. Ag;* **39**: 367-76.

Misra AK and Bhatnagar SP (1995). *Phytomorphology* **45**: 47-55.

Mittal RK, Singh HN, Singh RR and Singh JB (1974). Combining ability in tomato. *Haryana Journal of Horticultural Sciences* **3**:168-76.

Mittal RK, Singh SN and Singh HN (1976). Genetics of some characters in brinjal (*Solanum melongena* L.). *Vegetable Science* **3**: 79-86.

Mittal SP and Thomas TA (1961). A functional sterile mutant in tomato variety 'Merruti'. *Indian Journal of Horticulture* **18**: 150-51.

Mittal SP, Thomas TA and Singh HB (1962). Use of male sterility in the exploitation of Hybrid vigour in tomato. *Indian Journal of Horticulture* **19**:50-60.

Mohamed MF (1998). *Acta Horticulturae* **457**: 243-53.

Mohanty BK and Mishra RS (1998). Manifestation of heterosis for yield and yield- contributing characters in pumpkin. *Vegetable Science* **25**: 58-62.

Mohanty BK and Mishra RS (1999). Studies on heterosis for yield and yield attributes in pumpkin (*Cucurbita moschata* Duch. Ex. Poir.). *Indian Journal of Horticulture* **56**: 173-78.

Mohr HC (1956). Mode of inheritance of the bushy growth characteristics in watermelon. *Proceedings of Association of Southern Agriculture* **53**: 174.

Mohr HC, Blackhurst HT and Jenson ER (1955). F_1 hybrid watermelon from open pollinated seed by use of genetic marker. *Proceedings of Symposium on the Mutational Process Prague*, pp. 105-7.

Moon SS, Verma VK and Munshi AD (2002). Gene action of quality traits in muskmelon (*cucumis melo* L.). *Vegetable Science* **29**: 134-36.

More TA (1993). Development of exploitation of tropical genotypes lines in hybrid varieties of cucumber. *In: Golden Jubilee Symposium on Horticultural Research, Changing Scenario,* Horticultural Society of India, Bangalore, 24-28 May.

More TA and Seshadri VS (1988). Development of tropical gynoecious lines in cucumber. *Reports on Cucurbit Genetics Cooperative* USA, No. 11, pp 17-18.

More TA and Sheshadri VS and Sharma JC (1980). Monoecious sex expression in muskmelon. *Cucurbit Genetics Crop Report* **3**: 32-33.

More TA, Joshi S and Verma TS (1991a). Self-incompatible cabbage hybrids for the plains of India, *In: Golden Jubilee Symposium*, Organised by Indian Society of Genetics and Plant Breeding, New Delhi, pp. 607.

More TA, Sharma SC and Mishra JP (1991b). Per significant estimates performance of gynoecious muskmelon hybrids. *In: Golden Jubilee Symposium*, Organised by Indian Society of Genetics and Plant Breeding, New Delhi, 607 pp.

More TA, Sharma TC and Sheshadri VS (1982). A note on pollination studies in muskmelon. *Science and Culture* **48**: 177-179.

More, TA and Seshadri, VS (1980). Studies on heterosis in muskmelon (*Cucumis melo* L.). *Vegetable Science* **7**: 27.

More, TA and Sheshadri, VS (1987). Maintenance of gynoecious muskmelon with silver thiosulphate. *Vegetable Science* **14**: 138-42.

Morelock TE (1974). Influence of protoplasm source expression of male sterility in carrot (*D. carota). Ph. D. Thesis,* Wisconsin University, USA.

Morelock TE, Simon PW and Peterson CE (1996). Wisconsin Wild: Another petaloid male sterility cytoplasm for carrot. *HortScience* **31**:887-88.

Morues P, Thoquet P, Oliver J, Laterrot H and Grimsly N (1998). Genetic mapping of pH_2, a single locus controlling partial resistance to *Phytophthora infestans* in tomato. *Molecular Plant-Microbe Interaction* **11**: 259-69.

Morurgues F, Brisset MN and Chevreau E (1998). Strategies to improve plant resistance to bacterial diseases through genetic engineering. *Trends in Biotechnology* **16**: 203-10.

Motten AF, Campbell DR, Alexander DE and Miller HL (1981). Pollination effectiveness of specialist and generalist visitors to a North Carolina population of *Claytonia virginica. Ecology* **62**:1278-87.

Moyer JW, Kennedy GG and Romanow LR (1985). Resistance to watermelon mosaic virus-2 multiplication in *Cucumis melo. Phytopathology* **75**: 201.

Muchlabawer FJ (1983). Eight germplasm lines of pea resistant to pea seed borne mosaic virus. *Crop Science* **23**: 1019.

Muller CH (1940). A revision of the genus *Lycopersicon. US Department of Agriculture Miscellanous Publication* 382.

Munger HM (1942). The possible utilization of 1st generation muskmelon hybrids and an improved method of hybridization. *Proceedings of American Society of Horticultural Science* **40**: 405-10.

Munshi AD and Sirohi PS (1993). Studies on heterosis in bitter gourd (*Momordica charantia* L.). *Vegetable Science* **20**: 147-51.

Munshi AD and Verma VK (1997). Studies on heterosis in muskmelon (*Cucumis melo* L.). *Vegetable Science* **24**: 103-6.

Murfett J, Atherton TL, Mou B, Gasser CS, McClure BA and Clarke AE (1994). *S-Rnase* expressed in transgenic *Nicotiana* causes *S-allele* specific pollen rejection. *Nature* **367**: 563-66.

Murphy JF, Sikora EJ, Sammons B and Kaniewski WK (1998). Performance of transgenic tomatoes expressing cucumber mosaic virus *CP* gene under epidemic conditions. *HortScience* **33**: 1032-35.

Musmode AM and Kale PN (1986). Heterosis and combining ability in cucumber (*Cucumis sativus*). *Vegetable Science* **13**: 60-68.

Myres CE (1942). The Penn State Ball-Head Cabbage. Some problems encountered in its development. *Bulletin*, 430, Pa Agriculture Experiment Station, pp. 52.

Mythili JB and Thomas P (1999). *Scientia Horticulturae* **79**: 87-90.

Nagai K and Kida M (1926). An experiment with some variety of crosses of eggplants. *Japanese Journal of Genetics* **4**: 10-30.

Nair PK, Kapoor SK and Verma D (1974). A study of protoplasmic streaming in pollen grains. *New Botanist* **1**: 134-41.

Nalsari C, Sohan D and Srinivas P (1986). A study on inheritance of some economically important fruits in four cultivars of eggplant. *Kesetsart Journal* **20**: 239-48.

Nandpuri KS and Brar JS (1966). Studies on floral biology in muskmelon (*Cucumis melo* L.) *Journal of Research* **3**: 395.

Nandpuri KS and Tyagi DD (1976). Inheritance of some character in tomato. *Journal of Research* **13**:80-84.

Nandpuri KS, Kanwar JS and Singh S (1974 a). Line 'X' tester analysis for some quantitative characters in tomato. *Punjab Horticultural Journal* **14**: 137-46.

Nandpuri KS, Kanwar JS and Singh S (1974 b). Genetic variability and genotypic 'X' environmental interactions in tomato (*L. esculentum* Mill.). *Journal of Research* **11**: 242-246.

Nandpuri KS, Kanwar JS and Singh S (1975). General versus specific combining ability on a set of top-crosses of male sterile lines 'X' pollinators in tomato (*L. esculentum* Mill.). *Journal of Research* **12**: 34-39.

Nandpuri KS, Kanwar JS, Singh S and Saimbhi MS (1975 b). Performance of tomato varieties under low temperature conditions. *Haryana Journal of Horticultural Science* **4**: 46-50.

Nandpuri KS, Kumar JC and Dhillon GS (1974). Heterosis in watermelon. *Punjab Horticultural Journal* **14**: 75-83.

Nandpuri KS, Singh S and Lal T (1974 b). Study on the comparative performance of F_1 hybrids and their parents in muskmelon. *Journal of Research* **11**: 230-38.

Nandpuri KS, Singh S and Lal T (1983). Line 'X' Tester analysis of combining ability in muskmelon. *Journal of Research* **20**: 29-38.

Nandpuri, KS and Tyagi DD (1976). Inheritance of some characters in tomato. *Journal of Research,* **13** : 80-4.

Nandpuri KS, Singh S and Lal T (1974a). Combining ability studies in muskmelon. *Journal of Research* **11**: 225-229.

Nandpuri, KS, Singh S and Lal T (1982). Punjab Hybrid- A new variety of muskmelon. *Progressive Farming* **18**: 3-4.

Naralluh ME and Wallace DH (1968). The influence of modifier genes on the intensity and stability of self-incompatability in cabbage. *Euphytica* **17**: 495.

Narasimha Prasad BC, Madhavi Reddy K and Sadashiva AT (2003). Heterosis studies in chilli (*Capsicum annum* L.). *Indian Journal of Horticulture* **60**: 69-74.

Narasimha Rao N (1979). The barriers to hybridization between *Solanum melongena* and some other species of *Solanum. In: The Biology and Taxonomy of the Solanaceae,* Hawkes, JG, Lester, RN and Skelding, AD (Eds). Academic Press.

Narayanan Kutty C and Sirohi PS (2003). Exploitation of hybrid vigour in radish (*Raphanus sativus* L.). *Indian Journal of Horticulture* **60**: 64-68.

Narsinghani VG and Kumar S (1978 a). The effect of gibberellic acid on chromosomal abberations induced by EMS and MMS in peas. *Cytologia* **43**: 377-81.

Narsinghani VG and Rao VSN (1978). Genotype environment interaction in pea. *JNKVV Research Journal* **12**: 40-41.

Narsinghani VG and Singh SP (1979). Heterosis in mutant pea types. *Indian Journal of Agricultural Sciences* **49**: 39-42.

Narsinghani VG, Pal BS and Singh SP (1980). Note on the reaction of pea genetics of rust. *Indian Journal of Agricultural Sciences* **50**: 453.

Narsinghani VG, VSN Rao and Singh SP (1982). Diallel cross analysis for quantitative traits in mutant pea types. *Indian Journal of Agricultural Sciences* **52**: 364-67.

Nath P and Dutta OP (1970). Heterosis in watermelon. *Indian Journal of Horticulture* **27**: 176.

Natrajan S (1993). Heterosis in tomato under moisture stress. *South Indian Horticulture* **41**: 245-47.

Naulsri C, Sohan D and Srinivas P (1986). A study on inheritance of some economically important traits in fruit cultivars of eggplant. *Kesetsart Journal* **20**: 239-48.

Navratilova B, Bujek J Siroky J and Havvanek P (1997). Construction of intergeneric somatic hybrids between *Brassica oleracea* and *Armocaria rusticana. Biology Plantarum* **39**: 531-41.

Nee M (1990). The domestication of *Cucurbita* (Cucurbitaceae). *Economic Botany* **44**: 56-8.

Nelson RS, McCormick SM, Delannay X, Dubre P, Layton J, Anderson EJ, Kanieska M, Proksch RK, Horsch RB, Rogers SG, Fraley RT and Beachy RN (1988). Virus tolerance, plant growth and field performance of transgenic tomato plants expressing coat protein from tobacco mosaic virus. *Biotechnology* **6**: 403-9.

Netra Pal, Singh N and Saimbhi MS (1973). Genetic variability and correlation studies in onion. *Indian Journal of Horticulture* **30**: 391-93.

Netra Pal, Singh N, Choudhury B and Bhagchandani PM (1988). Isolation of male sterile and maintainer lines in Indian cultivars of onion (*Allium cepa* L.). *South Indian Horticulture* **34**: 189-90.

Newman PO, Krishnaraj S and Saxena PK (1996). *Journal of Plant Sciences* **157**: 554-60.

Newton KJ (1988). Plant mitochondrial genomes: Organization, expression and variation. *Annual Review of Plant Physiology and Plant Molecular Biology* **39**: 503-32.

Nienhuis J and Singh SP (1988a). Genetics of seed yield and its components in common bean (*Phaseolus vulgaris* L.) of middle-American origin. I. General combining ability. *Plant Breeding* **101**: 143-54.

Nienhuis J and Singh SP (1988b). Genetics of seed yield and its component in common bean (*Phaseolus vulgaris* L.) of middle-American origin. II. Genetic variance, heritability and expected response from selection. *Plant Breeding* **101**: 155-63.

Nieuwhof M (1961). Male sterility in Brussels sprout cauliflower and cabbage. *Euphytica* **10**: 351-56.

Nieuwhof M (1963). Hybrid breeding in early spring cabbage. *Euphytica* **12**: 189.

Nieuwhof M (1969). Cole crops. *World Crops Book.* Leonard Hill (Book), N.D.

Nieuwhof M and Garretsen F (1975). Hybrid breeding in early spring cabbage (*Brassica oleracea* var. *capitata* L. f. *alba* DC). *Euphytica* **24**: 87.

Nijs APM and Den Visser DL (1979). Silver compounds inducing male flowers in gynoecious cucumbers. *Reports of Cucurbit Genetic Cooperative* **2**: 14-5.

Nikanishi T and Hinata K (1975). Self seeds production by CO_2 gas treatment in self-incompatability cabbage. *Euphytica* **24**: 117.

Nikanishi T, Esashit and Hinata K (1969). Control of self-incompatability by CO_2 gas in *Brassica. Plant Cell Physiology* **10**: 925.

Nilkamp HJJ, Overbeeke N and Kool AJ (1983). Genetic manipulation of cytoplasmic plant genes. *Proceedings of 10th Congress European Association for Research on Plant Breeding,* EUCARPIA, Wageningen, the Netherlands, 19-24 June 1983, eds; W. Lange, AC Zeven and NG Hogenboom, pp. 239-243.

Nishi, S (1955). The commercial seed production of F_1 varieties of vegetables in Japan. *Proceedings of 14th International Horticulture Congress*, pp 468-78

Nishi, S (1967). F_1 seed production in Japan. *Proceedings of 17th International Horticulture Congress*, 231-57.

Nishio T, Mochizuki H and Yamakawa K (1984). Interspecific crosses between eggplant and related species. *Bull. Veg. Ornam. Crops. Res. Sta*; **12**: 57.

Nitch JP, Kurtz EB, Liverman JI and Went FW (1952). The development of sex expression in cucurbit flowers. *Proceedings of American Journal of Botany* **30**: 32-43.

Norman JC (1973). An evaluation of Japanese hybrid cultivars of cabbage in Ghana. *Journal of Kamauya University of Science and Technology* **10**: 18-20.

Novak FJ (1974). Induction of a haploid callus in anther cultures of *Capsicum* spp; Z. *Pflanzenzuecht* **72**: 46.

Nunome T and Hirai M (1998). A linkage map of eggplant (*Solanum melongena*) based on RAPD markers. *Plant and Animal Genome Conference* 18-22 January San Diego, CA.

Obra GIRiner ME and Scot DH (1945). Experimental production of hybrid tomato seed. *Proceedings of American Society of Horticultural Sciences* **46**: 269-76.

Ockendon DJ (1984). Anther culture in Brussels sprouts (*Brassica oleracea* var. *gemmifera).* I. Embryo yields and plant regeneration. *Annals of Applied Biology* pp. 105-285.

Ockendon DJ (1985). Anther culture in Brussels sprouts (*Brassica oleracea* var. *gemmifera).* II. Effect of genotype on embryo yields. *Annals of Applied Biology* pp. 101-107.

Odland ML and Noll CJ (1948). Hybrid vigour and combining ability in eggplants. *Proceedings of American Society of Horticultural Sciences* **51**: 417-22.

Odland ML and Noll CJ (1950). The utilization of cross compatibility and self-incompatibility in the production of F_1 hybrid seed in cabbage. *Proceedings of American Society of Horticultural Sciences* **55**: 391-402.

Oeller PW, Ming-Wong L,Taylor LP, Pike DA and Theologis A (1991). Reversible inhibition of tomato fruit senescence by antisense *RNA. Science* **254**: 437.

Ogunwolu O and Idowu O (1994). Potential of powdered Zanthoxylum zanthoxyloides (Rutaceae) root bark and *Azadirachta indica* (Melliaceae) seed for control of cowpea seed bruchid in Nigeria. *Journal of African Zoology* **108**: 251-58.

Ogura H (1968). Studies of the male sterility in Japanese radish with special reference to the utilization of this sterility towards the practical raising of hybrid seeds. *Meun. Fac. Agriculture Kagoshima University* **6**: 39-78.

Otsuka Y, Eberhart SA and Russel WA (1972). Comparision of prediction formulae for maize hybrids. *Crop Science* **12**: 325-31.

Owen KW, Peterson CE and Tolla GE (1980). Introduction of perfect flowers on gynoecious muskmelon by silver nitrate and amino ethoxynylglycine. *HortScience* **15**: 654-55.

Ozaki Y, Tashiro T and Okubo H (2000). Use of allozyme variation for evaluating genetic purity in asparagus (*Asparagus officinalis* L.) cultivars. *Journal of Horticultural Science and Biotechonology* **75**: 105-10.

Pal AB, Diojode SD and Biswas SR (1983). Line 'X' tester analysis of combining ability in bitter gourd (*Momordica charantia* L.). *South Indian Horticulture* **31**: 72.

Pal AB, Shivanandappa DT and Vani A (1984). Manifestation of heterosis in bottle gourd (*Lagenaria siceraria* (Mol.) Standl.) *South Indian Horticulture* **32**: 33-38.

Pal AK, Ram D, Maurya AN and Rajput CBS (2002). Heterosis over mid and better parents under Line 'X' tester design in cowpea. *Indian Journal of Horticulture* **59**: 279-287.

Pal BP (1945). Studies in hybrid vigour. II. Notes on the manifestation of hybrid vigour in gram, sesamum and chillies. *Indian Journal of Genetics and Plant Breeding* **5**: 106-21.

Pal BP and Sikka SM (1956). Exploitation of hybrid vigour in the improvement of crop plant, fruits and vegetables. *Indian Journal of Genetics and Plant Breeding* **16**: 98-104.

Pal BP and Singh HB (1943). A note on the economic possibilities of the cross *Lycopersicon esculentum* x *L. pimpinellifolium*. *Indian Journal of Genetics and Plant Breeding* **3**: 115-20.

Pal BP and Singh HB (1946). Studies in hybrid vigour. II. Note on the manifestation of hybrid vigour in the brinjal and bitter gourd. *Indian Journal of Genetics and Plant Breeding* **6**: 19-33.

Pal BP and Singh HB (1949). Hybrid brinjal give increased yields. *Indian Farming* **10**: 78-80.

Pal BP, Singh HB and Swarup V (1952). Taxonomic relation and breeding possibilities of okra (*Abelmoschus esculentus*) species, *Indian Journal of Genetics and Plant Breeding* **45**: 261-270.

Pal G and Rajki E (1966). *Solanum melongena* x *Solanum gilo* hybrids. *Asta Agron Hung* **15**: 37-44.

Palloix A, Herve Y, Knox RB and Dumas C (1985). Effect of carbon dioxide and relative humidity on self-incompatability in cauliflower, *Brassica oleracea*. *Theoretical and Applied Genetics* **70**: 628.

Panda PK and Singh KP (1995). Line 'X' tester analysis for combining ability studies in okra (*Abelmoschus esculentus* (L.) Moench). *Haryana Journal of Horticultural Sciences* **24**: 281-86.

Pandey Sudhakar and Ram D (2001). Bower system: Productivity of cucurbits. In: *XIXth Group Meeting of AICCP on Vegetable Crops*, held from 15-18 January, p.23.

Pandey SC and Kalloo G (1975). Line 'X' tester analysis for study of heterosis and combining ability in muskmelon, *In : Advances in Plant Sciences*, Kalyani.

Pandey SC and Kalloo G (1976). Line 'X' tester analysis for the study of heterosis and combining ability in muskmelon. *Symposium on Advances in Plant Science on Genetic Research and Education: Current Trends and Next Fifty Years (ab ISGPB),* New Delhi, 12-15 Feb., pp. 610-611.

Pandey SC, Pandita ML and Dixit J (1978). Heterosis and combining ability in radish. *Haryana Journal of Horticultural Sciences* **7**: 197-202.

Pandey SC, Pandita ML and Dixit J (1981). Studies on heterosis in chilli (*Capsicum annum* L.). *Haryana Journal of Horticultural Sciences* **10**: 116-21.

Pandey SK, Srivastava BL, Srivastava BP and Srivastava JP (2004). Studies on heterosis and inbreeding depression for yield and its traits in bottle gourd. *Indian Journal of Horticulture* **46**: 146-50.

Pandey SK, Srivastava SBL, Singh SP and Srivastava JP (2004). Heterosis and combining ability for earliness in bottle gourd [*Laganaria siceraria* (mol.) Stand.]. *Vegetable Science* **31**: 11-16.

Pandey V, Ahmed Z and Narendra Kumar (2002) Heterosis and combining ability in diallel crosses of sweet peper (*Capsicum annuum* L.). *Vegetable Science* **29**: 66-67.

Pandey Vandana, Pandey HK, Ahmed Z and Kumar Narendra (2004). Heterosis for net head weight and vitamin C in cabbage (*Brassica oleracea* var. *capitata*). *Vegetable Science* **31**: 177-78.

Pang SZ, Nagpal P, Wang M, Slightem JL and Gonsalves D (1992). Resistance to heterologous isolates of tomato spotted wilt virus in transgenic tobacco expressing its nucleo capsid protein gene. *Phytopathology* **82**: 1223-29.

Panmariammal T and Vijendra Das LD (1996). Heterosis for fodder yield in cowpea. *Madras Agricultural Journal*, **83**: 658-59.

Papova D and Mikhailov L (1973). Heterosis effect in *Capsicum. Priroda, Bulgaria* **22**: 73-4.

Paran I, Aftergoot E and Shifriss C (1998). Variation in *Capsicum annuum* revealed by RAPD and AFLP markers. *Euphytica* **99**: 167-173.

Park SK, Yu IO and Choi CI (1975). Study on some characteristics of *Capsicum* hybrids. *Research Republic Office Rural Development* **17**: 43.

Parker FD and Haatley CL (1978). Onion pollination: Viability of onion pollen and pollen diversity on insect body hairs. *Proc. IVth International Symposium Pollination,* Maryland Agricultural Experimental Station Special Miscellaneous Publication **1**: 201-206.

Parmar BS and Godawat SI (1990). Heterosis and inbreeding depression in pea (*Pisum sativum* L.). *International Journal for Tropical Agriculture* **8**: 19-22.

Partap PS and Dhankar BS (1980). Heterosis studies in okra (*Abelmoschus esculentus* (L.) Moench). *Haryana Agricultural University Journal of Research* **18**: 336-40.

Paszkiewicz Z (1983). Methods of testing peas for resistence to the pea enation mosaic and pea leaf roll viruses, *Bull. Inst. Hodowlii Aklimatzaoji Roslen* **150**: 75.

Patel JA, Patel MJ, Patel AD, Acharya and Bhalala MK (2001). Heterosis studies over environment in chilli (*Capsicum annum* L.). *Vegetable Science* **28**: 130-32.

Patel, JA, Shukla MR, Joshi KM, Patel SB and Patel SA (1997). Hybrid vigour of quantitative traits in chilli (*Capsicum annum* L.). *Vegetable Science* **24**: 107-10.

Paterson PA (1958). Cytoplasmically inherited male sterility in *Capsicum. Amer. Nat.* **92**: 111-19.

Pathak CS, Singh DP and Deshpandey AA (1980). *Annual Report*, IIHR, Bangalore, pp. 34-36.

Pathak CS, Singh DP and Pandey AA (1987). A new type of cytoplasmic male sterility in onion (*Allium cepa* L.). *First All India Conference of Cytology and Genetics*. Bangalore University, Abstract, pp. 46.

Pathak CS, Singh DP, Despandey AA and Sreedhar TS (1986). Source of resistance to purple blotch in onion. *Vegetable Science* **13**: 300-3.

Pathak GN and Singh SN (1949). Studies in the genus *Luffa*. I. Cytogenetic investigation in the interspecific hybrid L. *cylindrica* x L. *acutangula*. *Indian Journal of Genetics* **9**: 18-26.

Patil AA and Patil SS (1988). Heterosis for certain quality attributes in tomato. *Journal of Maharashtra Agricultural University* **13**: 241.

Patil JA, Jadhav AS and Rane MS (1973). Male sterility in Maharashtra onion (*Allium cepa* L.). *Mahatma Phule Agricultural University Journal of Research,* **4**: 29-31.

Patil MM and Chaudhury AN (1986). Heterosis in French bean. *Journal of Maharashtra Agriculture* **11**: 72-73.

Patil RB and Shete MM (1986). Combining ability analysis in cowpea. *Journal of Maharastra Agricultural University* **11**: 293-95.

Patil RB and Shete MM (1987). Heterosis in cross of seven genotypes of cowpea. *Journal of Maharashtra Agricultural University* **12**: 51-54.

Patil RB and Shinde SR (1984). Heterosis in eggplant. *Journal of Maharashtra Agricultural University* **9**: 289-99.

Pawar VY, Poshiya VK and Dhankar HL (1999). Combining ability analysis in okra (*Abelmoschus esculentus* (L.) Moench). *Gujarat Agricultural University Research Journal* **25**: 106-9.

Payne RC (1993). *Handbook of Variety Testing*. ISTA, Zurich, pp. 142.

Pearson OH (1931). Methods of determining the solidarity of cabbage heads. *Hilgardia* **5**: 383-93.

Pearson OH (1931). The influence of inbreeding upon the season of maturity in cabbage. *Proceedings of American Society of Horticultural Science* **29**: 359.

Pearson OH (1981). Nature and mechanisms of cytoplasmic male sterility in plants: A review. *Hort.Science* **16**: 482-87.

Pelletier G (1983). Cytoplasmic hybridization in higher plants. *Proceedings of 10th Congress, European Association for Research on Plant Breeding,* EUCARPIA, Wageningen, the Netherlands, 19-24 June, W. Lange, Zeven AC and Hogenboom NG, pp. 244-247.

Pelletier G, Freault M, Lancelin D, Boulidard L, Dore C, Bonhomme S, Grelon M, and Budar F (1995). Engineering of cytoplasmic male sterility in vegetables of protoplast fusion. *Acta Horticulturae* **392**: 11-16.

Pelletier G, Raquih C and Simon G (1972). La culture *in vitro* d, Asperge (*Asparagus officinalis*), *C.R. Acad. Sci. Ser. D.* **274**: 848.

Perkins JM and Jinks JL (1970). Detection and estimation of genotype X environmental linkage and epistatic components of variability. *Heredity* **25**: 157-77.

Peter JE, Crocomo OJ, Sharp WR, Peddock EF, Tegenkamp I and Tegenkamp T (1977). Haploid callus cells from anthers of *Phaseolus vulgaris, Phytomorphology* **27**: 79.

Peter KV (1998) *In: Genetics and Breeding of Vegetables*, pp. 266-71. ICAR New Delhi.

Peter KV and Rai B (1980). Combining ability in tomato. *Indian Journal of Genetics and Plant Breeding* **40**: 25-31.

Peter KV and Singh RD (1973). Diallel analysis of economic traits in brinjal. *Indian Journal of Agricultural Sciences* **43**: 452-5.

Peter KV and Singh RD (1976). Combining ability, heterosis and analysis of phenotypic variation in brinjal. *Indian Journal of Agricultural Sciences* **44**: 393-399.

Peter KV, Kumar Sadhan, Mani Raj and Prasanna KP (1991). *Indian Horticulture* **36**: Cover page and p. 35.

Peterson CE (1960). A gynoecious line of cucumber. *Quarterly Bulletin Mich. Agriculture Experiment Station* **43**: 40-2.

Peterson CE and Andher L (1960). Induction of staminate flowers on gynoecious cucumbers with gibberellin (GA_3). *Science* **131**: 1673-4.

Peterson CE and Simon PW (1986). Carrot breeding. *In:* Bassett MJ (Ed.). *Breeding Vegetable Crops*, AVI Publ. Co; pp. 321-356.

Peterson CE and Weigle JL (1958). A new method for producing hybrid cucumber seed. *Quarterly Bulletin Mich. Agriculture Experiment Station* **40**: 960-5.

Peterson CE, Owens KW and Row PR (1983). Wisconsin 998 muskmelon germplasm. *HortScience* **18**: 116.

Peterson PA (1958). Cytoplasmically inherited male sterility in *Capsicum*. *Am. Nat.* **92**: 111-19.

Petrova TC (1985). Inheritance of some characters in crosses between Bulgarian and Mexican French bean varieties. *Rasteniv "dri Nauki"* **22**: 25-31.

Phippen C and Ockendon DJ (1990). Genotype, plant, bud size and media factors affecting anther culture of cauliflowers (*Brassica oleracea* var. *botrytis*). *Theoretical and Applied Genetics* **79**: 33-38.

Picard E, Hours C, Gregoire S, Phan TH and Meunier JP (1987). Significant improvement of androgenetic haploid and diploid induction from wheat plants treated with a chemical hybridization agent. *Theoretical and Applied Genetics* **74**: 289-297.

Pike LM and Peterson CE (1969). Inheritance of parthenocary in the Cuke (*Cucumis sativus* L.). *Euphytica* **18**: 101-5.

Pilrat M (1990). Gene list for *Cucumis melo* L. Cucurbits and Genet. *Crop. Rep.* **13**: 58.

Pinnel EL, Rinke EH and Hayes HK (1952). Gamete selection for specific combining ability. *In: Heterosis*, pp. 378-388. Iowa State College Press.

Pious T and Peter KV (1988). Heterosis in intervarietal crosses of bell pepper (*Capsicum annum* var. *grossum*) and hot chilli (*C. annum* var. *fesiculatum*). *Indian Journal of Agricultural Science* **58**: 747-50.

Pious T. (1985). Transfer of clusterness to bell peppers (*Capsicum annum* var. *grossum* Stand). *M.Sc. Thesis*, Kerala Agricultural University, Vellanikara.

Pitchaimuthu M (1991). Heterosis in bottle gourd (*Lagenaria siceraria* (Mol.) Standl.). *Ph. D. Thesis*, IARI, New Delhi.

Pitrat M and Lecow H (1984). Inheritance of Zuchini yellow mosaic virus resistance in *Cucumis melo* L. *Euphytica* **33**: 57.

Pochard E (1970). Obtaining three new male sterile mutants of pepper (*C. annum*) through application of mutagenes on monoploid material. *Eucarpia, Versailles,* France, pp. 93-95.

Poole CF (1937). Improving the root vegetables. *Year Book* pp. 300-25. US Department of Agriculture.

Popova D and Mikhailov L (1973). Heterosis effect in *Capsicum*. *Priroda Bulgaria* **22**(4): 73-74.

Popova DT, Petrov T, Murtazov KH and Dasklov S (1976a). New studies of heterosis in eggplant. *Grandinerskai Nauka* **13**: 44-48.

Popova DT, Petrov T, Murtazov KH and Dasklov S (1976b). Some manifestation of heterosis in eggplant. *Biolog. Zh. Armenia* **29**: 67-72.

Poshiya VK and Shukla PT (1986). Combining ability analysis in Okra. *Gujarat Agricultural University Research Journal* **2**: 25-8.

Poshiya VK and Vashi PS (1995). Heterosis in relation to general and specific combining ability in okra. *Gujarat Agricultural University Research Journal* **20**: 69-72.

Powell PA, Nelson RS, De BN, Hoffman N, Rogers SG, Fraly RT and Beachy RN (1986). Delay of disease development in transgenic plants that express the tobacco mosaic virus coat protein gene. *Science* **232**: 738.

Power L (1945). Relative yield of inbred lines and F_1 hybrids of tomato. *Bot.Gazette* **106**: 175-8.

Prabhu Shankar HR (1990). Genetic analysis of yields and yields components in tomato (*Lycopersicon esculentum* Mill.). *M.Sc. (Agri.) Thesis,* UAS, Dharwad.

Prakash, Shivashankar KT and Gowda PHR (1993). Line 'X' tester analysis for hybrid vigour in brinjal (*Solanum melongena* L.). *Progressive Horticulture* **25**: 123-29.

Prasath D, Natrajan S and Thamburaj S (2000). Line 'X' tester analysis for heterosis in brinjal (*Solanum melongena* L.). *Orissa Journal of Horticulture* **28**: 59-64.

Pratap PS and Dhankar BS (1982). Heterosis studies in okra (*Abelmoshus esculentus* (L) Monech.). *Haryana Agricultural University Research Journal* **18**: 336-41.

Pratiha, Singh YV and Gupta Amarjeet (2004). Heterosis in brinjal (*Solanum melongena* L.). *Progressive Horticulture*, **36:** 335-338.

Pray LA and Good night CJ (1995). Genetic variability and inbreeding depression in the flour beetle, *T. castaneum*. *Evolution* **49**: 179-88.

Pritam Kalia and Yudhuir Singh (1996). Heterosis and combining ability studies for some quantitative traits in pea. *Haryana Journal of Horticultural Sciences* **25**: 54-56.

Providente R and Alconero R (1988). Inheritance of resistance to a third pathotype of pea seed borne mosaic virus in *Pisum sativum*. *Journal of Heredity* **79**: 76.

Przywara L, White DWR, Sanders PM and Maher D (1989). Interspecific hybridization of *Trifolium repens* with *T. hybridum* using in ovule embryo and embryo culture. *Annals of Botany* **64**: 613-24.

Pujari CV and Kale PN (1994). Heterosis studies in tomato. *Journal of Maharashtra Agricultural University* **19**: 83-85.

Purewal SS and Randhawa GS (1947). Chromosome and pollination studies in okra. *Indian Journal of Agricultural Sciences* **17**: 129-36.

Radkov P (1979). Intervarietal hybrids of French bean. *Priroda Bulgaria* **28**: 78-80.

Raghuveer P, Khetarpal S and Patil P (1994). *Seed Research* **22**: 112-13.

Rahman H, Hazarika GN, Alam S and Thakur AC (1996). Genetic variability and heterosis in a few crosses of chilli. *Proceedings of the Seminar on Problems and Prospects of Agricultural Research and Development in North-East India,* held at Assam Agricultural University, Jorhat, 27-28, November, pp. 78-82.

Rai M, Singh Major, Kumar S, Pandey S, Singh B and Ram D (2004). Hybrid techonology in vegetable crops: A over view. *In: National Symposium on Harnessing Heterosis in Crop Plants*, held at IIVR, Varanasi, 13-15 March, pp.110-118.

Rai Mangala (2001). Challenges and prospects in horticulture in the changing post GATT scenario. *In: Emerging Scenario in Vegetable Research and Development.* Kalloo G and Singh Kirti, (Eds). Research Periodical and Book Publishing House, pp.32-44.

Rai N (1992). Genetics of yield and quality traits in tomato (*Lycopersicon esculentum* Mill.). Ph.D.Thesis, Banaras Hindu University, Varanasi.

Rai N and Joshi AK (1994). Genetic studies for earliness in tomato. *Vegetable Science* **21**: 154-57.

Rai N and Joshi AK (1994). Genetics of blotchy ripening resistance in tomato. *Vegetable Science* **21**: 140-44.

Rai N, Syamal MM and Joshi AK (1996). Heterosis for certain quality traits in tomato (*Lycopersicon esculentum* Mill.). *Mysore Journal of Agricultural Sciences* **30**: 240-43.

Rai N, Syamal MM and Verma SK (1996). Inheritance of radial fruit cracking in tomato. *Progressive Horticulture* **28**: 6-11.

Rai N, Syamal MM, Joshi AK and Ghosh PK (1997). Diallel analysis for pericarp thickness and storability in tomato (*Lycopersicon esculentum* Mill.). *Annals of Agricultural Research* **18**: 71-75.

Rai N, Syamal MM, Joshi AK and Rajput CBS (1997). Genetics of yield and yield components in tomato (*Lycopersicon esculentum* Mill.). *Indian Journal of Agricultural Research* **31**: 46-50.

Rai N, Syamal MM, Joshi AK and Vijay Kumar (1998). Heterosis and inbreeding depression in tomato (*Lycopersicon esculentum* Mill.). *Indian Journal of Agricultural Research* **32**: 21-27.

Rai N (2005). Interspecific hybridization in brinjal. *Annual Report,* ICAR Research Complex For NEH Region Meghalaya (Press).

Raina SK and Iyer RD (1973). Differentiation of haploid plant from pollen callus in anther cultures of *Solanum melongena L. Z Pflazenzucht* **70**: 275-80.

Raj SK, Aslam AM and Singh BP (1991). Elimination of eggplant mottled crinkle virus using virzzole in explant cultures of *Solanum melongena. Indian Journal of Experimental Biology* **29**: 594-95.

Rajadhav SB and Kale P N (1985). Heterosis in root-knot nematode resistant tomato lines. *Journal of Maharashtra Agricultural University* **10**: 265-68.

Rajadhav SB, Choudhuri KG, Kale PN and Patil RS (1996). Heterosis in tomato under high temperature stress. *Journal of Maharashtra Agricultural University* **21**: 229-31.

Rajani B, Manju P, Nair PM and Saraswathy P (2001). Combining ability in okra (*Abelmoschus esculentus* (L.) Moench). *Journal of Tropical Agriculture* **32**: 98-101.

Rajendran R (1961). Studies on the exploitation of hybrid vigour in bottle gourd [*Lagenaria siceraria* (Mol.) Standl.]. *M. Sc. Thesis*, IARI, New Delhi.

Rajput YS, Rai N, Yadav RK and Asati BS (2003). Line 'X' Tester analysis in tomato. *Journal of Assam Science Society* **44**: 32-39

Ram B and Lal G (1989). Heterosis and inbreeding depression in chilli (*Capsicum annum* L.). *Progressive Horticulture* **21**: 364-72.

Ram D and Rajput CBS (1999). Heterosis studies in vegetable French bean (*Phaseolus vulgaris* L.). *Haryana Journal of Horticultural Sciences* **28**: 86-88.

Ram D, Kalloo G. and Singh Major (1997). Heterosis in bitter gourd (*Momordica charantia* L.). *Vegetable Science* **24**: 99-102.

Ram D, Rai M., Verma AK and Pandey S (2004). Heterosis and combining ability in sathputhia (*Luffa hermaphordita*). *Vegetable Science* **31**: 129-31.

Ram D, Singh SN, Chauhan YS and Singh ND (1981). Heterosis in brinjal. *Haryana Journal of Horticultural Sciences* **10**: 201-6.

Ram Harihar (1997). Squash and Pumpkin. *In: Vegetable Breeding: Principles and Practices,* pp. 349-50. Kalyani Publisher, New Delhi.

Ram, D and Rajput CBS (1998). Studies on combining ability in French bean (*Phaseolus vulgaris* L.) for green pod yield and its attributes. *Haryana Journal of Horticultural Sciences* **27**: 125-30.

Raman KR (1964). Further studies on hybrid vigour in brinjal (*Solanum melongena* L.). *Madras Agriculture Journal* **51**: 79.

Raman KR and Ramu N (1963). Studies on intervarietal crosses and hybrid vigour in bhindi. *Madras Agricultural Journal* **50**: 90-91.

Ramirez MR and Ochoa AN (1996). *Plant Cell Reporter.* **16**: 226-31.

Randhawa JS (1987). Genetic assessment of floral biology and productivity in brinjal (*Solanum melongena* L.). Ph. D. Thesis, Punjab Agricultural University, Ludhiana.

Rao B, Venkata Rao P and Muralimohan Reddy B (2000). Heterosis in ridge gourd (*Luffa acutangula* (Roxb) L.). *Haryana Journal of Horticultural Sciences* **29**: 96-98.

Rao PS, Ganapathi TR, Suprasanna P and Bapat VA (1996). Synthetic seed technology as a method of plant propagation and delivery of tissue cultured plants. *In: Trends in Plant Tissue Culture and Biotechnology,* pp-47-52. Agro. Botanical Publishers. Jaipur.

Rao PV and Singh B (1991). Plant regeneration from encapsulated somatic embryos of (*Solanum melongena* L.) *Plant Cell Reporter* **10**: 7-11.

Rao RGS Kumar S, Singh M and Rai M (2004). Genetic purity testing of hybrid seeds. *In: Souvenir of National Symposium on Harnessing Heterosis in Crop Plants*, held at IIVR, Varanasi, 13-15 March, pp.32-38.

Rao VR and Rilley KW (1994). The use of biotechnology for conservation and utilization of plant genetic resources. *Plant Genetic Resources Newsletter* **97**: 3-20.

Rao VSN and Narsinghani VG (1987). Hybrid performance of divergence peas (*Pisum sativum* L.). *Indian Journal of Genetics and Plant Breeding* **47**: 137-44.

Rastogi K and Saini SS (1984). Inheritance of seed shape and resistance to Ascochyta blight in pea (*Pisum sativum* L.). *Journal of Agricultural Sciences* **103**: 532.

Rattan RS and Bindal Arvind (2000). Development of okra hybrids resistant to yellow mosaic virus. *Vegetable Science* **27** : 121-125.

Rattan RS, Saini SS and Singh M (1978). Breeding tomatoes for soluble solid contents. *Vegetable Science* **5**: 78-82.

Ravindra Mulge, Anand A and Mukesh C (2001). Dry matter heterosis in bell pepper (*Capsicum annum* L.). *Vegetable Science* **28**: 172-74.

Reddy VVP and Reddy BMM (1994). Heterosis for fruit characters in tomato. *Journal of Maharashtra Agricultural University* **19** : 312-14.

Redenbaugh K (1990). Application of synthetic seeds to tropical crops. *HortScience* **25**: 251-55.

Redenbaugh K (1993) Synseeds: *Application of Synthetic Seeds to Crop Improvement.* CRC Press.

Redenbaugh K, Fujii J and Slade D (1988). Encapsulated plant embryos. *In: Biotechnology in Agriculture*, Mizrahi A, and Liss AR, (Eds). New York, pp. 225-248.

Redenbaugh K, Fujii J, Slade D, Viss P and Kossler M (1991). Artificial seeds encapsulation of somatic embryos. *In: Biotechnology in Agriculture and Forestry,* **17**. High Tech Micropropagation I. Bajaj YPS (Ed.) Springer Verlag Berlin, pp. 395-416.

Redenbaugh K, Nichol J, Kossler M and Paasch B (1984). Encapsulation of somatic embryos for artificial seed production. *In vitro Cell Development Biology* **20**: 256.

Redenbaugh K, Paash BD, Nichol JW, Kossler ME, Viss PR and Walker KA (1986). Somatic seeds: encapsulation of asexual embryos. *Biotechnology* **4**: 797-801.

Renbing Zhang Yang Yu Ke Haiying Zhang (2004). A genetic linkage map for watermelon derived from recombinant inbred lines. *Journal of American Society of Horticultural Sciences* **129**: 237-43.

Richey FD (1927). The convergent improvement of selfed lines of corn. *American National* **61**: 430-49.

Rick CM and Boynton JE (1967). A temperature-sensitive male-sterile mutant of the tomato. *American Journal of Botany* **54**: 601-11.

Rick CM and Fobes JF (1977). Association of an allozyme with nematode resistance. *Rep Tomato Genet Coop* **24**: 25.

Rick CM, Deverna JW, Chetelat RT and Stevens MA (1986). Meiosis in sesquidiploid hybrids of *Lycopersicon esculentum* and *Solanum lycopersicoides*. *Proceedings of National Academy of Sciences,* USA **83**: 3580-83.

Risser G (1984). Correlation between sex expression and fruit shape in muskmelon. *In: Proceedings of 3rd Eucarpa Meeting on Cucumis and melons,* Plovdiv.

Robbins MA, Witsenboer H, Michelmore RW, Laliberte JF and Fortin MG (1994). Genetic mapping of turnip mosaic virus resistance in *Lactuca sativa*. *Theoretical and Applied Genetics* **89**: 583-589.

Robbins WJ (1941). Growth of excised roots and heterosis in tomato. *American Journal of Botany* **28**: 216-25.

Robbins WJ (1952). Hybrid nutritional requirements. *In: Heterosis*. pp. 114-23. Iowa State College Press.

Robertson D, Palmer JD, Earle ED and Mutschler MA (1987). Analysis of organelle genomes in somatic hybrid derived from cytoplasmic male sterile *Brassica oleracea* and atrazine resistant *B.compestris. Theoretical and Applied Genetics* **74**: 303.

Robinson RW, Munger HM, Whitaker TW and Bohn GW (1976). Genes of cucurbitaceae, *HortScience* **11**: 554-68.

Robinson RW, Whitaker TW and Bohn GW (1970). Promotion of pistillate flowering in *Cucurbita* by 2-chloroethylphosphoric acids. *Euphytica* **19**: 180-83.

Roggen HP, Van Dijk AJ and Dorsman C (1972). Electric aided pollination: A method of breaking incompatibility in *B. oleracea* L. *Euphytica* **21**:181.

Roggen HP, Van Dijk AJ and Dorsman C (1973). Electric aided and bud pollination, which method to use for self-seed production in cole crops (*B. oleracea). Euphytica* **22**: 260.

Rogozinska J and Drozdowska L (1996). *Journal of Applied Genetics* **37**: 357-66.

Rom M, Bar M, Rom A, Pilowsky M and Gieloni O (1995). Purity control of F_1 hybrid tomato cultivars by RAPD markers. *Plant Breeding* **14**: 188-190.

Rosa JT (1923). The inheritance of flower types in *Cucumis* and *Citrullus*. *Hilgradia* **3**: 233.

Rotino GL, Perri E, Zottini M, Sommer and Spena A (1997). Genetic engineering of parthenocarpic plants. *Nature Biotechnology* **15**:1398-1401.

Rowe PR (1969). The genetics of sex expression and fruit shape, staminate flower induction and F_1 hybrid feasibility of a gynoecious muskmelon. Ph D Thesis, Michigan State University.

Rudich J, Kedar N and Halvey AH (1970). Changed sex expression and possibilities of F_1 hybrid seed production in some cucurbits by application of ethrel and alar (B-995). *Euphytica* **19**: 47-53.

Ruenbenbauer T (1962). The phenomenon of heterosis in relation to the plasmogenoma interaction-A theoretical study. *Genetica Polonica* **3**: 125-36.

Rundfeldt H (1960). Unter suchungen Zur Zuchtung des Koptohls (*B. oleracea* L. var. *capitata*). *Z. Pflanxzenz* **44**: 30-42.

Rybnokova VA and Rudikova AA (1990). Forms of pea resistant to fusarium root rot, *Selektsiya I Semenovodstvo (Moskva)*, **4**: 23.

Sachan SCP and Nath P (1974). Studies on hybrid vigour in watermelon (*Citrullus lanatus* (Thumb.) Mansf). *Proceedings of 19th International Horticultural Congress held at* Warsaw, 676.

Sadhu MK and Das PC (1978). Effect of ethrel (ethephon) on growth, flowering and fruiting of three cucurbits. *Journal of Horticultural Sciences* **53**: 1.

Saha MG, Hossain AKMA, Hoque KR and Bhowmik A (1991). *Annals of Bangladesh Agriculture* **1**: 91-97.

Saito Y, Komari T Masuta C, Hayashi Y, Kumeshiro T and Takanami Y (1992). Cucumber mosaic tolerant transgenic tomato plants expressing a satellite RNA. *Theoretical and Applied Genetics* **83**: 679-83.

Salehuzzaman M and Alam MS (1983). Genetic analysis of yield and its components in eggplant. *SABRAO Journal* **15**: 11-15.

Sambandan CN (1962). Heterosis in egg plant (*Solanum melongena* L.). Prospects and problem in commercial production of hybrid seeds. *Economic Botany* **16**: 71-76.

Sampson DR (1966). Linkage of genetic male sterility with a seedling marker and its use in producing F_1 hybrid seed of *Brassica oleracea* (Cabbage, broccoli, kale *etc.*). *Canadian Journal of Plant Sciences* **46**: 703.

Sanders PR, Sammons B, Kanieski W, Haley L, Layton J, LaVallee BJ, Delannay X and Tumer NE (1992). Field resistance of transgenic tomatoes expressing the tobacco mosaic virus or tomato mosaic virus coat protein gene. *Phytopathology* **82**: 683-90.

Sandha MS and Tarsem Lal (1999). Heterosis breeding in muskmelon: A review. *Vegetable Science* **26**: 1-5.

Sandhu JS, Thakur JC and Nandpuri KS (1977). Investigation on hybrid vigour in cauliflower (*Brassica oleracea* var. *botrytis* L.). *Indian Journal of Horticulture* **34**: 430.

Sanford JC, Smith FD and Russel JA (1993). Optimising the biolistic process for different biological application. *Methods Enzymology* **217**: 483-504.

Sanford, JC (1988). The Biolistic Process *Trends in Biotechnology* **6**: 299-302.

Sanguineti MC, Coltelli C and Conti S (1985). Heterosis and combining ability in eggplant (*Solanum melongena* L.). *Genet Agra* **39**: 345.

Sangwan RS and Lodhi GP (1995). Heterosis for grain characters in cowpea (*Vigna unguiculata* L. Walp.). *Legume Research* **18**: 75-80.

Santos JB and Venocovsky R (1986). Genetic control of some component of plant introduction in French bean. *Perquisa Agropecuaria* **21**: 957-963.

Santos PC, Cardoso AA, Viera C and Silva JC (1986). Heritability and correlation between yield and its components in two *Phaseolus* crosses. *Rovista Ceres* **33**: 404-12.

Sarkissian IV and McDaniel RG (1967). Mitochondrial polymorphism in maize. I. Putative evidence of a correlation of de novo orogin of hybrid-specific mitochondria. *Proceedings of National Academy of Science* USA **52**: 1262-66.

Sarkissian N (1972). Mitochondrial polymorphism and heterosis. 2. *Pblanzenuecht* **67**: 53-54.

Savithriamma DL and Lath J (1999). Heterosis for yields traits in cowpea (*Vigna unguiculata* (L) Walp). *ACIAR Food Newsletter* **29**:7-10.

Savova N (1985). Genetical studies of F_1 and F_2 hybrids of *Phaseolus vulgaris* L. x *Phaseolus coccineus. Genetika i. Selektsiya* **18**: 383-95.

Sawant DS, Birari SP and Jadhav BB (1994). Heterosis in cowpea. *Journal of Maharashtra Agricultural University* **19**: 89-91.

Sawant SV, Desai UT, Kale PN and Joi MB (1992). Heterosis studies in brinjal. *Journal of Maharashtra Agricultural University* **17**: 394-96.

Sawhney VK (1983). Temperature control of male sterility in a tomato mutant. *Journal of Heredity* **74**: 51-54.

Sawhney VK and Shukla A (1994). Male sterility in flowering plants: Are plant growth substances involved *American Journal of Botany* **81**: 1640-47.

Saxena JK (1977). Genetic diveristy and combining ability in peas (*Pisum sativum* L.). *Ph.D. Thesis*, Kanpur University.

Scala A, Bethiini P, Buiathi M, Bogani P, Pellegrini G and Togononi F (1984). *In vitro* analysis of tomato-*Fusarium oxyosporum* system and selection experiments. *In: Plant Tissue and Cell Culture: Application to Crop Improvement*, Novak, FJ *et al.,* (eds.), pp, 361-62.

Schnell FW and Cockerham CC (1992). Multiplicative *vs* arbitary gene action in heterosis. *Genetics* **131**: 461-69.

Schroeder HE, Gollash S, Moore A, Tame LM, Craig S, Hardie DC, Chrispeels MJ, Spencer D and Higgins TJV (1995). Bean α amylase inhibitor confers resistance to the pea weevil (*Bruchus pisorum*) in transgenic peas (*Pisum sativum* L.). *Plant Physiology* **107**:1233-39.

Schroeder WT and Providenti R (1966). *Plant Disease Reporter* **50**: 337.

Schroeder WT and Providenti R (1977). A common gene for reistance to bean yellow mosaic virus and watermelon mosaic virus in *Pisum sativum. Phytopathology* **61**: 846.

Schuler TH, Popy GM, Kerry BR and Denholm I (1998). Insect resistant transgenic plants. *Trends in Biotechnology* **16**: 168-175.

Schumacher H, Genal M and Theres K (1995). Genetic and physical mapping of the lateral suppressor (Ls) locus in tomato. *Molecular and General Genetics* **246**:761-66.

Scott DH and Riner ME (1946). Inheritance of male sterility in winter squash. *Proceedings of American Society of Horticultural Sciences* **47**: 375-77.

Scott JW, Agrama HA and Jones JP (2004). RFLP- based analysis of recombination among resistance genes to fusarium wilt race 1, 2 and 3 in tomato. *Journal of American Society of Horticultural Sciences* **129**: 394-400.

Sedgley M (1981). Ultrastructure and histochemistry of watermelon stigma. *Journal of Cell Science* **48**: 137.

Sedgley M (1982). Anatomy of the unpollinated watermelon stigma. *Journal of Cell Science* **54**: 341.

Sedgley M and Scholefield PB (1980). Stigma secretion in the watermelon before and after pollination. *Botan Gaz* **141**: 428.

Segal G, Sarfatti M, Schaffer MA, Ori N, Zamir D and Fluhr R (1992). Correlation of genetic and physical structure in the region surrounding the IZ *Fusarium oxysporum* resistance locus in tomato. *Molecular and General Genetics* **231**: 179-85.

Selvaraj V and Annappan RS (1983). A study of heterosis in cowpea (*Vigna unguiculata* L. Walp.). *In: Proceedings of Heterotic Systems*, Coimbatore.

Selvaraj V, Annapam RS (1983). A study of heterosis in cowpea (*Vigna unguiculata* (L) Walp.). *In: Proccedings of Scientific Meeting on Genetics and Improvement of Heterotic Systems,* Coimbatore.

Sen B and Srivastava SN (1957). Utilization of cytoplasmic male sterility in production of hybrid onion seeds. *Proceedings of 44th Indian Science Congress*, part 2 pp., 225.

Seo YK, Paek, KY, Hwang JK and Cho YH (1991). *Journal of Korean Society of Horticultural Science* **32**: 137-145.

Serquen FC, Bacher J and Staub JE (1997). Mapping and QTL analysis of horticultural traits in a narrow cross in cucumber (*Cucumis sativus* L.) using random-amplified polymorphic DNA markers. *Molecular Breeding:* 257-268.

Seshadri VS (1996). Hybrid Technology in Vegetable Crops: Present Status and Future Outlook. *In: Training on Vegetable Hybrids and their Seed Production* held at IARI, New Delhi, 15 March to 14 June, pp.23-27.

Seshadri VS (1998). BR Sharma Memorial Lecture on Breeding for Disease Resistance in Vegetable Crops in India. *National Symposium on Emerging Scenario in Vegetable Research and Development,* held at IIVR, Varanasi from 12-14 December, pp. 25-40.

Seshardri VS, Sharma JC and Choudhury B (1972). Pusa watermelon varieties assure good yield and quality. *Indian Horticulture* **16**: 21-32.

Shahin EA and Spivey R (1986). A single dominant gene for *Fusarium* wilt resistance in protoplast derived tomato plants. *Theoretical and Applied Genetics* **73**: 164-69.

Shamel AD (1905). The effect of inbreeding in plants *Year Book* US Department of Agriculture, pp. 377-392.

Shankaraiah V and Rao RR (1990). Studies on heterosis for growth characters and earliness in brinjal. *Vegetable Science* **17**: 56-62.

Shannan S and Robinson RW (1979). The use of ethephon to regulate sex expression on summer squash for hybrid seed production. *Journal of American Society of Horticultural Sciences* **104**: 674-77.

Sharma BR and Mahajan YP (1978). Line 'X' tester analysis of combining ability and heterosis for some economic characters in okra. *HortScience* **9**: 111-18.

Sharma C (1965). Studies in hybrid vogour in bhindi (*Abelmoschus esculentus* (L.) Monech.), *Ph. D. Thesis*, Punjab Agricultural University, Ludhiana.

Sharma DK, Chaudhary DR and Sharma PP (1999). Line 'X' tester analysis for study of combining ability of quantitative traits in tomato. *Indian Journal of Horticulture* **56**: 163-68.

Sharma DK, Choudhury DR and Pandey DP (2001). Studies on hybrid vigour in tomato (*Lycopersicon esculentum* Mill.). *Haryana Journal of Horticultural Sciences* **30**: 236-38.

Sharma Neeraj, Malik YS and Sharma NK (2004). Heterobeltiotic effects in long-fruited bottle gourd. *Haryana Journal of Horticultural Sciences*, **23**: 109-112.

Sharma NK, Dhankar BS and Tewatia AS (1993). Heterosis in bottle gourd (*Lagenaria siceraria* (Molina) Standl.). *Haryana Agricultural University Journal of Research* **23**: 814.

Sharma P and Rajam MV (1995). *Journal of Experimental Botany* **46**: 135-41.

Sharma RK (1992). Heterosis breeding in bell pepper (*Capsicum annum* L.). Ph. D. Thesis University of Horticulture and Forestry, Solan, p. 213.

Sharma RP and Kumar PA (1995). Revolutionizing plant breeding through biotechnology, sustaining crop and animal productivity. *The Challenge of the Decade*, Deb DL (Ed). Associated Publishing Co, New Delhi, pp. 147-156.

Sharma SR, Gill HS and Kapoor KS (1988). Heterosis and combining ability studies in late cauliflower (*Brassica oleracea* var. *botrytis* L.). *Vegetable Science* **15**: 55-63.

Sharp WR, Dougall DK and Paddock EF (1971). Haploid plantlets and callus from immature pollen grains of *Nicotiana* and *Lycopersicon*. *Bull Torrey Botan Club* **98**: 219-22.

Sharp WR, Raskin SR and Sommer EF (1971). Use of nurse culture in development of haploid clones in tomato. *Planta:* **104**: 357.

Sheehy RE, Kramer M ad Hiatt WR (1988). Reduction of polygalacturonase activity in tomato fruit by antisense RNA. *Proceedings of National Academy Science* USA **85**: 8805-09.

Sheshadri VS, Mishra JP, Sharma JC and More TA (1983). Heterosis breeding in melons: Problems and prospects. *Symposium,* held in Bidhan Chandra Krishi Viswavidyalaya, Kalyani.

Shifriss C (1977). Male sterility in pepper (*Capsicum annum* L.). *Euphytica* **93**: 83-88.

Shifriss C and Frankel R (1969). A new male sterility gene in *Capsicum annum. Journal of American Society of Horticultural Sciences* **94**: 385-387.

Shifriss O (1945). Male sterilities and albino seedlings in cucurbits. *Journal of Heredity* **36**: 47-52.

Shifriss, C and Rylski I (1972). Comparative performance of F_1 hybrids and open pollinated bell pepper varieties (*C. annum* L.) under sub-optimal temperature-regimes. *Euphytica* **22**: 530-4.

Shimotsuma M (1962). Cytological studies in the genes *Citrullus*. VI. Investigation on the cytology of induced autotriploid *C. vulgaris* and its progeny. Seiken Jino Republic. *Kihara Institute Biological Research* **13**:39.

Shimotusma M (1963). *Seiken Ziho* **15**: 24-34.

Shoba N and Aruangam R (1991). *South Indian Horticulture* **39**: 350-52.

Shrivastiva LS and Sachan SC (1975). Note on heterosis in pea. *Indian Journal of Agricultural Sciences* **45**: 278.

Shukla A and Sawhney VK (1993). Metabolism of dihydrozeatin in floral buds of wild type and a genic male sterile line of rapeseed (*Brassica napus*). *Journal of Experimental Botany* **44**: 1497-1505.

Shukla AK and Gautam NC (1990). Heterosis and inbreeding depression in okra (*Abelmoschus esculentus* (L.) Moench). *Indian Journal of Horticulture* **47**: 85-88.

Shukla AK, Gautam NC, Tiwari AK and Chaturvedi AK (1989). Heterosis and combining ability in okra. *Vegetable Science* **16**: 191-96.

Shukla MR, Patel JA, Joshi KM and Patel SA (1999). Line 'X' tester analysis of combining ability in chilli (*Capsicum annum* L.). *Vegetable Science* **26**: 45-49.

Shull GH (1908). The composition of field maize. *Report of American Breeders Association* **4**: 296-301.

Shull GH (1909). A pure line method in corn breeding. *Report of American Breeders Association* **5**: 51-59.

Shull GH (1910). Hybridization methods in corn breeding. *American Breeders Magazine*, **1**: 98-167.

Shull GH (1952). Beginnings of the heterosis concept. *In: Heterosis*, Gowen JD (Ed). pp.14-48. Ames: Iowa State College Press.

Sibi M Dumas de Vaulx R and Chambonnet D (1980). Androgenese *in vitro* chez le piment (*Capsicum annum* L.): Impact des pretraitements sur la taux de plantes regenerees, *in Application de la Culture in Vitro d l Amelioration des Plantes Potageres,* CR Eucarpia, Versailles, France.

Sibi M, Dumas de Valux R and Chambonnet D (1979). Obtention de plantes haploids par andogenese *in vitro* chez le piment (*Capsicum annuum* L). *Ann. Amelior Plant* **29**: 583.

Sidhu AS and Brar JS (1977). Heterosis and combining ability of yield and its components in watermelon (*Citrullus lanatus* (Thumb.) Mansf). *Journal of Research* **14**: 52.

Sidhu AS and Brar JS (1985). Genetic divergence and hybrid performance in watermelon. *Indian Journal of Agricultural Sciences:* 459-61.

Sidhu AS, Dixit J, Kalloo G and Bhutani RD (1981). Heterosis and combining ability in pear shaped tomato (*L. esculentum* Mill.). *Haryana Agricultural University Journal of Research* **11**: 1-7.

Sidhu AS, Kalloo G and Pandita ML (1980). Studies on some important aspects of floral bilogy in vegetable crops: A review. *Haryana Journal of Horticultural Sciences* **9**.

Sidhu AS, Singh S and Verma MM (1993). Studies on heterosis and divergence in tomato. *Symposium on Heterosis and Inbreeding in Crop Plants-Theory and Application*, held during 23-24 February, Ludhiana, pp. 64-65.

Simon PW (1986). *Daucus carrota* L. *In: Genetic Improvement of Vegetable Crops*, Kalloo G. and BO Bergh (Eds), Pergamon Press, pp. 479-484.

Simon PW (1993). *In: Genetic Improvement of Vegetable Crops* (Eds). Kalloo G. and Bergh BO, Pergamon Press.

Sindle N, Costache M and Sociu S (1977). Greenhouse egg plants hybrids. *Acta Horticulturae* **58**: 75-87.

Singh A and Kao TH (1992). Gametophytic self-incompatibility: Biochemistry, molecular genetics and evolutionary aspects. *In : Sexual Reproduction in Flowering Plants,* Russel SD and Dumas C., (Eds). Academic Press, pp. 449-483.

Singh A and Singh HN (1973). Genetic divergence and heterosis in chilli (*Capsicum annum* L.). *Haryana Journal of Horticultural Sciences* **6**: 155-60.

Singh A, Singh HN and Mittal RK (1973). Heterosis in chilli. *Indian Journal of Genetics* **33**: 398.

Singh A, Singh PK, Dixit J and Gautam JPS (1995). Heterosis and inbreeding depression in tomato. *Horticultural Journal* **8**: 125-29.

Singh Ajeet, Singh PK, Dixit J, Gautam JPS, Singh DN and Singh A (1996). Heterosis and inbreeding depression in tomato. *Birsa Agricultural University Journal of Research,* **8**: 89-90.

Singh AK and Pandey UB and Singh M (2000). Studies on heterosis in bitter gourd (*Momordica charantia* L.). *Vegetable Science* **27**: 158-61.

Singh AK and Saini SS (1983). Studies on heterosis and inbreeding depression for primary components and yield components in French bean (*Phaseolus vulgaris* L.). *Indian Horticulture* **34**: 187-95.

Singh AK and Sood S. (1999). Heterosis and inbreeding depression in okra. *Indian Journal of Horticulture* **36**: 67-78.

Singh AK, Pan RS, Rai M. and Prasad K. VSR (2004). Heterosis for yield and its contributing attributes in brinjal (*Solanum melongena* L.). *Vegetable Science* **31**: 146-48.

Singh B and Khanna KR (1974). Heterosis in chromosome behaviour of eggplant. *Current Science* **43**: 5-6.

Singh B and Kumar N (1988). Studies on hybrid vigour and combining ability in brinjal (*Solanum melongena* L.). *Vegetable Science* **15**: 72-78.

Singh B, Gupta VP and Gupta PK (1986) Heterosis studies in radish (*Raphanus sativus* L.). *Indian Journal of Horticulture* **43**: 242-247.

Singh B, Joshi S and Kumar N (1978). Hybrid vigour in brinjal (*Solanum melongena* L.). *Haryana Journal of Horticultural Sciences* **7**: 95-99.

Singh B, Rawat PS and Pant PC (1979). *Indian Journal of Agricultural Sciences* **44**: 807.

Singh B, Singh Sanjay, Pal AK and Rai Mathura (2004). Heterosis for yield and yield components in okra (*Abelmoschus esculentus* (L.) Monench). *Vegetable Science* **31**: 168-71.

Singh BD (1983). Plant Breeding : Principles and Methods. Kalyani Publishers, New Delhi.

Singh BD (2002). Plant Breeding. Kalyani Publishers, New Delhi.

Singh Brahma, Subodh Joshi and Narendra Kumar (1978). Heterosis and combining ability in tomato (*Lycopersicon esculentum* Mill.). *Vegetable Science* **3**: 91-96.

Singh BS (1979). Genetical studies in okra (*Abelmoschus esculentus* (L.) Monech.). *Ph. D. Thesis*, Punjab Agricultural University, Ludhiana.

Singh D (1983). Biometrical and genetical studies in okra, *(Abelmoschus esculentus* (L.) Monech.). Ph. D. Thesis, Punjab Agricultural University, Ludhiana.

Singh D and Sharma RR (1990). Genetics of seed number and fruit characters in okra. *Haryana Journal of Horticultural Sciences* **19**: 353-57.

Singh D, Shrivastava PS and Singh SP (1975). Heterosis and inbreeding depression in table pea (*Pisum sativum* L.). *Indian Journal of Agricultural Sciences* **45**: 278.

Singh DK and Gautam NC (1991*). Vegetable Science* **18**: 177-83.

Singh DP and Krishna Prasad VSR (1995). Heterosis in eggplant (*Solanum melongena* L.). *Indian Journal of Horticulture* **52**: 291-295.

Singh DP, Swarup V and Chatterjee SS (1976). Genetic studies in Indian cauliflower (*Brassica oleracea var. botrytis* L.). II. Heterosis and combining ability in maturity group II. *Vegetable Science* **3**: 41.

Singh G and Nandpuri KS (1974). Combining ability studies in tomato *(L. esculentum* Mill.) cultivars with functional male sterile lines. *Journal of Research,* Ludhiana **11**: 26.

Singh H, and Nandpuri KS (1975). The genetic analysis of acidity and total soluble solids in tomato. *Journal of Research,* **12**: 367-72.

Singh IJ (1993). Assessment of heterosis and combining ability in bottle gourd. (*M Sc.*) *Thesis*, Punjab Agricultural University, Ludhiana.

Singh J (1980-81). A note on hybrid vigour in chillies (*Capsicum annum* L.). *Punjab Vegetable Grower*, **15-16**: 31-33.

Singh J and Kaur S (1986). Present status of hot pepper breeding for multiple resistances in Punjab. In: *Proccedings VIIth Meet. Genet. Breed. Capsicum and Eggplant, Zaragoza,* Spain, pp. 111-14.

Singh J, Singh JP and Singh RD (1996). Studies on heterosis and inbreeding depression in tropical cauliflower. *Indian Journal of Horticulture* **53**: 135-140.

Singh J, Singh T and Khurana DS (1992). Heterosis studies in chillies (*Capsicum annum* L.). *Vegetable Science* **19**: 161-165.

Singh JP, Gill HS and Tewari RN (1966). Induction of closed anther character in tomato cultivars Pusa Ruby, Money Maker and Sioux. *Current Science* **35**: 252.

Singh JP, Gill HS and Tewari RN (1971). Studies on hybrid vigour in radish. *Indian Journal of Horticulture* **27**: 67-75.

Singh Kirti (1998). *In: Silver Jubilee National Symposium on Emerging Scenerio in Vegetable Resaerch and Development,* held at IIVR, Varanasi during 12-14 December, p 2.

Singh M. Chakraborty S., Srivastava Kiriti and Kalloo G (2001). Biotechnological advances in vegetable crops. *In: Emerging Scenario in Vegetable Research and Development.* Kalloo G and Singh Kirti, (Eds). Research Periodical and Book Publishing House, pp.58-84.

Singh M, Kumar S, Singh B and Banerjee MK (2002). *In: International Conference on Vegetables*, PNASF, Bangalore (India), 11-14 November.

Singh N, Arora SK, Ghai TR and Dhillon TS (1995). Gene action in okra (*Abelmoschus esculentus* (L.) Moench). *Vegetable Science* **22**: 98-100.

Singh N, Arora SK, Ghai TR and Dhillon TS (1996). Heterobeltiosis studies in okra (*Abelmoshus esculentus* (L) Monech.). *Punjab Vegetable Growers* **31**: 6-9.

Singh NJ, Khattra AS and Thakur (1997). Heterobeltiosis studies in bitter gourd. *Haryana Journal of Horticultural Sciences* **26** : 211-14.

Singh P (1983-84). Unpublished data.

Singh PK, Kumar C and Sharma R (2000). Genetic estimate (gene action) in long fruited bottle gourd. *Vegetable Science* **27**: 162-64.

Singh PK, Kumar JC and Sharma JR (1998). Heterosis studies in long fruited bottle gourd. *Vegetable Science* **25**: 55-57.

Singh Rajender and Hundal JS (2001). Manifestation of heterosis in chilli (*Capsicum annum* L.). *Vegetable Science* **28**: 124-26.

Singh Ramveer, Verma TS, Thakur PC and Yadav HL (1999). Heterotic response in maternal seeds crossing in cucumber. *Haryana Journal of Horticultural Sciences* **28**: 223-25.

Singh RD and Rai B (1990). *Vegetable Science* **17**: 180-83.

Singh RK and Mandal G (1993). Studies on heterosis in okra (*Abelmoshus esculentus* (L) Monech.). *Annals of Agricultural Research* **14**: 429-33.

Singh RK and Singh VK (1993). Heterosis breeding in tomato. *Annals of Agricultural Research* **14**: 416-20.

Singh RR and Singh HN (1980). Note on variance components in tomato. *Indian Journal of Agricultural Sciences* **50**: 36-3.

Singh S (1988). *Current Science* **13**: 730-32.

Singh S and Nandpuri KS (1970). Utilization of male sterile line for hybrid tomato production. *Indian Journal of Horticulture* **27**: 32-35.

Singh S and Sawhney VK (1992). Cytokinins in a normal and the Ogura (*ogu*) cytoplasmic male sterile line of rapeseed (*Brassica napus*). *Plant Science* **86**: 147-54.

Singh S, Dhaliwal MS, Cheema DS and Brar GS (1998). Diallel analysis of some processing attributes in tomato (*Lycopersicon esculentum* Mill.). *Indian Journal of Genetics and Plant Breeding* **52**: 265-69.

Singh SJ (1979 a). Studies on pumpkin mosaic virus. *Ph. D. Thesis*, Banaras Hindu University, Varanasi, pp.379.

Singh SJ (1981b). Studies on causing mosaic disease of pumpkin (*Cucurbita maxima* Duch.). *Phytopathology Mediterranea* **20**: 184-88.

Singh SN (1957). Studies in the floral biology of *Luffa*. *Horticulture Advance* **1**: 86.

Singh SN (1984). The magnitude of GCA in relation to heterosis and inbreeding depression in eggplant. *Haryana Journal of Horticultural Sciences* **13**: 175-81.

Singh SN, Singh D, Chauhan YS and Katiyar RP (1974). Genetic variability, heretibility and genetic advance in brinjal. *Progressive Horticulture* **6**: 15-18.

Singh TP and Singh KB (1970b). Heterosis in field peas. *Indian Journal of Genetics* **30**: 619-24.

Sirohi PS (1994). Heterosis in pumpkin (*Cucurbita moschata* Duch, ex. Poir). *Vegetable Science* **21**: 163-65.

Sirohi PS (2000a). Pusa Hybrid-1, a new bitter gourd hybrid. *Indian Horticulture* **45**: 30-31.

Sirohi PS (2000b). Pusa Hybrid-1, a new pumpkin hybrid. *Indian Horticulture* **45**: 32-34.

Sirohi PS (2000c). Pusa Hybrid-3, a new bottle gourd hybrid. *Indian Horticulture* **45**: 33-34.

Sirohi PS (2004). Enhancement of heterosis in cucurbitaceous vegetable crops. *In: National Symposium on Harnessing Heterosis in Crop Plants*, held at IIVR, Varanasi, 13-15 March, pp.121-29.

Sirohi PS and Choudhury B (1977). Combining ability in bitter gourd (*Momordica charantia* L.). *Vegetable Science* **5**: 15-22.

Sirohi PS and Choudhury B (1978). Heterosis in bitter gourd (*Momordica charantia* L.). *Vegetable Science* **6**: 106-112.

Sirohi PS and Choudhury B (1983). Diallel analaysis for variability in bitter gourd (*Momordica charantia* L.). *Indian Journal of Agricultural Sciences* **53**: 80-88.

Sirohi PS, Kumar TS and Choudhury B (1985). Studies on heterosis in pumpkin (*Cucurbita moschata* Duch ex Poir.). *South Indian Horticulture* **33**: 297-301.

Sirohi PS, Kumar TS and Choudhury B (1986). Studies on combining ability in pumpkin (*Cucurbita moschata* Duch, ex. Poir.). *Indian Journal of Horticulture* **43**: 98-104.

Sirohi PS, Kumar TS and Choudhury B. (1986). Diallel analysis for variability in pumpkin (*Cucurbita moschata* Duch. ex. Poir.). *Indian Journal of Agricultural Sciences* **56**: 171-73.

Sirohi PS, Reddey YS and Behera TK (2002). Heterosis for yield and its contributing traits in pumpkin (*Cucurbita moschata* Duch. ex. Poir.). *Vegetable Science* **29**: 178-179.

Sirohi PS, Sivakami N and Choudhary B (1985). Heterosis in long fruited bottle gourd. *Annals of Agricultural Research* **6**: 210-14.

Sirohi PS, Sivakani N and Choudhury B (1987). Studies on exploitation of heterosis in bottle gourd. *Vegetable Science* **14**: 37-41.

Sit AK (1999). Heterosis, combining ability and gene action in bottle gourd (*Lagenaria siceraria* (Mol.) Standl.). Ph. D. Thesis, IARI, New Delhi.

Slack CR and Hatch MD (1967). Comparative studies on the activity of carboxylase and their enzymes in relation to the new pathway of photosynthetic carbon dioxide fixation in tropical grasses. *Biochemical Journal* **103**: 660-65.

Smith CJS, Watson CF, Ray J, Bird CR, Morris PC, Schuh W and Griesson D (1988). Antisese RNA inhibition of polygalacturonase gene expression in transgenic tomatoes. *Nature* **334**: 724- 726.

Smith, JSC and Register, JC (1998). Genetic purity and testing technologies for seed quality: A company perspective. *Seed Science Research* **8**: 285-93.

Sob, AM and Weyman LJ (1966). Intergenic complementation at the diploid level in *Neurospora.* Cited by Srb (1966). *In: Plant Breeding* Iowa USA, pp. 372-76.

Sohi HS and Sohni SS (1974). Behaviour of okra varieties of damping off, powdery mildew and cercospora blight. *Indian Phytopathology* **27**: 91-2.

Song MX and Wei SL (1996). On the heterosis of hybrid radish and the strategy for parent choice. *Acta Agricultuae Zhejiangensis* **8**: 46-49.

Song PL, Pen WZ and Yang YH (1988). *Acta Sci. Nature Univer. Hunan* **11**: 340-45.

Song R and Messing J (2003). Gene expression of a gene family in maize based on non-collinear haplotypes. *Proceedings of National Academy of Sciences*, USA **100**: 9055-60.

Songstad DD, Somers DA, Griesbach RJ (1995). Advances in alternative DNA delivery techniques. *Plant Cell Tissue Org Culture* **40**:1-15.

Sonone HN, Deore BP and Patil SK (1981). A preliminary estimation of hybrid vigour in tomato. *Journal of Maharashtra Agricultural University* **6**: 250-52.

Sood and Kalia Pritam (2001). Heterosis and combining ability studies in some quantitative traits in okra (*Abelmoschus esculentus* (L.) Moench). *Haryana Journal of Horticultural Sciences* **30**: 92-94.

Sopory SK and Rogan PG (1976). Induction of pollen division and embryoid formation in anther cultures of some dihaploid clones of *Solanum tuberosum. Z. Pflanzenphysicol* **80**: 77.

Sopory SK, Jacobson E and Wenzel G (1978). Production of monohaploid embryoids and plantlets in cultural anthers of *Solanum tuberosum. Plant Science Letter* **12**: 47.

Sorvari S *et al.* (1997). Polysacharide and galactomannan gelling agents for synthetic seeds. *Journal of American Society of Hort.Sciences* **122**: 878-883.

Souza CM, Pinto JEBP, Rodrigues BM, Furtado DF, Morais AR, Arrigoni BM and Morais AR (1998). *Ciencia-Agrotecnologia* **22**: 52-56.

Spetsidis N, Sapountzakis G and Tsaftaris AS (1996). *Cucurbit Genetic Cooperation. Republic* **19**: 63-65.

Sprague GF (1983). Heterosis in Maize: Theory and Practice, *In : Heterosis: Reappraisal of Theory and Practice Frankel R.* Springer-Verlag, Berlin, pp. 47-70.

Srb Ames and Weyonan LJ (1966). Intergenic and intragenic complementation at the diploid level in *Neurospora*. Cited by Srb (1966), In: *Plant Breeding:* Annes; Iowa USA, pp. 372-376.

Sretenovic RT, Vinterhalter D, Vinterhalter B and Lazic B (1997). *Acta Horticulturae* **462**: 589-93.

Srivastava D (1982). Studies on pollen biology of certain cultivated malvaceae. *In : Advances in Pollen Spore Research*, pp. 18-44. Today and Tomorrow Printers and Publishers, New Delhi.

Srivastava HK (1975). Heterosis and homeostasis in mitochondria of wheat hybrid. *Second International SABRAO, Congress*, New Delhi **12**: 30-39.

Srivastava HK (1981). Intergenomic interaction, heterosis and improvement of crop yield. *Advances in Agronomy* **34**: 117-195.

Srivastava HK (1983). Heterosis and intergenomic complementation: Mitochondria, Chloroplast and nucleus. *In: Frankel R (ed) Heterosis: Reappraisal of theory and practice. Monograph, Theoretical and Applied Genetics,* **6**: 260-86.

Srivastava LS and Sachan SC (1975). Note on heterosis in pea. *Indian Journal of Agricultural Sciences* **45**: 278.

Srivastava VK (1970). Studies on hybrid vigour, combining ability and inheritance of some quantitative characters in bitter gourd (*Momordica charantia* L.). *Ph D Thesis*, University of Udaipur, India.

Stadler LJ (1944). Gamete selection in corn breeding. *Journal of American Society of Agronomy* **36**: 988-89.

Stalker DM McBridge KE and Malyj LD (1988). Herbicide resistance in transgenic plants. Expressing bacterial detoxification gene. *Science* **242**: 419.

Stanley RG and Linskens, HF (1974). *In: Pollen: Biology, Biochemistry and Management.* Springer.

Stark DM and Beachy RN (1989). Protection against polyvirus infection in transgenic plants. Evidence from broad spectrum resistance. *Biotechnology* **7**: 1257-62.

Stebbins GL (1958). The inviability, weakness, and sterility of interspecific hybrids. *Advances in Genetics* **9**: 147-215.

Stenitz B, Wolf D, Josef TM and Zelcer A (1999). *Capsicum and Eggplant Newsletter* **18**: 9-15.

Stevens MA and Rick CM (1986). *In: Genetics and Breeding. Tomato Crop*, Athenton J.G. and Rudich J., (Eds). Chapman and Hall Ltd, London. pp.35-109.

Stommel JR Tousignnant ME, Wai T, Pasini R and Kaper JM (1998). Viral satellite RNA expression in transgenic tomato confers field tolerance to cucumber mosaic virus. *Plant Diseases* **82**: 391-96.

Strathof TP, Van Tuyl JM, Keijzer CJ, Wilms HJ, Kwakkenbos AAM and Van Dien MP (1987). Overcoming post-fertilization barriers in *Lilium* by ovary-and ovule culture. *Plant Cell Incompatibility Newsletter* **19**: 69-74.

Strel'nikova TR, Mashtakova AK and Pogoda LP (1984). Cucumber hybrids with combined resistance. *Kartofel Ovoshchi* **4**: 24.

Stringfield GH (1950). Heterozygosis and hybrid vigour in maize. *Agronomy Journal* **42**: 145-52.

Strittmatter G, Janssens J, Opsomer C and Botterman J (1995). Inhibition of fungal disease development in plants by engineering controlled cell death. *Biotechnology* **13**: 1085-89.

Stuber CW (1994). Heterosis in plant breeding. *Plant Breeding Review* **12**: 227-51.

Stuber CW (1999). Biochemistry, molecular, biology and physiology of heterosis. *In: Genetics and Exploitation of Heterosis in Crop Coors.* JG and Pandey S, (Eds). pp. 173-83.

Stuber CW, Lincilin SE, Wolff DW, Helentjaris T and Lander ES (1992). Identification of genetic factors contributing to heterosis in a hybrid from two elite maize inbred lines using molecular markers. *Genetics* **132**: 823-839.

Stuber GW, WilliamsWP and Moll RH (1973). Epistasis in maize (*Zea mays* L): III. Significance in preditions of hybrid performance. *Crop Science* **13**: 195-200.

Stuckey HP (1916). Transmission of resistance and susceptibility to blossom end rot in tomatoes. *Georgia Agriculture Experiment Station Bulletin* **121**: 83-91.

Studentsove LI (1973). Choice of pairs of red pepper in breeding for heterosis, Nauchn. *Tr. Maikop. Abst. St. VNII Rastenivod* **7**: 96.

Stuiver MH, Tigelaar H, Molendijk L, Toost van Deventer E, sela-Burlaage MB, Storms J, Plooster L, Sifbolts F, Custers J, Apotheker-de Groot M and Melchers LS (1996). Broad-spectrum resistance in transgenic carrot plants. Stacey G, Mullin B and Greshoff PM (eds), *8th International Congress. Molecular Plant-Microbe Interaction*, pp. 13-93.

Subashri S and Natarajan S (1999). Studies on residual heterosis for yield and fruit characters in F_2 generation of chilli (*Capsicum annum* L.).

Suneal Kumar T (1984). Studies on heterosis in pumpkin (*Cucurbita moschata* Duch. ex. Poir.). M Sc Thesis, IARI.

Suresh Babu KV (1987). Cytological studies in okra (*Abelmoschus esculentus* (L.) Moench.). *Ph. D. (Hort.), Thesis,* UAS, Bangalore.

Suresh Kumar, Banerjee MK, Pratap PS and Kumar S (1995). Heterosis study for fruit quality and its component in tomato. *Annals of Agricultural Research* **16**: 212-17.

Swamy KRM and Mathai PJ (1978). Heterosis in tomato (*L. esculentum* Mill.). *South Indian Horticulture* **26**: 108-15.

Swamy KRM and Mathai PJ (1982). Note on combining ability in tomato. *Indian Journal of Agricultural Sciences* **52**: 193-6.

Swarup *et al.* (1992) *In: Compedium on Vegetable Hybrids and their Seed Production Training,* March 15 to 14 June 1996, at Division of Vegetable Crops, IARI, New Delhi, p. 25.

Swarup V (1991). *In: Breeding Procedures for cross-pollinated vegetable crops*. ICAR, Publication.

Swarup V and Chatterjee SS (1974). Heterosis in Indian Cauliflower. *Proceedings of 19th International Horticulture Congress.*

Swarup V and Gill HS (1964). The use of marker gene in hybrid seed production in cabbage. *Current Science* **33**: 315.

Swarup V and Pal AB (1966). Gene effects and heterosis in cauliflower. *Indian Journal of Genetics and Plant Breeding* **26**: 269.

Swarup V, Gill HS and Singh D (1963). Studies on hybrid vigour in cabbage. *Indian Journal of Genetics and Plant Breeding* **23**: 90.

Synge AD (1947). Pollen collection by honey bees. *Journal of Animal Ecology* **16**: 122-38.

Szymanska M and Molas J (1995). *Acta Agrobotanica* **48**: 75-82.

Tanskley SD and Bernatzky R (1987). Molecular markers for the nuclear genome of tomato. *In : Tomato Biotechnology*. Nevinss DJ and Jones RA (Eds), Alan R Liss Inc, New York, pp. 37-44.

Tanskley SD and Jones RA (1979). Application of alcohol dehyrogenase allozymes in testing the purity of F_1 hybrids of tomato. *HortScience* **16**: 179-81.

Tanskley SD and Orton TJ (1983). *In: Isozymes in Plant Genetics and Breeding*, Elsevier.

Tanskley SD and Rick CM (1980). Isozymic gene linkage map of the tomato. Application in genetics and breeding. *Theoretical and Applied Genetics* **57**: 161-70.

Tanskley SD, Bernatzky R Lapitan NL and Prince JP (1988). Conservation of gene repeotoire but not gene order in pepper and tomato. *Proceedings of National Academy of Sciences* USA, 6419-23.

Tanksley SD, Ganal MW, Prince JP, de Vicente MC, Boinerable MW, Broun P, Fulton TM, Giovanonni Wing RA, Wu W and Young ND (1992). High-density molecular linkage maps of the tomato and potato genomes: Biological inference and practical applications. *Genetics* **132**: 1141-60.

Tanksley SD, Grandillos, Fulton TM, Zamir D, Eshed Y, Petiard V, Lopez J and Beek-Bunn T(1996). Advanced back cross QTL analysis in a cross between an elite processing line of tomato and its wild relative L. *pimpinellifolium. Theoretical and Applied Genetics* **92**: 213-24.

Tanskley SD, Rick CM and Vallejos CE (1984). Tight linkage between a male sterile locus and an enzyme marker in tomato. *Theoretical and Applied Genetics* **68**: 109-13.

Tendulakar SK (1997). Heterosis, combining ability and gene action in bitter gourd (*Momordica charantia* L.). *Ph. D. Thesis*, IARI.

Tepedino VJ (1981). The pollination efficiency of the squash bee (*Peponapis pruinosa*) and the honey bee (*Apis mellifera*) on summer squash. *Journal of Kansas Entomololical Society* **54**: 359-77.

Tewari Deepali and Ram, Hari Har (1999). Heterosis in bitter gourd (*Momordica charantia* L.). *Vegetable Science* **26**: 27-29.

Tewari SP (2004). IPR- related issues in hybrid technology. *In: National Symposium on Harnessing Heterosis in Crop Plants*, held at IIVR, Varanasi, 13-15 March, pp.18-21.

Tewatia AS and Kalloo G (1992). Partial diallel analysis for the study of heterosis and combining ability in garden pea. *Haryana Journal of Horticultural Sciences* **21**: 76-82.

Thakur BS, Korla BN and Spehia (2001). Heterosis studies in late group cauliflower. *Haryana Journal of Horticultural Sciences* **30**: 233-35.

Thakur JC, Singh SJ and Kattra AS (1995). Heterobeltiosis studies in main season cauliflower. *Journal of Research* **32**: 143-50.

Thakur MR (1976). Inheritance of yellow vein mosaic in cross of okra species *A. esculentus* x *A. manihot* spp. *manihot. SABRO Journal* **8**: 69-73.

Thakur MR and Arora SK (1986). Okra. *In: Vegetable Crops in India,* Bose, TK (Eds). Naya Prokash, pp. 606-22

Thakur MR, Singh K and Singh J (1968). Hybrid vigour studies in brinjal. *Journal of Research* **5**: 490-95.

Thakur PC (1987). Gene action. An index for heterosis breeding in sweet pepper. *Capsicum Newsletter* **6** : 41-42.

Thakur PC, Joshi S, Verma TS and Verma HC (1998). 'Pusa Depti' new pp.35-109. *Capsicum* hybrid. *Indian Horticulture* **43**: 6.

Thomas BR and Pratt D (1981). Efficient Hybridization between *Lycopersicon esculentum* and L. *peruvianum via* embryo callus. *Theoretical and Applied Genetics* **59**: 215-19.

Thomas CM, Jones DA, Parniske M, Harrison K, Balint-Kurti PJ, Hatzixanthis, K and Jones JD (1995). Characterization of the tomato *cf-4* gene for Cladosporium fulvum identifies sequences that determine recognitional specificity in *cf-4* and *cf-9*. *Plant Cell.* **9**: 2209-24.

Thomas HA (1932). Vegetable breeding at California. *Proceedings of American Society of Horticultural Sciences* **29**: 572.

Thomas P and Peter KV (1988). Heterosis in intervarietal cross in bell pepper and hot chilli. *Indian Journal of Agricultural Sciences* **54**: 747-50.

Thomas TA, Kamala V, Dabas BS and Bisht IS (1990). Genetic resources of onion and garlic, their evaluation and maintenance. *National Symposium on Onion and Garlic: Production, Marketing and Export*, held at Solan, India.

Thompson KF (1959). Breeding better kales. *Agriculture* **65**: 487-91.

Thomson JD (1986). Pollen transport and deposition by bumblebees in Erythronium: Influences of floral nectar and bee grooming. *Journal of Ecology* **74**: 329-41.

Tikka SBS, Jaimini SN, Asawa BM and Mathur JR (1977). Genetic variability in inter-relationship and discriminant function analysis in cowpea (*Vigna unguiculata* (L) Walp.). *Indian Journal of Heredity* **9**: 1-9.

Tikoo SK and Anand N (1980). Development of tomato genotypes with exserted stigma and a seedling marker for use as female parents to exploit heterosis. *Current Science* **49**: 326-27.

Timin NI (1979a). Tr. VN II Selektsii I semenovod ovosshch, Kultur **9**: 46-50.

Timin NI (1979b). Tsitplazmatich, muzhsk, sterl'nost'I seleksiya rast. Kiev, Ukrainian.

Timin NI and Vasilevskii (1995). Carrot lines for heterosis breeding using cytoplasmic male sterility (CMS). *Kartofeli Ovoschchi*, **2**: 27-28.

Tinker NA and Mather DE (1995). Methods of QTL analysis with progeny replicated in multiple environments. *JQTL* **1**:1-22.

Tissereau H, Florin B, Meschin MC, Thierry C and Petiard V (1994). Cryopreservation of somatic embryos: A tool for germplasm preservation and commercial delivery of selected plants. *Annals of Botany* **74**: 547-55.

Tiwari CV and Kale PN (1994). Heterosis studies in tomato. *Journal of Maharashtra Agricultural University* **19**: 83-85.

Tiwari RD (1966). Studies on hybrid vigour in *Solanum melongena* L. *Journal of Indian Botanical Society* **45**: 138-49.

Tokmakov Yu G. and Rabunetz NA (1967). *Sel'skohoz Biology* **2**: 563-6.

Tolla GE and Peterson LE (1979). Comparison of gibberellin GA_{4+7} and silver nitrate for induction of staminate flowers in a gynoecious cucumber line. *HortScience* **14**:542-44.

Tomassoli L, Iiardi V, Barba M and Kaniewski W (1999). Resistance of transgenic tomato to cucumber common mosaic virus under field conditions. *Molecular Breeding* **5**: 121-30.

Towill LE (1985). Low temperature and freeze-(vacuum) drying preservation of pollen. In: *Cryopreservation of Plant Cells and Organs,* Kartha KK (Ed). pp.171-98. FL: CRC Press.

Toyoda H, Shimizu K, Chatani K, Kita N, Matsuda Y and Onchi S (1989). Selection of bacterial wilt resistant tomato through tissue culture. *Plant Cell Reports* **8**: 317-320.

Trivedi DP (1983). Morphological and genetical studies in bitter gourd (*Momordica charantia* L.). *Thesis*, Kanpur University.

Trivedi RN and Nath P (1973). Cytogenetics of *Momordica charantia* and its polyploids. *Cytologia* **38**: 317-25.

Trivic S and Leskovac V (1987). Biochemical study of heterosis in maize: mitochondrial heterosis in Novi Sad hybrids of maize. *Arh Poljopr Nauke* **48**: 91-98.

Troyer AF (1991). Breeding corn for the export market. *Proceeding, 46th Annual Corn and Sorghum Research Conference,* American Seed Trade Association **46**: 165-176.

Troyer AR (1990). A retrospective view of corn genetic resources. *Journal of Heredity* **81**: 17-24.

Tsaftaris SA (1995) Molecular aspect of heterosis in plants. *Physiology of Plants* **94**: 362-70.

Tsaftaris SA, Kafka M and Polidoros A (1997). Epigenetic modification of total genomic maize DNA. The role of growth conditions. *In: Genetics, Biochemistry and breeding of Maize and Sorghum* Tsaftaris A.S. (Eds.). The Royal Society of Chem. Press Cambridge.

Tumer NE, O' Connell KM, Nelson RS, Sanders PR, Beachy RN, Fraley RT and Shah DM (1987). Expression of alfalfa mosaic virus coat protein gene confers cross-protection in transgenic tobacco and tomato plants. *EMBO Journal* **6**: 1181-88.

Tyagi ID (1973). Heterosis in bottle gourd. *Indian Journal of Horticulture* **30**: 394-400.

Ugale SP, Patil RC and Khupse SS (1976). Cytogenetic studies in the cross between *A. esculentus* and *A. tetraphyllus*. *Journal of Maharashtra Agricultural University* **1**: 106-10.

Uhcman R, Zel J and Ravnikar M (1998). *Scientia Horticulture* **73**: 193-202.

Ustimova EJ (1949). A description of the interspecific hybrids of *H. esculentus* x *H. manihot. Nature* **6**: 58-60.

Vaeck M, Reynaerts A, Hofte H, Jansens S, De Beuckeleer M, Dean C, Zabeau M, Var Montagu M and Leemans J (1987). Transgenic plants protected from insect attack. *Nature* **327**: 33-37.

Vagera J and Hairane KP (1985). *In vitro* induction of androgensis in *Capsicum annuum* L and its genetic aspects. *Biol. Plant* **27**: 10.

Vahab MA (1989). Homeostatic analysis of components of genetic variance and inheritance of fruit colour, fruit shape and bitterness in bitter gourd (*Momordica charantia* L.). *Ph D Thesis,* Kerala Agricultural University.

Vaid K and Singh RM (1986). Genetic variablility in dry bean (*Phaseolus vulgaris* L.). *Research and Development Reporter* **3**: 55-58.

Vallejos CE and Tanksley SD (1983). Segregation of isozyme markers and cold tolerance in an interspecific backcross of tomato. *Theoretical and Applied Genetics* **66**: 241-247.

Van T Tuyl JM, Van Dien MP, Van Creij MGM, Van Kleinwee TCM, Franken J and Bino RJ (1991). Application of *in vitro* pollination, ovary culture, ovule culture and embryo rescue for overcoming incongruity barriers in interspecific *Lilium* crosses. *Plant Science* **74**: 115-26.

Van Tunen AJ, Van der Meer IM and Moe JNM (1994). Flavonoids and genetic modification of male fertility. *In: Genetic control of self-incompatibility and reproductive development in flowering plants*, Williams EG, Clarke AE and Knox RB, (Eds). pp.423-42. Kluwer Academic Publ.

Vandemoortele JL, Billard JP, Boucaud J and Gaspar T (1999). *In vitro Cell Development Biology Plant* **35**: 13-17.

Varma PM (1955). Ability of the white fly to carry more than one virus simultaneously. *Current Science* **24**: 317-18.

Varnell SH (1965). Cytogenetics of the vegetable crops. *IV Legumes Botanical Review* **31**: 247-330.

Vasudeva RS and Lal TB (1943). A mosaic diseases of bottle gourd. *Indian Journal of Agricultural Sciences* **13**: 182-91.

Veeragavathama D and Irulappan J (1990). Genetic analysis in okra [*Abelmoschus esculentus* (L.) Moench]. *South Indian Horticulture* **38**: 75-82.

Velich I (1985). Use of marker genes in heterosis breeding of muskmelon. *Zoldsegtermesztesi Kutato Intezet Bull.* **18**: 59.

Venkataramani KS (1952). A preliminary study on some crosses and hybrid vigour in *Hibiscus esculentus* (L.). *Journal of Madras University* **22**: 183-200.

Verma MM and Chahal GS (1993). Genetic basis of heterosis. *In: ProceedingsofSymposium on Heterosis Breeding in Crop Plants: Theory and Application*. Verma MM, Virk DS, Chahal GS and Dhillon BS (Eds). Crop Improvement Society of India, Ludhiana, pp. 1-18.

Verma S (1986). Combining ability studies in brinjal (*Solanum melongena* L.). *Progressive Horticulture* **18**: 111-13.

Verma TS, Joshi S and Thakur PC and More TA (1990). Hybrid vigour in cabbage (*Brassica oleracea* var. *capitata* L.). *Haryana Journal of Horticultural Sciences* **19**: 177-81.

Verma TS, Yadav HL, Joshi S, Thakur PC and Singh RV (1999). Heterosis studies in Capsicum. *Haryana Journal of Horticultural Sciences* **28**: 226-28.

Verma VK (1997). Hybrid technology and seed production in gourds, pumpkin and summer squash. *In: Compendium of Hybrid Seed Production Technology of Vegetable Crops,* held at IARI, New Delhi, April 7-26, pp. 143-148.

Verma VK, Sirohi PS and Choudhury B (1984). Note on the response to chemical seed treatment on sex expression and fruiting in bitter gourd. *Indian Journal of Horticulture* **41**: 113.

Vijay OP and Manohar MS (1986). Heterosis in okra (*Abelmoschus esculentus* (L.) Moench). *Indian Journal of Horticulture* **43**: 252-59.

Vijaya Kumari P, More TA and Seshadri VS (1993). Heterosis in tropical and temperate gynoecious hybrids in cucumber. *Vegetable Science* **20**: 152-157.

Vijayaraghavan C and Warrier VA (1946). Evaluation of high yielding hybrid bhindi [*Hibiscus esculentus* (L.)]. *Proceedings of 33rd Indian Science Congress* **33**: 165.

Virdelwala HA, Nandpuri KS and Singh Surjan (1981). Heterosis and combining ability in tomato. *Vegetable Science* **8**: 120-29.

Vishwanathan TV (1973). Hybrid vigour in brinjal. *Proceedings of Indian Academy of Science* **77**: 176-80.

Visser DL (1977). The effect of alternating temperatures on the self-incompatibility of some clones by Brussels Sprout *B. oleracea* L. var. *gemmifera* (DC) Schulz. *Euphytica* **26**: 273.

Viswanath P (1999). Genetic analysis of some physiological traits in cowpea [*Vigna unguiculata* (L)]. Walp. *Mysore Journal of Agricultural Sciences* **33**:234-36.

Vos P, Rene H, Bleeker M, Reijans, M Lee, T, Hornes M, Fritters A, Pot J, Peleman J, Kuiper, M and Zabean N (1995). AFLP: a new technique for DNA fingerprinting. *Nucleic Acid Research* **23**: 4407-14.

Walkey DGA, Webb MJW, Bolland CJ and Miller A (1987). Production of virus free garlic (*Allium sativum* L.) and shallot (*A. ascalonicum* L.) by meristem-tip culture. *Journal of Horticultural Science* **62**:211-20.

Wang Y, Sun C, Wang C and Chien N (1973). Induction of pollen plantlets of Triticale and *Capsicum annuum* from anther culture. *Science Sin* **16**: 147-57.

Wang YH, Thomas CE and Dean RA (1997). A genetic map of melon (*Cucumis melo* L.) based on amplified fragment length polymorphism (AFLP) markers. *Theoretical and Applied Genetics* **95**: 791-98.

Wankhade RV, Kale PS and Dod V (1995). Genetics of earliness, yield and fruit characters in okra. *MPKV Research Journal* **19**: 117-20.

Ward GM (1964). Observation of root growth in the green house tomato. *Canadian Journal of Plant Science* **44**: 492-94.

Watts LE (1965). Investigation into the breeding system of cauliflower. II. Adaptation of the system to breeding. *Euphytica* **14**: 67.

Weatherhead MA and Henshaw GG (1979). The production of homozygous diploid plants of *Solanum verrucosum* by tissue culture techniques. *Euphytica* **29**: 47.

Weide R, van Wordragen MF, Lankhorst RK, Verkerk R, Hanhart C, Liharska T, Pap E, Stam P, Zabel P and Koornneef M (1993). Integration of the classical and molecular linkage maps of tomato chromosome. *Genetics* **135**: 1175-86.

Weinand U and Saedler (1987). Plant transposable elements: Unique structure for gene tagging and gene cloning. Hohn Th and Schell J (eds). Plant DNA infections agents. Springer-Verlag, pp. 205-227.

Weiring D (1961). The breeding of new cauliflower varieties for summer culture, *Zaadhelangen* **15**: 306.

Welch JE and Grimball JrEL (1947). Male sterility in carrot. *Science*, **106**: 594.

Wellington R (1912). Influence of crossing in increasing the yield of tomato. *Bulletin, 346,* N.Y. Agriculture Station, pp. 67-76.

Wellington R (1922). Comparison of first generation tomato crosses and their parents. *Technical Bulletin.* Ministry of Agriculture Experiment Station, **6**: 25.

Whaley WG (1952). Physisology of gene action in hybrids. *In: Heterosis.* Iowa State College Press, pp. 98-128.

Whitaker TW (1933). *Bot Gaz* **94**: 780-90.

Whitaker TW (1974). Cucurbita. *In: Handbook of Genetics,* Plenum Press, pp. 135-44.

Whitaker TW and Bemis WP (1976). Cucurbits. *Evolution of Crop Plants,* Simmonds NW (Ed.) Longman Group Ltd., pp. 64-69.

Whitaker TW and Davis GN (1962). Cucurbits. Interscience, NewYork, pp 250.

Whitaker TW and Davis GN (1962). *Cucurbits: Botany, Cultivation and Utilization.* Leonard Hill (Books) Ltd, London.

Wiering D (1958). Artificial pollination in cabbage. *Euphytica* **7**: 223-27.

Wietsma WA, DeJong KY and Van Tuyl JM (1994). Overcoming pre-fertilization barriers in interspecific crosses of *Fritillaria imperialis* and *F. raddeana* plant cell incompatibility. *Newsletter* **26**: 89-92.

Williams JK, Kubelik AR, Livak KJ, Ratalski JK and Tingu SV (1990). DNA polymorphisms amplified by arbitrary primers are useful as genetic markers. *Nucleic Acid Research* **18**: 6531-35.

Williams PH, Walker JC and Pound GS (1968). Hybelle and Sanibel multiple disease resistant F_1 hybrid cabbage. *Phytopathology* **58**: 791.

Williamson VM, Ho JY, Wu FF, Miller N and Kaloshian I (1994). A PCR based marker tightly linked to the nematode resistance gene, Mi in tomato. *Theoretical and Applied Genetics* **87**: 757-763.

Witsenboer H, Kesseli RV, Fortin MG, Stanghellini M and Michelmore RW (1995). Sources and genetic structure of a cluster of genes for resistance to three pathogens in lettuce. *Theoretical and Applied Genetics* **91**: 178-88.

Worrall D, Hird DL, Hodge R, Wyatt P, Draper J and Scott R (1992). Pre-mature dissolution of the microsporocyte callose wall causes male sterility in transgenic tobacco. *Plant Cell* **4**: 759-71.

Wright S (1968). *Evaluation and Genetics of populations*, University of Chicago Press.

Xu B, Grun P, Kheyr-Pour A and Kao T-h (1990a). Identification of pistil specific proteins associated with three *S-alleles* in *Solanum chacoense. Sex Plant Reproduction* **3**: 54-60.

Xu B, MU J, Nevins DL, Grun P and Kao T-h (1990b). Cloning and sequencing of cDNA encoding two self-incompatibility associated proteins in *Solanum chacoense. Molecular and General Genetics* **224**: 341-46.

Yadav ED, Kale PN and Wavhal KN (1988). Genetic analysis of fruit size and shape in tomato. *Journal of Agricultural Research* **55**: 141-52.

Yadav ED, Wavhal KN and Kale PN (1991). Genetic analysis of pericarp thickness in tomato. *Journal of Maharashtra Agricultural University* **16**: 181-182.

Yadav RD, Kale PN and Wavhal, KN (1988). Genetic analysis and fruit dry matter in tomato. *Vegetable Science* **15**: 49-54.

Yaghoobi J, Kaloshian I, Wen Yand Williamson VM (1995). Mapping a new namatode resistance locus in *Lycopersicon peruvianum. Theoretical and Applied Genetics* **91**: 457-64.

Yahab MA (1989). Homeostatic analysis of components of genetic variance and inheritance of fruit colour, fruit shape and bitterness in bitter gourd (*Momordica charantia* L.). *Ph.D. Thesis,* Kerala Agricultural University.

Yamamoto T (1994). *In: 22nd Annual Meeting of The International Plant Propagators' Society*, Wellington, New Zealand, pp. 361-65.

Yanagisawa N and Hosono M (1951). The development of the Shinasahi watermelon. *Japan Journal of Breeding* **1**: 66-70.

Yasseen MY, Splittstoesser WE and Litz RE (1994). *Plant Cell Tissue and Organ Culture* **36**: 243-47.

Ye ZB, Li HX and Zhou GL (1994). Jornnal *of Huazhong Agricultural University* **13**: 291-95.

Yoder JI, Palys J, Alpert K and Lassner M (1988). *Ac* transposition in transgenic tomato plants. *Molecular and General Genetics* **213**: 291-96.

Yoneyama K and Anzai H (1991). Gene technological study on disease control by the inactivation of pathogenic toxins in transgenic plants. *Journal of Pest Science* **16**: 291.

Yordanov M (1983). Heterosis in the tomato. *In: Heterosis, reappraisal of theory and practice*, (Ed.) R. Galun. *Monographs on Theoretical and Applied Genetics*, **4**:189-19.

York TL, Munger HM and Wallace DH (1957). Experimental release of the cabbage inbred. Department of Plant Breeding. Mimeogr. Publ Cornell University, pp. 152-53.

Young ND and Tanksley SD (1989). RFLP analysis of the size of chromosomal segments retained around the Tm-2 locus of tomato during backcrossing breeding. *Theoretical and Applied Genetics* **77**: 353-59.

Yu SB, Li JX, Xu CG, Tan YF and Gao YJ (1997). Importance of epistasis as the genetic basis of heterosis in an elite rice hybrid. *Proceedings of National Academy of Science,* USA, **94**: 9226-31.

Yun JS, Hwang SG, Song IG, Lee CH, Yun T, Jeong IM, Paek KY (1998). *RDA-Journal of Horticultural Science* **40**: 14-19.

Yurnia OV (1977). New cucumber hybrid for mechanical harvesting. *Kartofelt Ovashchi* **9**: 28-9.

Zamir D, Ekstein- Michelson I, Zakay Y, Navot N, Zeidan M, Sarfatti M, Eshed Y, Harel E, Pleban T, Oss H Van, Kedar N, Rabinowitch HD and Czoshek H (1994). Mapping and introgression of a tomato yellow leaf curl virus tolerance gene, TY-1. *Theoretical and Applied Genetics* **88**: 141-46.

Zaveri PP, Patel RK, Yadavendra JP and Shah RM (1983). Heterosis and Combining ability in cowpea. *Indian Journal of Agricultural Sciences* **53**: 703.

Zeng ZB (1993). Theoretical basis for separation of multiple linked gene effects in mapping quantitative trait loci. *Proceedings of National Academy of Science*, USA, pp. 10972-976.

Zhang Y and Stommel JR (1998). Identfication of carotenoid loci in tomato with RAPD and AFLP markers. *Plant and Animal Genome Conference*, 18-22 January, 1898, San Diego, CA.

Zhao JP, Bi KH, Jiang XM and Bai XF (1998). Plant Physiology. *Community* **29**: 435-437.

Zheng HR, Shen MJ, Zhong WJ, Zhang ZQ and Zhou Y (1998). *Acta Agricultural Shanghai* **14**: 33-38.

Zhou ZX, Zhang ZJ, Jiu YJ, Jiang QY and Mi JJ (1994). *Acta Agrcultural Boreali Sinica* **9**: 59-63.

Index

A

A. altaicum 178
A. manihot 49, 177, 295, 297, 301
A.tetraphyllus 177
Abelmoschus esculentus 48
Aberrations 73
Abiotic stresses 1, 194, 243, 247, 298, 357, 358, 364
Abscisic acid 75
Abscission 103, 104, 262, 313, 314, 317, 322, 399
Accessions 340, 341, 348, 356, 363
Acetone 108
Ach-15 286
Achievements of tissue 192, 193
Aconium delphonifolium 119
Additional returns 250
Additive effects 26
Additive genes 9
Adoptive research trials 127
Adoptive trials 125, 126, 127, 138
Advantages 1, 9, 21, 43, 51, 117, 143, 147, 149, 153, 150, 155, 184, 198, 200, 214, 218, 222, 231, 245, 247, 391
Advantages of triple-cross 310
Agrobacterium 160, 161, 194, 195
Agrobacterium tumefaciens 160, 194
Agroclimatic zones 126
Agronomic practices 127
Alata 89, 90
Alcohol dehydogenase 20
Alexander fleming 157
Alfalfa 112, 113, 114, 218, 219, 221
All India Coordinated 10, 327
Alleles 3, 6, 9, 15, 16, 17, 18, 19, 20, 25, 26, 27, 71, 86, 88, 89, 90, 91, 92, 93, 94, 97, 98, 101, 305, 309, 310, 338
Allium cepa 178
Allium Fistulosum 178
- Species 178
Allogamous plants 17
Allogamy 69
Allopolyploidy 183
Alstroemeria 104, 105
Alternaria blight 314, 405
- dauci 60
Amaranthus hybridus 207
Amino acid 73, 90, 91, 147
Ammonium chloride 75
Amphibolis 68
Amplified Fragment Length Polymorphism (AFLP) Mark 153
Analogous 21, 27, 103, 220
Androecious 66, 353, 354
Andromonoecious 54, 55, 66
Andromonoecy 66
Anemos 69
Angular leaf spot 53, 327
Anigozanthos 105
Animal biotechnology 158
Antagonists 74, 75
Anther 66, 67, 69, 71, 72, 73, 74, 76, 78, 79, 95,163,164,190, 191, 192, 193, 206, 262, 275, 284, 287, 290, 298, 299, 359, 360
Anther culture 74, 305
Anthesis 77, 91
Anthracnose 12, 53, 138, 260, 288, 289, 293, 297, 327, 341, 382, 393
Anthracnose resistant 260
Antioxidant 1
Antipodal 68
Antirrhinum majus 114
Antisense 90, 205, 206
Aphids 173, 303, 304, 335, 336
Apis mellifera 112, 116
Apogamy 65
Apomixis 5, 65, 66

Application of 2n-Gametes 105
Applications 75, 112, 114, 151, 152, 241,242, 382
Approaches 157, 163, 170, 188, 194, 201, 207, 213, 214
Arabidopsis thaliana 190, 209
Arachis 105, 184
Arbh-1 277
Area under hybrids 247
Arginine 73
Arka Abhijit 268
- Shreshtha 268
- Vardan 268
- Vishal 269
Armenica vulgaris 108
Arth 269, 271
Artificial Pollination Alone 32, 329, 337, 342
Artificial Seed Technology 164, 218
Artistic creations 226
Asexual 65
- reproduction 65
Asiatic 105, 177, 358
Asparagines 73, 91
Asparagus 34, 192, 221, 222, 251, 255
Aspartic acid 73
Aspergillus oryzae 90
Atrazine 82, 193, 207
Atta texana 119
Attractant sprays 112, 115
Attractiveness of competing crops 115
Autogamous 17
Auxin antagonists 75
Auxins 74, 75
Axillary bud 167

B

B.carinata 97
B.mixtus 119
Bacillus thuringensis 161
Backcross method of breeding 121
Backcrossing 56, 85, 122, 212, 285, 286, 332, 333, 367
Bacterial cancker 260
Bacterial leaf Spot 283, 382
- resistance 203
- wilt 12, 137, 173, 193, 376, 379, 382
- wilt Resistant 260, 273, 283, 327
Bacterium 161, 196, 206
Balsam Pear 317
Banana 42
Barriers 99, 100, 101, 102, 103, 104, 105, 179, 181, 182, 184, 201
- hybridization 98
Basic requirements 31
Basmati grain 240
Beet root 129
Bell pepper 41, 42, 43, 129, 133, 134, 135, 175, 209, 290
Benefit analysis 255, 257
Benefits from IPR 226
Better parent 4, 7, 8, 9, 35, 36, 39, 41, 42, 43, 45, 46, 49, 51, 52, 53, 57, 58, 59, 61, 62, 63
Better stability 5
Binary 195
Biochemical changes 73, 107
Biodiversity 238, 240, 246
Biolistics 198
Biometrical basis of heterosis 26
Biosafety regulations for transgenic hybrids 241
Biosystematic 182
Biotechnology 4, 65, 70, 77, 106, 157, 158, 162, 163, 182, 201, 212, 218, 234, 240, 241, 243
Biotic stresses 12, 125,163, 335
Bisexual 31, 252, 254, 262, 358
Bitter gourd 4, 49, 50, 66, 130, 131, 132, 135, 317, 319, 320, 391, 392, 393, 394, 395, 396, 397, 400, 401, 404
- mosaic 393
Black leg 312, 388
- rot 12, 303, 304, 312, 314, 388, 405
Bombus 117, 120
- *flavifrons* 120
Borer 12, 274, 277, 278, 297, 348, 377, 380, 383, 385, 410

Botany and floral biology 261, 274, 284, 290, 298, 304, 314, 318, 323, 328, 337, 341, 349, 354, 364
Bottle gourd 3, 32, 51, 66, 130, 131, 132, 135, 136,251, 253, 254, 256, 257, 395, 396, 397
Branching type 330
Brassica 5, 65, 66, 72, 82, 91, 97, 98, 104, 105, 113, 115, 120, 153, 179, 183, 190, 192, 193, 194, 195, 199, 207, 209, 221, 303, 311
- *campestris* 113, 190, 191
- *cybrids* 82
- *oleracea* 66, 72, 91, 98, 179, 183, 192, 193
Brassicaceae 91
Breaking hybridization 182
Breeder 16, 125, 126, 132, 138, 140, 146, 183, 214, 229, 232, 233, 236, 243, 361
Brinjal 1, 2, 3, 4, 5, 9, 10, 11, 12, 66, 129, 130, 131, 132, 133, 135,136,165,166, 174, 205, 212, 220,247, 249, 251,252, 253, 254, 256, 257, 276
British biotechnologist 158
Broadcast 373, 404
Broccoli 32, 33, 34, 130, 131, 132, 153, 179, 251, 256
Bt gene 161, 212
Buckeye rot 375
Buckwheat 86
Bud-pollination 304, 306, 307, 310, 311
Buffering capacity 2
Bulb crop 58, 168, 170
Butanol 157, 158

C

C. anguria 177, 182, 327, 335
C. annuum 175, 177, 281, 289
C. callosus 327, 336
C. cardenasii 175, 282
C. chacoense 177
C. eximium 175
C. metuliferus 327
C. microcarpum 177
C. pendulum 177
Cabbage 1, 2, 3, 4, 9, 10, 12, 13, 31, 32, 33, 34, 35, 43, 44,92, 93, 94, 95, 98, 118, 129, 130, 131, 132, 133, 134, 135, 136, 137, 138, 148, 149, 153, 156, 163, 164, 168, 172, 179, 212, 247, 249, 251, 256, 258, 303, 304, 305, 307, 310, 311, 314, 316, 386, 387, 388, 389, 390, 391
Cabbage butterfly 304, 389
- yellow 303, 388
Caenorthabditis elegans 162
Calcium phosphate precipitation 201
Callose 72, 74, 92, 96, 100, 101, 201, 298
Camelia japonica 108
Cancellation effects 9
Capsicum 3, 4, 9, 12, 13, 32, 34, 43, 75, 130, 131, 132, 135, 136, 137, 138, 166, 167, 175, 176, 192, 193, 209, 249, 251, 252, 253, 254, 258, 281, 283, 289, 291
- *annum* 75, 193, 209, 291
Carnation 160, 172
Carpillate 66
Carrot 1, 3, 5, 10, 13, 31, 32, 33, 34, 59, 60, 129, 133, 201, 209, 220, 222, 249, 251, 256, 257, 357, 358, 359, 360, 361, 362
- yellows 405
Cataloguing 147, 236
Cauliflower 1, 10, 12, 13, 129, 130, 131, 132, 133, 135, 137, 138, 168, 179, 196, 212, 249, 251, 258, 303, 313, 314, 315, 389, 390, 391
Causal 375, 377, 379, 382, 384, 387, 393, 405, 409
Celery 219, 220
Cercospora carotae 60, 357
- leaf spot 297, 405
Certification 31, 141,143, 232, 252, 271, 279, 280, 287, 288, 293, 300, 301, 312, 316, 320, 321, 326, 334, 339, 346,347, 352, 356, 361, 362, 371
Character 5, 7, 8, 16, 22, 29, 35, 55, 95, 123, 128, 144, 267, 307, 310, 322, 343
Characterization of hybrids and parents 239

Chemical hybridizing agents 70, 74, 78, 79, 292
- sex expression 325, 355
- treatments 103
Chemically induced male sterility 71
Chilli 12, 31, 32, 41, 43, 129, 130, 131, 132, 133, 134, 135, 136, 137, 138, 166, 175, 191, 212, 249, 251, 252, 253, 257, 258, 281, 282, 283, 284, 285, 286, 288, 289, 293, 381, 382, 383
Chinese cabbage 32, 34, 148, 153, 156, 251
Chitinase genes 203
Chloroplast genes 83
Chromosome doubling 105, 182
- elimination 180
Citrus ovules 159
Cladosporium cucumerinum 53
Classification of barriers 99
Clostridium acetobutylicum 157
Coat protein 161, 202, 212
Coefficients 27, 29, 176, 213
Cointegrate 195
Cold tolerant 259
Cole crops 32, 33, 34, 43, 44, 65, 98, 168, 179, 257, 303, 314
Colletotrichum 53, 293, 382
Colony replacement 114
Combined gene product 20
Combining ability of inbred line 33
Commercially managed 112
- managed pollinators 112
Committee 125, 141, 132, 138, 139, 141, 235, 236, 241, 242
Competent authority 228, 241
Competitive ability 7, 111
Complementary epistasis 27, 28
Complementary genes 16, 28, 121, 123
Complementation 21, 24, 25
Complete dominance 4, 93
Concept of Heterosis 3, 4, 20
Consortium approach 239
Control measures 375, 377, 379, 380, 382, 383, 384, 385, 387, 388, 391, 392, 393, 394, 396, 397, 399, 400, 401, 402, 403, 405, 406, 409, 410
Controller of patents and designs 225
Convergent improvement 121, 122
Cooperative society 11, 12
Copper chelators 77
Corn 2, 3, 32, 33, 34, 35, 81, 83, 108, 207, 255, 256
Cost 248, 249, 250, 251, 252, 253, 254, 255, 256, 257, 258, 264, 265, 266, 267, 268, 277, 285, 286, 305, 307, 310, 315, 324, 329, 345
Cost of emasculation 250, 251, 253
Cost of emasculation and pollination 250, 251, 253
- hybrid seed production 250, 253, 265
- land rent 250
- parental lines seed 250
- post-harvest operations 251, 252
- seed production 250, 305
- seed sowing 250
Cotyledon 165, 166, 167, 168
Cowpea 34, 62, 63, 204, 376
Crepis foetida 91
Critical difference 9
Crop 1, 3, 5, 6, 8, 10, 11, 12, 13, 16, 23, 24, 25, 31, 33, 41, 43, 49, 58, 65, 66, 67, 68, 71, 73, 79, 82, 83, 86, 99, 100, 106, 111, 112,113, 114, 115, 116, 123, 130, 134, 135, 138,230, 234, 244, 245, 246, 247, 248, 250, 251, 252, 253, 255, 256, 257, 259, 261, 268, 269, 270, 272, 275, 276, 280, 282, 283, 287, 288, 290, 292, 293, 295, 298, 299, 301,304, 312, 313, 317, 319, 321, 322, 324, 326, 327, 328, 333, 334, 337, 340, 346, 347, 353, 356, 357, 362, 363, 364, 365, 370, 372
- odours 113, 116
Cropping intensity 2
Crossability barriers 99, 103, 182
Crossability polygon 178
Cross-breeding 2, 3
Cross-fertilization 2

Cross-pollination 2, 69, 82, 96, 115, 245, 267, 275, 286, 291, 299, 324, 325, 332, 351, 355, 357, 359, 360, 365
Cryopreservation 106, 109, 110, 193, 219
Cucumber 2, 3, 4, 5, 9, 10, 12, 13, 32, 33, 34, 35, 51, 52, 53, 61, 66, 130, 131, 132, 133, 135, 136, 167, 177, 182, 202, 249, 251,253, 254, 256, 257, 258, 332, 396, 397, 398
- mosaic 327, 336, 340
- mosaic virus 61, 202, 336, 340, 397
Cucumis anguria 327
- *hardwickii* 327, 334
- *sativus* 51, 167
Cucurbit phyllody 393
Cucurbita maxima 57
Cucurbita pepo 117, 177, 193, 351
Cucurbitaceous crops 32, 49, 317, 401
- vegetables 167
Cucurbits 11, 31, 32, 33, 34, 66, 162, 167, 168, 177, 251, 253, 254, 256, 324, 401
Culture 157, 159, 160, 163, 164, 165, 170, 173, 181, 182, 183, 184, 187, 188, 190, 191, 192, 193, 194, 195, 196, 197, 207, 208, 219, 221, 223, 305
- hybrid Cells 188
Curl virus 12, 138, 173, 260, 283, 297, 300, 376, 377, 382
Cybridization 83
Cyclamen 104
Cytokinin 159, 160, 166, 195
Cytological Changes 72, 191
Cytoplasmic Basis of Heterosis 24
- genetic male sterility 71, 83
- incompatability 180
- male sterile line 44, 85
- male sterility 70, 71, 79
- orgenelles 24

D

D. dentatus 60
Damping 296, 373, 374, 375, 379, 382, 386, 407, 408
Damping off 296, 373, 375, 379, 382, 407, 408
Date of expiry 252
Date of packing 252
Datura 2,102, 159, 160, 164, 179, 188, 190, 191
Daucus carota 159, 190, 357, 358
- *siliculus* 60
Dehiscence 69, 262, 275, 284, 290, 291, 298, 319, 324, 337, 342, 359, 370
Deleterious effects 15
Demand of a particular crop 250
Densely pubescent 319
Dentatus 60
Desiccated 68, 107, 108, 221
Desirable genotypes 220
Development of self-incompatible parental lines 305
Diamond back 12, 304, 314, 389
- back moth 304, 314
Dianthus barbatus 172
- *caryophyllus* 172
Dioecious 34, 222, 251, 255, 335
Diploid parthenogenesis 65
Diploid secondary 68
Direct DNA uptake 197
Disclosure 226, 227, 236
Distant crosses 172, 181, 183
Distant hybridization 172, 182, 183, 184
Distinctiveness 230
Distribution 15, 16, 27, 31, 69, 89, 111, 138, 214, 226, 252, 295, 335, 345, 348, 357, 397, 403
DNA based markers 146, 147
Documentation 236
Domesticated 6, 22, 322, 327, 347
Dominance 15, 16, 17, 18, 21, 26, 27
Dominance effects 26
Dominance of linked genes hypothesis 16
Dominant marker gene 305, 311
Double cross 3, 18, 34, 96, 121, 122, 123, 124, 307, 309, 310
- fertilization 70, 101
- mutant 22

Double-top-crosses 31
Downey mildew 12, 137, 314, 387, 393, 409
Dramatic 18
Drought tolerance 209, 259, 274
Duplicate 27, 28

E

Earliness 35, 37, 39, 41, 42, 43, 44, 45, 47, 50, 52, 53, 54, 55, 56, 57, 59, 64, 178, 287, 314, 319, 335, 341, 358
Early blight 12, 173, 212, 260, 271, 375
- determinate 327
- flowering 46, 47, 49, 50, 53, 277, 296
- group 45, 390
Early-maturing 2, 60, 260, 282, 297, 320, 333, 336, 349, 363, 387
Ecological balance 226
Economic status 250
Economically 1, 6, 82, 83, 97, 111, 112, 114, 218, 247, 250, 251, 263, 324, 355
Economics 247, 252, 253, 268
Effect in two different tissues 20
Electrical fusion 187
Electrophoresis 20, 73, 147, 148, 149, 154, 155, 160, 176, 218
Electrophoresis of proteins 147, 148
Electroporation method 199
Emasculation 10, 12, 65, 194, 207, 250, 251, 252, 253, 254, 255, 256, 257, 263, 359, 365
Embryo culture 159, 163, 164, 173, 183
Embryo Sac 65, 68, 70, 101, 179
Endosperm Abortion 180, 181
Engineering for abiotic stress 209
- disease resistance 202
- male sterility 206
- post-harvest traits 205
- salt tolerance 209
Entomon 69
Enzymatic balance 24
Enzymes 19, 20, 23, 24, 25, 73, 74, 79, 107, 110, 120, 150, 152, 157, 160, 164, 185, 208, 223
Epistases 27
Epistasis hypothesis 18
- interactions 28
Erwinia carotovora 312
Erysiphe herachi 60
Erythronium 120
Ethephon 55, 320, 351
Ethrel 56, 75, 79, 248, 256, 257, 325, 355
Ethylene 77, 206
Euheterosis 5, 6, 7
European economic community 228
European types 315, 358, 403
Evaluation 44, 76, 125, 126, 127, 212, 214, 216, 227, 296, 313, 327, 348, 354, 363
Exploiting self-incompatibility 97
Export 1, 11, 12, 13, 247, 258, 283, 298, 364
Expression of GMS 71
Expression of Heterosis 3, 5, 23, 24, 56

F

F. raddeana 103
F_1 hybrids 3, 4, 9, 11, 15, 17, 21, 24, 26, 32, 35, 39, 44, 46, 49, 52, 54, 57, 58, 59, 60, 61, 63, 101, 247, 304, 305, 315, 316, 360
Factors 3, 5, 21, 28, 66, 71, 92, 100, 101, 115, 117, 120, 123, 148, 154, 181, 214, 227, 250, 262, 263, 264, 271, 277, 280, 293, 301, 312, 361, 362, 371, 377
- affecting production 250
Failure of zygote development 179, 180
- formation 179
FAO resolution 232
Farmers right 143
Farmyard manure 373, 374, 378, 381, 384, 387, 390, 391, 395, 397, 399, 400, 401, 404, 407, 408
Fecundity 32
Fermentation 157, 264, 345
Fertility 31, 32, 17, 72, 74, 75, 76, 78, 79, 82, 83, 84, 85, 105, 108, 115, 121, 194, 206, 207, 291, 292, 299, 329, 337, 344, 354, 359, 361, 366, 368, 374, 389
- restoration 82

Fertilization 65, 69, 70, 86, 87, 88, 96, 99, 100, 101, 103, 104, 111, 115, 145, 159, 170, 179, 205, 264, 342
Field inspection 271, 280, 287, 293, 300, 312, 320, 326, 334, 339, 346, 352, 356, 371
Filament 66, 71, 77, 79, 359
First hybrid 3, 5, 261, 352
First-generation Hybrids 53
Fixed heterosis 7
Flavr savr 4, 206
Flesh weight 56
Floral biology 65, 261, 274, 284, 290, 298, 304, 314, 318, 323, 328, 337, 341, 349, 354, 358, 364
Flowering 31, 35, 39, 45, 46, 47, 48, 49, 50, 53, 62, 63, 65, 66, 68, 75, 76, 99, 105, 106, 112, 113, 114, 116, 118, 139, 181, 184, 251, 255, 266, 271, 274, 277, 280, 282, 284, 285, 287, 288, 289, 292, 293, 296, 298, 299, 300, 301, 305, 306, 310, 312, 316, 319, 320, 321, 325, 326, 330, 334, 339, 340, 342, 344, 346, 347, 352, 355, 356, 359, 361, 362, 364, 365, 369, 370, 371, 372, 383, 391, 392, 395, 397, 399, 400, 403
Fluchloralin 378, 381, 387, 396, 399, 408
Fluorescent activated cell 190
Focusing 89, 149
Food security 226, 244
Foreign seeds 258
Free insect 32, 257, 304, 305, 316
Free insect-pollination 32, 304, 305
Free radicals 1
French bean 34, 63, 64
Fritillaria imoerialis 103
Fruit and seed-setting 111
- fly 392, 394
- fly resistance 318
- girth 39, 41, 317
- length 39, 40, 41, 42, 43, 46, 49, 50, 51, 317, 320
- rot 174, 260, 274, 283, 289, 291, 297, 379, 382, 396
Fruit and setting 43, 55, 75, 96, 112, 123, 263, 264, 266, 267, 275, 284, 285, 291,320, 324, 325, 330, 339, 343, 345, 350, 351, 355, 373, 374, 381, 392, 396, 399
- yield/plant 41, 43, 317
Fruit-borer 274, 278, 297, 377, 380, 383, 385
Full-diallel 305
Fumigation 373, 398
Functional male sterility 265, 267
Fungal resistance 203
- toxin 170
Fusarium oxysporum 304, 379, 388
- *wilt* 137, 260, 269, 273, 297, 311, 323, 335, 336, 341, 376, 379, 382
Fusion of protoplasts 185, 187
Future strategies 11

G

Gamete selection 121, 122, 123
Gametogenesis 67, 307
Gametophytic 72, 86, 89, 90, 92, 94, 95, 96, 97
- system 86, 95, 96, 97
Garlic 65, 169, 170, 363
Geitonogamy 69, 70
Gels ssed for encapsulation 221
Gemini virus 196
Gene 3, 5, 6, 8, 9, 10, 16, 18, 19, 20, 21, 22, 25, 160, 161, 163, 170, 172, 178, 180, 181, 182, 184, 194, 195, 196, 198, 199, 200, 201, 203, 204, 205, 206, 207, 208, 209, 212, 213, 214, 215, 216, 218, 222, 223, 232, 238, 242, 243, 244, 245, 246,266, 267, 285, 286, 291, 322, 338, 343, 351, 353, 366, 367, 370
- tagging 212
Gene-cytoplasmic male sterility 58
General Areas 11
- features 5
- requirements 271, 288, 293, 301, 312, 321, 326, 334, 346, 352, 356, 339, 362, 371
Genes systems 121

Genesis 78
Genetic Basis of Heterosis 15, 19
- constitution 24, 65, 366, 367, 369
- diversity 5, 33, 61, 143, 151, 317
- engineering 71, 83, 158, 163, 194, 202, 205, 241, 242
- engineering approval committee 242
- male sterility 71, 83, 84, 264, 267, 285, 290, 291, 305, 310
- manipulation 83, 212, 241
- polymorphism 24
- purity 249, 272, 280, 281
- purity testing 143, 148, 153, 155
- resources 143, 232, 234, 238, 241, 259, 272, 281, 282, 289, 303, 313, 317, 322, 327, 335, 340, 347, 353, 357, 363
- transformation 162, 163, 168, 197
Genetically dissimilar 4
Genomic DNA 150, 152, 153, 155
Genomics 157
Genotype of plant 86, 93
Genotype X environmental interactions 29
Genotypic disharmony 179, 180
Geographical indications and IPR 240
Gerbera 160
Germination 46, 47, 53, 74, 75, 88, 91, 92, 95, 96, 98, 99, 100, 101, 102, 104, 110, 111, 120, 145, 159, 181, 184, 244, 246, 252, 262, 268, 272, 280, 288, 293, 301, 312, 321, 326, 334, 340, 345, 347, 353, 356, 362, 372, 373, 375, 386, 391, 393, 395, 396, 400, 403, 404, 407
Germplasm exchange 106, 220
Germplasms 232, 259, 272, 282, 296, 317, 322, 323, 327, 348, 363
Gibberellic acid 66, 248, 331, 332, 333
Gibberellins 75, 103
Gingko biloba 108
Gladiolus 105
Glandular hairs 262
Glucanase thionin 203
Glycerol 158
Glycine max 184, 192
GM parental lines 238
Gold dust 68
Gossypium 81, 115
Grafting 157, 181
Granular flesh texture 56
Grape 230
Growth chamber test 146
Growth regulators 19, 66, 76, 103, 166, 167, 205, 324, 355
Gynoecious 10, 13, 52, 53, 55, 56, 66, 136, 256, 332
- *line* 55, 327, 331, 332, 333
Gynomonoecious 331, 353, 354

H

H. pungens 46
H. tetraphyllus 46
Half-diallel 305
Hand emasculation 252, 253, 254, 255, 256, 262, 263, 276, 284, 285, 291, 299, 365
Hand emasculation and hand pollination 41, 252, 315
Hand pollination 10, 41, 164, 251, 252, 253, 254, 255, 256, 276, 285, 291, 299, 304, 305, 315, 325, 331, 355, 365
Handling 110, 118, 152, 220, 242, 248, 252, 254, 359, 405
Haploid 65, 66, 67, 68, 88, 90, 106, 159, 160, 164, 180, 182, 190, 191, 192, 193
Hard ripe 377
Harvesting, curing and seed extraction 330
Heat tolerant 259
Heat-sensitive inhibitors 102
Helianthus 119
Helminthosporium maydis 160
Hermaphrodite 34, 55, 66, 328, 329, 331, 332, 333, 337, 338, 341, 342, 353, 354, 355
Hermaphrodite lines 55
Heterobeltiosis 4, 8, 26, 39, 49, 57, 61
Heterogeneity 24, 33
Heterokaryon 22
Heteromorphic system 86
Heteromorphy 95

Heterosis 1, 2, 3, 4, 5, 6, 7, 8, 9, 10, 13, 15, 16, 17, 18, 19, 20, 21, 22, 23, 24, 25, 65, 66, 70, 166, 264, 266, 271, 292, 360
Heterozygote 16, 18, 19, 20, 22, 24, 27
Hibiscus manihot 46
High ascorbic acid 260
High heterosis 35, 36, 39, 41, 42, 43, 44, 45, 46, 47, 50, 52, 53, 54, 56, 57, 60, 61, 62, 64
High Lands 361
- TSS 13, 36, 56, 57, 130, 260, 336, 363
- volatile 260
- yield 33, 34, 37, 40, 41, 47, 50, 51, 56, 59, 60, 259, 273, 287, 289, 303, 314, 317, 327, 336, 341, 348, 357, 363, 374, 378, 396, 398, 408
High-density planting 26
High-yielding 1, 43, 44, 53, 55, 56, 57, 60, 61, 122, 137, 232, 250, 261, 289, 282, 297, 303, 313, 313, 322, 325, 333, 336, 346, 349, 352, 363
High-yielding hybrids 364
Higher magnitude 17
Himalayan region 361
Hints for cultivation 373
History 2, 111, 112, 157, 172, 225, 228, 225, 243, 282
Hoeing 374, 381, 390, 404
Homozygotes 6, 16, 17, 19, 20, 22, 27, 91
Homozygous 5, 8, 15, 17, 20, 29, 32, 33, 66, 70, 92, 163, 285, 286, 307, 310, 331, 332, 355, 361
- *autodiploid* 66
- *genotypes* 29
Hormone antagonists 74
Hover fly 120
Human genetic map 158
Human genome project 158, 161
Hybrid seed 163, 164, 194, 206, 240, 241, 242, 244, 247, 248, 249, 250
- seed production 10, 12, 13, 31, 32, 33, 34, 54, 55, 70, 82, 96, 97, 135, 247, 248, 249, 250, 251, 252, 253, 254, 255, 256, 257, 258, 299, 308, 332, 365, 366
Hybrid seed production and IPR 240
- substance 20, 22, 23
- technology 135, 225, 238, 239, 240, 242, 247, 248, 250
- Testing Centres in India 130
- varieties 1, 9, 11, 12, 96, 121, 122, 125, 133, 244, 261, 268, 361
- vegetables 11
- vigour 2, 3, 4, 19, 24, 25, 33, 34, 35, 41, 45, 46, 49, 52, 58, 61, 62, 164, 247, 261
Hybridization 2, 3, 12, 43, 46, 65, 75, 77, 78, 99, 102, 105, 150, 153, 155, 157, 160, 161, 172, 173, 177, 178, 182, 183, 184, 185, 188, 193, 244, 263, 264, 276, 285, 291, 315, 319, 360
Hybrids 12, 65, 81, 96, 100, 102, 104, 105, 106, 107, 108, 121, 124, 125, 126, 127, 129, 130, 132, 133, 135, 136, 137, 138, 139, 141, 143, 144, 146, 148, 155, 178, 186, 189, 190, 229, 231, 238, 239, 240, 241, 242, 243, 244, 261, 266, 267, 268, 271, 274, 277, 279, 280, 281, 283, 284, 285, 286, 288, 289, 291, 292, 297, 298, 299, 301, 303, 304, 306, 307, 309, 310, 314, 315, 316, 364, 366, 370
Hydrophilous flowers 69
Hypothesis 15, 16, 17, 18, 25, 26, 89, 213

I

Identification 3, 12, 54, 124, 125, 132, 133, 134, 135, 138, 143, 146, 147, 148, 151, 154, 155, 215, 244, 252, 255, 256, 264, 265, 266, 285, 305, 311
Identification of hybrids 132
Identification of superior F_1 combinations 305
Imidazolinones 207
Immature 130, 169, 197, 198, 318, 377, 380, 403, 408, 409
Impatiens 108
Import 242, 247, 258
In vitro pollen germination 110, 111
Inbreeding 2, 3, 8, 9, 15, 16, 17, 18, 19, 24, 25, 32, 33, 39, 44, 61, 63, 70, 96, 97, 305, 337, 361, 369

Inbreeding depression 8, 9, 16, 17, 19, 25, 39, 63, 70, 361
Incompatibility 3, 10, 70, 86, 87, 88, 89, 90, 91, 92, 93, 94, 95, 96, 97, 98, 100, 102, 106, 112, 181, 182, 190, 248, 251, 252, 257, 305, 306, 307, 309, 313, 316
Increasing pollinator visitation rates 112
Indexing 236
India seed sector 258
Indian Patent Act 225, 228
Indian Patent and Design Act 225
Indian Trade and Merchandise Marks Act 225
Industrial application and usefulness 227
Inert matter 272, 280, 288, 293, 301, 312, 321, 326, 334, 340, 347, 353, 362, 372
Ingenuity 225
Inheritance 18, 80, 265, 360
Inhibition in stigma and style 101
Inhibition of the ovary 101
Inhibitor gene 59
Inhibitors of pollen fertility 78
Initial Evaluation Trials 126, 127
Innovativeness 225
Insect body hairs 117, 118
Insect-pollination 32, 304, 305, 342, 343
Insect-proof cage 369
Insect-proof wire nets 307
Institutional Biosafety Committees 241
Intellectual inputs 225
Intellectual property 225, 226, 227, 228, 238, 243, 244
Intellectual property rights 225
Interacting genes 28
- pairs 27
Interaction 8, 15, 18, 21, 22, 24, 25, 27, 28, 69, 70, 89, 95, 96, 100, 116, 117, 157, 180, 310, 360, 366
Interaction between foraging insects 116, 117
Interagenic complementation 21
Intergeneric crosses 65, 81, 99, 100, 183
Intergeneric hybridization 99, 105, 160, 172, 173, 177, 184
Interspecific hybrids 45, 46, 48, 49, 105, 106, 172, 178, 179, 180, 181, 182, 184
Intraspecific crosses 81
Irrigation 11, 114, 115, 138, 145, 248, 249, 250, 258, 374, 378, 381, 384, 387, 391, 392, 396, 397, 399, 400, 401, 403, 404, 406, 407, 408
Isoenzymes 147
Isolation of protoplast 185
Isoleucine 73, 207
Isozyme genes 44
Isozyme marker 149, 213

J

Japanese biotechnologist 158
Jassids 380
Juglans regia 108

K

Klebsiella pneumoniae 208
Knol-khol 34, 129
Koh-bleach test 144
Korea 258, 285, 286

L

L. auratum 105, 106
L. candidum 105
L. cheesmani 173
L. cheesmani 259, 260
L. chilense 88, 173, 174, 181
L. esculentum 100, 173, 181, 182, 193, 194, 260
L. henryi 105, 106
L. hirsutum 88, 173, 181, 260, 261
L. longiflorum 105
L. pennellii 173
L. peruvianum 88, 89, 98, 173, 181, 182, 193, 209, 260
L. pimpinellifolium 38, 173, 181, 259, 260
Labelling 252
Laboratory examinations 144
Lagenarium 53, 341
Large fruit 129, 260
Lasthenia 118

Late blight 260, 375
- blight resistant 260
- group 45, 390
- maturing 260, 386, 387, 389
Leaf curl virus 138, 173, 260, 283, 297, 300, 382
- spot 53, 173, 174, 271, 283, 297, 327, 382, 385, 405
Leafhopper 385
Legume crops 61
Lepidoptera 118
Lethal genes 101, 179, 180, 181
Lettuce 34, 73, 148, 207, 209, 220
Liable heterosis 7
Liberalized economic 11
Ligation of oligonucleotide adapters 153
Lilium 96, 104, 105, 106, 108
Linkage between interacting genes 28, 151
Linum 119, 159, 164
Liposome mediated DNA 201
Liposomes 201
Loci 16, 18, 19, 25, 26, 27, 71, 83, 88, 89, 90, 95, 100, 147, 148, 150, 151, 154, 155, 213, 214, 223
Long duration fruiting 296
- fruited 252, 296, 378
- styled flower 276
- term storage of Pollen 109
Low-yielding lines 26
Lupinus 120, 159
Luxuriance 5, 6
Lycopersicon 72, 89, 96, 100, 104, 105, 165, 173, 181, 182, 183, 190, 192, 193, 208, 209, 260, 261, 265

M

M. cochinchinensis 317, 321
M. cymbalaria 317
M. denundata 317
M. dioica 317, 321
M. macrophylla 317
M. tuberosa 317
Major constraints 10
Male sterile anthers 73
Male line 10, 13, 41, 44, 54, 58, 59, 60, 79, 83, 85, 162, 163, 164, 262, 264, 265, 266, 267, 271, 277, 284, 285, 286, 291, 299, 304, 315, 324, 325, 360, 366, 367
- mutants 54, 71, 73, 265, 360
Male sterility 70, 71, 72, 73, 74, 75, 76, 77, 78, 79, 80, 81, 82, 83, 84, 121, 251, 252, 254, 257, 264, 265, 266, 267, 284, 285, 290, 291, 292, 304, 305, 310, 311, 360, 361, 367
Malus pumila 108
Malvaceous crops 46, 295
Management of IPR 227
Management of pollination insects 111
Manihot 49, 177
Marker systems 147, 148, 149
Marketing 11, 143, 229, 243, 389, 394, 406, 410
Mass production of hybrid seed 248
Maternal effects 28
Matromorphic seeds 66
Matromorphy 65
Mature green 377
Mean error variance 9
Measurement of heterosis 7
Mechanism of self-incompatibility 89, 95, 106
Medicago sativa 207
Medium 22, 56, 60, 103, 104, 105, 110, 119, 146, 159, 164, 165, 166, 167, 168, 170, 181, 185, 187, 188, 189, 190, 191, 196, 201, 208, 221, 222, 264, 269, 270, 275, 276, 277, 287, 292, 297, 300, 369, 370, 383, 390, 396
Megachile rotundata F. 113
Megagametogenesis 68
Meloidogyne incognita 261, 357
Mentor pollen 102
Meristem culture 159
Meristems 197, 198, 219, 314
Metabolic interaction 21
Methods of gene transfer 195
- hybrid seed production 252, 276, 284, 285, 291, 299, 304, 315, 329, 337, 342, 350, 355, 359
Methods of storage of pollen 108

Methods of Testing 144
Microgametogenesis 67
Microgametophyte 67
Microinjection 197, 198, 199
Microjectiles 198
Micromanipulator 198
Microorganisms 157, 158
Micropropagation 163, 164, 165, 166, 167, 168, 170
Micropyle 67, 101, 103
Microsatellite or minisatellite 154
Microspores 67, 73, 74, 77, 79, 81, 164
Microsporogenesis 67, 72, 73, 74, 81
Mid-late group 45, 390
Mid-parent 3, 4, 9, 25, 35, 36, 41, 46, 52, 57, 62, 63
Mid-parent value 25, 63
Mid-season group 390
Minikit trials 125, 127, 128
Mite 173, 283, 297, 335
Mitochondria 24, 79, 82, 83, 185
Mitochondrial genome 83
Mode of distribution 252
Model of single gene heterosis 21
Moisture 37, 68, 107, 109, 115, 301, 312, 321, 326, 334, 340, 345, 347, 353, 356, 362, 364, 372, 373, 374, 384, 386, 387, 389, 391, 392, 396, 399, 400, 403, 404, 407, 408
Moisture-holding 373
Molecular basis of heterosis 24, 25
- cloning 215, 216
- markers 16, 143, 146, 147, 148, 155, 162, 212, 239
- techniques 155, 155, 156
Momordica charantia 317
Monoclonal antibodies 158
Monoecious 3, 248, 251, 254, 255, 256, 317, 318, 319, 322, 323, 324, 328, 329, 331, 332
- character 55
- lines 55
Monogenic 17, 184, 265
More number of fruits/plant 277
Morphological test 144
Morphological variations 259, 272
Morphology of GMS mutant 71
Mother root production stage 362, 371, 372
Multicellular 25, 162, 192, 262, 342
Multilocation trials 125
Multiple effects 25
Multiplication of lines 369
Multiplicative 18
Murashige 104, 159
Muskmelon 10, 11,12, 13, 32, 33, 34, 53, 54, 55, 66, 130, 133, 135, 136, 137, 148, 177, 252, 254, 256, 276, 285, 335, 336, 337, 338, 359, 360, 398, 399
Muslin 365
Mustard aphids 389
Mustard sawfly 388
Mutation 6, 70, 73, 81, 149, 152, 218
Mutational euheterosis 5, 6
Myoinositol 209

N

$NANO_3$ treatment 185
NAOH test 144
National authority 232
National released hybrids 268, 277, 286, 292, 299, 311, 316, 320, 325, 333, 346, 352
National trials 125, 127
Native pollinators 112, 115
New hybrid technologies 239
Nicotiana 18, 81, 88, 89, 90, 104, 160, 186, 188, 189, 190
- *alata* 89, 90
- *glauca* 188
Nitrate reductase 23
No dominance 4
Nomia melanderi 113
Non-adaptability attribute 122
Non-alleles 18
Non-reproducibility 5
Non-segregating generations 28
Normal anthers 72
North India 13, 335, 381, 389, 391, 396, 398, 400, 401, 403, 407

Novelty 143, 227, 230
NSC 10, 249
Nucellus 65, 159
Nucleosides 79
Nursery raising 250, 381, 407
Nursery-bed 373, 386, 387, 407

O

Objectionable weed seeds 301, 321, 334, 340, 347
Oenothera 96, 119
Off types 271, 280, 288, 293, 301, 312, 370, 371, 372
Off-season 11, 12, 13, 145
Okra 1, 4, 10, 12, 34, 46, 47, 49, 66, 118, 129, 130, 131, 132, 135, 136, 137, 138, 162, 177, 249, 251, 252, 253, 257, 376, 383, 384, 385
Okra stem fly 385
Oligogenes 150, 183
Onion 1, 2, 3, 4, 5, 10, 12, 13, 130, 131, 132, 133, 135, 136, 170, 193, 199, 256, 363, 364, 365, 366, 367, 371, 370, 406, 407, 408, 409, 410
Onion thrips 410
Open-pollinated 9, 10, 55, 60, 122, 123, 244, 245, 247, 248, 256, 257, 345, 361
Orange 270, 276, 284, 290, 329, 337, 338, 339, 350, 357, 394, 409
Organelle polymorphism 24
Organism 4, 24, 150, 162, 375, 377, 379, 382, 384, 387, 393, 405, 409
Ornamental plants 157
Osmia spp. 113
Osmotin 203
Other crop seeds 272, 280, 288, 293, 301, 312, 321, 326, 334, 340, 347, 353, 356, 362, 372
Outcrossing 17, 97, 112, 267, 282
Ovary 66, 67, 101, 102, 103, 104, 105, 262, 264, 298, 315, 319, 323, 327, 329, 343, 350, 354, 359, 364
- culture 104
- manipulations 102
Ovary slice culture 104
Over ripe 377
Overcoming F_1 sterility 105
Over-dominance 15, 16, 17, 18, 19, 21, 26, 28, 39, 46, 49
- dominance hypothesis 15, 16, 17
Ovule culture 104, 193

P

P. glaberrima 120
Packaging 119, 240, 252, 277, 285, 326
P-aminobenzoic acid 19, 22
Papaver rhoeas 159
Parent 3, 4, 6, 7, 8, 9, 25, 28, 31, 34, 35, 36, 39, 41, 42, 43, 44, 45, 46, 49, 51, 52, 53, 54, 55, 57, 58, 59, 60, 61, 62, 63, 71, 76, 79, 80, 82, 85, 95, 100, 103, 123, 124, 148, 162, 163, 173, 174, 175, 177, 178, 179, 181, 182, 183, 244, 251, 252, 253, 254, 255, 256, 257, 266, 271, 276, 277, 280, 285, 286, 287, 291, 299, 307, 311, 314, 319, 324, 325, 329, 331, 332, 335, 338, 339, 344, 346, 348, 350, 351, 355, 360, 361, 365, 366, 367, 369, 370
Parental lines 78, 97, 162, 163, 164, 189, 214, 216, 238, 239, 240, 242, 324, 350
Parental lines of hybrids 146, 163, 214, 216, 238, 277
Parental values 4, 7
Parthenium argentarum 91
Parthenocarpic 66, 175, 205
Parthenocarpic-fruits 175, 176, 205
Parthenogenesis 65
Partial dominance 4, 28
Particle gun 198
Patent 225, 226, 227, 228, 238, 239, 240, 242, 243, 275, 284, 290, 304, 319, 323, 328, 349, 350
- requirements 227
Patent ability 227
Patents and designs protection act 225
PCR analysis of SSR loci 154
PCR polymerase 157
Pea 34, 62, 61, 66, 129, 148, 193
Pearlmillet 81

Pendimethalin 374, 378, 381, 384
Penicillin 157, 158
Penicillium notatum 157
Pepper 66, 75, 148, 151, 156, 175, 212, 256, 257, 281, 282, 285, 286, 289, 290, 291, 292, 293, 383
Pericarp thickness 35, 36, 37, 38, 278
Period protection 230
Peroxidase test 145
Petunia 83, 88, 89, 90, 91, 96, 160, 188, 201, 202, 207
- *hybrid* 90
- *inflata* 89, 90, 91
Pharmaceuticals 194, 228
Phaseolus 64, 104, 105, 183, 192
Phenol test 145
Phenylamine 73
Philein 69
Phloem mobility 79
Phlox pilosa 120
Phomopsis blight 273, 274, 278, 280, 379
Photoperiod 13, 71, 72
Physiological basis of heterosis 19
Phytochemical 1
Phytophthora blight 382
Pin 86, 95, 173
Pinching of male flowers 253,254,256,324,355
Pink 61, 123, 262, 329, 350, 364, 377
Pistil 67, 69, 70, 88, 89, 90, 92, 94, 98, 99, 100
Pistillate 32, 66, 117,182,319, 320, 325, 328, 329, 331, 337, 339, 341, 342, 350, 351, 354, 355
Plant Biotechnology 106, 163, 212, 218
- breeders rights 229
- breeding 7, 193, 212, 228, 231
- hormones 74
- Patents Act 226
- regeneration 65, 164, 190, 197
- tissue culture 159, 163, 192, 209
- Variety Protection Act 226
- viruses 196
Plasmids 82, 83, 195, 201
Poinsett 52, 327
Polar nuclei 68
Pollen biology 65, 68, 69
- biotechnology 65, 70, 182
- development 69, 70, 72, 73, 74, 75, 77, 79
- dispersion 69
- grain traits 118
- losses 117, 119, 120
Pollen movement 117, 118
- removal 117, 118, 119, 120
- stainability 176
- storage in organic solvents 108
- viability 106, 107, 108, 109, 110, 111, 120, 319
- viability after storage 110
Pollen-Grains 66, 67, 68, 69, 70, 74, 79, 86, 88, 91, 92, 93, 95, 96, 100, 101, 103, 106, 107, 108, 117, 118, 120, 190, 191, 192, 197, 315, 319, 342, 344, 397
- pistil interaction 69, 70
- producing capacity 255
- stigma interaction 95
Pollination 65, 66, 67, 69, 70, 81, 82, 86, 88, 91, 92, 96, 97, 98, 99, 102, 103
- constraints 111
- operation 330, 343
Pollinator 60, 79, 85, 112, 113, 114, 115, 116, 118, 119, 251, 257, 271, 275, 280, 305, 332, 360, 361
- populations 113
Pollinators 111, 112, 113, 114, 115, 116, 117, 118, 120, 251, 256, 267, 298, 324, 325, 331, 337, 355, 365
Poly cross 33
Polyacrylamide gel electrophoresis 149, 154
Polyembryony 65
Polyethylene glycol 187, 197
Polyethylene glycol (Peg) 187, 197
Polygalacturonase 205, 209
Popularization of hybrid seeds 250
Population improvement 121, 122
Positive and negative heterosis 6
Positive heterosis 6, 35, 36, 59, 62, 63
Post-fertilization barriers 101, 104

Post-harvest industry 11
Post-harvest operations 250, 251, 252
Potassium 79, 115, 144, 183, 220, 221
Potato 34, 65, 66, 86, 149, 151, 152
- virus 152
Potentialities of synthetic 219
Preclude 3
Pre-fixed prices 252
Pre-selective amplification 153
Primula 86, 95
Private sectors 133, 242, 274
Problems from IPR 226
Procedure of patenting 228
Processing 3, 31, 37, 38, 59, 129, 194, 248, 261, 268, 287, 298, 406
Production cost analysis 252
- objectives 261, 274, 283, 289, 318, 323, 328, 336, 341, 349, 354, 364
- of balanced 21
- hybrid seeds 248, 255, 370
- hybrid substance 22, 23
- inbred lines 32
Production of tetraploid lines 344
Progamic 70, 121, 122
Progeny 8, 33, 55, 79, 80, 81, 86, 111, 150, 164, 170, 180, 194, 196, 216, 265, 267, 279, 331
Proline 73, 79, 209
Promising crosses 52, 53
Propagation 7, 33, 65, 97, 164, 166, 170, 219, 220, 229, 248
Proposed PBR in india 231
Protection of plant varieties 225, 234, 238, 240
Proteomics 157, 160
Protoplasts 159, 160, 161, 164, 185, 187, 188
Prunus persica 108
- *salicina* 108
- *triloba* 108
Pseudoheterosis 5
Pseudomonas ascochyta 53
- *syringae* 203
Pseudoperonospora cubensis 55
Pumpkin 32, 33, 34, 57, 58, 66, 130, 167, 254, 256, 257, 394, 401, 402
Pumpkin and squashes 57
- mosaic virus 402
Pure seed 272, 280, 288, 293, 301, 312, 321, 326, 334, 340, 347, 353, 356, 362, 372
Purple blotch 363, 364, 409
Pyrus ussuriensis 108

Q

QTL mapping 213, 214, 215, 216
Quality components 54
Quality test 125, 128
Quantitative 16, 18, 22, 25, 73, 78, 110, 184, 213
- trait 150, 151, 213
- traits loci 25, 213

R

Radish 1, 5, 32, 33, 34, 61, 73, 98, 129, 148, 172, 249, 251, 257, 258, 310, 311
Random amplified polymorphic DNAS 152
Raphanobrassica 172, 183
Raphanus sativus 61, 98
Recessive 6, 8, 15, 16, 17, 19, 20, 21, 32, 54, 55, 56, 70, 95, 122, 170, 184, 265, 267, 285, 286, 291, 310, 311, 338, 343, 351, 366, 367
- alleles 15, 17, 20, 338
- marker 265, 285, 311
Reciprocal crosses 28, 181, 307, 354
Recombinant DNA 70, 157, 160, 241
- DNA advisory 241
Recombination 6, 24, 27, 83, 123, 183, 184
Recurrent parent 79, 80, 85
Red Pumpkin Beetle 58, 322, 336
Reduced amount of a single gene product 19
Regeneration of hybrid plants 188
Registration of germplasm 238
Registry 234, 237, 238
Release and notification 125, 139
- of hybrids 125, 138
- procedures 125
Removal of male buds and use of insect pollinators 331

Remunerative 1, 11, 257
Re-patterned 25
Reproducibility 5, 152, 201
Reproduction 5, 19, 65, 66, 68, 229
Resistant 1, 11, 12, 37, 40, 44, 48, 53, 55, 56, 60, 61, 69, 82, 83, 90, 137, 160, 161, 162, 163, 170, 173, 177, 185, 194, 202, 203, 204, 205, 207, 208, 209, 211, 212, 303, 304, 311, 312, 314, 316, 318, 323, 327, 335, 336, 341, 346, 349, 363, 364, 370, 371, 376, 377, 380, 382, 385, 388, 410
Restoration system 31, 194
Restorer gene 84, 85
Restriction digestion 153, 155
Review committee on genetic manipulation 241
Rhizoctonia solani 375, 382
Rhizopus niveus 90
Ribosome-inactivating protein 203
Ridge and smooth gourds 58
Ridged gourd 66, 402
Ridged gourd and smooth (sponge) gourd 353
Root 21, 23, 59, 60, 61, 129, 140, 152, 159, 172, 173, 185, 195, 221, 357, 361, 358, 360, 362, 375, 376, 377, 379, 380, 381, 387, 388, 390, 403, 404, 408
- crops 59
- rot 173, 283, 382
- knot nematode 212, 270, 377
Rosa carolina 117
Rotation 114, 235
Rotation 375, 376, 388, 405, 409
Round melon 66
Round-fruited 57

S

S. chacoense 182
S. gilo 174, 175, 176, 274
S. indicum 175
S. khasianum 174, 175, 273, 274
S. melongena 175
S. officinarum 172
S. pinnatisectum 182
S. polyadenium 182
S. tuberosum 90, 182
S. xanthocarpum 174, 175, 273
Salix babylonica 108
S-allele 86, 91, 100, 305, 309, 310
Salmonella typhimurium 207
Salt tolerance 209, 259
Sandalwood 220
Sarton 327
Sca effects 52
Scab 53, 392, 405
Seed certification standards 252, 300, 320, 326, 334, 339, 346, 352, 356, 361, 371
Seed germination 46, 104, 246, 345, 373, 386, 391, 395, 396, 400, 403
Seed import and export 258
- organizers 249
- production 247, 248, 249, 250, 251, 252, 253, 254, 255, 256, 257, 258, 262, 263
- standard 288
- yield 75, 78, 249, 257, 263, 310
Seedling development 181
Seedlings 148, 163, 219, 344, 345, 371, 373, 374, 375, 378, 379, 381, 382, 386, 387, 389, 390, 404, 407, 408
Segregating generations 5, 28, 264
Segregation 3, 214
Selective heterosis 7
Self-fertilization 2
Self-incompatability 96, 305
Semi-loopers 406
Sepal 264, 276
Separate gene products 19
Septoria spot 376
Sex expression 34, 55, 252, 320, 324, 325, 337, 339, 355
Sexual 53, 65, 66, 68, 75, 106, 184, 194, 196, 239
- progeny 194, 196
- reproduction 65, 66, 68
Share of profit 250
Shipping quality 53, 56
Shoot and fruit-borer 274, 278, 385
Sibmating 32, 96

Single gene 19, 21, 88, 91, 102, 205, 310
Single stem type 330
Single-cross 18, 25, 33, 34, 244, 267, 306, 307, 308, 310
Single-cross hybrid 25, 33, 34, 244, 306, 307
Slimy masses 264
Smooth gourd 66, 354, 355, 402
Sodium alginate 220, 221
Soft rot 312, 405
Solanaceous crops 35, 259
Solanum chacoense 89, 90
- *melongena* 175, 176, 193, 222
- *phureja* 88
- *tuberosum* 89, 190, 192
Soluble proteins 73, 149
Somaclonal variation 161, 170, 171, 193, 194
Somatic hybrid cells 188
Somatic hybridization 161, 183, 184, 185, 188
Some success stories 9
Sorbitol 185, 209
Soybean 198, 220, 226
Spatial 89, 99, 114
Spatial isolation of parental species 99
Specific combiners 49
Specific combining ability 5, 266
Specific requirement 334
Sperms 67, 68, 179
Sphaerotheca fuliginea 55, 323
Spinach 32, 33, 34, 35, 251, 255, 256
Sporophytic 87, 92, 93, 307
- self-incompatibility 92, 97
- system 87, 91, 92, 93, 94, 95, 96, 97
Sprouting broccoli 179, 251
Sprouts 33, 34, 98, 179, 251
Stability 5, 201, 230, 290, 297, 355
Staminate 66, 319, 325, 329, 331, 337, 342, 350, 351, 355
Staminodes 75, 319, 350
Standard heterosis 7, 8, 49, 51
State 4, 12, 20, 229, 230, 234, 235, 248, 249, 286, 361
State seed corporations 249
Station trials 125
Status 1, 9, 26, 65, 78, 97, 106, 129, 179, 198, 241, 250, 261, 262, 263, 278, 383
Stem 298, 300, 318, 322, 328, 330, 337, 339, 341, 348, 350, 354, 358, 364, 375, 376, 377, 379, 382, 384, 385, 388, 393, 398, 409
Stemphylium blight 364, 409
Stenotomum 88
Step ladder method of breeding 121, 124
Steps for production 220
Steps in development of F_1 hybrids 305
Stigma 66, 67, 69, 70, 86, 87, 90, 91, 92, 93, 95, 96, 98, 100, 101, 102, 103, 107, 110, 111, 119, 298, 299, 304, 315, 319, 320, 324, 331, 337, 342, 350, 354, 359
Stigmatal 102
Stimulation of divergent alleles 16
Stylar barriers 102
Style 66, 67, 70, 86, 87, 88, 89, 93, 95, 96, 98, 101, 102, 103, 179, 181, 262, 309, 310, 365
Sulfonmide 22
Sulphonylurea 207
Surinam 327
Sweet pepper 43, 257, 281, 286, 289, 291
Sweet potato 65, 86, 222
Symmetrical distribution in F_2 15, 16
Symptoms 78, 375, 377, 379, 380, 382, 384, 385, 387, 394, 397, 398, 402, 405, 409
Synchronization 31, 44, 140
Synergids 68
Synseed 220
Synthetic 46, 76, 164, 218, 219, 220, 221, 222
Synthetic seed technology 218, 219, 220

T

T. hybridum 105
Tablegreen 327
Tagetes 105
Tall plant type 296
Temporal isolation 114
Terminator technology 242, 243, 244, 245, 246
Test crossing 305
- of significance 9

Tetraploid line 56, 344
Thailand 258
Theobroma cacao 96
Theoretical insight 23
Thrips 12, 363, 364, 383, 410
Ti Plasmid 160, 161, 195, 196
Timing of pollinator introduction 113
Tip burn 44, 357
Tissue culture 97, 159, 163, 164, 165, 170, 192, 193, 194, 208, 209
Tobacco 159, 161, 163, 164, 173, 191, 197, 198, 201, 202, 204, 207, 209, 211, 244, 271, 283, 398
Tomato 1, 2, 3, 4, 5, 9, 10, 11, 12, 13, 16, 21, 31, 32, 33, 34, 35, 36, 129, 130, 131, 132, 133, 135, 136, 137, 138, 165, 206, 247, 249, 251, 252, 253, 254, 255, 256, 257, 258, 262, 263, 265, 338, 341, 373, 374, 375, 377, 378, 379, 381, 382
- fruit-borer 377
- leaf 376
- mosaic 260
Top-cross 33, 305, 361
Top-parent 39, 52, 54, 58, 60
Total weed seeds 301, 321, 334, 340, 347, 362
Transformation by ultrasonication 201
- using pollen 201
Transgenic chimeras 197
- parental lines 163
- plants 193, 194, 198, 201, 202, 203, 205, 206, 207, 208, 209, 220, 245
- tomato 204, 208, 212
Transmission 24, 130, 194, 252
Transportation 35, 260, 261, 269, 339, 405, 409
Transposon tagging 71, 218
Trials for quality test 125, 128
Trials on disease and insects 125
Trichosanthus dioica 167
Trifolium 88, 97, 105, 113
Trifolium repens 105
Trigenic 28, 117
Trigona iridipensis 117
Trimonoecious 66
Triple fusion 70
Triple-cross 34, 307, 309, 310
Triple-test cross analysis 28
Triploid watermelon 342, 343
Triticale 172, 183
Tropaeolum 159
Tumour 195
Turnip 129, 183, 251
Type of F_1 combinations 305
Types of hybrids 31

U

Undesirable complementary genes 123
Unfertilized gametes 65
Unicellular 262, 127, 129
Uniformity 33, 34, 43, 44, 45, 51, 60, 99, 111, 143, 230, 360, 361
Unilateral incompatibility 100, 182
UPOV 229, 230, 231, 243
Use insect pollination 338
Use of cytoplasmic-nuclear 286
- functional sterility 277
- gynoecious lines 55, 338, 355
- hynoecious plants 331
- gynomonoecious lines 331
- male sterile lines 13, 262, 324, 366
- male sterility 41, 311, 338, 351
- marker genes 311, 342, 343
- molecular markers 212
Useful heterosis 7
Utility Patents Act 226

V

Vacuum-drying 109
Vagans 117, 118
Valine 73, 207
Vegetable hybrids 9, 11, 129, 130, 135, 163, 247
- improvement 7, 10, 33, 65, 125, 126, 127, 130, 132, 133, 141, 157, 327, 352
Verminator technology 243
Verticillium wilt 173, 260, 274

Viable pollen 71, 72, 79, 86, 264, 360, 397
Vicia 120, 192
Vigorous 15, 16, 17, 18, 46, 61, 122, 182, 311, 312
Vinegar 157
Virus resistance 4, 182, 202
Vitis 104

W

Watermelon 2, 9, 12, 13, 31, 32, 33, 34, 35, 56, 66,130, 131, 132, 133, 135, 136, 137, 212, 256, 327, 336, 340, 343, 344, 347, 351
Watermelon mosaic 327, 336, 402
Water-soluble aniline 100
Weed seeds 272, 280, 288, 293, 301, 321, 334, 340, 347, 356, 372
Weedicide 384
Weeding 251, 373, 378, 381, 384, 397, 399, 401, 403, 404, 408
Weight of seeds 57, 252
Well-aerated 373, 378, 381, 383, 401, 403
Western plains 270, 317
White fly 38, 261, 377
Wide crosses 172, 177
Wilt 12, 40, 61, 137, 138, 173, 194, 297
Wind pollination 32, 255, 256, 257
Wire stem 388
Witches broom 394
World Trade Organization (WTO) 228, 229

X

Xerophytes 209

Y

Yellow vein mosaic 12, 137, 138, 177, 402
- vein mosaic of pumpkin 402
Yomaki 327

Z

Zea mays 108
Zostera 68
Zygote development 179, 180

FSC
www.fsc.org
MIX
Papier | Fördert gute Waldnutzung
FSC® C083411

Zeitfracht Medien GmbH
Ferdinand-Jühlke-Straße 7
99095 Erfurt, Deutschland
produktsicherheit@kolibri360.de